THE DECOMMISSIONING HANDBOOK

Sponsored by
ASME Environmental Engineering Division
ASME Nuclear Engineering Division
U.S. Department of Energy
American Nuclear Society

In cooperation with

Edited by
Anibal L. Taboas
A. Alan Moghissi
Thomas S. LaGuardia

ASME
Three Park Avenue ♦ New York, N.Y. 10016

ASME Press ISBN: 0-7918-0224-8
American Nuclear Society ISBN: 0-89448-041-3

Cover image: The cover was adapted from the inscription on a historical marker. This rock marks "Site A" within Red Gate Woods, a part of the Cook County Forrest Preserve District, near Chicago, IL. The full inscription reads:

THE WORLD'S FIRST NUCLEAR REACTOR
WAS REBUILT AT THIS SITE IN 1943 AFTER
INITIAL OPERATION AT THE UNIVERSITY OF CHICAGO.
THIS REACTOR (CP-2) AND THE FIRST HEAVY WATER
MODERATED REACTOR (CP-3) WERE MAJOR FACILITIES
AROUND WHICH DEVELOPED
THE ARGONNE NATIONAL LABORATORY.
THIS SITE WAS RELEASED BY THE LABORATORY IN 1956,
AND THE U.S. ATOMIC ENERGY COMMISSION
THEN BURIED THE REACTORS HERE.

Photo credits: R. R. Freeman and A. L. Taboas, May 2004.

PREFACE

Practitioners have long recognized the need for an authoritative reference in nuclear decommissioning. Most of us, however, have been too heavily immersed in our respective activities to undertake its preparation. But now we've finally done it.

Rapid change and innovation have become constants, as a now global community revisits nuclear power and technology. A key factor affecting resources is the level of communication and positive industry-wide relationship among professionals. The purpose of The Decommissioning Handbook is to facilitate the forum needed to build a network of consistent approach, practices, and results. These approaches will apply not only to decommissioning existing facilities, but crossing the traditional line between operations and reuse, they will allow us to rethink the construction of new ones. When we build, we should already have a plan for returning the site to a desired end-state condition. The last twenty years have given industry the decommissioning experience it needs to plan for reusing any site. It is time to reap the benefits of that experience by securing the complete life-cycle of any nuclear facility. The Decontamination Handbook is a first step toward that goal.

Recognizing the need for an authoritative and comprehensive document, we assembled a team representing policy, engineering, and science. The goal was to produce both a full introduction for those new to the field and a current desk reference on regulations, resources, and experience. Part 1 summarizes major elements of decommissioning, placing the various aspects of the process in perspective. Novices, students, and decision makers not familiar with the subject would benefit from reading this Part. Part 2 presents regulatory requirements and related issues as risk analysis and stakeholder participation. The third and fourth Parts detail the actions necessary to decommission a facility while Part 5 describes projects and positive and negative lessons learned.

A persistent concern was the addition of newer technologies and process improvements. The valuable lessons that continue to emerge from past and ongoing projects should influence planning, construction, and operation of future facilities. Some techniques used just a few years ago have already been shelved, sometimes for older, more traditional methods. Some technologies evolved only through site-specific situations and innovation. Providing all known technologies proved overly ambitious; instead we have focused on only the time-tested and proven technologies – some newer, some older -- and postponed for a subsequent volume the discussion of how to incorporate decommissioning-friendly features into new facility designs for power generation, fuel fabrication, and/or conversion of surplus nuclear material.

Given the rapidly evolving field, how could The Decontamination Handbook be thorough as a reference and yet flexible to hold the latest information on regulation, technology, and resources? We opted for a living resource -- to supplement the printed text with a compact disk and provide updates via Internet access. Confident that you'll discover shortcomings, we applaud your sharp perception in advance, and invite constructive comments and additions as we prepare our first updates or supplements.

Anibal L. Taboas
A. Allan Moghissi
Thomas S. LaGuardia
Contributing Editors

ACKNOWLEDGMENTS

Completing this handbook took the dedicated effort of individuals and organizations -- foremost of which is the U.S. Department of Energy (DOE). The effort was led by Anibal L. Taboas (DOE), then the incumbent Chair of the Environmental Engineering Division (EED) of the American Society of Mechanical Engineers (ASME). His initial goal was to update the DOE's internal Decontamination and Decommissioning Guide and to broaden its applicability to government, industry, and research. The EED' Executive Committee endorsed the initiative, and two of its members, A. Alan Moghissi, President of the Institute for Regulatory Science (RSI), and Tom S. LaGuardia, President of TLG Services, Inc. (TLG), joined Taboas in forming an editorial team with the appropriate diversity of experience for a balanced presentation of the industry.

Many other dedicated professionals contributed to The Decommissioning Handbook, including:

Aker, Rock E.	*New Horizon Scientific*	**Loomis, Jr., John**	*DOE*
Austin, William E.	*Westinghouse S.M.S*	**Love, Betty**	*RSI*
Baker, Richard	*DOE*	**Maxted, Maxcine**	*DOE*
Bossart, Steven	*DOE*	**McCune, Mary**	*DOE*
Breault, Andrea	*Booz, Allen, Hamilton*	**McNeil, James**	*Consultant*
Brownstein, Martin	*Duratex*	**Meservey, Richard H.**	*Bechtel BWXT Idaho*
Cloutier, Jr., William	*TLG Services*	**Newey, John**	*SMUD*
Cummin, Christopher	*Duratek*	**Petroff, Dale M.**	*Westinghouse S.M.S.*
DeGregory, Jamie	*Ithaca College*	**Petrosky, Tim**	*Consumers Energy*
Dietz, William	*TLG Services*	**Pettit, Paul**	*Fluor Fernald*
Doolittle, Henry G.	*Bechtel Hanford*	**Remark, John**	*Framatome*
Endsley, Stephen	*TLG Services*	**Rolfston, Greg**	*Entergy Nuclear*
Fricke, Victor R.	*Westinghouse S.R.*	**Siry, Glen**	*Westinghouse S.R.*
Gardner, Dennis	*SMUD*	**Snyder, Tom**	*RWE NUKEM*
Hubbard, Kathleen	*TLG Services*	**St. Onge, Richard**	*Southern Cal. Edison*
Johnesee, James A.	*Bechtel BWXT Idaho*	**Straja, Sorin**	*RSI*
Kahn, Bernard	*Georgia Tech.*	**Szilagyi, Andrew**	*DOE*
Kaufmann, Joshua	*The De Mink Group*	**Thayer, Jay**	*Entergy Nuclear*
Kirkland, Michael C.	*RSI*	**Wallis, Bruce R.**	*Portland General Electric*
Lackey, Michael B.	*Fluor Hanford*	**Wilson, Greg**	*Connecticut Yankee*
Lahr, Jeffrey S.	*Georgia Tech.*	**Young, Dan**	*RIO*

In keeping with a long tradition of service by professional societies, the EED joined with ASME's Nuclear Engineering Division, and with the American Nuclear Society (ANS) for publication. Most contributors belong to one or more of these organizations.

Completing an extensive handbook covering such a complex topic would have been impossible without an even greater number of un-named but dedicated people ranging from technology developers to peer reviewers, and the staff at ASME. The Decommissioning Handbook results from voluntary public service in relation to official government business, and intentionally remains in the public domain. Thanks to all who have brought this book to publication.

CONTENTS

PART 1   INTRODUCTION

Chapter 1

DECOMMISSIONING OVERVIEW

1.1 INTRODUCTION

Decommissioning brings to closure, or terminates, the mission or useful life of a radiological or nuclear facility. The owner or licensee normally decides when a facility is to permanently cease operations. It must then show that a proposed decommissioning project plan can be conducted safely, and that at completion the facility will comply with regulatory requirements. This chapter presents a road map of the decommissioning process, highlighting the major issues to address in planning, implementing, and completing a project, and including references for further detail.

1.1.1 Background: The number of radiological and nuclear facilities that have reached the end of their useful life has increased steadily over the last decade. Decommissioning is now, more than ever, a major activity involving regulators, owners and operators, and the public. Nuclear safety at most facilities in the U.S. is under the regulatory purview of the Department of Energy (DOE) or the Nuclear Regulatory Commission (NRC). The Environmental Protection Agency (EPA) promulgates generally applicable standards for the protection of the environment and human health; these standards are then implemented through NRC and DOE rules.

For the purpose of decommissioning, the NRC and/or DOE rules are comprehensive enough to be considered the most relevant, appropriate, and applicable guides. In general, procedures applicable to nuclear facilities (where there may be criticality considerations) are more comprehensive than for radiological facilities (such as x-ray or gamma sources), and require higher formal documentation. Radiological and/or occupational safety can have a separate set of applicable regulations, such as for accelerators, medical applications, and food processing devises. There also may be regulatory consideration by the nature of the material (e.g., naturally occurring, or activated), defined origin (e.g., as source, byproduct, or special nuclear material (SNM)), and even by application (e.g., sealed sources for logging petroleum wells, or used in agriculture). Regulations are intricate, and their applicability determined by factors such as ownership, date generated, location, and constituents exhibiting hazardous chemical properties.

Radiological and occupational safety at non-DOE facilities is regulated by the Occupational Safety and Health Administration (OSHA), which may delegate authority directly to approved "State Plan" states, or by the NRC to "Agreement States." Individual states have also promulgated a wide variety of statutes addressing occupational safety, transportation and disposal, and/or implementing related environmental management authorities. Civilian facilities include educational research reactors, accelerators, medical facilities, irradiation (foods, diamonds, wood, medical product sterilization, and others), enrichment and fuel fabrication, power generation, and such other applications as sealed lighting sources and space-based thermionic generators. The DOE self-regulates most of its facilities, which pursue a diverse mission of research, development, demonstration, production, and testing for defense and civilian applications.

To deactivate and decommission the global backlog of facilities without harming human health and safety or the environment requires using technology that can cost-effectively reduce risk and meet regulatory requirements. It is a basic premise of this book that remediation is not complete until all the goals of decommissioning are met and no further action is required (excluding long-term monitoring and maintenance). Except for the time cost of money, the major planning and implementation factors for commercial and federal decommissioning projects are very similar. The critical importance of clearly defining the facility end-state criteria in terms of radioactive and chemical hazards cannot be over-emphasized. This factor alone can drive cost feasibility since it can increase the amount of material to dispose of at restricted disposal sites by tenfold. In general, the planning process includes:

1. Identification of the facility to be deactivated, including a decommissioning decision basis. Candidate facilities for decommissioning are identified after evaluating their potential for reuse and recovery of material and equipment.

2. A Deactivation (or Transition) Plan that addresses:
   - Site specific environment, safety, and health (ES&H) considerations
   - Training
   - Regulatory drivers
   - Waste minimization and disposal
   - Cost and schedule
   - Equipment removal
   - Work readiness review.

3. An Historical Site Assessment (HSA) to gather process and historical knowledge.

4. Decisions on how to deal with people long associated with facility operations. Decommissioning presents a very real and unique challenge from a human resource perspective. Although some staff can contribute significantly to the planning process, decommissioning always involves a change in the needed labor and skills mix, and may involve complete relocation or release of staff if a Decommissioning Operations Contractor (DOC) is chosen.

5. A Characterization Plan that addresses:
   - The detailed determination of radioactive and hazardous chemical materials present, including quantities and locations
   - The conduct of characterization surveys, including process
   - The criteria used to conduct the characterization surveys, and how determinations are made.

6. A Decommissioning Plan that addresses:
   - Facility description and history
   - Project scope and objectives
   - Characterization data summary
   - Specific decommissioning methods
   - Radiological criteria, including cleanup criteria
   - Health and safety plans
   - Safety analysis and review
   - Risk assessment
   - Site release criteria
   - Waste generation estimates and waste disposal procedures

- Project plan
- Key decision points
- Alternative considerations.
- Project reporting
- Work readiness review

When a decommissioning project plan is developed, it is based on:

- Adequately protecting public and occupational safety and health
- Potential environmental and ecological impacts
- Compliance with statutory, contractual, and regulatory requirements
- Effective project management, including selection among viable alternatives based on risk, cost, and desired facility end-state
- Human capital management, consistent with future site utilization plans

7. Decommissioning Operations -- the actual hands-on activities to accomplish the safe decommissioning of the facility

8. A Closure Plan, including long-term monitoring, that addresses:
   - Information review
   - Field data
   - Sample analysis
   - Data comparison of methods
   - Instrumentation plans
   - Health and safety measures
   - Reports and recommendations
   - Release criteria, methods and approaches
   - Closure documentation, such as facility history, contamination and decommissioning activity, waste volumes, release survey and status, independent verification contractor surveys/report, characterization report, and a closure certification statement.

1.1.2 Decommissioning Phases: Facility decommissioning follows deactivation (i.e., after shutting down operations and removing high risk large quantities of readily accessible radioactivity such as spent fuel and sealed sources, and highly hazardous reactive chemicals such as bulk quantities of sodium). The decommissioning process can be described in four principal phases, each involving distinct, specific activities. Successful project planning means understanding these phases and the activities that must be performed to complete them.

1. The *pre-planning* phase of the project, which can begin early, even before the facility is permanently shut down, involves the preliminary assessment of decommissioning options, conceptual cost estimates and schedules, waste generation and disposition estimates, and exposure estimates to workers and the public. The objective is to define major options and a funding approach that will meet owners' needs and satisfy regulators.

2. The *detailed engineering* and evaluation phase covers detailed planning and specific engineering feasibility evaluation. It normally begins after the facility is permanently shut down and when remaining radiological and hazardous material is stable and can be inventoried by measurement and calculation. This phase includes interaction with regulators and stakeholders for acceptance of the approach, particularly the proposed facility end-state. This phase can generate high risk, as shortcuts in the interest of visible progress can result in safety blunders and underestimated costs.

3. The *operations* phase is the actual hands-on activity for decommissioning. It may also involve removal, packaging, shipping, and disposal of systems and structures to meet end-state objectives. This is normally the hardest part of the project to manage, and takes an especially nimble and responsive team to keep everything on track.

4. The *closeout* phase consists of the final disposition of all systems and structures, restoration of the site to the desired end-state (as verified by a final survey), and formal documentation of no further regulatory action.

1.2 EVALUATION OF DECOMMISSIONING ALTERNATIVES

In anticipation of facility decommissioning, management should address the available alternatives for achieving the planned final disposition of the site. The evaluation process is iterative: a preliminary assessment based on existing information, regulatory guidance, economic factors, and management objectives yields, as more detailed information becomes available, to the selected alternative, which will itself likely be modified to suit changing conditions and objectives.

Although there are some differences, the work process for commercial operations is generally similar to that used for government facilities. A principal difference is in how the project is financed. Most commercial endeavors use funds set aside throughout the facility's operating life. However, most government facility projects assume eventual decommissioning via multiyear planning for federal funding, resulting in annual allocations that may or may not be enough to fund the plan. Thus, the planner must divide the project into discreet tasks that may or may not be accomplished within a budget year, always preserving an exit option if funding is depleted.

A second significant difference is that DOE projects often have significant on-site quantities of hazardous chemicals that must be cleaned up and disposed of as hazardous waste. Many DOE projects involve the cleanup and disposal operations of both hazardous chemicals and radiological contaminations.

A third difference is that the NRC has described, in various detailed nuclear regulatory guides (NUREGs) and other publications, the specific documents to be produced and sent for review, the contents of these documents, and the general procedures to be followed by the licensee. The DOE is much less prescriptive and allows much more latitude in its overall decommissioning process.

1.2.1 Potential NRC Decommissioning Alternatives: The NRC has identified three immediate choices:

1. DECON: The equipment, structures, and portions of the facility and site that contain radioactive contaminants are removed or decontaminated to a level that permits license termination shortly after operations stop.

2. SAFSTOR: The facility is placed in a safe, stable condition and maintained in that state until it is later decontaminated to levels that permit license termination. During SAFSTOR, a facility remains intact, but the fuel has been removed from the reactor vessel and radioactive liquids have been drained from systems and components, and processed. Radioactive decay during SAFSTOR reduces potential occupational dose

and the quantity of material that must be disposed of during decontamination and dismantling, leading to eventual decommissioning.

3. ENTOMB: Entombment involves encasing radioactive structures, systems, and components in concrete or other structurally long-lived substance. An entombed structure is appropriately maintained and kept under continued surveillance until the radioactivity level allows license termination. Internal contamination of most power reactors will exceed the limits for unrestricted use even after 100 years; and because current regulations require that decommissioning be completed within 60 years, this option cannot be implemented without a special official NRC review, resolution of issues, and the promulgation of appropriate decision and implementation documents.

1.2.2 DOE Process: Consistent with implementing legal requirements, the DOE has developed a hierarchy of instructions in internal directives. A DOE Order on Life Cycle Asset Management (DOE 1998a), includes requirements for transition, deactivation, decommissioning, and surveillance and maintenance at the end of a facility's life cycle. It also requires that a value-added, quality-driven graded approach be applied to ensure that performance measures are established in accordance with best industry practices, and that formal, comprehensive planning and integration result in controlled methods for facility disposition. A framework for removal actions that is not time-critical is encouraged, and the end state of the project can encompass any of the NRC end states described above. Specific facility disposition requirements include:

- Compliance with National Environmental Policy Act evaluations (NEPA 1982).
- Integration of safety management practices
- Collection, maintenance, and update of data to produce a detailed written summary of physical, chemical, and radiological hazards present, including estimated quantities of specific hazardous materials and their physical locations
- Surveillance and maintenance activities, as required by facility conditions
- Methods for identifying, and evaluating alternatives, and documenting a preferred alternative process to identify activities necessary to achieve specific facility end-points, including a final survey to determine that the end-points were achieved
- Methods for detailed engineering planning and plan documentation
- Update of corporate facility database
- Development of a final report

1.2.3 Alternative Selection: The decommissioning alternative decision process is site-specific and iterative. The process must allow for changes to the decision to account for evolutionary factors that become more evident as greater detailed engineering and planning are performed. If a facility is still in operation at the time the evaluation is performed, the radiological characterization is subject to change by the time the facility is permanently shut down for decommissioning. Planners must take this element of change into consideration in their evaluation.

The decision process to select the right alternative to use in decommissioning a facility can be quite complex. Generally, economic factors dominate the process. For an economic model to provide accurate decommissioning cost estimates, the following input is required for each alternative:

- Project delays
- Fund growth rates in investments

- Projected escalation rates
- Projected interest rates
- Funding sources – whether an external trust fund or annual budget allocations
- Projected availability of waste disposal sites and associated costs

Other indirect economic factors influencing the decision process may include:

- Extent of residual radioactivity and hazardous materials
- Potential changes to regulations
- Availability of waste disposal sites
- Availability of knowledgeable plant staff for deferred decommissioning
- Storage of spent nuclear fuel until transfer to a federal repository

Life cycle factors involve the timing and/or conditions for shutting down a facility. Power reactors are typically licensed for 40 years following construction. However, recent programs to extend the license for an additional 20 years are becoming more common, and greatly affect the decision as to selection of a decommissioning alternative. Extended operation of a reactor will generate more spent nuclear fuel and may increase the radioactive inventory to be removed. The availability of knowledgeable plant personnel familiar with the early years of plant operation is an important consideration in determining when to accomplish decommissioning, and delay may potentially suffer the loss of these valuable personnel. Delay beyond 40 years will provide additional time for trust fund balances to grow, thereby rendering annual accruals unnecessary in the later years. Current NRC regulations specify that decommissioning activities must be completed within 60 years of permanent shutdown unless special waivers are granted. This time restriction significantly limits the options for the methods and procedures used for commercial reactor decommissioning.

The life cycle decision factors for government owned facilities are generally more difficult. A facility may have been shut down for many years following a technology demonstration or termination of a program, and may be heavily contaminated with both radiological and hazardous chemicals. The cleanup of contaminated soil in DOE facilities is often a major consideration of the project (in addition to the contaminated structures). Life cycle factors usually focus on how long the facility may need to remain in safe storage condition until adequate funding is appropriated. In some cases, the threat of spreading contamination becomes the driver as to timing.

Subsequent to decontamination and appropriate modifications, multiple users may use certain sites. The driving force to promptly decontaminate and decommission (but not dismantle) a facility lies within the needs of these commercial entities to take over and operate the facility on an economically advantageous basis. Commercial power reactor owners have also identified specific applications for the reutilization of some structures for power generation, office space, or warehousing facilities. The end-use needs of these companies drive the decision-making process. Research reactors located in universities, hospitals, and government facilities may undergo decommissioning to make room for new non-nuclear research facilities, teaching classrooms, laboratories, and administration facilities. The disposition of such reactor facilities might include salvaging the building and some of the support equipment.

Many stakeholders have legitimate interest in the conduct of decommissioning. Some may have conflicting objectives, and satisfying all may be virtually impossible. They may include:

- Residents of neighboring communities
- Company shareholders and management
- State and federal regulatory agencies
- Local and national public interest groups

Decommissioning planners should address stakeholder concerns early to prevent intervention that could jeopardize the schedule and budget for decommissioning Compromises are usually necessary for progress.

1.2.4 Risk Analysis: A well managed risk program can save significant costs in the management of major projects. This is also valid for a decommissioning project. Accidents, real or perceived risk, and financial risk are closely related, and environmental risk assessments can both determine and support cost saving alternatives.

Accident risks associated with specific decommissioning activities include those related to decontamination, component and structure removal, packaging, transportation, and disposal. Major expenditures may arise from accidents involving worker injury or death, injury to the public (during transportation accidents), spills at a disposal facility, or damage to unrelated structures. Serious accidents can result in a temporary or permanent job shutdown. Even relatively minor worker injury (reportable under OSHA regulations) requires investigation, root cause analysis, and corrective action. These require the expenditure of valuable management attention. These activities detract from the function of controlling the project to keep it on budget and schedule. Overall, the incidence of accidents during decommissioning has been remarkably low.

The technical risks for a project may be unique challenges that require unique or unproven solutions and must be factored into the overall project risk profile. In addition, as with most nuclear projects of any kind, there may be regulators or stakeholders with a perception of risk. The real technical risks must be aggressively managed by a sound engineering plan developed to address and mitigate these risks. The perception of risk must likewise be managed in the form of education and communication to ensure that the perception is addressed and managed.

Financial risks arise from inadequate cost estimating and funding, and the corresponding cost overruns. This type of risk involves both the facility owner and the contractors performing the work. The owner attempts to identify all the costs associated with decommissioning to ensure that adequate funds will be available for safe completion. The NRC requires the licensee of a commercial power reactor to establish an independent decommissioning trust fund (DTF) to ensure that all the funds will be available at the start of the project. In addition, licensees must re-evaluate the adequacy of the fund and report to the NRC every two years. However, once into an established project, costs exceeding available funds may arise from unexpected changes in work scope, large increases in the cost for waste disposal, or changes in the selected project alternative.

Unlike commercial projects, environmental risks are a valid cost-benefit consideration at DOE decommissioning sites. The NRC has regulated the general radiological dose end-points that must be met (e.g., a pathways dose of 25 mrem/y to an individual) at a decommissioned site for its unrestricted release. Hence, compliance with this regulation is relatively straight forward. However, the DOE may elect to leave larger or smaller quantities of radioactivity in place provided that it can be shown by a detailed environmental risk analysis that the probability of dose to the public (as stipulated by appropriate DOE, NRC, or state regulation) will not be exceeded. Typically, the DOE's environmental risk analysis also includes a quantitative assessment of other factors, including the risk of hazardous chemicals to remain at the site.

1.3 DECOMMISSIONING PRE-PLANNING

Pre-planning is a key aspect of a successful decommissioning since it requires certain major decisions. These include resolution of financial issues, evaluation and selection of technologies, establishment of conditions for license termination, and explorations of waste disposal options. Pre-planning also includes the development of worker protection plans, quality assurance, security, and public/stakeholder participation. The overriding issue in most of these decisions is compliance with numerous regulatory requirements. This phase often includes plans for deactivation -- the transition from an operating facility to a decommissioned facility.

1.3.1 Regulatory Requirements: Regulations are issued by federal, state, and local agencies. The decommissioning planner must develop a plan that satisfies all within this framework yet remains within the boundaries of cost and schedule. It is not possible to cover all the federal, state and local regulations since they vary considerably on a site-specific basis. The planner must work through these site-specific requirements, armed with a sufficient knowledge of the federal regulations.

The primary statute governing most decommissioning activities is the Atomic Energy Act (AEA), and its subsequent amendments (AEA 1954). For commercial facilities, the NRC is the agency responsible for the promulgation, implementation, and enforcement of regulation. The disposition of DOE facilities is administered through a system of DOE Orders, which consists of directives to establish requirements, optional guidance, and technical standards in the form of notices, procedures, and manuals that may reflect commercial practices. The hierarchy of these regulatory mandates is based on the DOE's self-regulation authority under the AEA.

1.3.1.1 Nuclear Regulatory Commission: The NRC, authorized by the AEA, has promulgated regulations dealing with decommissioning in the Code of Federal Regulations (CFR), Chapter 10, Parts 20, 30, 50, 61, 71, and 72 (NRC 2003a-f). Consequently, the NRC is responsible for protection of public health and safety during all facets of the decommissioning process.

The NRC issues licenses to nuclear power plant and research reactor owners, among others. In recent years, the NRC has imposed regulations specifically addressing decommissioning activities. In general, they define the process to be implemented to satisfy the regulatory requirements, leaving the details of implementation to the licensee. For commercial power plants, detailed rules for decommissioning are found in 10 CFR 50.52, Termination of License.

The NRC provides guidance in implementation of its regulations in the form of NUREGs to identify specific methods for meeting its regulations. These guides do not preclude other methods for implementation of the regulations as long as they meet the same criteria. However, the licensee must then demonstrate to the satisfaction of the NRC that the alternative method is acceptable. Some NUREGs are reports prepared by NRC contractors to provide additional decommissioning information and guidance to licensees.

NRC regulations require that a commercial plant licensee submit a Post-Shutdown Decommissioning Activities Report (PSDAR) and a License Termination Plan (LTP), and that a licensee of a research reactor or other nonpower nuclear facility submit a Decommissioning Plan (DP). In general, these documents contain:

- A description of the site conditions
- The planned decommissioning activities
- A description of the methods used to ensure protection of workers and the public against radiation hazards
- A description of the planned final radiation survey
- An updated cost estimate
- A comparison of the cost estimate with funds set aside for decommissioning
- A plan for ensuring the availability of adequate funds for completing the project

Subsequently, the NRC and the licensee interact to begin the process of decommissioning.

1.3.1.2 Department of Energy: The DOE promulgates DOE Orders dealing with a wide range of facility operations. These Orders are contractually enforced as regulation for internal use in DOE facilities. Initially, the DOE imposed the same Orders for decommissioning that were imposed for operations, a process that was burdensome for the facility operator to follow because of the tight regulatory control. Beginning in the early 1990s, the DOE has allowed a "graded approach" to implementation to more closely match the Orders to the degree of potential hazard associated with the decommissioning activity. The DOE Orders also implement the EPA's requirements for remediation during decommissioning. This regulatory process is more complicated and prescriptive, and the legal implications of failure to meet these requirements have more far-reaching litigious impacts to both the facility operator and on-site contractors.

Similarly, there are DOE Orders relating to the transportation of hazardous (including radioactive) materials. These follow the requirements for transportation of such materials as promulgated by the Department of Transportation (DOT), and those of the NRC dealing with packaging of radioactive materials. There are several other related statutory and regulatory requirements that differ from the commercial sector, such as indemnification provisions under various amendments of the Price Anderson Indemnification Act (PAA 1994), and the payment of prevailing union labor wages as determined under the Davis Bacon Act (DBA 1964).

1.3.1.3 Environmental Protection Agency: The EPA is responsible for promulgating regulation, and in certain cases implementation and enforcement of a number of laws, including the:

- Clean Air Act (CAA 1970)
- Clean Water Act (CWA 1972)
- Safe Drinking Water Act (SDWA 1974)
- Resource Conservation and Recovery Act (RCRA 1976)

Other relevant and applicable laws include the:

- Toxic Substances Control Act (TSCA 1976)
- Comprehensive Environmental Response, Compensation, and Liability Act

(often called "Superfund") (CERCLA 1980)
- Superfund Amendment and Reauthorization Act of 1986 (SARA 1986)

1.3.1.4 Occupational Safety and Health Administration: The Occupational Safety and Health Act authorizes its administration to develop, implement and, in certain cases, enforce regulations dealing with occupational safety and health. OSHA regulations are codified in 29 CFR 1910 and 1926 (OSHA 2003a and b). The safety issues covered include:

- Design of buildings
- Mechanical guarding machine operation
- Electrical equipment
- Occupational disease prevention
- Noise exposure

General occupational requirements for hazardous waste are established in 29 CFR 1910.120, and include:
- Provision for characterization and analysis
- Personal protective equipment controls
- Training
- Medical surveillance
- Engineering
- Administration
- Site control
- Site monitoring
- Information programs
- Drum and container handling
- Decontamination
- Personal protective equipment controls

By agreement between the OSHA and NRC, the lead responsibility for occupational safety, including radiological and industrial safety, rests with the agency performing the regulatory oversight of nuclear safety (including the DOE). Similarly, the EPA defers exercising its authority (under Superfund) for most facilities under NRC authority. Individual states, however, may have different combinations of agencies to regulate aspects of occupational safety, environmental concerns, and public health.

1.3.1.5 Department of Transportation: Based on the Hazardous Material Transportation Act of 1990 (HMTA 1990), the DOT develops, implements, and enforces the transportation of mixed waste (materials that are both hazardous and radioactive). The DOT regulations are codified in 49 CFR 171-180 (DOT 2003 a-j) and establish criteria for safe transport of hazardous and radioactive materials on public roads, railways, waterways, and aircraft. The DOT regulations include specifications for acceptable shipping containers. Radioactive material specifications require that containers meet the NRC packaging regulations provided in 10 CFR Part 71.

1.3.1.6 Individual State Requirements: State governments can also exercise control over decommissioning activities, including material release criteria, storage of radioactive materials, packaging and transportation of waste shipments, and disposal of hazardous waste. States may be authorized to develop regulations dealing with hazardous waste as long as they are at least as restrictive as those of the EPA (EPA 1999). A significant complicating factor is the deviation of interpretation of a specific regulation among various states. Consequently, the relevant state agency must be appraised of the proposed decommissioning activities so that a productive and cooperative relationship can be established to facilitate decommissioning. Some states have generated their own limited statutory jurisdiction for limited activities, such as accelerators, or have some delegated authority from the NRC as an "Agreement State" or from the OSHA as a "State Plan" state. These regulatory delegations, by state, are listed in Table 1.1 (NRC 2003g)

Table 1.1 NRC and OSHA Regulatory Delegations

Jurisdiction	NRC	OSHA	Jurisdiction	NRC	OSHA
Alabama	Y		Montana	N	
Alaska	N	Y	Nebraska	Y	
Arizona	Y	Y	Nevada	Y	Y
Arkansas	Y		New Jersey	N	Y*
California	Y	Y	New Hampshire	Y	
Colorado	Y		New Mexico	Y	Y
Connecticut	N	Y*	New York	Y	Y*
Dstr. of Columbia	N		N. Carolina	Y	Y
Delaware	N		N. Dakota	Y	
Florida	Y		Ohio	Y	
Georgia	Y		Oklahoma	Y	
Guam			Oregon	Y	Y
Hawaii	N	Y	Pennsylvania	I	
Idaho	N		Puerto Rico	N	Y
Illinois	Y		Rhode Island	Y	
Indiana	N	Y	S. Carolina	Y	Y
Iowa	Y	Y	S. Dakota	N	
Kansas	Y		Tennessee	Y	Y
Kentucky	Y		Texas	Y	
Louisiana	Y		Utah	Y	Y
Maryland	Y	Y	Vermont	N	Y
Massachusetts	Y		Virgin Islands		Y
Maine	Y		Virginia	N	Y
Michigan	N	Y	Washington	Y	Y
Minnesota	I	Y	Wisconsin	I	
Mississippi	Y		Wyoming	N	Y
Missouri	N				

Y = Status as NRC Agreement State or OSHA State Plan State.
N = NRC non-agreement status..
I = NRC letter of intent for Agreement State status.
* = OSHA plan covers public sector (State & Local Government) employees only.

1.3.2 Financial Requirements: The availability of funds to accomplish a decommissioning project is one of the principal factors in selecting the disposition alternative for the facility. Accordingly, an accurate and detailed cost estimate and schedule are needed to determine the amount of funding required, how the funds will be generated, and the administrative controls needed to keep the rate of spending consistent with the budgetary limitations.

The financial requirements are usually established early in the project by the preparation of a preliminary cost estimate and schedule. As additional information becomes available and decisions are made as to the approach to be taken, the cost estimate and schedule are refined and ultimately established as the baseline cost and schedule. This baseline is used as a basis for funding the project and to monitor and control progress during implementation.

1.3.2.1 NRC Decommissioning Trust Fund: Facilities licensed under the NRC have specific requirements as to the minimum funding amount that must be provided before the approval process to initiate decommissioning activities may start. Licensees are required to demonstrate how the funds will be accumulated and how the fund balances will be applied to ensure that the project will not experience funding shortfalls. Licensees are required under 10 CFR 50.75(f)(2) to prepare a preliminary decommissioning cost estimate at about five years prior to the projected end of operations at the facility. However, licensees typically prepare these estimates earlier since the state Public Utility Commission (PUC) usually requires a detailed estimate for the proposed decommissioning, and it is the PUC that allows the licensee to collect funds from ratepayers for the electricity generated. The NRC requires these funds to be held in an external DTF.

Within two years following permanent cessation of operations, the NRC requires the licensee to submit a site-specific cost estimate, usually submitted as part of the PSDAR. It also requires an updated site-specific estimate of remaining costs as part of the LTP, at least two years before termination. States also may impose restrictions on withdrawals from the DTF. California, for example, requires the licensee to formally request the amount needed to decommission the plant two years prior to the start of the project.

1.3.2.2 DOE Budgetary Process: In contrast to the above, funding for DOE facilities is normally based on an annual congressional appropriation. If a specific project is identified as a line item in the budget, funding is normally ensured for completion (often referred to as a Closure Project).

If no such designation is given, funding will be reviewed on an annual basis and authorized as needed, based on an assessment of need for the entire DOE complex. However, although a budget for decommissioning may have been established under and approved for use by the DOE, approval to spend those funds by contractors requires a specific authorization. Accordingly, although a facility may already be in decommissioning, its funding may be withheld so that the funds can be applied to another facility of higher priority.

The environmental management process is clearly dependent on congressional imposed budgetary conditions. It is not unusual to find a site facility being remediated under a defined set of conditions, while a similar facility at the same site is on a completely different timetable. Some of the discrepancy results from external factors, such as regulatory regime (e.g., CERCLA, RCRA, or voluntary action) or availability of funds in the specific sponsoring program.

1.3.3 Clearance Criteria: The term clearance criteria, sometimes referred to as free release criteria, refers to the radiological level deemed acceptable to regulatory agencies to allow equipment, materials, or facilities to be returned to service as clean, scrapped, or abandoned in place. The specified limits for these radiological criteria are established by the EPA, NRC, and DOE. The NRC and EPA have somewhat different site release criteria, in the form of computer model pathways doses to an individual per year. For example, the NRC value is 25 mrem/y while the EPA value is somewhat less, and the EPA has a separate drinking water dose limit. The NRC and DOE have very detailed criteria for the release of radiologically contaminated and hazardous materials from facilities. Procedures employed by the NRC and DOE may differ significantly depending on circumstances. The matter is complex and is discussed elsewhere in this handbook.

1.3.4 License Termination: Termination of a license under NRC regulations requires the licensee to demonstrate that the residual radiological levels (after decontamination, or on clean systems and structures) meet the NRC criteria for clearance.

1.3.5 Evaluation of Technologies: The pre-planning and assessment phase of decommissioning should include an evaluation of the technologies available and proven to perform the desired activities where they will be used. This is a critical step in the planning process, as decisions made to implement the project depend on the success of these technologies. Failure of these technologies can severely affect the budgetary limitations and schedule constraints for a timely completion. The planner is cautioned not to sidestep this important aspect. Some facilities avoid application of innovative technology as a mechanism to reduce risk.

1.3.5.1 Facility Characterization: One of the most important steps in decommissioning planning is characterization of the facility for radioactive and other hazardous materials. Characterization sets the scope and extent of the processes needed to decontaminate, remove, package, transport, and dispose of systems, structures, soils, and tooling. There has been a great deal of work in recent years to develop the methodology and instrumentation to characterize facilities for decommissioning. Initial efforts were directed at extensive and often unnecessary characterization, at significant cost and radiation exposure to the workers. More recent work has refined the process to characterize the facility to a level sufficient for safely planning and implementing the work, and to add characterization data as needed to support specific activities immediately before they are attempted.

1.3.5.2 Decontamination: Related decontamination technologies have improved significantly in the last ten years. Most use chemical, electrochemical, and mechanical processes.

Older chemical processes were replaced by processes such as low oxidation metal ion, difluoroboric acid, and the Siemans CORD. Chemical processes can be applied to reduce radiation levels sufficiently such that work crews can cut and remove surface contaminated equipment through contact handling. Mechanical processes are effective in further reducing worker exposure during removal and disposal. Chemical processes, however, are seldom able to remove radioactivity to levels that would permit free release of materials.

Electrochemical processes (also called electropolishing) can result in free release of materials by removing the surface of the contaminated component. In the U.S., most of these processes have not been used frequently because of the relatively low cost for direct disposal of contaminated materials. However, the desire to recycle materials is likely to change the current practice.

Decontamination of concrete structures is generally accomplished by scarification, either by a grinding process or by a percussion process. The technologies are well developed, and considerable experience is available for review by the planner.

1.3.5.3 Removal/Dismantling: The technologies for removal/dismantling of components have also made major improvements in application and speed of cutting. Mechanical methods consisting of sawing or milling of the component, with the associated generation of swarf (metal removed) from the kerf (cut opening). Thermal cutting methods involve melting and, in some cases, vaporizing the metal in the kerf to cut through the material, and the gaseous effluent must be controlled. These techniques are generally slower than thermal methods, yet they do minimize the gaseous effluent. Innovative materials have been developed, such as ultra hard, diamond-like materials with the promise to be significantly more versatile and less expensive than diamond, saw, or wire. Dismantling concrete structures is generally accomplished by a hydraulic ram mounted on a backhoe, by wrecking ball, or by controlled blasting. Generally, the extent of dismantling requires a tradeoff involving operator exposure. Each technique must be evaluated in terms of the cost and benefit for the application.

1.3.5.4 Waste Management: Waste disposal, a major cost element for decommissioning, is a universal problem. With the exception of spent nuclear fuel and waste containing concentrated transuranics, nearly all radioactive waste from a decommissioning is considered low level waste (LLW). The cost of disposing LLW has been escalating rapidly over the last 20 years, far exceeding indicators such as the consumer price index. Accordingly, planners must pay special attention to reducing the quantity of LLW to reduce overall cost. Several approaches have been developed, but each must be evaluated for cost and benefit to optimize the expenditure of labor to decontaminate and survey, versus package, transport, and disposal by burial. Exceptional care should be given to minimize waste with commingled radioactive and hazardous chemical characteristics since it escalates both the regulatory regime and cost.

There are a number of factors that must be considered in optimizing the waste management process, including minimization, processing, and disposal. The first step in waste minimization consists of a detailed characterization of potentially clean materials to ensure that they do not contain residual contamination exceeding free release levels. Due to the cost of characterization, the planner must assess the amount of labor required versus the amount of material that can potentially be free-released. The second approach is to minimize the amount of clean materials that are brought into the radiological controlled area. In the U.S., this practice is established early (during plant operations) and is therefore an easy administrative control to extend to decommissioning.

Several vendors offer services to package clean, potentially clean, and contaminated components, and to transport them to off-site facilities where labor costs are substantially lower. The activities of surveying, free-releasing for sale as clean scrap, decontaminating, packaging, transport, and disposal at one of the commercial waste disposal facilities are usually far more economical than on-site processing. Most commercial projects have reviewed, if not contracted, these vendors.

Sometimes waste quantities can be reduced by on-site decontamination of components, using mechanical, chemical, or electrochemical processes. The approach requires detailed characterization to establish realistic expectations. However, in certain cases, the quantity of secondary waste generated exceeds the amount of material that can be released. Electrochemical decontamination removes the contamination by removing micro-pores from the surface. In this manner, the cleaned surface cannot trap contamination and the component can probably be free-released. For each of the processes, the planner should evaluate the labor, equipment, and material cost to perform the decontamination, and the cost of disposing secondary waste. Any component would still require survey for free-release, at additional cost.

Under some circumstances it is cheaper to assume that all waste is contaminated, and to package and transport directly for burial. DOE can usually dispose of waste at considerably lower cost than commercial facility owners. Options include the Nevada Test Site, Hanford, WA, and Oak Ridge, TN. The DOE also uses several commercial disposal sites. Generally, the lowest cost of disposal is associated with materials that can be released without condition because of hazardous radiological or chemical characteristic. Bundling of disposal volumes has demonstrated cost advantage in procuring transportation and disposal capacity.

1.3.6 Radiation Exposure: Control of radiation exposure to workers and the public is a cornerstone of decommissioning planning. Depending on the facility being decommissioned, radiation exposure can arise from surface contamination and material activation, as well as from unplanned sources. The planner must deal with the nature and quantity of all radioactive materials.

Workers may be exposed to radiation during field surveys, decontamination, removal, packaging, shipping, and disposal. The planner must use appropriate methods of exposure reduction to the workers, including time, distance, and shielding. Planning must consider the decay rate of various radioactive materials, and take advantage of radionuclides with a short half-life. Processing of these at a later date will ensure a reduced exposure. Various regulations require limiting exposure to As Low as Reasonably Achievable (ALARA) below regulatory maximum limits. This approach helped reduce exposures to a small fraction of what had been incurred in earlier projects.

Radiation exposure to the public at the site boundary is likely to be no greater than that during plant operations, particularly after spent nuclear fuel has been removed from the site. The major potential exposure to the public may occur during transport.

1.3.7 Decommissioning Management Team: The management team represents one of the largest costs of a project and is the factor most associated with the cost of delay. A carefully selected team of skilled and experienced managers is vital for efficiency. There are often two elements of the management team: the facility owner (DOE or utility licensee) and the DOC. The owner usually conducts a competitive bidding process to select a decommissioning operations contractor who will manage fieldwork. The owner retains primary responsibility for public health and safety, budget, security, quality assurance, and general oversight of the work. The DOC has primary responsibility for selecting and managing crew performance, scheduling work, overseeing occupational health and safety, waste management, and obtaining necessary equipment.

Former utility operating personnel are frequently available for decommissioning activities at commercial power reactors. Such personnel are familiar with the facility, the operating technical specifications (later modified for decommissioning), health and safety program (modified as new and different activities are involved), security, and quality assurance. They normally control the allocation of funds from the DTF. This group usually has primary oversight of the DOC work scope, and reviews programs, procedures, and techniques to be used.

Since the original site operators within the weapons complex are generally not available due to the typical delay in decommissioning, DOE usually conducts a competitive bidding process to select a DOC (or facility program manager) to provide the oversight for a specific facility or an entire site. The DOC usually takes over all activities, including the infrastructure of maintenance activities, until each building or facility is ready for decommissioning. After identifying the work scope for each facility, the DOC evaluates competitive bids and awards further subcontracts for specific portions of the fieldwork. The DOC also controls the budget allocation, overall health and safety, security, quality assurance, and provides oversight for the subcontractor.

The DOC for the utility/licensee is contractually responsible for performing the decommissioning activities, but this does not diminish the regulatory responsibilities of the owner/licensee. Usually, the DOC forms a team of subcontractors with capabilities in specific technical areas of the project, such as project management, waste management, reactor vessel and internals segmentation, and radiological health and safety. In some cases, the DOC may select additional subcontractors to fulfill specific project responsibilities.

1.3.8 Quality Assurance Requirements: Quality Assurance (QA) requirements for decommissioning follow the same principles as for operations, albeit on a graded approach commensurate with the safety implications involved with each project. As for any major project, a Quality Assurance Program Plan (QAPP) is developed for the project (and sub-projects, as applicable), and a QA staff is maintained to implement the QAPP. Decommissioning activities involve design of special equipment, heavy lifting, radiological exposure, packaging, transportation, and disposal. These activities must be controlled in the same manner as for plant operations. QA activities blend well with an integrated safety management system (ISMS).

The QAPP should cover all the major project activities and address the principles of:

- Organization and responsibilities
- Qualified vendor selection and vendor surveillance
- Material receipt inspection
- Design reviews
- Document control and record keeping
- Record closeouts
- Audits

QA staffing requirements begin with a QA Manager who should report directly to the highest ranking officer on site. For a major project, the manager will have several QA engineers, and in some cases Quality Control (QC) specialists to assist in implementing the program. The number of QA/QC personnel will depend on the size of the project, and will vary as the needs of the project change. It is not possible to provide specific guidance for the number of QA personnel required, as each project size, complexity, and safety implications vary. The staffing level is reduced as safety-related activities diminish. To ensure that issues are not overlooked, QA responsibilities are formally documented to flow down to subcontractors performing specialized elements of work.

1.3.9 Security Program: The security program for decommissioning is essentially the same as for operations. DOE and commercial industry have similar requirements for the protection of spent fuel and special nuclear materials. Security program requirements must be implemented as long as there is any material vulnerability. An important distinction, however, is that as soon as special materials are to be removed, significant changes must be planned for the security program at the site. The security force is routinely responsible for:

- Personnel background checks

- Personnel access control
- Guarding against intruders
- Monitoring access for service and vendor vehicles
- Guarding against theft of materials, such as tools, equipment, and SNM

The number of guards and their requirements at a site with spent fuel or SNM is the same as for an operating site. The NRC has specific guidelines for the numbers of trained, armed responders for such sites. Access control is tightly observed, and routine training drills are conducted to maintain the proficiency of the guard force. Once all spent nuclear fuel and SNM is removed from a site, the security force need not monitor personnel and vendor vehicles or require armed responders to prevent intruder access. The force may be reduced to reflect the reduced concerns for the remaining activities, eventually to the more economical security guard or night watchman.

1.3.10 Public and Stakeholder Participation Program: Experienced managers are increasingly recognizing the importance of appropriate public and stakeholder participation. Stakeholders are individuals and organizations with some vested interest in how the decommissioning will be conducted. The group includes those who are:

- Personally impacted
- Administratively impacted by virtue of election or appointment (such as state and federal regulatory agencies)
- Concerned by virtue of their interest, philosophy, or ideology (such as local and national public interest groups who observe the appropriate role of stakeholders)

Planners should address public relations and stakeholder concerns early in the planning phase to ensure that interventions do not jeopardize the project's schedule and budget.

1.3.11 Transition Planning: Transition refers to an operating facility being turned over for decommissioning. It is usually a traumatic time for personnel whose jobs are no longer required. The first step in transition planning usually is to identify key personnel who are needed to support the project. A retention bonus program is usually developed to entice them to remain until their special expertise is no longer needed. The bonus is forfeited if the employee leaves before the stated activity is completed. Several facilities have effectively used completion bonuses. Similarly, personnel whose expertise is no longer needed and who cannot be trained easily to perform a comparable role in decommissioning are placed on the severance list. These people are given earned severance packages, and in some cases (e.g., premature shutdown of the facility) an extra sum to accommodate special circumstances.

Another early step in transition is to establish the status of systems and structures in preparation for turnover for decommissioning. This may identify certain activities to clean out legacy wastes from operations, isolate systems from potential accidental releases, shut down systems not required for support, and identify clean equipment that can be sold immediately as scrap or salvage. This process may require a year or more to complete, and the planner must be involved in these key decisions to ensure that essential systems or structures are not shut down prematurely.

Maintaining a comprehensive knowledge of the operational history of the nuclear facility is a valuable resource for developing the DP. The operational history of the nuclear facility is typically maintained in extensive volumes of documentation and records. The NRC requires that specific design, construction, and operational records be generated and maintained as a means to monitor and evaluate the safety-related aspects of a licensed facility. These records and other regulatory and operational management records are valuable to develop operational plant and facility history.

1.3.12 Records: Construction records may provide detailed information relative to the locations of construction joints and critical stress/tensioning apparatus, and construction materials and reinforcement patterns. Photographs showing construction progress are useful as a means to determine unique or critical assemblies and structural configurations, heavy lift sequencing and embedded structural members, and piping installations. Similarly, design records are helpful to identify configuration of equipment, piping, and component interface.

Engineering drawings for nuclear facilities are most valuable in decommissioning. Design and installation drawings help determine the as-built condition of the facility. System descriptions and technical and functional specifications will be important if the decontamination process is to use existing equipment, or circulate chemicals in process equipment and piping. Specific documentation representative of specifications, equipment drawings, vendor data, and maintenance manuals for installed equipment will provide information for many purposes, such as the details of internal configurations. Weight lifting points and foundation details will be useful if the equipment is being salvaged. Operating manuals, spare part lists, and maintenance requirements are valuable if the equipment is to be used during decommissioning.

Facility operations and maintenance records provide significant insight into the various conditions that may have occurred during the operating life of the facility. Contamination occurs during both routine and abnormal events. Records of these events can provide significant information regarding the extent and nature of contamination. Operating and maintenance procedures can be used as protocol for identifying and accommodating differing conditions that may not have been considered during the design. As an example, process or contaminated fluids may have been dumped to a sump or transferred to another tank or system if the original design did not make adequate provisions. The reason for maintaining operating logs is to record operational occurrences. These logs may identify sources of contamination, length of time of spill, and cleanup measures, if any. Survey reports are frequently the best source for mapping the extent of contaminated areas. Exposure records, ALARA studies, and other information generated by health physics (HP) personnel provide significant details of radiological history. Waste and chemical inventories help predict the types and quantities of hazardous material contamination that might be present.

Safety-related documents provide potential information specific to fire hazards, emergency plans, and accident investigations. The documentation and regulatory notifications of incidents, spills, and accidents are addressed in weekly, monthly, and annual reports that summarize operational occurrences and describe potential dangerous conditions. Licensing documents include those regulatory records and reports that pertain to hazards to workers and to the public. The licensing process begins with a detailed description of the facility design that addresses possible accidents and upset conditions, and presents design features that control the hazards or mitigate their effects. The Preliminary Safety Analysis Report (PSAR) describes the intended design and the basis for granting a construction permit. This report is ultimately revised and reissued as a Final Safety Analysis Report (FSAR) to describe the facility in its "as-built" condition.

1.4 DECOMMISSIONING CHARACTERIZATION PLANNING

A proposed plan of action must be based on a defensible, comprehensive characterization of the facility. Such an effort is a prerequisite for evaluating the health and safety, technical, waste management, and financial requirements that must be included in definitive estimates of costs and disposal options.

Characterization is based on the definitive collection of all information needed to describe the following in detail:

- The hazards present at the facility
- The condition of the facility structures as it may affect occupational health and safety
- The extent, nature, and concentration of radiological and hazardous chemical contamination
- The institutional, legal, and technical restraints on alternatives

A detailed facility characterization is generally a major effort. Except for simple facilities, a written plan is a key document if the characterization is subcontracted, and is generally required by the waste disposal facility to ensure that the waste is identified and documented sufficiently to meet specified acceptance criteria. There are several software packages commercially available to optimize this effort, such as to help stage activities in order to minimize worker exposure.

An initial scoping survey is required to identify personnel hazards and to develop the DP. Detailed characterization data is collected to determine the type and magnitude of contaminants prior to beginning any activities. Assessments and surveys are performed during the decontamination phase as a means to evaluate facility conditions. The QA for characterization by surveys requires a documented program of QC activities that measure performance relative to specified standards. This program must be described in a formal QAPP that requires:

- Periodic calibration and maintenance of monitoring instruments
- Checking of data entries (measurements and locations) by selective replication
- Tracking of records with regard to instrument identification, operator, location, and the date the instrument was used

A reliable background reference must be established for radioactive and other contaminants. The major contaminants in a structure are readily identified at many facilities. Following a guide such as can provide a simple path to unambiguous characterization of areas acceptable for release and those that need decontamination. The Multi-Agency Radioactive Site Survey and Investigation Manual (MARSSIM) (NRC 2000) is a widely accepted tool that was prepared jointly by the NRC, the DOE, and the U.S. Department of Defense (DOD).

The sample collection plan must be able to identify the spread and level of contaminants. Sample collection techniques for must meet the project specifications, and sample preservation must meet criteria for maintaining radioactive and other contaminants in their original chemical and physical forms. A chain of custody must track the sample from collection through analysis and reporting. This information will then form the basis of the facility's final status of the required and a potentially required confirmatory survey to document the end-state status of the facility. As no single technology can address the full spectrum of requirements, multiple characterization methods are applied for facility and material characterization. Decommissioning managers must identify the appropriate suite of characterization methods necessary to characterize the facility.

Characterization methods are often combined to take advantage of the unique strengths and compensate for the limitations of each method. Some characterization programs attempt to go to extreme details to identify specific levels of contamination within systems, structures, and soils. However, the needs are significantly less sophisticated for cost estimating, when there is only the need to determine clean versus contaminated and to identify the principal radionuclides.

The radioactive contamination in and around most nuclear facilities is normally classified as LLW, with the exception of spent fuel or transuranic waste having a concentration in excess of 100 nCi/g of alpha activity and mixed waste that contains radioactive material along with hazardous chemical constituents. Areas that require more extensive investigation are those with components that are either potentially clean or highly contaminated. In such cases, a more thorough investigation may be needed to determine that potentially clean components can be free-released or that very highly contaminated components might require special decontamination procedures prior to removal to reduce occupational exposure.

1.5 DETAILED DECOMMISSIONING PLANNING

This phase of the project is when specific technological evaluations are performed, and detailed working assumptions are made. It normally begins after the facility is permanently shut down, when the plant radioactive and other hazardous material inventory is no longer changing and can be determined by measurement and calculation. It involves interaction with regulators and stakeholders to satisfy regulatory rules and guidelines, and acceptance of the approach. Experience shows that shortcuts at this stage can result in risks to safety and budget overruns. In this phase, all final planning and engineering activities are defined. This includes:

- Preparing the DP or PSDAR
- Defining the approach to managing the project
- Selecting the techniques and methodologies for decommissioning activities
- Preparing the activity/work/task specifications
- Preparing the detailed baseline cost estimate and schedule

The information developed during pre-planning and characterization is re-examined with respect to the assumptions made, accuracy of the cost and schedule estimates, regulatory restrictions on each approach evaluated, and feasibility of implementing each technique or method. The planners decide which techniques provide the most cost-effective methods to achieve the decommissioning objectives and use this information to set the project path and prepare appropriate regulatory documentation. As necessary, additional engineering evaluations are performed to refine and update the previous work, and to provide sufficient detailed analyses to support a decision to select a course of action. Often these engineering evaluations continue through the operations phase.

1.5.1 Preparation of the Decommissioning Plan: The DP sets out the detailed approach to be taken on the project. Its major function is to inform regulators and other stakeholders of the course of action the owner/licensee chooses to follow. Commercial nuclear plants are required to submit a PSDAR, which describes the required activities performed prior to shutting down the reactor, what has been done to defuel, and what operating technical specifications were changed to match the reduced status safety requirements. Research reactors and DOE facilities need only to identify the removal of radioactive materials, and license termination if appropriate. The DP typically provides a summary characterization of the facility, the management organization to be used, license termination activities, waste disposal options, and an overview of the task organization to remove components, systems, and structures.

1.5.1.1 Decommissioning Project Plan: Sometimes an owner/operator will generate an internal decommissioning plan to describe how the project will be managed. It identifies the project management organization, and the relationship and responsibilities of each functional position. The value of such a plan is to eliminate the uncertainty of who has responsibility for each aspect of management. It avoids lapses, and overlapping or duplication in responsibilities. It should also identify an integrated safety management program to ensure safety during each phase of the project.

1.5.1.2 Identification of Detailed Activities: At this stage the planners have sufficient information to:

- Identify how to accomplish each of the major activities
- Elect the techniques for in-situ decontamination of systems and structures
- Establish general guidelines for the precautions, limitations, and operating parameters
- Determine methods for dealing with secondary waste from decontamination
- Identify the techniques for removal activities for systems and structures
- Evaluate and select he methods for removal of the reactor vessel and internals
- Determine necessary support equipment and services, such as electrical power supplies, ventilation, laydown areas, and interior transport routes of casks

At this stage the planners can also identify the type of acceptable waste disposal containers. These might include specialized casks and individual liners for vessel and internals, specification containers for piping, systems, and components, and bulk material containers for concrete rubble and soil. Staging areas must be built and designated for the temporary storage of empty containers; restricted areas must be established for temporary storage of loaded containers awaiting shipment.

The selection of transport options is typically based on access to major roadways, railroads, barge facilities, or any combination of these. Most shipments are made by trucks, which have the greatest flexibility as to destination, but may also have such routing limitations as bridge widths and wide loads. Therefore, rail shipments may be more desirable and cost-effective for larger shipments of casks for the reactor vessel and internals. When adequate facilities are available, barge transport is often preferred for very large loads, such as steam generators, pressurizes, and intact reactor vessels.

The options available for disposal of radioactive and other hazardous materials are reviewed in detail with respect to each facility's waste acceptance criteria. For each facility, the criteria identify:

- Types of wastes accepted
- Form of the waste
- Limits of concentrations and total quantity of radionuclides
- Limitations of the maximum amount of radioactive materials that may be received and stored on site until transported for burial

1.5.1.3 Project Scope: An assessment of the site's historical operations data prepares the decommissioning planner with a preliminary overview of the general status. After the HSA is established, specific information is integrated into the planning for the decommissioning process. At this stage, operations can be translated into technical objectives that are incorporated into concise planning documents and a detailed project-specific work scope. The work scope must define numerous discreet elements of the project execution in terms of management and technical approach, types and levels of resources, and the hierarchy of tasks. Explicit and quantified work scope definition is critical for projects with a significant level of regulator and stakeholder involvement. Additionally, changes in the scope that might result in changes to the FSAR could significantly impact the baseline project budget and schedule.

First, structuring the work scope helps to ensure that all operational activities required to complete the planned project are captured and facilitates assignment to responsible organizations (i.e., managers, plant forces, contractors). In most projects of this nature, there are multiple project participants or contractors to perform tasks; they must all be coordinated and integrated to ensure a focus on safe operations and to minimize conditions that might cause work slowdowns.

A Work Breakdown Structure (WBS) defines the scope of work for the project, displays the project in graphical form, and categorizes cost elements and work scope activities into logical groupings that have a direct or indirect relationship to each other and to the end product. The work groupings are usually related to the accounting system, or chart of accounts used for budgeting and tracking major elements of the project costs. The WBS allows for organizing tasks so that responsibilities can be assigned appropriately and identifying interrelated activities and decision points. Activity sequencing requires the determination and documentation of the relationships between discreet activities. Activity definition requires examination of the total work scope and using the WBS to develop discreet activities that are unique and can be associated with a deliverable. Typical scope elements of the project that may be included in the WBS include the following:

- Preliminary site characterization
- Safety/Licensing documentation
- Regulatory documentation
- Permits and approvals
- Site characterization
- Dismantling and demolition
- Waste treatment
- Transportation and disposal
- Inspection
- Conceptual design
- Preparation of the DP
- Preliminary design
- Detailed design
- Procurement
- Final site survey and documentation
- Independent verification and documentation
- Site restoration

Each activity may have a corresponding predecessor or successor activity. The activity flow model is the road map of the entire project as defined by the planned end-state and the activities required to achieving that state on schedule and within the projected budget. A detailed WBS is presented in the compact disk enclosed in the back of this handbook.

1.5.2 Detailed Cost Estimate and Schedule: At this stage, the planners can develop the detailed cost estimate and schedule for the project. These estimates are usually adopted as the baseline estimate for the project, against which project performance can be measured and funding requirements established. The cost estimates are broken down into phases of the project, and further refined to specific major tasks. Cost estimators and schedulers must work closely with the procurement group so that meaningful correlations can be made between how the work packages are to be combined and contracted, and how to measure their progress against the budget. Similarly, the schedule is carefully coordinated with the contracting practices in order to maintain continuity and properly balance resources throughout the project. Financial planning must include, as a minimum, the following:

- Evaluation of anticipated schedules
- Labor costs
- Radiological and hazardous material disposal options
- Availability of proven characterization, decontamination, dismantling, and waste management technologies

Based on the approach to the final facility end-state, (DECON, SAFSTOR, or ENTOMB) plans, budgets, and schedules specific to the selected approach will form the framework of the decommissioning project. As in any major multi-year project, it is imperative to ensure that all available financial and economic data is used to develop the cost estimates. Cost estimating approaches are best applied to an accurate inventory of facility equipment and quantities of anticipated demolition debris. Projects the size and complexity of most nuclear facilities often encounter unplanned (but preventable) delays that result in significant cost escalation.

1.5.2.1 Decommissioning Cost Planning: As discussed earlier, a thoroughly sound cost estimate and baseline resource-loaded schedule must be developed for the decommissioning option selected. Numerous examples (including cost modeling software) are available to use as validation tools to estimate costs of materials, labor, equipment, waste handling, and project management. The total cost depends on the sequence and timing of the various stages. A prudent approach, if feasible, is to ensure that estimates developed by incumbent staff are compared and validated against independently developed estimates. Overall project uncertainties, risks, and delay routinely cause project costs to be impacted by 15% to 35%, with an average of 25% (see Chapter 12).

Labor costs are a large component of any decommissioning project. Cost comparisons of using facility staff labor versus outsourcing to specialty firms with dedicated personnel and equipment should be evaluated. In the case of government projects, the Davis-Bacon Act affects the cost of labor. A specific project's scope and schedule will affect the extent of work performed under the provisions of that act since contractors and their subcontractors must pay workers employed at the site no less than the prevailing wages and fringe benefits paid on projects of a similar nature.

A site-specific detailed inventory of systems and structures should be generated in a format that ensures that the most accurate information is available for all future planning and estimating activities. This inventory identifies the scope of systems and structures to be in any estimate and should list specific information necessary to obtain component identification, location, and size (overall dimensions, volume, and weight). Design specifications of structural and civil components can also be valuable for dismantling planning. Whenever practical, heavy equipment demolition is generally less costly than the hands-on removal of systems and components followed by demolition with heavy equipment. Obviously, the contamination of the facility, system, or component is the critical factor in choosing removal methods.

In the case of either commercial or federal facility decommissioning, unit cost factor estimates (discussed further in Chapter 12) can be developed as a demonstrable basis for establishing reliable cost estimates. Such detailed line-item factoring of activity time, labor costs (by craft), and equipment and consumable costs ensures that cost elements have not been omitted. Generally, the planner should develop separate unit costs for both the types of equipment and the duration (in minutes) of specific activities required for removing a component. Work difficulty parameters should be applied to include specifics pertaining to work under ALARA. Conditions that increase the time necessary to complete specific operations include such factors as protective equipment, heat stress, confined space, and additional work breaks. Crew utilization and interface requirements also impact performance, based on composition and size. The total cost for labor, materials, and equipment should be summed and divided by the appropriate unit value, such as pipe length, size of valve, and weight of heat exchanger.

Although the major effort in decommissioning is the removal of radioactivity, other potentially hazardous contaminants must also be characterized. After an area is surveyed and the radiological conditions are established, biological cleanup and general housekeeping will commence. This includes removing loose bird/animal wastes and rubble, and sweeping and vacuuming the floors. Under some jurisdictions and waste management programs, non-friable asbestos can be left in a structure and removed by heavy equipment along with the demolition rubble. However, any asbestos-containing material, typically found in the following items, is usually removed:

- Vinyl floor tile
- Cement board, wall, and roof panels
- Thermal system insulation on steam and water piping
- Boiler insulation
- Roofing felt
- Sheet rock joint tapes and finishing plaster
- Tile and cove mastic
- Window and door putty

In addition to the nuclear complex structures, most nuclear facilities require large storage, maintenance, and repair facilities, laydown yards, and boundary perimeters. Because of the potential for hazardous material releases in these areas, all radiological and other hazardous materials should be considered, such as soil and water (surface and subsurface), adjacent waterways, gaseous effluents, and surface solids. Consideration must also be given to their possible pathways to the public and the environment, including drinking water, irrigation, agriculture, fish, and wildlife. Finally, the inventory of contaminated materials and their distribution should be included in the estimate.

1.5.2.2 Project Schedule: Proper management of a decommissioning project requires rigorous scheduling and ongoing control throughout the project, with appropriate change controls when necessary. The project schedule is the road map that determines how the process will move from conceptual engineering to completion and site restoration. During project execution, the schedule should be used as a forward-looking tool rather than a mechanism for reporting what has happened and when. As a tool, it supports different activities for different members of the project team, as follows:

1. For senior management, it provides a measure of project performance
2. For the stakeholder, it provides information on the work scope, the status of the decommissioning, and the potential effects of specific activities
3. For the project manager, it provides an overview of where the project is going, a means of assessing project performance, an understanding of differences between planned and actual progress, a means of allocating resources and adjusting resources, and the framework for assessing, managing, and controlling project risk
4. For the engineer or field supervisor, it provides information on what must be done and when, and what activities may affect the start and completion of the work

It is critically important that the project team implements the schedule and that senior management routinely enforces and supports its continued use. Senior management should attend project status meetings, actively ask questions, and challenge personnel on performance of work according to the schedule. If management either has or is perceived to have its own schedule, or allows for milestone dates that are in conflict with the team's schedule, the entire project team will quickly lose commitment and interest in support of the plan. Management must control but refrain from dictating the schedule.

Routine meetings with all project personnel should be held to review current status and evaluate subsequent activities that must be accomplished to support project goals. Team members should discuss problems that impact the schedule, resource requirements to support the schedule, significant hand-off issues (to ensure that personnel know when one activity ends and another begins), and evaluate risk issues and plan for contingencies.

1.6 DECOMMISSIONING OPERATIONS

The decommissioning operations phase entails the actual performance of activities. As noted earlier, it is usually the DOC who has the responsibility to implement the project for the owner/licensee. The DOC brings on its experienced staff of professionals to manage the project and subcontracts for the specific tasks of decontamination, removal, packaging, shipping, and disposal. Similarly, the DOC provides the survey of the decontaminated facility to meet the criteria for license termination, and subcontracts to experienced vendors for instrumentation, analyses, and reporting of the residual levels of contamination. Most DOC contracts include a prime contractor and several subcontractors. In some cases the subcontractors are imbedded in the DOC management; that is, they work as part of the single management team and there are no specific lines of demarcation where the DOC and subcontractor staffs interface.

1.6.1 Work Scope Activities: In most cases, subcontractors have a specific role for a portion of the work and, when that work is completed, the subcontractor is relieved of its responsibilities at the site. The major elements of subcontracted work are:

- Specialty decontamination methods
- Remote special tooling for vessel/internals removal
- Concrete scarification technologies
- Piping and equipment removal
- Demolition activities
- Packaging of materials
- Waste disposal facilities
- Transport company (truck/rail/ barge)
- Heavy rigging companies
- Waste processing vendors

Vendors may also be prepared to supply instrumentation for determining the levels of residual contamination for free-release, and for the experienced staff to oversee this element of the work.

It is not possible to detail the specific activities for each type of facility. Part 5 of this handbook includes specific case histories of decommissioning projects. These can be a valuable reference for actual experience in dealing with unique problems as they arise at each site and the planner is encouraged to review them for additional insights.

1.6.2 NRC License Termination Plan: The LTP, required two years prior to license expiration, identifies the specific activities required to remove the radioactivity to meet the free-release (or partial release) Criteria. It also includes the plan for the final site survey to certify that the free-release criteria have been met. Upon NRC approval of the LTP, decommissioning activities may proceed.

As contaminated systems, components, and structures are removed, the staff endeavors to isolate the cleaned areas to prevent recontamination from ongoing activities. Doors to cleaned buildings are locked and sealed closed, the area is posted, and in some cases a guard may be used to control access. In areas adjacent to active decommissioning work, plastic sheeting may be put up to seal the clean areas, and a temporary chain link fence or other barrier erected to prevent access. For these cleaned areas, the site radiological engineering and HP staff will survey to demonstrate that the residual radioactivity meets the free-release criteria. The NRC typically performs independent surveys; upon satisfactory verification, it will issue a License Termination Order.

DOE facilities do not have specific license termination procedures since they are generally not licensed by regulatory agencies. However, DOE projects typically have an independent verification survey conducted by an outside organization to determine that project cleanup criteria were met. Also, in the case of a DOE D&D project conducted under CERCLA at a site listed on the National Priorities List, the specific conditions and clean up criteria specified in the CERCLA Record of Decision for the project must be met. A DOE D&D project that is solely under RCRA rather than CERCLA may require that certain portions of the project involve cleaning areas containing hazardous chemical materials; the successful clean up of these areas must then be formally signed off by the EPA or its designated responsible state agency, as appropriate.

1.7 CLOSEOUT AND POST-DECOMMISSIONING

This phase is the final disposition of decontaminated or noncontaminated systems and structure, arid restoration of the site to the desired end-point of the facility. The DOC is usually retained to complete the process by demolishing the remaining nonradioactive structures and restoring the site. Using the DOC for this purpose is logical as the DOC is most familiar with the facility, has some, if not all of the necessary equipment on site, and has contracts in place to access the necessary crews.

Funding for demolition of nonradioactive structures and restoration of the site is not part of the NRC definition of decommissioning. Accordingly, no separate funding is required by the NRC to demolish these structures or restore the site. However, most Public Utility Commissions consider this phase of decommissioning as important as the removal of radioactivity. The concern is that the owner/licensee will abandon the site, leaving it unsafe or in an aesthetically unacceptable condition.

Accordingly, most allow utilities to collect and designate an amount to be used as part of the DTF. By comparison, such demolition and restoration are definite elements of any DOE project and are included in the D&D project plan, cost estimate, and schedule.

1.7.1 Demolition of Remaining Nonradioactive Structures: The remaining nonradioactive structures at a decommissioned site are usually candidates for total demolition. In most cases, the remaining buildings are not suitable for reuse without extensive refurbishment and upgrading to meet local codes and standards for reoccupation under a new charter. Chemical or hazardous building materials may be present, requiring removal or remediation prior to site restoration and stabilization. The buildings may have large access openings cut into them to facilitate removal of major equipment. They may be structurally unsafe for abandonment, as the original means of egress may have been changed to facilitate decommissioning. These buildings may be deemed unsafe structures under any one of the state codes formulated by the Building Officials and Code Administrators (ICC 1997) and/or the Uniform Building Code (BOCA 1996), thereby requiring either demolition or indefinite maintenance for safety. The local building official has the authority to make that determination. Site restoration can take many forms, including leaving it as an industrial site, returning it to near pre-construction conditions, or a combination of these options.

1.7.2 Site Restoration for NRC-Licensed Facilities: The degree of site restoration to be implemented is generally up to the utility/licensee, but often stakeholders are influential in guiding the decision process as to how best to meet the local community's needs. The most desirable restoration is to return the site to a condition whereby the buildings are removed to below grade, the site is back-filled and covered with clean soil, and the ground is seeded to stabilize the soil. In some cases it may be acceptable to leave the site as an industrial facility, using gravel to control runoff and the growth of weeds and other unwanted vegetation. The specific approach adopted for each site should incorporate stakeholder input and yet make financial sense for the owner/licensee.

As previously mentioned, site restoration is an official part of an overall DOE project unless it is specifically excluded in project documents such as the Project Plan. For example, site restoration such as the clean up of soil beneath a building that is being decontaminated may be deferred (as explicitly stated in the project documents) until a later date when the building structure itself and its foundations will be demolished and the rubble removed. Likewise, certain soil decontamination activities of low-level soil contamination may be delayed until a later more suitable date. Also possible groundwater contamination resulting from radioactive leakage from the original facility may be remediated under a separate project from the facility decommissioning project.

1.7.3 Demobilization: Upon completion of each major phase of the project, some demobilization of equipment and manpower resources can be anticipated. Specialized equipment (such as reactor vessel cutting equipment) is packaged and shipped off site as it is no longer needed. The type of human resource required will change as the more highly skilled craftsmen needed for decontamination, special cutting, concrete demolition, etc., are replaced with less critical employees to complete the decommissioning activities. The crews should receive timely notification as to when these demobilization activities are to take place so that they can plan their future work assignments, just as in the case for initial site transition.

The DOC must ensure that the proper resources will be available to complete the project safely and that the crew does not try to extend the project until they can find replacement work. This is an important part of DOC management, and continued safety programs must be implemented to reduce the risk of real or imagined injuries to workers, thereby minimizing the opportunity for them to claim Worker's Compensation coverage to bridge any gaps in employment. It is the DOC's responsibility to balance crews and work scope to the last day of the project.

Chapter 2

THE NUCLEAR AND RELATED INDUSTRY

2.1 INTRODUCTION

During the more than half-century history of the U.S. weapons programs, a vast research, production, and testing network was developed that came to be known as the "nuclear weapons complex.." The DOE and its predecessor agencies have been responsible for most of these facilities. DOE-owned plants and equipment include:

- Facilities for uranium milling and refining
- Gaseous diffusion plants for producing weapons-grade uranium
- Facilities for fabricating and testing weapons
- Nuclear reactors for testing materials and equipment components
- Nuclear reactor prototypes
- Cyclotrons
- Hot Cells
- Research laboratories

They also included reactors for making tritium, plutonium, and radioactive and stable isotopes of many elements; and chemical separation facilities for extracting uranium, plutonium, and other radionuclides.

In 1954, Congress passed the Atomic Energy Act, directing the Atomic Energy Commission "to encourage widespread participation in the development and utilization of atomic energy for peaceful purposes." A logical extension of nuclear science and engineering was the development of commercial nuclear power. Generating electricity by nuclear power was found to be commercially viable by the early 1950s. By 1960, as a result of cooperation between the commercial industry and the government, two nuclear reactors primarily designed to make electricity had begun operations: Dresden Unit 1, a boiling water reactor (BWR), and Shippingport, a pressurized water reactor (PWR). The development of U.S. commercial reactors continued through the 1980s, and still continues internationally.

There is extensive literature on the history of nuclear technology, the establishment of various facilities, and the events that led to both the development of nuclear weapons and the peaceful application of nuclear energy. The manager of the program known as the Manhattan Project describes in detail the formation of national laboratories in Los Alamos, Oak Ridge, Idaho Falls, Richland, and elsewhere (Groves 1962). Compton explores scientific issues, particularly those faced by Fermi and others during the initial reactor development (Compton 1956). The early history of nuclear technology is also described by Rhodes (1986) and Hewlett (1969). Descriptions of the weapons complex and the legacy of wastes from weapons manufacturing operations come from the DOE (1995a). DOE (1997) published what is probably the first comprehensive analysis of the sources of waste and contamination generated from nuclear weapons production, and later (DOE

1998b) later gave a site-by-site, project-by-project projection of the technical scope, cost, and schedule required to complete its remaining cleanup projects (DOE 1998b).

2.2 DEFINITION OF DECOMMISSIONING AND RELATED TERMS

The industry's decommissioning plans are at various stages of development. Some nuclear facilities have either ceased operations or are nearing the end of their operating lives; others have already achieved a free-release from all regulatory stakeholders. As of this date, there is no agreed-upon definition of activities related to decommissioning. Nevertheless, regardless of variations in terminology and definitions, all agencies and technical communities agree that the ultimate objective is to reach a condition that affords protection of the environment, including the health and safety of workers and the general public.

2.2.1 Deactivation: The process of placing a DOE facility in a stable condition is often called deactivation (DOE 2000). The principal purpose of deactivation is to significantly reduce potential hazards. The process includes removing radioactive, chemical, and other hazardous materials to ensure adequate protection of workers, the general public, and the environment. Actions include removing fuel and stored radioactive and other hazardous materials, draining and/or de-energizing nonessential systems, and related actions. Deactivation doesn't necessarily include removing contamination remaining in fixed structures and equipment. Often the objective of deactivation is to limit the long-term cost of surveillance and maintenance.

2.2.2 Decontamination: There is a high level of agreement on the definition of decontamination -- the complete or partial removal of contamination by a deliberate physical, chemical, or biological process. It may be accomplished by:

- Treating the surface to remove or decrease the contamination
- Allowing the material to stand so radioactivity decreases over time through natural decay
- Covering the contamination to shield or attenuate emitted radiation

2.2.3 Disposition: This term is specific to the DOE and consists of those activities that follow completion of program mission, including but not limited to surveillance and maintenance, deactivation, and decommissioning.

2.2.4 Decommissioning: According to the DOE, the ultimate goal of decommissioning is the unrestricted release or restricted use of the site (DOE 1995b). Consequently, decommissioning takes place after deactivation and includes surveillance and maintenance, decontamination, and/or dismantling. These actions are taken with adequate regard for the health and safety of workers and the public and protection of the environment.

The NRC defines decommissioning as the process of closing down a facility followed by reducing residual radioactivity to a level that permits the release of the property for unrestricted use and termination of the license (NRC 2002a).

The IAEA defines decommissioning in its Safety Glossary (IAEA 2000) as follows:

1. Administrative and technical actions taken to allow the removal of some or all of the regulatory controls from a facility (except for a repository, which is not decommissioned, but closed). The use of the term decommissioning implies that no further use of the facility (or part thereof) for its existing purpose is foreseen.

   The actions will need to be such as to ensure the long-term protection of the public and the environment and typically include reducing the levels of residual radionuclides in the materials and the site of the facility so that the materials can be safely recycled, reused or disposed of as exempt waste or as radioactive waste, and the site can be released for unrestricted use or otherwise reused.

   Decommissioning typically includes dismantling the facility (or part thereof), but in the Agency's usage this need not be the case. It could, for example, be decommissioned without dismantling and the existing structures subsequently put to another use (after decontamination).

2. All steps leading to the release of a nuclear facility, other than a disposal facility, from regulatory control. These steps include the processes of decontamination and dismantling.

2.3 DOE DECOMMISSIONING

The legacy of more than five decades of weapons production manifests itself in the thousands of nuclear weapons complex facilities no longer in use. There are no operating facilities currently dedicated to producing key weapon ingredients, such as tritium or plutonium. Although some portions of the weapons complex are continuing to operate, its current mission is to clean up the Cold War legacy. The 1995 inventory includes more than 130 sites in 33 states and Puerto Rico, comprising a combined area of about 3,300 mi^2 (DOE 1995a). In its diversity, the nuclear weapons complex ranged from tracts of isolated desert in Nevada, where weapons were tested, to warehouses in downtown New York City that once stored uranium. Its national laboratories in New Mexico and California designed weapons for production in Colorado, Florida, Missouri, Ohio, Tennessee, Texas, and Washington. Even now, long after some of the sites have been turned over for other uses, the DOE owns 2.3 M acres of land and 120 M ft^2 of buildings. The growing inventory of aged, degraded, and contaminated buildings requiring decommissioning totals over 7,000. Contained within these structures and elsewhere in the weapons complex are nearly 35,300,000 ft^3 of radioactive waste and materials that need treatment, storage, or disposal after removal.

However, the DOE has also been involved in decades of research, development, demonstration, and application of energy technologies for civilian applications, as well as basic and applied sciences. Most of this work is accomplished through a complex of national laboratories, but a significant portion of the portfolio involves other institutions, such as universities and commercial facilities. Although many facilities require eventual decommissioning, such as activation from a high energy physics accelerator, or alpha-gamma hot cells used for examination of nuclear fuel failure, their cumulative impact is considered much smaller than that of the remediation backlog of the weapons facilities.

2.3.1 Weapons Production: The U.S. weapons program involved many facilities, each performing a discrete step in the production process. The following is a general summary of the types of facilities that will require decommissioning.

2.3.1.1 Uranium Mining, Milling, and Refining: Uranium ore was mined (primarily in western states) and processed into uranium oxide, known as yellow cake, in the milling stage. Refining involved chemically converting uranium concentrates into purified forms suitable as feed material for the next step in the process. Wartime refining was performed in several facilities at Fernald, OH and Weldon Spring, MO. Although mining, milling, and refining operations would account for only a fraction of the radioactivity in the byproducts of weapons production, the sheer volume of ore mined and processed resulted in mill tailings accounting for a larger share in waste volumes.

2.3.1.2 Enrichment: This process consisted of separating naturally occurring isotopes of uranium into low enriched, highly enriched, and depleted uranium. Uranium was converted to UF_6 to be processed in a gaseous diffusion plant. The first enrichment facilities were located in Oak Ridge, TN; later plants were built in Portsmouth, OH and Paducah, KY.

2.3.1.3 Fuel and Target Fabrication: This step consisted of the foundry and machine shop operation required to convert uranium feed material, principally metal, into fuel and target elements used in nuclear materials production reactors. Uranium (as UF_6) was converted into metal at Weldon Spring and Fernald. After uranium was coated with aluminum or zirconium metal, it was assembled into fuel and targets for the reactors. The fuel and target assembly areas were located at the reactor sites in Hanford, WA and Savannah River, SC. Tritium was also produced at SR. Eventually, Fernald and Weldon Spring took over part of the fabrication mission.

2.3.1.4 Reactor Operations: This step included fuel and target loading and removal, reactor maintenance, and the operation of the reactor itself. Hanford's nine reactors all consisted of large cubes of graphite blocks with horizontal channels cut in them for the uranium fuel and cooling water. SR's five reactors had a large tank of heavy water (D_2O) in which highly enriched fuel and separate depleted uranium targets were submerged. Because only a small fraction of the uranium in fuel and targets was converted to plutonium during each cycle through a reactor, workers at Hanford and SR processed large quantities of uranium. Almost the entire radioactivity in the environmental legacy of nuclear weapons production was created by reactor operations.

2.3.1.5 Chemical Separations: This process consisted of dissolving spent nuclear fuel and targets, and isolating and concentrating the plutonium, uranium, and other nuclear materials they contained. Because of potential high levels of radiation exposure, the process was performed remotely behind lead glass shielding and thick concrete walls. In the U.S., eight of these chemical separation plants, called canyons because of their long narrow shape, were operated for recovering plutonium and uranium until the late 1980s. For example, the Plutonium-Uranium Extraction (PUREX) facility at the Hanford site in Washington operated from 1956 to 1972 and resumed operation from 1983 through 1988. Plants were also operated for decades in South Carolina and Idaho. The canyons are highly contaminated with fission products and transuranic elements. The plutonium metal from the reprocessing plants generally was formed into "buttons" for temporary storage and shipments.

2.3.1.6 Component Fabrication: Weapons component fabrication included the manufacturing, assembly, inspection, bench testing, and verification of specialized nuclear and other parts and major subassemblies. Most plutonium from the reprocessing plants went to the Rocky Flats Plant in Colorado to be machined into warhead components. It was usually in the form of a metal, but liquid and powdered forms of plutonium were also produced. In addition, the weapons laboratories used some plutonium to make and test prototype designs for weapons. Because of the high toxicity of plutonium, workers were required to use glove boxes equipped with safety and ventilation systems. This stage of the process resulted in contamination of facilities by solvents and heavy metals, explosives, and plutonium.

2.3.1.7 Weapon Operations: These operations included the assembly, maintenance, and dismantling of nuclear weapons. A variety of weapon components were manufactured at multiple locations and shipped to several locations for final assembly into weapons. Most recently, the Pantex Plant in Texas was the location for this final assembly process. Weapons operations resulted in relatively small waste quantities and contamination as compared to other weapons production steps because radioactive materials handled in this step were generally in the form of sealed weapons components.

2.3.2 Classification of Radioactive Wastes at DOE Facilities: Unwanted radioactive materials were produced in every step of the weapons production process. In addition, a number of other hazardous materials, including toxic chemical compounds, were used during the production process. Because the regulatory requirements for radioactive waste and other waste were developed independently, several inconsistencies emerged for treatment of wastes posing comparable human or ecological risk. As a general rule, waste resulting from decommissioning activities of facilities can be classified into the following groups.

2.3.2.1 High-Level Waste: As defined in federal regulations (NRC 1991a), high-level waste (HLW) consists of spent reactor fuel and those wastes resulting from reprocessing. It is the most radioactive byproduct from reprocessing and contains most of the radioactivity originally found in the spent fuel. The DOE currently stores about 100 M gallons of HLW. The waste, stored in tanks, contains a variety of radioactive liquids, solids, and sludge. The Idaho Chemical Processing Plant converted radioactive liquid from processing naval reactor fuel into a dry calcine form.

Five facilities were built for reprocessing operations in Hanford, WA, and two in SR. Four of the Hanford canyons and one of SR's were primarily devoted to plutonium extraction. Two others (in SC and ID) were used for extracting highly enriched uranium from spent fuel. The fifth Hanford canyon was briefly used to recover uranium from HLW tanks. In addition, a demonstration plant for reprocessing commercial spent fuel was built and operated briefly in West Valley, NY. The HLW from this plant is also a DOE responsibility.

2.3.2.2 Transuranic Waste: This waste stream contains alpha emitting isotopes of elements heavier than uranium (therein the "transuranium" or "TRU" designation) with half-lives greater than 20 years, at a concentration greater than 100 nCi/g (DOE 2001). TRU waste result from processes such as reprocessing spent nuclear fuel, weapons production, handling some kinds of SNM (such as enriched uranium targets used in pulsed neutron sources), and from protective clothing, contaminated equipment, and solvents contaminated primarily with plutonium. The denoted concentration level is within the range conservatively deemed acceptable for routine disposal without having to consider permanent isolation, such as in a geologic repository.

Both accidents and normal operations have generated significant quantities of TRU waste. The Rocky Flats Plant experienced two major fires in 1957 and 1969, and numerous small fires in its production line. Fire-fighting and subsequent decontamination efforts generated a significant number of waste-filled drums, most of which were shipped to Idaho for storage. The site of the previous Rocky Flats Plant is transitioning into a wildlife preserve.

Throughout the nuclear weapons complex, the TRU waste inventory in storage totals about 3,530,000 ft^3, or the rough equivalent of 500,000 55-gallon drums. As in the case of spent fuel and HLW, much of the TRU material was placed in temporary storage until a permanent repository became operational. The Waste Isolation Pilot Plant (WIPP) in Carlsbad, NM has since opened and is currently receiving TRU waste for disposal.

2.3.2.3 Low-Level Waste: Low-level waste (LLW) is primarily defined by what it is not. LLW is radioactive waste that is not HLW, TRU, byproduct material (as defined in section 11e(2) of the Atomic Energy Act), or naturally occurring radioactive material (DOE 2001). DOE is self-regulating for its LLW, but is subject to NRC regulations when commercial treatment and disposal facilities are used (NRC 2002b).

LLW was generated from a diversity of activities at the DOE sites, ranging from wastewater treatment and equipment maintenance associated with research, development, and weapons production activities to activation of materials from reactor and accelerator operations. Some LLW was also derived from the pretreatment of HLW and the management of chemical separation facilities. Decommissioning has and is likely to continue to produce significant quantities of LLW; it is disposed of at both commercial and DOE facilities.

2.3.2.4 Hazardous Waste: Many solvents, effluents from industrial processes, and other wastes are regulated by the EPA, based on requirements of the Resource Conservation and Recovery Act (RCRA 1976). The principal implementing regulation appears in 40 CFR 270, entitled EPA Administered Permit Programs: Hazardous Waste Permit Program (EPA 1976). These materials are considered to be hazardous waste if they are either listed in the regulations as a hazardous waste or exhibit a specific characteristic as corrosive, ignitable, reactive, or toxic. According to RCRA, states may develop their own hazardous waste regulations as long as they are at least as restrictive as EPA regulation. Consequently, several states have promulgated their own regulations in addition to, or different from, EPA regulation. Therefore, a waste management practice that may be acceptable in one state may not be acceptable in another.

Like other large industrial operations, the nuclear weapons production complex created vast quantities of hazardous waste. DOE examples include various acids, nitrates, oils, heavy metals, fluorides, and explosives. Except for simple on-site treatment processes, such as stabilization and neutralization, most hazardous waste is managed at commercial facilities because of the existing treatment and disposal experience, and the economy of scale.

2.3.2.5 Mixed Low-Level Waste: Mixed LLW contains both radioactive and hazardous waste properties. Consequently, management of mixed waste attempts to be consistent with the requirements for separately managing each component. However, as requirements were developed independent from each other, their implementation is often complicated. Recent cooperative activities between the NRC and EPA have reduced problems associated with the management of mixed waste.

Mixed LLW is generated during a broad spectrum of processes and activities, including equipment maintenance, materials production, cleaning, environmental restoration, decommissioning, and handling and treatment of radioactive substances. Mixed LLW generally is not disposed of at DOE sites. Except for specialized facilities such as the TSCA Incinerator in Oak Ridge, mixed LLW is often treated and disposed at commercial facilities.

2.3.3 Summary of DOE Sites and Facilities Requiring Decommissioning: There are several active and planned programs to decommission DOE facilities. DOE programmatic responsibilities for decommissioning are evolving; however, the vast majority of decommissioning performed over the last decade and for the foreseeable future is through the DOE Office of Environmental Management. Future organization responsibilities for decommissioning of facilities beyond those currently in the Environmental Management inventory may reside with the National Nuclear Safety Administration, the Office of Science, and the Office of Nuclear Energy, as "landlords" for specific sites. Internal organizational responsibility for future liabilities is evolving.

The following is a brief description of the range of activities and substantial challenges facing the DOE at its major sites (Bossart 1999). A more current description of decommissioning plans may be found at the DOE Environmental Management's Internet web site.

2.3.3.1 Active Weapons Laboratories and Facilities: Los Alamos National Laboratory, Sandia National Laboratory, Lawrence Livermore National Laboratory, Y-12, and the Pantex Plant have continuing missions for nuclear weapons stockpile stewardship. Los Alamos has 89 structures that require decommissioning, including a plutonium processing facility, a phase separator pit, tritium facility, facilities contaminated with highly explosive materials, and the Omega West Reactor. Pantex has excess facilities that contain highly explosive contamination and traces of radioactive contamination. The Y-12 decommissioning includes the Alpha IV (Building 9201-4) Lithium Colex Process and a 600,000 ft^2 transite-covered structural steel and concrete facility with three floors and a sub-basement.

2.3.3.2 Active Science and Nuclear Energy Facilities: The science and nuclear energy facility decommissioning includes a suite of accelerator, research reactor, laboratory, and hot cell facilities at Brookhaven National Laboratory, Argonne National Laboratory (Illinois and Idaho sites), Oak Ridge National Laboratory, and Lawrence Berkeley National Laboratory. Oak Ridge projects include the decommissioning of research reactors, tanks, auxiliary buildings and equipment, isotope processing buildings, surface facilities and wells associated with hydrofracture operations, and other radioactively contaminated structures. Remaining work at Argonne includes the Zero Power and Juggernaut Reactor facilities and the Building 301 Hot Cell facility. Decommissioning of the Graphite Research Reactor is ongoing at Brookhaven, and the High Flux Beam Reactor should be completed over the next few years.

2.3.3.3 Inactive Small Sites: These sites include a range of facilities that were part of nuclear energy research, weapons development, and radiological research. Example facilities include:

1. Energy Technology Engineering Center, with its sodium disposal and radioactive materials handling facilities;
2. Laboratory for Energy-Related Health Research, Building 23 (housing a 7,400 ft^2 hot cell facility) at General Atomics in San Diego, CA (excess facilities include five buildings, underground tanks, and waste storage areas);
3. Glove box and hot cell at General Electric in Vallecitos, CA;
4. Separations Process Research Unit in Schenectady, NY, with its plutonium contaminated pilot separations units; and
5. Various sites in Ohio, such as the RMI Extrusion facility, Battelle-Columbus locations on King Avenue and West Jefferson, West Valley, Mound, and Fernald sites. Remediation of the RMI facility involves deactivation of 26 buildings and associated equipment, decontamination of 4 of the 26, and demolition of the remaining 22 buildings. The West Jefferson site includes large hot cells in Building JN-1; Building JN-2, a two-story office and laboratory building with a high bay; and Building JN-3, which houses a pool-type reactor.

2.3.3.4 Larger Environmental Management Sites:

1. Idaho National Environmental and Engineering Laboratory (INEEL): INEEL has operated facilities for reprocessing spent nuclear fuel from naval reactors, fuel storage, tank farm, and calcining HLW. Over 200 contaminated excess facilities and structures remain at INEEL. Current decommissioning projects focus on risk reduction and accelerating cleanup so that the site's large surveillance and maintenance (S&M) mortgage can be reduced. The main areas for decommissioning are the Test Reactor Area, Test Area North, Idaho Chemical Processing Plant, Central Facilities Area, Power Burst Facilities, Auxiliary Reactor Area, Radioactive Waste Management Complex, and the Experimental Breeder Reactor Area. Projects are grouped as surveillance and maintenance, deactivation, decontamination, and decommissioning. Initial S&M activities focus on portions of the chemical processing plant that contain fissile material and numerous process cells, the Power Burst Facilities reactor, and the 120,000 gallon canal of the Materials Test Reactor that holds spent fuel elements.

2. Hanford Reservation: Major decommissioning projects involve production reactors, spent fuel storage facilities, and plutonium production facilities.

   The 300 Area Fuel Supply Shutdown Complex produced fuel for the Hanford Single Pass Reactor and the N-Reactor. It consists of several buildings and contains 1,200 MT of un-irradiated low-enriched fuel, and 668 MT of low-enriched uranium trioxide powder.

   The Plutonium Finishing Plant converted Pu-bearing solutions into oxide and metal, and also reclaimed SNM from scrap. 10 of its 12 major facilities will be deactivated.

   The 324/327 Facility Transition Project includes the Waste Technology Engineering (324) and the Post-Irradiation Test (327) Laboratories. 324 includes controlled areas with potential dose in excess of 0.10 Gy/h. 327 houses 11 high density metal-shielded hot cells, two unlined concrete water basins, and a dry storage cell. This facility was used for inspection and analysis of fuel elements and materials, primarily from Pu production reactors.

The K-Basin Deactivation Project includes 18 facilities designed for wet storage of spent nuclear fuel. The Advanced Reactors Transition project includes decommissioning of the Plutonium Recycle Test Reactor, non-nuclear facilities associated with the Liquid Metal Fast Breeder Reactor, the Fast Flux Test Facility, and Fuels and Materials Examination Facility. Hanford also has eight production reactors remaining (in the 100 Area) and five fuel reprocessing canyons (in the 200 Area) that will eventually undergo decommissioning. The reactor facilities will most likely be placed in an interim safe storage condition.

In addition to these excess facilities, Hanford has many facilities with an operating mission. One such facility of particular significance is the Waste Encapsulation and Storage Facility, with 150 M Ci in 2,000 Sr and Cs capsules.

3. East Tennessee Technology Park (ETTP): The effort includes two projects. The first focuses on centrifuge equipment removal -- radiological characterization, stripping of components inside buildings, decontamination, and the demolition of auxiliary support facilities. Facilities not demolished may be leased after proven free of contamination. The second project focuses on decontamination and demolition of radioactively contaminated facilities. Additional decommissioning will include two buildings representing the process of highly enriching uranium. Specific efforts will include utility deactivation, strip-out of equipment and piping, decontamination/demolition, and waste disposal.

4. Rocky Flats Environmental Technology Site (RFETS): This site fabricated and processed plutonium metal for the DOE Weapons Complex. In 1992, when the site's mission was terminated, nearly 1,000 glove boxes and miles of process pipe and ventilation ducts contaminated with highly fissile material remained in place. RFETS cleanup is divided in "clusters." One cluster (Bldg. 371) includes several buildings used to recover and convert Pu into high purity metal buttons, and Am into dioxide. Remediation will require disposing 300 MT of stored waste, 40,000 MT of structural materials, 6,900 MT of equipment, and 58,000 g of chemicals. A second cluster (Bldg. 771/774) involves tanks and piping systems, including removal of liquids, sludge, and raschig rings. The third cluster (Bldg. 707/750) involves several buildings, including one where SNM was stored without packaging. The fourth cluster (Bldg. 776/777) involves facilities associated with nuclear weapon manufacturing and assembly; waste operations (size reduction and incineration), and Pu recovery. The fifth cluster (Bldg. 779) consists of 18 structures used for research and development on Pu processes, and contains about 100 glove boxes. Remediation of this site is through a particular performance incentive contract, which is likely to become a conceptual model for future remediation activities.

5. Savannah River Laboratory: The Savannah River Site (SR) produced Pu, tritium, and other materials supporting defense programs. Facilities include fuel fabrication, reactors, and reprocessing canyons. Most decommissioning work is scheduled for after 2006. Two spent fuel reprocessing facilities, known as the F and H canyons, are being deactivated. These large, remotely operated, heavily shielded structures were built in the early 1950s to provide materials (initially ^{239}Pu) for the country's nuclear arsenal. They are made of reinforced concrete, and each is about 850 ft long and 125 ft wide.

Deactivation projects also include the 235-F plutonium facility, and the P, C, R, K, and L reactors. The 235-F facility was used to fabricate and test Pu^{238} heat source fuels, and the reactors irradiated target materials for tritium and plutonium production. Other deactivation projects include a storage facility known as the Receiving Basin for Off-Site (irradiated) Fuels, the D-Area heavy water production and M-Area reactor fuel fabrication facilities.

6. Ohio Facilities: The Mound Plant and the Fernald Site are the larger environmental remediation sites in Ohio. Mound supported the tritium program for over four decades. The decommissioning of three buildings (the Semi-Works, Research, and Technical buildings) will involve over 1,000 lineal ft of tritium contaminated glove boxes, oil, and mercury, and miles of process piping and ductwork, uranium beds, molecular sieves, and other equipment. In addition, several other buildings (Building 58, H-Building, B-stack, C-Building, and E Annex) will be demolished or decontaminated for reuse. The SM/PP Hill project involves decommissioning 22 structures consisting of 116,000 ft^2 of processing and laboratory facilities, facility support structures, a water tower, and an air emission stack. The Test Fire Valley project will demolish 37 facilities and transition an additional 24 for reuse. Fernald received raw uranium ore and processed the ore; its decommissioning activities deal with the above-ground portion of contaminated structures. Most of the decommissioning activities at Fernald are scheduled for early completion, with waste disposition remaining as an issue.

7. Nevada Test Site (NTS): Decommissioning at NTS includes the Test Cell A and C facility, the Reactor and the Engine Maintenance Assembly and Disassembly facilities, and others previously engaged in weapons testing and assembly/disassembly.

2.3.3.5 Other DOE Decommissioning: Numerous diverse facilities continue in operation or have not yet been declared excess to DOE needs. In addition to special application nuclear laboratories and reactors, facilities range from alpha-gamma hot cells and intense pulsed neutron sources, to major accelerators. They will require eventual decommissioning, which is likely to be performed through a dedicated DOE legacy management organization.

2.4 TRAINING, RESEARCH, AND TEST REACTORS

There are 194 training, research, and test reactors in the U.S. Of those, 140 have been shut down and await restoration plans, while an additional 33 have been decommissioned to some degree. There are no reactors currently under construction or being planned. Research reactors have been used since the beginning of the atomic age and many are more than 40 years old. Therefore, these represent a significant portion of the reactors that will soon require decommissioning decisions.

The purpose and mission of the research reactors largely determine the radioactive contaminants that will be encountered during decommissioning. The newer ones located on university campuses are expected to generally contain the same contaminants as commercial reactors. However, older ones and those located at government facilities are expected to contain a wide variety and larger fractions of some wastes that more closely resemble facilities used in the nuclear weapons complex. A contributing factor to the variety and quantity of wastes that will be encountered is failed reactor fuel. Early fuel manufacturing operations were based on limited experience and consequently there were significant fuel failures during some of the early reactor operations. Many radionuclides, including isotopes of transuranic elements, are likely to be encountered during decommissioning.

2.5 COMMERCIAL NUCLEAR POWER REACTORS

In the U.S., nuclear powered electrical generation began in the 1950s. Current power reactors are designed to primarily be a heat source that ultimately produces steam to drive a turbine generator, thereby producing electricity. There is a rich history of evaluating several types of reactors. Light water reactors are dominant throughout the world. Pressurized and boiling water reactors are in this group are. Figure 2.1 shows diagrams of these two reactor types.

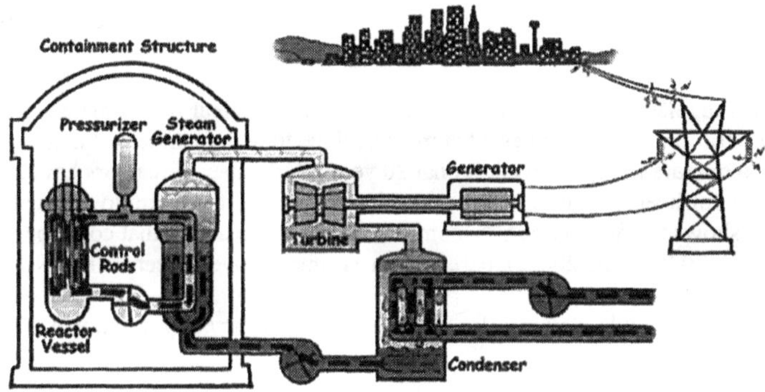

BWR

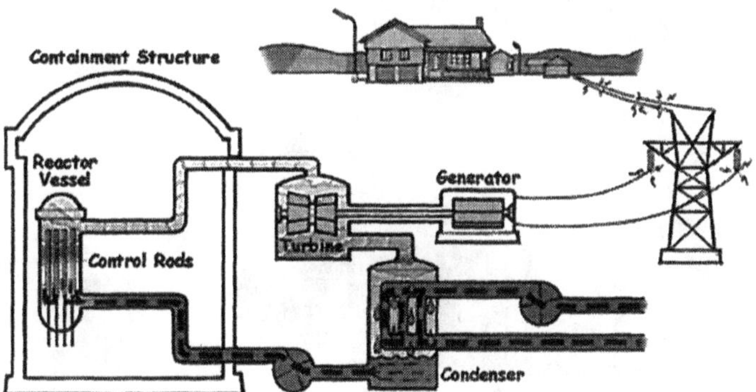

Figure 2.1 Typical Light-Water Reactors

The term "gas-cooled" includes both the high temperature gas-cooled reactor and advanced gas reactor designs. They usually use graphite as moderator and helium as coolant. The pressurized heavy water reactor is similar to light water reactors except that the moderator is heavy water (deuterium oxide). The Canadian CANDU reactor is prominent in this group. The Russian RBMK being an example of light water-cooled, graphite-moderated reactor. Other reactors include the fast breeder reactor and liquid metal fast breeder reactor.

NRC licensed nuclear reactors should face a wave of decommissioning activity. However, several key factors have altered the time when power reactors will permanently shut down, as follows.

1. Many jurisdictions have completed or are in the process of deregulating the electric power utilities. Utility deregulation, following the practice within the airline, telecommunications, and rail industries in the U.S., typically provides for increased customer choices in their provider of electricity, and the promise of reduced cost. This transition forces each electric generation asset to be viewed as an individual profit center, rather than on the historic regulated basis of cost of operation plus regulated profit. When placed under this economic microscope, some commercial power reactors, particularly single-unit sites, did not withstand the test and were permanently shut down prior to the expiration of the 40 year license.

2. NRC issued regulations dealing with license renewal of nuclear power plants in 10 CFR 54 (NRC 1991b). This rule describes what must be done to demonstrate that a power reactor can continue to operate for up to an additional 20 years. A number of reactors have submitted applications for license extension and others have indicated the intent to pursue this option. Figure 2.2 shows the number of U.S. commercial nuclear reactors that could be shut down, per year, through 2035 (NRC 2002c), based on an assumed 40-year license life (NRC 1991c).

Tables 2.1 and 2.2 provide statistics for operating and permanently shutdown commercial power reactors in the U.S. as of the end of 1998.

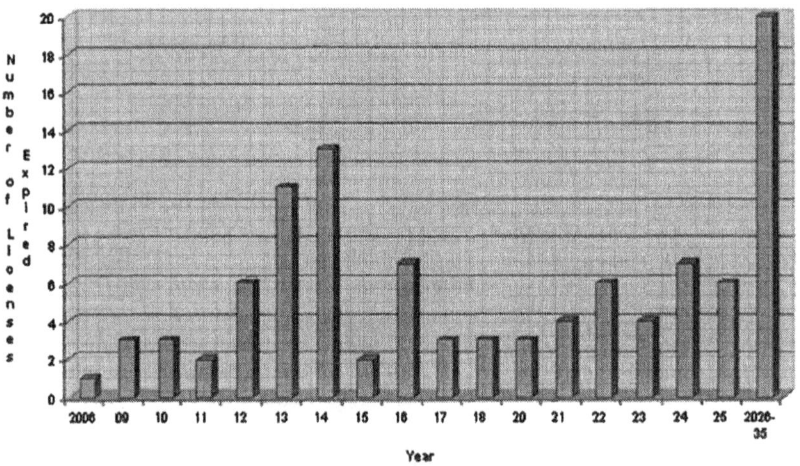

Figure 2.2 Commercial Nuclear Reactors to Be Shut Down Through 2035

Table 2.1 Reactors Operating, Shut Down, or Decommissioned

Reactor Type	Operating	Shut Down or Decommissioned
PWR	69	11
BWR	35	10
HTGR		2
FBR		2
Other types		4

Table 2.2 Permanently Shut Down or Decommissioned Commercial Power Reactors

Name	Rx Type	MWt	Shutdown Year
Haddam Neck (CT Yankee)	PWR	1825	1996
Indian Point 1	PWR	615	1974
Maine Yankee	PWR	2700	1997
Rancho Seco	PWR	2772	1989
San Onofre 1	PWR	1347	1992
Shippingport	PWR	236	1982
Three Mile Island 2	PWR	2770	1979
Trojan	PWR	3411	1992
Yankee Rowe	PWR	600	1991
Zion 1	PWR	3250	1998
Zion 2	PWR	3250	1998
Big Rock Point	BWR	240	1997
Dresden 1	BWR	700	1978
EBWR	BWR	100	1967
Elk River	BWR	58	1968
GE VBWR	BWR	50	1963
Humboldt Bay 3	BWR	200	1976
LaCrosse	BWR	165	1987
Millstone 1	BWR	2011	1998
Pathfinder	BWR	190	1967
Shoreham	BWR	2436	1989
Ft. St. Vrain	HTGR	842	1989
Peach Bottom 1	HTGR	115	1974
EBR II	FBR	62	1994
Fermi 1	FBR	200	1972
CVTR	PHWR	65	1967
Hallam	Sodium Graphite	240	1964
Hanford N	LGR	4000	1987
Piqua	Organic Cooled/Moderated	46	1966

At the international level, the situation is somewhat similar to the situation in the U.S. However, because power reactors became operational somewhat later in other countries than in the U.S., there are fewer reactors shut down or decommissioned. Table 2.3 provides the statistics for power reactors worldwide. Current statistics on both operating and permanently shutdown power reactors and their basic operations are maintained by many organizations. They may be found within the International Nuclear Safety Center at Argonne National Laboratory, the DOE's Energy Information Agency, the International Atomic Energy Agency (IAEA), and the Nuclear Energy Institute (NEI).

Table 2.3 Reactors Operating and Either Permanently Shut Down or Decommissioned Internationally

Reactor Type	Operating	Shut Down or Decommissioned
PWR	184	12
BWR	57	9
PHWR	37	4
LGR	18	5
GCR	20	18
AGR	14	1
Other	5	11

2.5.1 Commercial Pressurized Water Reactor: As shown in Figure 2.1, in a commercial PWR the primary (or reactor coolant) system transfers the heat from the fuel via primary coolant pumps and coolant piping to the steam generator. The steam formed in the steam generator is transferred by the secondary system to the main turbine generator, where it is converted into electricity. The steam then condenses and is cooled by circulating water in a tertiary or condensate system. Condensate cooling water is provided by a nearby ocean, sea, lake, river or cooling tower. Only the primary reactor coolant system is expected to contain radioactivity. Both the primary and secondary systems are closed-loop systems and the water is recirculated within each system. Additionally, there are approximately 100 systems that support the operation of the primary and secondary systems.

The reactor coolant system consists of the reactor vessel, the steam generators, the reactor coolant pumps, a pressurizer, and the connecting piping. A reactor coolant loop consists of a reactor coolant pump, a steam generator, and the piping that connects these components to the reactor vessel. There are multiple reactor coolant loops at each reactor (varying from 2 to 4 loops, depending on the reactor manufacturer and model). The primary function of the reactor coolant system is to transfer the heat from the fuel to the steam generators. A second function is to contain any fission products if they were to escape the fuel. The reactor coolant system is located inside the containment building. The containment serves to prevent contamination outside allowable limits from reaching the environment. Containment buildings are designed to withstand the pressures and temperatures that would accompany an accidental primary coolant or steam release into the building.

2.5.2 Commercial Boiling Water Reactor: As shown in Figure 2.1, the reactor recirculation system in a commercial BWR transfers the heat from the fuel via reactor recirculating pumps and piping to the steam separator and drier located in the upper region of the reactor vessel. The steam is then transferred to the main turbine generator, where it is converted into electricity. This entire system is radioactively contaminated. Similar to the PWR, the steam then condenses and is cooled by circulating water in a condensate system. The condensate cooling water is provided by a nearby ocean, sea, lake, river or cooling tower.

The primary (or reactor recirculation) system consists of the reactor vessel, the steam separator - dryer, the main pumps, and the connecting piping. The reactor vessel and recirculation piping and the main steam line piping, up to and including the main steam line isolation valve, are located inside the primary containment, which also includes a large pool of water called the suppression pool. In the event of accidental break in the main coolant circulation line, the primary containment is designed so that the main steam line isolation valve would close and the steam (produced by the reactor and from recirculation cooling water flashing to steam as it escapes through the break) is routed by the primary containment piping and structural features into a large pool of suppression water. Here the steam is condensed to water, thereby reducing the pressure in the containment. The reactor containment structure is located within a reactor building, thereby providing secondary containment for the small amount of steam and gas leakage that comes out of primary containment. The secondary containment feature is provided by an iodine removal system that filters the reactor building exhaust under accident conditions, significantly reducing the potential release of radioactive iodine to the environment.

The BWR primary and secondary containment features serve to prevent contamination beyond the allowable limits from reaching the environment. The primary containment structure is designed to withstand the pressures and temperature that would accompany a design basis accident. The reactor building (with its iodine removal system) does not become pressurized in the event of an accident since the gas and steam leakage from the primary containment is quite small. For these reasons, this type of containment system design (in general use for all U.S. BWR's) is generally referred as a "pressure suppression" type of containment. A typical U.S. PWR containment is a "dry" one (no suppression pool of water). It is designed to have a much larger containment volume than the BWR in order to limit the pressure buildup by having a much greater volume for the steam to expand into. Thus, the suppression pool concept allows a containment to be much smaller in size for a given reactor power level.

2.6 THE NUCLEAR FUEL CYCLE

The nuclear fuel cycle consists of several steps that lead to the preparation of uranium for use as reactor fuel, and several additional steps necessary to safely manage the highly radioactive spent nuclear fuel. The foundation of nuclear technology is the fission of the isotope of an element, notably ^{235}U and ^{239}Pu. During the ^{235}U fission, the more abundant ^{238}U adsorbs neutrons and eventually results in ^{239}Pu. The nuclear fuel cycle starts with the mining of uranium, a naturally occurring element containing a small fraction of ^{235}U. Either naturally occurring uranium or uranium enriched in ^{235}U was the fuel used in the early development of nuclear power in the U. S. Generally, the front end of the nuclear fuel cycle is separated into the following steps.

2.6.1 Exploration, Mining, and Milling: Uranium ore normally is mined by methods similar to those used for mining other metals. Uranium ores in the U.S. typically range from about 0.05% to 0.3% uranium oxide (U_3O_8). Mined uranium is normally processed by grinding the ore to a uniform particle size and then extracting the uranium by chemical leaching. The milling process commonly yields natural uranium in a dry powdered form known as yellowcake.

2.6.2 Conversion and Enrichment: Yellowcake must be converted to UF_6, required by most uranium enrichment facilities. A solid at room temperature, UF_6 is a gas at higher temperatures. The concentration of the fissionable isotope ^{235}U in natural uranium is only 0.72%, considerably less than that required for sustaining a nuclear chain reaction in light water reactors (typically 4 - 5%). Gaseous diffusion and gas centrifuge are the commonly used uranium enrichment technologies. The gaseous diffusion process consists of passing the natural UF_6 gas feed under high pressure through a series of diffusion barriers (semi-porous membranes) that permit passage of the lighter $^{235}UF_6$ molecules at a faster rate than the heavier $^{238}UF_6$ molecules. This differential treatment, applied across a large number of diffusion stages, progressively raises the product stream concentration of ^{235}U relative to ^{238}U. In the gaseous diffusion technology, the separation achieved per diffusion stage is relatively low, and a large number of stages are required to achieve the desired level of isotope enrichment. Because this technology requires a large capital outlay for facilities and it consumes large amounts of electrical energy, it is relatively cost-intensive. In the gas centrifuge process, the natural UF_6 gas is spun at high speed in a series of cylinders. This acts to separate the $^{235}UF_6$ and $^{238}UF_6$ molecules based on their slightly different atomic masses. Gas centrifuge technology involves relatively high capital costs for the specialized equipment required, but its power costs are below those for the gaseous diffusion technology.

2.6.3 Fuel Fabrication: For use as nuclear fuel, enriched UF_6 is converted into uranium dioxide (UO_2) powder that is then processed into pellet form. The pellets are then high temperature fired in a sintering furnace to create hard, ceramic pellets of enriched uranium. The pellets are stacked (according to each nuclear core's design specifications) into tubes of corrosion resistant metal alloy. The tubes, called fuel rods, are sealed to contain the pellets. The finished rods are grouped into "assemblies" that are then used to build up the fuel core of a power reactor. Figure 2.3 presents this fabrication cycle.

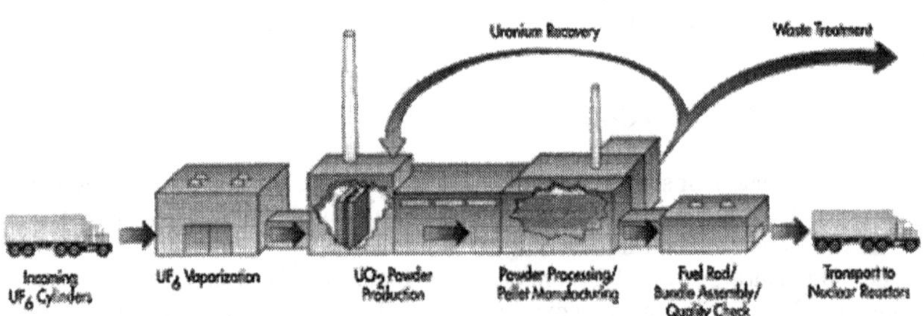

Figure 2.3 Schematic of Nuclear Fuel Fabrication

2.6.4 Interim Storage of Spent Fuel: The reactor is shut down for refueling after its operating cycle (typically one to two years). The fuel discharged at that time (spent fuel) is typically stored in on-site pools for a minimum number of years to allow for radioactive and thermal decay. If on-site pool storage capacity is exceeded, it may be necessary to store aged fuel away from the site or in an on-site modular dry storage facility known as an independent spent fuel storage installation (ISFSI).

2.6.5 Reprocessing: Much like the process described for weapons production, spent fuel discharged from light water reactors can also be processed to recover fissile (^{235}U, ^{239}Pu), fertile (^{238}U) and other radioactive materials. These materials can be chemically separated and recycled. Commercial fuel reprocessing halted in the U.S. in 1977 but is still performed in other countries.

2.6.6 Waste Disposal: The DOE has responsibility for the development of the country's waste disposal system for spent nuclear fuel and HLW. Current plans call for the ultimate disposal in solid form, placed deep within Yucca Mountain, NV, in licensed, stable, geologic structures.

2.6.7 Power Reactor Decommissioning Wastes: Similar to the DOE complex, decommissioning of nuclear power plants will result in the generation of differing types of radioactive waste. Residual radioactive contamination in a nuclear power plant will typically be found on the surface and near-surfaces of buildings and components, limited exterior soil surface, subsurface contamination, and piping interiors. In addition to surface contamination, surfaces and components adjacent to the reactor itself will likely contain radioactive atoms due to neutron activation from fission products. Table 2.4 notes the nuclides of interest for personnel protection at a decommissioning (EPRI 1999). Nuclides of interest over all periods include: ^{3}H, ^{14}C, ^{54}Mn, ^{55}Fe, ^{59}Ni, ^{63}Ni, ^{58}Co, ^{60}Co, ^{65}Zn, ^{90}Sr, ^{94}Nb, ^{134}Cs, ^{137}Cs, ^{152}Eu, ^{154}Eu, ^{238}Pu, ^{239}Pu, ^{241}Am, and ^{242}Am. Note that the abundance of transuranic nuclides in nuclear power plants is generally small.

Table 2.4 Nuclides of Interest from Nuclear Power Plant Decommissioning

Time Period after Shutdown	Nuclides of Interest
Within 10 to 20 years	^{60}Co, ^{55}Fe, ^{137}Cs
From 20 to 30 years	^{63}Ni, ^{137}Cs, ^{60}Co, ^{90}Sr
Greater than 100 years	^{238}Pu, ^{239}Pu, ^{240}Pu, ^{241}Am, ^{242}Cm, ^{244}Cm

There are widely ranging numbers for estimates of the volume of radioactive materials to be disposed of from the decommissioning of a commercial reactor facility. Examples of radioactive waste volumes from decommissioning are shown in Table 2.5.

Table 2.5 Quantity of Radioactive Waste Generated after Shutdown of a Reactor

Facility	Rx Type	MWe	Radioactive Waste Volume (ft^3)	(m^3)
Saxton (with contaminated soil)	PWR	7	~73,000	1,971
Oyster Creek	BWR	610	550,000	14,850
Yankee Rowe	PWR	167	146,000	3,942
NRC Reference PWR	PWR	1175	647,500	17,482
NRC Reference BWR	BWR	1155	670,300	18,098

PART 2

REGULATORY REQUIREMENTS AND POLICY ISSUES

From the initial shutdown decision to the completion of the planned end-state, decommissioning activities, for both federal and commercial facilities, are subject to various requirements, including laws, regulations, and policy issues. The Nuclear Regulatory Commission (NRC) and Department of Energy (DOE) are the primary agencies overseeing these requirements. Although nuclear facilities undergoing decommissioning are also subject to many environmental laws and regulations, much of this authority has been delegated to the states. Consequently, requirements for a specific action may be different among various states. Issues of concern include worker safety and the fundamentals of risk analysis, including risk communication. Finally, a successful project includes public involvement and stakeholder participation. The appropriate interrelationship among the regulators, stakeholders, and the decommissioning team is a key to its overall success.

Nuclear facilities in the United States are subject to numerous federal requirements. The number of laws and their respective implementing regulations for both federal and commercial facilities is so large that it cannot be described in one chapter or even one volume. Instead, this part gives an overview of applicable requirements. A key principle included in all regulations dealing with radioactive materials is keeping the exposure to ionizing radiation as low as reasonably achievable (ALARA). This principle implies that a balance must be found between reducing worker and public exposure and the costs to reduce such exposure.

The federal agency responsible for regulating the decommissioning of commercial nuclear facilities is the NRC. In addition to Title 10 of the CFR, the NRC has published numerous other documents in the Federal Register and in the form of nuclear regulatory guides (NUREGS). The DOE's decommissioning activities follow the requirements of the Atomic Energy Act (AEA) and in most cases employ orders (comparable in status to regulations) for internal use by both federal staff and contractors.

Chapter 3

NRC REQUIREMENTS FOR DECOMMISSIONING

3.1 INTRODUCTION

This chapter focuses on decontamination activities at nuclear and radiological facilities regulated by the NRC. If a state is an NRC Agreement State, certain nuclear-related facilities (excluding federal facilities and nuclear power reactors) may be regulated through the relevant individual state agency.

The NRC normally accomplishes its regulatory function by issuing and then enforcing the provisions of a license. It licenses and regulates nuclear reactors (both power and nonpower generating) and over 6,000 radioactive material licensees (NRC 2003a). The NRC also regulates several DOE facilities, either by mutual agreement or by specific statute. A relevant example of statute regulation is the licensing of the permanent geologic repository at Yucca Mountain for the disposal of spent nuclear reactor fuel and high-level waste (HLW).

3.2 NRC REGULATORY FRAMEWORK

The NRC regulatory framework is codified within Title 10 of the Code of Federal Regulations (CFR), specifically Parts 0 through 199. These regulations apply to all individuals and organizations with a license to hold or use radioactive materials, or operate nuclear facilities. In addition, the NRC has developed documents providing general guidance and also to describing acceptable methods of implementing particular regulations. These documents include Regulatory Guides that delineate methods acceptable to the NRC for meeting its specific regulations, and NUREG documents, sponsored reports covering regulatory approaches, administrative information, research results, and other topics. The NRC Telephone Directory, for example, is a NUREG document (NUREG/BR-0046). A Regulatory Guide typically describes one method of meeting a regulatory requirement that is deemed acceptable to the NRC. Other methods of compliance may be used, but they must be reviewed and accepted by the NRC.

In practice, it is often more cost-effective to conform to the recommendations of the Regulatory Guides than to take the actions necessary to satisfactorily demonstrate an alternative as equally protective of public health and safety. Therefore, in many cases Regulatory Guides become *de facto* regulations. A NUREG document, on the other hand, may contain guidance but has no implied or *de facto* enforcement status.

3.3 DECOMMISSIONING ALTERNATIVES

The NRC defines decommissioning as the process of closing down a facility followed by reducing residual radioactivity to a level that permits the release of the property for unrestricted use and termination of the license (NRC 2002a). While the NRC decommissioning regulations generally address only radiological contaminants, nonradiological contaminants come under other laws administered by other state and federal agencies and may require remediation of affected media, such as soil and water that may be located at a significant distance from the source facility. The removal, handling, and disposal of hazardous materials within the facility are also subject to various laws and regulations.

Decommissioning alternatives are discussed in detail in Chapter 9 of this handbook. The NRC's Generic Environmental Impact Statement (GEIS) evaluated the environmental impact of three methods for decommissioning (NRC 2002b): DECON, SAFSTOR, and ENTOMB. The first two may be combined, by, for example, conducting some preliminary dismantling and cleanup (DECON), followed by a long storage period (SAFSTOR), and later resuming of DECON through to project completion. This GEIS terminology is fairly precise, restrictive, and may not be directly applicable to DOE decommissioning projects; that is, the NRC and the DOE may use these terms with a somewhat different definition in mind.

3.3.1 DECON: In this alternative, the equipment, structures, and portions of a facility and site containing radioactive contaminants are removed or decontaminated to a level that permits eventual termination of the license. In the case of a power reactor, this generally involves a physical dismantling of the facility, various equipment and structure decontamination operations, and shipment of radioactive wastes to licensed disposal areas. A partial DECON may be followed by a SAFSTOR to allow for further radioactive decay, and sometimes the accumulation of sufficient funding. The final stage of DECON completes the remediation effort, with all residual radioactivity being reduced to acceptable environmental levels. The GEIS found DECON to be an acceptable decommissioning method.

3.3.2 SAFSTOR: In this alternative, the facility is placed in a safe stable condition and maintained as such until it undergoes DECON. It may be left intact during this period, but fuel must first be removed from the reactor vessel, and radioactive liquids drained from systems and components and appropriately processed. Radioactive decay occurs during the SAFSTOR period, an approach intended to reduce levels of contamination and potentially reduce the quantity of material to be shipped off site during the subsequent DECON phase. In accordance with 10 CFR 50.82, decommissioning must be completed within 60 years of the permanent cessation of operations by nuclear power plants. The NRC would approve completion beyond 60 years only when necessary to protect public health and safety. Hence, any combination of the DECON and SAFSTOR alternatives must achieve termination of the NRC license within sixty years. Given this provision, the GEIS found SAFSTOR to be an acceptable decommissioning method.

3.3.3 ENTOMB: According to the GEIS, ENTOMB involves encasing radioactive structures, systems, and components in a structurally long-lived substance, such as concrete. The entombed structure is appropriately maintained, and continued surveillance is carried out until the radioactivity decays to a level that permits license termination. However, most commercial power reactors will have radionuclides in materials in concentrations exceeding the NRC limits for unrestricted release even after 100 years have elapsed (due to the presence of relatively long-lived radionuclides, such as ^{137} Cs and ^{90}Sr with half-lives of about 30 years). Thus, the ENTOMB option will generally not be feasible for commercial nuclear power plants due to the 60-year requirement. However, the NRC states that this option might be acceptable for facilities that could demonstrate decay to unrestricted-use levels in about 100 years. NRC document SECY-02-0191 (NRC 2002c) states that, unless directed otherwise, staff will defer rulemaking to permit entombment as an option for power reactors until completion of research on its viability issues (e.g., structure performance, source term development, and flow and transport modeling). Therefore, the ENTOMB alternative currently requires a highly specific review and approval on a case by case basis.

3.4 DECOMMISSIONING REGULATIONS

The NRC established decommissioning regulations on a variety of issues. These include:

- Technical and financial criteria for decommissioning licensed nuclear facilities (NRC 1988)
- Recordkeeping requirements for decommissioning
- Establishing timeframes and schedules for decommissioning licensed nuclear facilities (NRC 1994a)
- Financial assurance requirements for decommissioning (NRC 1995 and 1998)
- Decommissioning procedures for nuclear power reactors (NRC 1996a)
- Radiological criteria for license termination, also known as the License Termination Rule (NRC 1997a)

The License Termination Rule establishes criteria for both unrestricted and restricted use of the licensed facility. In accordance with 10 CFR 20.1402, Radiological Criteria for Unrestricted Use, a site will be considered acceptable if the residual radioactivity distinguishable from background radiation has been reduced to ALARA levels and has a total effective dose equivalent (TEDE) that does not exceed 25 mrem/y for an average member of the critical group. These criteria are also required for all groundwater sources of drinking water. Determination of ALARA levels must consider any potential detriments that might result from decontamination and waste disposal.

In accord with 10 CFR 20.1403, Criteria for License Termination under Restricted Conditions, a site will be considered acceptable if the licensee:

1. Has submitted a decommissioning plan (DP) or license termination plan (LTP) to NRC indicating intent to decommission in accordance with Parts 30.36(d), 40.42(d), 50.82(a and b), 70.38(d) or 72.54, and specifying the intention to decommission by restricting use of the site. The licensee must also document how the advice of local stakeholders has been sought and incorporated, as appropriate, following analysis of that advice.

2. Can demonstrate that further reduction in residual radioactivity necessary to comply with the provisions of Part 20.1402 would result in public or environmental harm, or was not made because the residual levels associated with restricted condition are ALARA.

3. Has made provision for legally enforceable institutional controls that provide reasonable assurance that the TEDE from residual radioactivity will not exceed 25 mrem/y.

4. Has provided sufficient financial assurance to enable an independent third party, including a governmental custodian of a site, to assume and carry out responsibilities for any necessary control and maintenance of the site.

5. Can ensure that residual radioactivity has been reduced so that if institutional control were not in effect, there would be reasonable assurance that the TEDE is ALARA and would not exceed either 100 mrem/y, or 500 mrem/y while meeting certain additional criteria.

Although all NRC licensees must meet essentially the same ultimate objective for license termination, somewhat differing regulatory requirements apply for decommissioning various facilities, depending on the type of license. Nuclear reactor licensees generally follow one regulatory path (discussed in the next section); other licensees, including fuel cycle licensees, materials licensees, and sites in the NRC's Site Decommissioning Management Program, generally follow the path discussed in Section 3.6.

3.5 DECOMMISSIONING OF NUCLEAR POWER REACTORS

Regulations for decommissioning nuclear power reactors are included in 10 CFR Parts 2, 50, and 51. Decommissioning activities for power reactors may be divided into three phases: 1) initial activities, 2) major decommissioning and storage activities, and 3) license termination activities. Regulatory Guide 1.184, "Decommissioning of Nuclear Power Reactors" (NRC 2000a), describes methods and procedures acceptable for implementing the rules relating to phases 1 and 2. License termination activities are further discussed in Section 3.8

Decommissioning action starts with a decision for a power reactor to permanently cease operations. After the final shutdown, a written certification of permanent cessation of operations must be submitted within 30 days as described in 10 CFR 50.82(a)(1)(i). The next milestone is the permanent removal of all fuel, after which a separate certification must be made as described in 10 CFR 50.82(a)(1)(ii). There is no time limit for removing the fuel. Once these two certifications are submitted, reactor operation and movement of fuel into the reactor vessel are prohibited, but limited decommissioning work may begin.

At this point certain operational requirements may be eliminated, including emergency core cooling systems and other systems involved in reactor operation, which may allow for a reduction in the annual license fee. No more than 3% of the decommissioning funds outlined in 10 CFR 50.75(c) may be used until 30 days after the public meeting following the submittal of a Post Shutdown Decommissioning Activities Report (PSDAR). This report includes specific decommissioning plans and a high level schedule, cost estimate, and an evaluation of environmental impacts associated with site-specific activities. It must be submitted to the NRC and the affected state(s) within two years after the permanent cessation of operations, as discussed in 10 CFR 50.82(a)(4)(i). Note that licensed reactors that had already filed a decommissioning plan before the publication of the current ruling were not required to file a PSDAR. Furthermore, an LTP, also submitted within two years of license expiration, requires specific information and formatting not required for a PSDAR. Guidance for PSDAR development is presented in Regulatory Guide 1.184 (NRC 2000a) and Regulatory Guide 1.185 (NRC 2000b).

Submittal of the PSDAR opens a forum for public participation, including the opportunity to make comments and participate in a public meeting, typically held within 90 days of submittal. Upon completion of the required submittals, and a 90-day waiting period, three major decommissioning activities may begin:

- Permanent removal of major radioactive components, such as the reactor vessel, steam generators, or other components of comparable radioactivity
- Permanent changes to the containment structure
- Dismantling components resulting in "greater than Class C" waste

A site-specific decommissioning project cost estimate is due within two years following permanent cessation of operations. Though this estimate enables full use of the decommissioning fund, state public utility commissions may require specific protocols to access funds under their jurisdiction. Another required report describes how spent nuclear fuel will be managed and how the funding for fuel management will be obtained. Funding and management are required until the fuel is transferred to the federal government. Without a specific NRC waiver, the site must be decommissioned within 60 years. Decommissioning will then proceed under one of the scenarios outlined earlier.

Although regulations applicable to a permanently shutdown plant in decommissioning are reduced, they are not eliminated. License conditions, technical specifications, quality assurance, emergency plans, security, and many other requirements for a power reactor will still apply. Also, although the hazards to the public from a plant in decommissioning are substantially less than from an operating plant, the stored spent nuclear fuel remains a substantial hazard to be properly and safely managed.

3.6 DECOMMISSIONING OF NRC NONREACTOR FACILITIES

Regulations for licensing and decommissioning power and nonpower reactors, and production and utilization facilities in general, reside in 10 CFR 50. Parts 30, 40, 50, 70, and 72 cover the licensing of byproduct material, source material, special nuclear material, and storage of spent nuclear fuel in independent storage facilities. This diversity of facilities requires that the NRC stipulate a variety of decommissioning rules, regulations, and guidance.

Before September 2003, the primary guidance documents for decommissioning nonreactor facilities were NUREG -1727, the Decommissioning Standard Review Plan (NRC 2000c), and NUREG/BR-024, the Handbook for Decommissioning Fuel Cycle and Materials Licensees (NRC 1997b). Both documents were replaced in September 2003 by the three volumes of NUREG-1757, Consolidated Nuclear Materials Safety and Safeguards Decommissioning Guidance (NRC 2003b):

- Decommissioning Processes for Materials Licensees
- Characterization, Survey, and Determination of Radiological Criteria
- Financial Assurance Recordkeeping and Timelines

Using a risk-informed approach, the guidance document divides facilities into seven groups, based on the risk posed by the site conditions and the needed decommissioning operations. Although detailed guidance applies to each group, some information, e.g., radiological survey details, is generic to reactor, fuel cycle, and materials licensee facilities.

3.7 SITE MANAGEMENT DECOMMISSIONING PROGRAM

The NRC's Site Decommissioning Management Program (SMDP) was created in 1990 to help ensure the timely cleanup of sites making only limited progress in completing remediation and license termination. Sites in this program met one or more of the following criteria:

- Problems with a viable responsible organization (e.g., the inability to pay for, or unwillingness to perform, decommissioning)
- Large amounts of soil contamination or unused settling ponds or burial grounds that would entail extensive disposition issues
- Long-term presence of contaminated, unused facility buildings
- A license previously terminated
- Contamination or potential contamination of the groundwater from on-site wastes

Additional information on the SMDP program can be found in NUREG-1444, Supplement 1 (NRC 1994b). Initially, 49 sites were identified in the program. In 1998, the criteria to add a site were revised to include restricted-use sites or complex unrestricted-use sites needing detailed, site-specific dose modeling. The criteria also included sites subject to significant public, state, or congressional interest, or that have questionable financial viability. The status of decommissioning activities, including the SMDP program, is periodically updated (NRC 2002d).

The NRC provided oversight of the cleanup of such contaminated sites as those identified under the Oak Ridge National Laboratory's Terminated License Review Project. The project reviewed documentation on previously terminated licenses to ensure that any residual radioactive contamination would meet the current criteria for license termination. As a result of this review, 39 formerly licensed sites were found to have residual contamination levels that exceeded unrestricted release criteria. Sixteen of these sites were re-released after successful remediation.

3.8 LICENSE TERMINATION

All decommissioning activities ultimately support the final objective of license termination. Final status surveys, data analysis, and documentation must demonstrate compliance with agreed-upon release criteria. All licensees must formally request license termination. Power reactors require submission of an LTP. The LTP indicates the methods the licensee will use to show that the site/facility is sufficiently decontaminated or remediated so as to meet the radiological dose and other requirements of 10 CFR 20 Subpart E (e.g., 20 CFR 1402, restricted release; and 20 CFR 1403, unrestricted release). The LTP, required at least two years before the requested termination date, must include:

- A site characterization
- Identification of remaining dismantling activities
- Plans for remaining site remediation
- Detailed plans for the final site survey of residual contamination
- A description of the site's end-use (if restricted use is proposed, a description of institutional controls and maintenance and surveillance programs is needed)
- An updated site-specific estimate of remaining decommissioning costs
- A supplement to the environmental report describing any new information or significant environmental change associated with the licensee's proposed decommissioning activities

Two guidance documents support development of LTPs:

- Regulatory Guide 1.179 - Standard Format and Content for License Termination Plans for Nuclear Power Reactors (NRC 1999)

- NUREG 1700 - Standard Review Plan for Evaluating Nuclear Power Reactor License Termination Plans (NRC 2003c)

Although specifically written for nuclear power reactors, these references give valuable insights into the overall preparation for license termination. After receipt of a license termination plan, NRC will post a notice in the Federal Register. The LTP is available for public comment, and a meeting is scheduled near the facility for discussion with the public. Upon approval of an LTP, the focus shifts to demonstrating compliance with its provisions. At the end of the LTP process, if the NRC determines that the remaining decontamination and dismantling has been performed in accordance with the approved LTP, and if the final radiation survey and associated documentation show that the facility and site are suitable for release, then the NRC will terminate the license and the decommissioning process will be considered complete.

Chapter 4

ENVIRONMENTAL AND RELATED REQUIREMENTS

4.1 INTRODUCTION

This chapter summarizes the major environmental laws relevant to decommissioning. Whereas the Atomic Energy Act (AEA 1954) is the primary law governing the management of radionuclides, the Resource Conservation and Recovery Act (RCRA) is the primary law governing hazardous waste. Other statutes also address the release and disposal of radioactive and hazardous materials. Although the DOE, the NRC, and the Environmental Protection Agency (EPA) are the primary agencies implementing these requirements, much of the authority to develop environmental regulations and enforce them has been delegated to the states. It is not practical to address the individual state environmental requirements in this handbook, many of which are similar to the federal requirements.

4.2 THE ATOMIC ENERGY ACT

The Atomic Energy Act (AEA) was enacted to ensure the proper management of source, special nuclear, and by-product material. The AEA and its amendments delegate the control of nuclear energy primarily to the DOE and the NRC. The final regulations implementing the AEA are found in various parts of Chapter 10 of the CFR, which includes various aspects of radiation protection and radioactive waste management

The AEA allows the NRC to delegate portions of its regulatory authority to the states. This authority permits the states to license and regulate by-products, source materials, and small quantities of special nuclear materials. The NRC delegation of authority to a state requires compatibility between its radiation control program and NRC requirements. The state must also establish its authority to enter into an agreement, usually by enacting appropriate enabling legislation.

10 CFR 50 (NRC 2003d) presents the NRC-established regulations addressing:

- Radiological criteria for decommissioning
- Requirements for, and the detailed contents of, decommissioning plans and related required licensing submittals
- Timeliness for submittal of decommissioning plans and related documents
- Decommissioning funding plans.

In support of these regulations, the NRC has developed a wide range of guidance documents to assist licensees in complying with the regulations, including standard format and content, standard review plans, and technical guidance. Its regulations are largely governed by the ALARA principle, a standard it applies to regulate radiation exposure limits and as criteria to terminate a license (NRC 1997a), a prerequisite for decommissioning. The NRC has provided a Decommissioning Standard Review Plan (NRC 2000c). The standard takes into account the state of technology, the economics of improvements in relation to benefits to the public health and safety, other societal and socioeconomic considerations, and the use of nuclear energy in the public interest.

A waste that contains radioactive constituents regulated under AEA and hazardous waste constituents regulated under RCRA is referred to as mixed waste. The DOE and NRC retain authority for the actual radionuclides in by-product material (under the AEA) and the EPA (or authorized states) regulates hazardous waste components (under RCRA).

4.3 RESOURCE CONSERVATION AND RECOVERY ACT

The Resource Conservation and Recovery Act (RCRA) authorizes the EPA to regulate the life cycle of hazardous waste. The RCRA addresses the generation, storage, treatment, transportation, and disposal of hazardous wastes. Under the RCRA, generators of hazardous waste are responsible for that waste from "cradle to grave." Decommissioning efforts are subject to the RCRA if they include hazardous waste. Examples of RCRA hazardous wastes that could result from the decommissioning process include debris contaminated with solvents; oils; waste stored in drums and tanks; and cleaning materials used during decontamination. Source material, special nuclear material, and by-product material defined by AEA are exempt from RCRA requirements unless mixed with hazardous wastes.

The RCRA requires that all treatment, storage, and disposal facilities obtain an operating permit (40 CFR 264, 265). Further, facilities must comply with general and specific requirements. General requirements include providing security, planning for emergencies, properly characterizing waste, planning for facility closure, and proper record keeping and reporting. All transportation, treatment, storage, and disposal activities carried out as part of decommissioning are subject to standards, permits, inspections, and enforcement actions of the EPA or authorized state regulatory agency.

Specific requirements apply to individual components, such as special requirements for waste storage tanks (40 CFR 280 & 281). Subtitle I, implemented by the Hazardous and Solid Waste Amendments creates a program to regulate underground storage tanks. Materials regulated by the AEA (i.e. source, special, and byproduct) are exempted from the tank regulations unless combined with other hazardous wastes.

The RCRA imposes corrective action requirements upon the release of hazardous waste from a solid waste management unit. The RCRA also includes facility closure requirements for all facilities with its permits; corrective action and closure requirements are usually incorporated into the permit.

The EPA may authorize states to develop more restrictive requirements than its own to enforce hazardous waste programs in lieu of the RCRA. These states may have authority to regulate decommissioning activities. Depending on a state's scope, regulations may include waste management, corrective action, and closure requirements.

For facilities that will be decommissioned under the corrective action provisions of an RCRA permit or order, EPA guidance encourages responsible state agencies, EPA regions, and facilities to provide public participation activities that are consistent with the public participation activities associated with a CERCLA response action.

4.4 HAZARDOUS MATERIALS TRANSPORTATION ACT

The Hazardous Materials Transportation Act (HMTA) regulates transportation of hazardous materials, including those covered by the AEA and RCRA. The RCRA transportation standards (40 CFR 263) have adopted most rules of the Department of Transportation (DOT) for transportation of hazardous wastes under the HMTA (HMTA 1990). Similarly, the NRC has adopted regulations for the transportation of radioactive materials from the DOT. Transportation requirements apply to all modes -- air, rail, highway, and water.

4.5 COMPREHENSIVE ENVIRONMENTAL RESPONSE, COMPENSATION, AND LIABILITY ACT

The passage of the Comprehensive Environmental Response, Compensation, and Liability Act (CERCLA), also known as the Superfund, was in response to growing national concern about the release of hazardous substances from abandoned waste sites. The CERCLA gives the federal government authority to regulate hazardous substances, to respond to hazardous substance emergencies, and to develop long-term solutions for the nation's hazardous waste problems. The Superfund Amendments and Reauthorization Act (SARA) expanded the federal government's response authority and clarified that federal facilities are subject to the same CERCLA requirements as private industry. Federal agencies and departments are responsible for sites within their jurisdiction.

Under CERCLA, the National Response Center must be notified of releases or threatened releases of hazardous substances above certain quantities (40 CFR 302) unless a federal permit authorizes the release. CERCLA's Community Right-to-Know requirements (40 CFR 350, 355, 370, 372) also mandate notification of state and local emergency response authorities.

CERCLA response efforts are guided by 40 CFR 300, the National Oil and Hazardous Substances Pollution Contingency Plan (EPA 1990), commonly referred to as the National Contingency Plan (NCP). The NCP describes the process that responsible parties (including federal agencies) must follow in response to releases of hazardous substances into the environment. It also establishes the criteria, methods, and procedures the EPA uses to determine which releases have priority for long-term evaluation and response. The NCP's goal is to select remedies that protect human health and the environment, maintain protection over time, and minimize waste generation.

Under the NCP, response actions include remedial and removal actions. Because the remedial action process, which typically includes extensive studies to support remedy selection, may take years, the EPA and DOE have found it generally inappropriate for decommissioning facilities. NCP's removal actions are categorized in three ways: emergency, time-critical, and non-time-critical. The NCP requires public involvement in the removal action planning process, through the administrative record process, public notices, and other mechanisms. The NCP evaluates the need for removal actions based on impacts on human health and the environment, materials in bulk containers that may leak, threat of migration in soil and water, and the threat of fire.

When feasible, the DOE's policy is to pursue non-time critical removals (40 CFR 300.410 and 300.415), which allows great flexibility in creating decommissioning plans and usually brings faster and more cost-effective benefits to worker safety, public health, and the environment than other remedial action options. An agreement between the EPA and the DOE established a tailored approach for decommissioning the DOE's contaminated facilities as non-time-critical removal actions. A Decommissioning Implementation Guide provides additional guidance for the planning of related projects (DOE 1999a).

The CERCLA also created a Hazardous Substance Response Trust Fund, supported by an excise tax on the chemical and petroleum industries and called on to pay for cleanup activities at abandoned waste sites. The 1980 law requires the parties responsible for the contamination to conduct or pay for the cleanup. If the EPA's cleanup enforcement actions are not successful, the federal government can clean up a site using the CERCLA Trust Fund. If the Superfund program conducts the cleanup, the government can take court action against responsible parties to recover up to three times the cleanup costs. However, funds from the Superfund are not generally used for the cleanup of federal facilities or for NRC decommissioning.

Under CERCLA [(105(a)(8)(B)], the EPA prepares the National Priority List (NPL) of highly contaminated sites. Candidate sites are selected after being ranked according to relative risk. NPL sites, published in the Federal Register for public comment, will undergo a remedial investigation/ feasibility study to determine clean up requirements. Once appropriately cleaned, sites will be removed from the list.

CERCLA requires the DOE and EPA to enter into an Interagency Agreement (IAG) to define cleanup requirements for NPL facilities. An IAG may also be forged for sites not on the NPL. EPA encourages incorporating RCRA requirements into IAG at CERCLA sites for facilities subject to both statutes so there is only one, comprehensive agreement. The state is usually a party to the IAG. Among other things, the IAG establishes the roles of the DOE, EPA, and the state in completing the remedial action. An IAG or an NRC-approved decommissioning plan may provide for public and stakeholder involvement in the decommissioning process. At facilities where CERCLA applies, the NCP mandates public involvement in the response action decision-making process. At facilities where RCRA applies, public involvement is governed by 40 CFR 270, EPA-administered permit programs (EPA 1976), or equivalent state regulations.

4.6 LOW-LEVEL WASTE POLICY ACT

The Low-Level Waste Policy Act (LLWPA) assigns the responsibility for disposal of low-level wastes (LLW) to the states. The LLWPA Amendments of 1985 directed states to develop commercial disposal facilities, either individually or by entering into agreements known as "compacts" with other states. The result has been the formation of disposal compacts among groups of states. There are currently ten regional compacts, but some states remain unaffiliated. Each compact assigns a host state the first tenure, typically twenty years, for disposing of LLW. Compacts can also enter into agreements with other compacts to dispose of their waste. As an incentive to form compacts, each compact may exclude the LLW of any non-member state from its disposal site. Unaffiliated states, however, have no authority to prohibit disposal of waste from other states.

At the time the act was passed, there were three commercial operational disposal sites in the country: Richland, WA, Beatty, NV, and Barnwell, SC. Since then, the Nevada site has closed, and one new site, the Envirocare facility in Clive, Utah, has opened for nationwide Class A waste. As explained above, Richland accepts waste only from its own compact and the Rocky Mountain compact. Barnwell is the only facility accepting Class A, B, and C waste from outside its compact. The NRC or an agreement state licenses commercial LLW facilities.

4.7 NUCLEAR WASTE POLICY ACT

The Nuclear Waste Policy Act (NWPA) of 1982 established the DOE's Office of Civilian Radioactive Waste Management to develop and manage a federal disposal system for all spent nuclear fuel from commercial nuclear reactors and HLW from atomic energy defense activities. The statute provides detailed direction for the scientific, technical, and institutional development of the system and requires that the NRC license waste management facilities. The NWPA Amendments of 1987 directed the DOE to characterize only the Yucca Mountain site in Nevada as a potential repository.

Under the NWPA, the costs of disposing of spent fuel and HLW are the responsibility of the generators. Fees levied on the owners and generators of commercial spent fuel are deposited in the Nuclear Waste Fund, a separate account in the U.S. Treasury that is managed and administered by the DOE. The Office of Civilian Radioactive Waste Management, however, can only expend funds from the Nuclear Waste Fund that are appropriated by Congress. Amounts not appropriated for current expenses are invested in U.S. Treasury securities and managed strategically to ensure that the long-term costs of disposal can be met. The NWPA also sets forth a multi-step process for the recommendation and approval of the Yucca Mountain site. The DOE formally initiated this process in May 2001. The current schedule calls for construction to begin in 2006 and for the repository to open in 2010.

4.8 CLEAN AIR ACT

The Clean Air Act (CAA) is the comprehensive federal law that regulates the quality of air. It has been amended many times to account for changing conditions in both the environment and industry. The CAA set emission standards for stationary and mobile sources, expanded local air pollution control programs, established air quality control regions, set standards and compliance deadlines, and initiated research on low-emission fuels and automobiles. The 1990 revision focused on air quality standards, car emissions and alternative fuels, hazardous air pollutants, acid rain, and ozone depletion. States have the primary responsibility for implementing and enforcing the CAA (Fleming 1990).

Several regulations have been promulgated under the CAA. The National Emission Standards for Hazardous Air Pollutants (40 CFR 61) is a body of federal regulations that set forth emission limits and other requirements, such as monitoring, record keeping, and operational requirements, for activities generating emissions of certain hazardous air pollutants. The national primary and secondary ambient air quality standards implemented by the states are concentration limits for specific pollutants considered harmful to public health and the environment (40 CFR 50). There are six: sulfur dioxide, carbon monoxide, nitrogen dioxide, ozone, lead, and particulate matter. Ambient air monitoring, carried out by state and local agencies, is a program designed to monitor these hazardous pollutants (40 CFR 58).

The many parts of the CAA are designed to work together to provide a comprehensive program to ensure acceptable air quality. All metropolitan and consolidated metropolitan statistical areas that do not comply with any of the air quality standards are considered non-attainment areas, consisting of those areas that do not meet primary or secondary ambient air quality standards. All urban counties within the statistical areas are considered part of the non-attainment area, regardless of their attainment status. Air quality standards are primarily implemented through state implementation plans; however, the EPA can rescind this authority if it finds a plan unsatisfactory.

The CAA prohibits construction of new major sources of non-attainment pollutant in metropolitan and consolidated metropolitan statistical areas. For example, if an area is classified as non-attainment for carbon monoxide, new major sources of carbon monoxide may be built only with a state-issued permit (40 CFR 52.24). However, it seems unlikely that a decommissioning project would require such a non-attainment permit.

The CAA forms a two-tiered set of air quality standards that states must meet. The primary standards established acceptable levels of air quality to protect the public's health with a built-in safety margin. Secondary standards state the level of air quality that must be maintained to protect the public, with a safety margin, from any known or suspected effects of an airborne pollutant. States primarily use data from air quality monitoring to track and trend national and local air quality.

The CAA also requires operating permits for all major sources (40 CFR 70). Some decommissioning activities (such as hazardous and mixed waste treatment, storage and disposal units) may require permits because of radionuclide emissions. Additionally, some activities may emit enough other regulated pollutants (e.g. carbon monoxide, nitrogen oxides, sulfur dioxide, particulate matter, volatile organic compounds, and lead) to qualify as major sources and therefore require permits. Major sources include any stationary source or group of stationary sources in a contiguous area that exceed emissions standards established by Section 112 (Hazardous Air Pollutants) of the CAA.

4.9 CLEAN WATER ACT

The Clean Water Act (CWA), enacted in 1948, has been amended several times to address growing pollution in the country's waterways. It is the principal law governing the country's streams, lakes, and estuaries. The CWA establishes a system of national effluent standards, water quality standards, a permit program to enforce these standards, provisions for special circumstances (e.g., hazardous chemicals and oil spills), and a construction loan program for publicly owned treatment works. The CWA definition of pollutant includes radioactive materials; however, the EPA's implementing regulations exclude radioactive materials governed by the AEA. Accordingly, CWA regulations apply to naturally occurring radionuclides, but not to source, byproduct, or special nuclear materials.

The EPA created the National Pollutant Discharge Elimination System (NPDES) in 1972 to implement the provisions of the CWA (40 CFR 122). This is a permit program that regulates the point source discharge of pollutants. Any point source that discharges a pollutant into surface water is required to have a NPDES permit. Point sources include, but are not limited to, pipes, ditches, channels, tunnels, wells, or containers that are or may be discharging pollutants. Specific decommissioning activities that could require a permit include discharge from wastewater treatment facilities and storm water from RCRA facilities and other industrial activities.

The NPDES program applies almost exclusively to industrial and municipal facilities. The EPA and state regulatory agencies monitor permit holders to ensure compliance. Permit holders test the water and submit the results to the monitoring agency, which may send inspectors to the site after reviewing the data. The monitoring agency has authority to penalize any permit holder that violates the permit restrictions. Violations can be addressed with either an administrative order from the responsible agency or with civil or criminal actions. Permits are public records and any member of the public may bring suit against delinquent permit holders, unless the EPA or state agency has already done so. Discharges to publicly owned treatment works do not require permits, but are subject to pre-treatment regulations (40 CFR 403). These regulations require removal or treatment of all pollutants that could pass through the system untreated or adversely impact the system's performance.

Section 404 of the CWA requires a permit for placing dredged or filler material into bodies of water (33 CFR 323). The Army Corps of Engineers administers this program and reviews all activities that involve filling wetlands not covered by a national or regional permit on a case by case basis. It has issued some national and regional permits allowing for certain placements of dredged or fill material (33 CFR 330).

4.10 SAFE DRINKING WATER ACT

The intent of the Safe Drinking Water Act (SDWA) is to protect the nation's drinking waters from both man-made and natural contaminants by setting standards for chemicals, human and animal wastes, pesticides, and other types of contaminants in drinking water supplies. States have primary enforcement authority, but the EPA may assume that authority from states that do not properly enforce the SDWA.

National Primary and Secondary Drinking Water Regulations are the major standards for establishing acceptable levels of many contaminants (40 CFR 141 - 143). The primary regulations set the enforceable Maximum Contaminant Levels (MCL) and non-enforceable Maximum Contaminant Level Goals (MCLG) for individual contaminants in drinking water, and states require methods for treating water to remove contaminants. MCLG are significant for decommissioning because they can be used as applicable or relevant and appropriate requirements under CERCLA. CERCLA [Section 121(d)] requires that remedial actions attain any applicable federal standard or requirement, or any more stringent state standard. Additionally, SDWA standards are used to establish groundwater protection regulations under other statutes, such as the RCRA.

4.11 NATIONAL ENVIRONMENTAL POLICY ACT

The National Environmental Policy Act (NEPA) was enacted to set overall direction for federal programs affecting the environment, including decommissioning projects under the jurisdiction of both the DOE and NRC. State, local, and private actions involving federal funding or permits may also trigger NEPA requirements. The NEPA requires the preparation of environmental assessments and, for any project that will have a major impact on the environment, an Environmental Impact Statement (EIS) (40 CFR 1500 B 1508). The EIS process requires that the public have opportunities to interact with the responsible agency. The purpose of an EIS is to ensure that environmental information concerning the impacts of a proposed action and appropriate alternatives are available and fully considered before decisions are made and actions taken. The EIS helps the agency avoid difficulties resulting from overlooked environmental impacts.

Individual actions, such as actions to decommission facilities, are to be evaluated to determine the level of NEPA review needed. The process usually starts with a determination of the potential environmental impacts from the proposed project. Its purpose is to determine if an EIS is required. The agency must assess the proposed action for direct, indirect, and cumulative impacts. No EIS is required if the assessment results in a finding of no significant impact, but the agency must explain why the proposed action does not significantly impact the human environment. If an EIS is required, the agency first issues a notice of intent and prepares a draft EIS. When the draft is complete, the agency allows a 45-day public comment period. Comments by the public and other federal agencies must be addressed in the final EIS, thereby requiring the responsible agency to completely evaluate its plan and any alternatives to the satisfaction of both the public and NEPA requirements. For either process, a comprehensive administrative record containing any documentation contributing to an agency's decision-making process should be collected. This record will serve as the evidentiary support to justify an agency's action against potential litigation arising from its decision.

Title II of the NEPA established the Council on Environmental Quality (CEQ) to oversee its administration and ensure compliance. The CEQ develops regulations for preparing an EIS, resolves lead agency disputes, trains federal agencies on NEPA, and mediates policy disputes between agencies. The EPA is responsible for reviewing the EIS and announcing it in the Federal Register. The EPA is also authorized to review any action of an agency that is not preparing an EIS when it believes one may be required.

4.12 ENDANGERED SPECIES ACT

The Endangered Species Act (ESA) was enacted to protect plants and animals in danger of extinction and to conserve their habitats. Threatened and endangered species are designated by the Secretary of the Interior and identified on the Endangered Species List. Whenever possible, critical habitat is designated at the time the species is listed. The Department of the Interior develops recovery plans for listed species. The Fish and Wildlife Service, the National Marine Fisheries Service, and/or the appropriate state agencies review all proposed federal actions that may affect threatened or endangered species. The Department of the Interior may require a biological assessment discussing listed species that could be affected by a proposed project. The Department then issues a biological opinion, and may offer alternatives that would avoid violations of the Act. If this occurs, the responsible agency (e.g., the Department of the Interior or the NRC) must alter or abandon the project, or apply for an exemption. The Endangered Species Committee grants exemptions if there are no reasonable alternatives, if the benefits of the action outweigh the benefits of the alternatives, or if the action is of regional or national significance.

4.13 TOXIC SUBSTANCES CONTROL ACT

The Toxic Substances Control Act (TSCA) was enacted to allow the EPA to control toxic substances, including the manufacture of various chemical agents. Sections of interest to decommissioning activities include those dealing with polychlorinated biphenyls (PCBs), lead, and asbestos; other TSCA-regulated chemicals may also be involved. Note that the disposal of PCBs and asbestos is regulated by several states under similar programs.

4.14 FEDERAL FACILITY COMPLIANCE ACT

Before the passage of the Federal Facility Compliance Act (FFCA), federal agencies maintained that the doctrine of "sovereign immunity" exempted them from fines and penalties prescribed by solid and hazardous waste laws. Congress enacted the FFCA to bring federal facilities into the same legal framework as the private sector; it has effectively waived sovereign immunity with respect to RCRA for federal facilities. The EPA is required to conduct annual RCRA inspections of all federal facilities. Authorized states also inspect to ensure compliance with the state hazardous waste programs [Section 104(3)]. To facilitate compliance, the DOE has entered into site management and cleanup agreements with the EPA and states pursuant to FFCA. Many DOE facilities are subject to these agreements and other binding administrative cleanup orders. The FFCA allows regulators to impose fines or penalties on entities that fail to comply with enforceable agreements.

4.15 NATIONAL HISTORIC PRESERVATION ACT

The National Historic Preservation Act (NHPA), enacted in 1966, is the most comprehensive national policy on historic preservation. The National Register of Historic Places was created under this Act. It is a list of cultural resources with national, regional, state, and local significance. Section 106 (36 CFR 800) is the primary provision that could potentially impact decommissioning. The NHPA requires federal agencies with jurisdiction over a federal, federally assisted, or federally licensed undertaking to consider the effects of the undertaking on properties included in or eligible for inclusion in the National Register of Historic Places, and to give the Advisory Council on Historic Preservation a reasonable opportunity to comment.

4.16 COASTAL ZONE MANAGEMENT ACT

The Coastal Zone Management Act (CZMA) is designed to preserve, protect, develop, and where possible, to restore or enhance the resources of the nation's coastal zone [16 U.S.C. 1452(1)]. The Act requires federal agencies that undertake development projects affecting the coastal zone to ensure that each project is, to the maximum extent practicable, consistent with the enforceable policies of the responsible state [16 U.S.C. 1456(c)(2)]. Implementing regulations have been adopted as 15 CFR Part 930. Under these regulations, federal agencies must (as soon as possible but at least 90 days before final approval of the proposed project) provide the state with a determination of consistency for any activity directly affecting the coastal zone. The Act also requires that applications for federal permits include a certification that the proposed project will comply with the enforceable policies of the state. It is unlikely that many decommissioning projects will be subject to the requirements of the CZMA. However, if it appears that it applies, the responsible person contacts a representative of the state agency responsible for the coastal zone management program, discusses the proposed project, and clarifies the enforceable policies of the state's program.

4.17 FISH AND WILD LIFE COORDINATION ACT

In addition to the Endangered Species Act, the Fish and Wild Life Coordination Act is the major federal statute designed to protect plant and animal resources from negative impact of development. Section 2 of the Act requires federal agencies to consult with the Fish and Wildlife Service and the appropriate state wildlife agency before proposing or approving the impounding, controlling, or diverting of any stream or other body of water [16 U.S.C. 662(a)]. It seems unlikely that the DOE will propose impounding, controlling, or diverting a body of water in association with decommissioning. Nevertheless, if such action occurs, DOE personnel should consult with the appropriate offices concerning recommendations they may have for enhancing the wildlife resource improvement and loss prevention plans.

4.18 STATE LAWS

States have enacted laws and implementing regulations concerning environmental protection and radiological control. Each state has its own unique body of regulations, subject to continual change. Facility decommissioning activities are subject to these state laws as well as to federal laws and regulations.

Chapter 5

INTERNAL REQUIREMENTS OF THE DOE

5.1 INTRODUCTION

In general, environmental laws, regulations, and other requirements for decommissioning of facilities under DOE management are interpreted and communicated through a directives system and are imposed on specific projects through contracts. The internal focus is on the final phases of the life cycle, including deactivation, decommissioning, surveillance and maintenance, and any potential transfer of ownership. The management system includes mandatory requirements, such as policy notices, orders, notices, and manuals, and guidance with some implementation flexibility, including the graded application of certain technical standards and guides. In addition, the management system is supported by optional handbooks, voluntary standards, and other documents illustrating established practices for meeting requirements. A thorough understanding of this system requires acceptance that the DOE is a mission-oriented federal agency with its share of statutory intricacies, diverse sites, and a diverse set of contractors.

5.2 ENVIRONMENTAL LAWS AND REGULATIONS

The primary authority for guiding facility disposition comes through the DOE's implementation of its authority granted by the AEA. In addition, numerous other laws and their implementing regulations apply to facility disposition, including the CERCLA, the RCRA, and various others. A few DOE facilities are under jurisdiction of the NRC, including the repository for disposal of commercial HLW at Yucca Mountain, NV. Under the AEA, the DOE has programmatic responsibility for facility disposition, including decommissioning and managing and disposing of radioactive waste. Furthermore, the DOE has lead federal agency responsibility for the disposition of former nuclear weapons production and research facilities. The DOE is responsible for project planning, budgeting, and executing facility disposition for these facilities, in appropriate coordination with regulators and stakeholders. It uses a system of directives, policies, guidelines, and manuals to control internal activities, including those of its implementing contractors. The hierarchy of the directives most relevant to decontamination or facility disposition is presented as Figure 5.1.

5.3 DIRECTIVES ESTABLISHING REQUIREMENTS

5.3.1 Life Cycle Asset Management: Order 430.1A and DOE O 430.1B (DOE 1998a), identify the requirements for managing the entire life cycle of a facility, are intended to ensure that an integrated and seamless process is in place for the management of physical assets from acquisition through operation and disposition. It therefore addresses specific requirements for the end of a facility's life cycle, including transition, stabilization, and the length of time before final disposition is complete (including deactivation, decommissioning, and residual surveillance or long-term maintenance and monitoring). The phases of the facility life cycle are discussed further in the sections corresponding to implementation guidance.

Environmental Laws/Regulations
Atomic Energy Act
Comprehensive Environmental Response, Conservation, and Liability Act
Resource Conservation and Recovery Act
Others

Directives Establishing Mandatory Requirements
Order 430.1A and Order 430.1B

Policy 450.4

Directives Providing Guidance
Guide 430.1-5
Guide 430.1-3
Guide 430.1-4
Guide 430.1-2
Guide 450.3-3
Guide O 430.1A Good Practice Guides
Guide O 5400.1

Handbooks and Standards
Deactivation Handbook
Decommissioning Handbook
Voluntary Standards

Figure 5.1 Hierarchy of DOE Facility Directives

Other requirements that apply to all phases of a facility's life cycle include:

- An integrated graded approach to record keeping, maintenance, and other aspects of asset management
- Performance measures based on industrial best practice
- Comprehensive methods for planning and control
- Demonstrating compliance with applicable federal, state, and local laws and regulations, as well as negotiated agreements, such as consent decrees and federal facility compliance agreements

For facilities currently in the operational phase, the requirements related to facility transition are specific to the implementation of actions to place a facility, systems, and materials in stable and known conditions and to ensure that hazards are known, evaluated, and identified pending disposition. For facilities that have ceased operations, requirements ensure that actions are taken to eliminate/mitigate hazards and that adequate protection is provided to workers, the public, and the environment.

Requirements specific to facility disposition include:

- Integrated safety management practices, to include an approved safety basis
- Collection, maintenance, and update of baseline data to support physical, chemical, and radiological characterization
- Surveillance and maintenance activities to correspond with facility condition
- An established method for identifying, assessing, and evaluating alternatives and documenting a preferred alternative
- A process applicable to deactivation and decommissioning that identifies the specific facility end-state and the activities to achieve it
- Standard, detailed, engineering planning and documentation
- Evaluation of a non-time-critical removal framework for decommissioning
- An up-to-date corporate database of the as-is facility condition
- Formal documentation of actions taken, and of meeting the release criteria for end-state condition

Further requirements in the planning and conduct of transition and disposition activities appear in various other directives. The federal project director and others involved in developing and/or executing a decommissioning project are required by Order 430.1A to be well versed on applicable federal, state, and local laws and enabling regulations. In addition, requirements addressing integrated safety management, conduct of operations, conduct of maintenance, radiological protection, and other requirements integral to the planning, development, and conduct of work are expected to be integrated into the program.

While the various phases of a facility's life cycle (specifically those occurring after operations are complete) are addressed, timing of any action is quite dependent on the priorities of the internal responsible program organization and its funding profile. As such, the sequence, scope, and end-state condition of similar facilities may differ significantly (e.g., deactivation followed by long-term surveillance and maintenance, integrated decommissioning for immediate reuse in a new mission, or simple decommissioning followed by entombment or demolition). The DOE Order was established to ensure that, regardless of the specific sequence, scope, or end state, each program element enables an integrated, seamless, and tailored approach to facility disposition. For example, for projects where no logical delineation between deactivation and decommissioning exists, the requirements are integrated to serve the overall project and completion objectives. In doing so, planning considers the possibility of future changes to priorities and identifies the conditions (end-points) where a project may be safely and efficiently slowed or accelerated as necessary in line with changing conditions and priorities.

5.3.2 Integrated Safety Management System: The basic requirement for incorporating an integrated safety management system (ISMS) into management and work practices is identified in DOE Policy 450.4, Safety Management System Policy (DOE 1996a). Throughout the policy statement, the term "safety" is used synonymously with environment, safety and health (ES&H) to encompass protection of the public, the workers, and the environment. The policy highlights components that facilitate the orderly development and implementation of a safety management system throughout the transition and disposition of an excess facility. Its objective is to ensure that safety management becomes an integral part of mission accomplishment through the integration of

ES&H considerations into all facets of work planning and execution in order to effectively protect the public, the workers, and the environment.

The guiding principles are fundamental policies that guide action, from development of safety directives to actual performance of work. The ISMS guiding principles are:

1. Line Management Responsibility for Safety: Line management is directly responsible for the protection of the public, the workers, and the environment.

2. Clear Roles and Responsibility: Clear and unambiguous lines of authority and responsibility for ensuring safety shall be established and maintained at all organizational levels within the Department and its contractors.

3. Competence Commensurate with Responsibilities: Personnel shall possess the experience, knowledge, skills, and abilities necessary to discharge their responsibilities.

4. Balanced Priorities: Resources shall be effectively allocated to address safety, programmatic, and operational considerations. Protecting the public, the workers, and the environment shall be a priority whenever activities are planned and performed.

5. Identification of Safety Standards and Requirements: Before work is performed, the associated hazards shall be evaluated and an agreed-upon set of safety standards and requirements shall be established. If properly implemented, these standards will provide adequate assurance that the public, the workers, and the environment are protected from adverse consequences.

6. Hazard Controls Tailored to Work Being Performed: Administrative and engineering controls to prevent and mitigate hazards shall be tailored to the work being performed and associated hazards.

7. Operations Authorization: The conditions and requirements to be satisfied for operations to be initiated and conducted shall be clearly established and agreed-upon.

The core functions provide the necessary structure for any work activity that could potentially affect the public, the workers, and/or the environment. These functions are applied as a continuous cycle, with the degree of rigor appropriate to address the type of work and its associated hazards. The ISMS core functions are:

- Define the Scope of Work: Missions are translated into work, expectations are set, tasks are identified and prioritized, and resources are allocated.
- Analyze Hazards: Hazards associated with the work are identified, analyzed, and categorized.
- Develop and Implement Hazard Controls: Applicable standards and requirements are identified and agreed-upon, controls to prevent/mitigate hazards are identified, the safety envelope is established, and controls are implemented.
- Perform Work within Controls: Readiness is confirmed and work is performed safely.

- Provide Feedback and Continuous Improvement: Feedback information on the adequacy of controls is gathered, opportunities for improving the definition and planning of work are identified and implemented, line and independent oversight is conducted, and if necessary, regulatory enforcement actions occur.

ISMS mechanisms are site-/facility-specific documents that define how its functions are performed. These mechanisms are based on policies and procedures to identify and analyze hazards, perform safety analyses, and set applicable safety standards. These mechanisms are established, as appropriate, in such documents as the Health and Safety Plan (HASP), Safety Analysis Report (SAR), Basis for Interim Operation (BIO), Chemical Hygiene Plan, and Process Hazard Analyses. Responsibilities for the ISMS, appropriate to the mechanisms used, must be clearly defined to satisfy each principle or function and to ensure identification of the appropriate qualified approval authority. Implementation involves specific definition of work and planning, hazards identification and analysis, definition and implementation of hazard controls, performance of work, development and implementation of operating procedures, and monitoring and assessment of performance for improvement.

A major mechanism for integrating safety and health into facility transition and disposition efforts is through the work planning process, during which existing safety documentation is evaluated and operations are identified and evaluated against existing controls. Since most of it is related to operations, the existing documentation frequently requires significant review and modification to accommodate a different kind of work -- the various aspects of decommissioning. Occupational safety documentation for older facilities often falls short of modern standards, and particularly requires tailoring for the planned sequencing of activities. Practices that are comparable to the levels established by the Occupational Safety and Health Act (OSHA) are contractually required.

Existing documentation from the facility's operational phase frequently serves as the starting basis for a project safety analysis. Revisions, comparisons, crosswalks, and other evaluation techniques can be used to determine those facility disposition actions that may be covered in existing documentation or may require supplemental coverage. Such evaluations are frequently more expedient than preparation of new safety documentation, particularly if they involve expert operational knowledge of the specific facility.

Early occupational worker involvement in all the levels of hazards identification and analyses, and the selection of hazard controls, required for the planning and execution of disposition is key to an efficient system. A comprehensive walkdown of the facilities/systems to be dismantled should be conducted by a team that includes the system engineer, the work planner, ES&H specialists, and applicable craft supervisors. Such a walkdown is an important step in identifying hazards, selecting the appropriate hazard controls, and preparing the work plans/procedures. Consulting with scientists and engineers that worked at the facility when it was operational can provide additional information on the materials and processes used at the facilities -- information that can aid in hazards identification for decommissioning activities. Such involvement is one of many leading factors for ensuring an effective safety system. Other such indicators include the ratio of unusual occurrences at other sites that actually trigger a lessons-learned review.

When selecting hazard controls and developing work plans/procedures, the planner should ensure a proper balance among the potentially conflicting safety considerations of the various ES&H

disciplines. For example, the ALARA considerations for working in a radiologically contaminated room would dictate accomplishing the work using the fewest possible number of workers. However, if the work involves using torches or other equipment producing flames/sparks, then precautions for fire protection versus working alone must be balanced with ALARA considerations, and both incorporated into the work planning.

Verification that hazard controls are in place must be a step in the process to check readiness prior to authorizing the start of work. Ensuring that the workers assigned to perform decommissioning tasks have the necessary experience and training must also be ensured prior to commencing the work. The effectiveness of the implementation of hazard controls while work is in progress, including checking that workers are using the necessary personal protective equipment, must be checked through line management and/or supervisory oversight.

An important aspect of the DOE's ISMS is the gathering and analyses of feedback information for the purpose of improving the safety performance of subsequent work. Decommissioning tasks/activities should be evaluated upon completion to determine if lessons can be learned that would improve subsequent work. Such lessons-learned information should be documented and disseminated to enable others to benefit. Lessons learned information, generated internally or obtained from external sources, should be applied as appropriate to improving work planning and performance. Feedback information that can lead to improvement of subsequent decommissioning work can come from line management and supervisory oversight, independent oversight, and input and self-assessment by the workers.

5.4 DIRECTIVES PROVIDING GUIDANCE

A number of guides (DOE 2001a, 1999a, 1999b, and 1999c) were developed to help facility managers implement the requirements of Order 430.1A. The goal of the processes described in these guides is to ensure a continuum of hazard mitigation and risk reduction throughout the transition and disposition phases, leading to a timely, cost-effective disposition of the facility. Note that guides described in this section are consistent with the principles and core functions of the DOE's ISMS policy.

5.4.1 Transition Implementation Guide: Guide 430.1-5 (DOE 2001a) provides guidance for transition activities between operations and disposition in a DOE facility's life cycle. Transition activities begin once a facility is declared excess to DOE needs. It includes placing the facility in a stable and known condition, identifying hazards, eliminating or mitigating hazards, and transferring programmatic and financial responsibilities from the operating program to the disposition program. Timely completion of transition activities can take advantage of facility operational capabilities before they are lost, eliminating or mitigating hazards in a more efficient, cost-effective manner. When preparing for the disposition phase it is important that material, systems, and infrastructure stabilization activities be initiated prior to the end of facility operations.

Due to the large number of activities required in transition from operation, this guide was prepared to aid in the development, planning, and implementation of requirements and activities during the transition phase at facilities declared as excess. The guide defines activities or actions that provide a sequenced risk reduction and seamless transition from operations to the selected disposition path and presents an eleven-step approach for implementing the transition requirements specified in Order 430.1A. These administrative and physical actions include:

- Officially declaring a facility excess
- Establishing an integrated transition team (with representatives from the operating and disposition organizations)
- Site characterization
- Identification, planning, and execution of both early stabilization actions (if any) and/or other stabilization actions consistent with the specified disposition path
- Integration of safety management, and adjusting the authorization basis as appropriate
- The conduct of a review to verify the satisfactory completion of stabilization
- The transfer of the facility from the operating to the disposition program

5.4.2 Deactivation Implementation Guide: Guide 430.1-3 (DOE 1999b) provides guidance for facility deactivation following transition and stabilization. The goal is to achieve a safe shutdown condition that is economical to monitor and maintain for an extended period, until eventual decommissioning. Deactivation of contaminated, excess facilities occurs as soon as reasonable and for as many facilities as possible. In this way, the resources can be applied to the greatest net safety gains, and to achieve stability in the shortest time. Deactivation should yield a low risk state with minimum surveillance and maintenance requirements.

Deactivation, analogous to the NRC's SAFSTOR, is characterized by the need to define end-points that are design specifications for identifying deactivation project completion. Specifying and achieving end-points is a systematic engineering method of proceeding from an existing facility condition to a stated final set of conditions in which a facility is safe and can be economically monitored and maintained.

5.4.3 Decommissioning Implementation Guide: Guide 430.1-4 (DOE 1999a) focuses on final facility disposition, when the facility is taken to its ultimate end-state through decontamination and/or dismantling to demolition or entombment. After decommissioning is complete, the facility or surrounding area may require institutional control.

In 1994, the DOE established by policy that decommissioning activities could be conducted as non-time-critical removal actions under CERCLA unless the circumstances at the facility made it inappropriate. Use of such removal actions effectively integrated state and stakeholder participation with the oversight responsibilities of the EPA and of the DOE as lead agency. These actions are defined and explained in the NCP found in 40 CFR 300. Although the Implementation Guide is modeled from the CERCLA non-time-critical removal action process, it is flexible enough to accommodate projects regardless of the statute, authority, or management decision that initiates them. A decommissioning project may be initiated by a variety of circumstances, including:

- Determination that a release or substantial threat of release to the environment is present, and a removal action under CERCLA Section 105 is appropriate
- Implementation of a decommissioning plan approved with the objective of termination of an NRC license
- Decommissioning in accordance with an RCRA permit or order

- A programmatic management decision to proceed with the disposition of a surplus facility to reduce operational costs through asset retirement

5.4.4 S&M during Facility Transition and Disposition: Guide 430.1-2 (DOE 1999c) guides the implementation for Surveillance and Maintenance (S&M) during facility transition and disposition. S&M activities are conducted throughout the facility life cycle, including when a facility is not operating or is not expected to operate again. It is important to ensure that S&M is adequate to maintain the facility safety envelope during the final stages of operations, and through a seamless transition to the final disposition. S&M includes periodic inspections and maintenance of structures, systems, and equipment to ensure, at a minimum, that there is adequate containment of any contamination and that the potential hazards to workers, the public, and the environment are eliminated or mitigated and controlled. The guide presents a six-step approach for implementing the S&M requirements specified in Order 430.1A. This approach includes:

- S&M throughout the transition and disposition phases (Step 1)
- A decision point to evaluate/reevaluate the program (Step 2)
- Collection and evaluation of data to support a revised program (Steps 3 and 4)
- Development of a plan to outline the program to be implemented (Step 5)
- Implementation of the program (Step 6)

5.4.5 Integration of ES&H into Facility Disposition Activities: Guidance for integrating and enhancing worker, public, and environmental protection during facility disposition activities is provided in DOE-STD-1120-98 (DOE 1998b). This standard is designed to support an ISMS consistent with the guiding principles and core functions contained in DOE Policy 450.4 and discussed in Guide 450.4-1. The standard includes expectations for:

- Clarification, integration, and reduction of applicable ES&H requirements
- An integrated and balanced approach to identification, analysis, and control of all types of hazards
- Implementing an effective ISMS, including those expectations related to subcontractor and transient workers
- A compilation of existing ES&H directives potentially applicable to facility disposition activities

5.4.6 Tailoring for ISMS Applications: The "graded approach" application of requirements to a particular project, activity, or facility is required by Order 430.1A. Implementation of the tailoring approach, as defined in Guide 450.3-3 (DOE 1997a), is an acceptable method of complying with this requirement. Tailoring is appropriate for all steps in transition facility disposition. It allows choices to be made from a variety of engineering and administrative controls to provide adequate protection for workers, the public, and the environment during the performance of work. Individual tasks are tailored so that each task has controls that fit the specific work and the hazards associated with it, and that are consistent with high-level performance expectations.

Tailoring permits the consideration of differences between facilities and provides a means to determine the extent to which actions are appropriate for a particular facility (or portions thereof). The detail required and the magnitude of resources expended for a particular management element is commensurate with the relative importance of that element to safety, environmental compliance, safeguards and security, the magnitude of any hazard identified, programmatic importance, financial impact, and/or other facility-specific requirements.

Tailoring is cost-effective because it does not demand a high level of analysis and/or planning for simple jobs already covered in established procedures. Worker involvement, as stated earlier, has also proven to be cost-effective because these employees are often those who have spent many years performing tasks during operations, and they may have a good understanding of the safety and performance requirements of the transition and facility disposition activities.

Tailoring an ISMS also offers a means to grade activities and processes in accordance with the different hazards associated with individual facilities. Tailoring is used to scale expectations and acceptable performance to the needs of the site, activity, facility, or work to be performed. When applied to the core safety management functions, tailoring promotes a work management system that is safe, efficient, and cost-effective.

5.4.7 Deactivation Methods and Practices Handbook: The Excess Facility Transition to Deactivation & Decommissioning Methods and Practices Handbook (DOE 2003d), describes practices, methods, and successful examples that can be adapted to other activities. It represents a third tier of information that follows from Order 430.1A and Guide 430.1-3. The Deactivation Management section includes the following topics:

- The stabilization/deactivation customer
- Project approach for a complex facility
- Establishing the overall end-state
- Viewing deactivation in two phases
- Early decisions
- Early deactivation tasks
- Facility-specific commitments
- Hazard reduction
- Detailed end-points
- Setup method and criteria
- Post-deactivation
- S&M plan
- Specify end-points
- Deactivation work plans
- Project management plan
- Decoupling from decisions
- Contiguous facilities
- Scheduling and estimating
- Use of independent reviewers
- Safety documentation
- Regulatory interaction
- Stakeholder interaction

An internal DOE Decommissioning Handbook (DOE 2000a), gives an excellent overview of how to implement the its decontamination and decommissioning framework.

Chapter 6

OCCUPATIONAL SAFETY AND HEALTH REQUIREMENTS

6.1 INTRODUCTION

Decommissioning of nuclear facilities requires attention not only to radiation safety but also to other safety requirements collectively referred to as occupational safety and health (OSH). As a general rule, a safety assessment is the initial phase of both radiation safety and OSH. The safety assessment identifies significant hazards during the decommissioning phase of a nuclear facility that are not normally encountered during the operational phase. Hazardous materials are major factors in the decommissioning of old nuclear facilities and represent a risk to the operators undertaking the work. Examples of common hazardous materials are lead, asbestos, PCBs, mercury, and beryllium. All require special disposal. The handling and disposal of mixed (hazardous and radioactive) wastes can pose special problems since, unlike radioactive materials with a relatively short half life, hazardous chemicals may pose a health hazard for a much longer (if not indefinite) time. If decommissioning will be long deferred, due regard should be given to the gradual deterioration of structures, systems, and components.

Upon permanent shutdown, a facility should undertake a critical review of all site records (paper, microfilm, and electronic formats) and should resist the initial urge to eliminate records used "for operations only." Records are critical assets for decommissioning planning; their maintenance and accessibility should preclude any decision as to reductions in related site staffing.

Decommissioning normally follows deactivation, and often is followed by long-term storage with S&M. A long time may pass between facility operation, deactivation, and final decommissioning. The S&M activities focus on monitoring and controlling any remaining hazardous substances or contamination and maintaining the facility's structural integrity. In some cases, facility operations may be temporarily suspended, then indefinitely shut down, establishing an S&M phase by default.

The duration of each facility phase (deactivation, decommissioning, and S&M) generally depends on hazard magnitude, project complexity, and funding availability. A major part of decommissioning is the removal of buildings, facilities, and other structures. Consequently, during the process, workers are exposed to physical hazards similar to those encountered during typical construction activities. Although potential risks to workers lessen as radiological or hazardous material inventories are removed, the potential risk of exposure to both radiological and hazardous materials during removal increases. Experience over the past few years highlights several key factors impacting the effectiveness of OSH programs:

1. The existing set of OSH requirements, promulgated primarily for the design and operation of facilities, are not always suitable for disposition activities. Although fundamental OSH principles and objectives remain similar, facility disposition activities entail unique work, hazard, and programmatic characteristics different from facility operation and therefore requiring clarification or modification.

2. Decommissioning activities introduce potential hazards and challenges not necessarily encountered during operations. Workers can be exposed to hazardous substances and physical hazards, including ionizing radiation, during a typical facility project. Addressing these hazards requires development and implementation of an effective and integrated hazards management system to identify, analyze, and establish controls.

A properly designed and executed OSH program must include the following key elements:

1. *Work Planning* to ensure that OSH is an integral part of planning and performing the work. This approach has shown to be most cost-effective in controlling hazards.

2. *Hazard Identification* and analysis to determine potential consequences.

3. *Integrated Hazards Analysis* as complex activities require integrated hazard elimination.

4. *Health and Safety Plan/Program* to ensure overall adequate worker protection.

5. *Establishing Controls for Hazards* to implement prevention and mitigation controls. Such engineering controls, work practices, personal protective equipment, or a combination, help prevent exposure to hazardous substances and ES&H hazards.

6. *Evaluating Readiness* to be completed before any work starts, to ensure that all hazards have been identified, appropriate OSH requirements have been met, and safety systems and controls are in place and capable of performing their intended function.

7. *Management of Change* to evaluate all proposed activities, changes, and discoveries that may affect facility or worker safety. A system for continual feedback and evaluation should be implemented.

6.2 OVERVIEW OF REGULATORY REQUIREMENTS

Despite the significant differences between facilities operated by the private sector and those operated by government agencies, notably the DOE, the two systems have much in common.

6.2.1 Requirements of The OSHA: The Occupational Safety and Health Administration (both the Act and the Administration are referred to as OSHA) is responsible for the implementation of requirements of the Occupational Safety and Health Act. This Act authorizes enforcement of its standards, assists and encourages the states in their efforts to ensure safe and healthful working conditions, and provides for research, information, education, and relevant training. The safety and health aspects of decommissioning activities at commercial or private nuclear facilities must also comply with OSHA regulations. Since many states have their own regulations, decommissioning must comply with the requirements of the respective state; however, they follow the requirements of OSHA in the majority of cases. Most of the regulations impacting decommissioning activities (Table 6.1) have been codified in 29 CFR 1910 (OSHA 2003a). In particular, 29 CFR 1910.120, "Hazardous Waste Operations and Emergency Response," (commonly known as HAZWOPER) requires a hazards management process (identification, analysis, and establishment of controls) and a health and safety program for worker protection, and is applicable to the restoration of sites contaminated by hazardous waste constituents.

Table 6.1 Relevant Regulations from the Code of Federal Regulations

Regulation	Agency	Title
10 CFR 20	NRC	Standards for Protection against Radiation
10 CFR 50	NRC	Domestic Licensing of Production and Utilization Facilities
10 CFR 830.120	DOE	Nuclear Safety Management: Quality Assurance Requirements
10 CFR 830.330	DOE	Nuclear Safety Management: Training and Qualification
10 CFR 835	DOE	Occupational Radiation Protection
29CFR1904.0	OSHA	Recording and Reporting Occupational Injuries and Illnesses - Purpose
29 CFR 1904.5	OSHA	Recording and Reporting Occupational Injuries and Illnesses - Annual Summary
29 CFR 1910	OSHA	Occupational Safety and Health Standards
29CFR1910.23	OSHA	Guarding Floor and Wall Openings and Holes
29 CFR 1910.25	OSHA	Portable Wood Ladders
29 CFR 1910.26	OSHA	Portable Metal Ladders
29 CFR 1910.28	OSHA	Safety Requirements for Scaffolding
29 CFR 1910.95	OSHA	Occupational Noise Exposure
29 CFR 1910.119	OSHA	Process Safety Management of Highly Hazardous Chemicals
29 CFR 1910.120	OSHA	Hazardous Waste Operations and Emergency Response
29 CFR 1910.134	OSHA	Respiratory Protection
29 CFR 1910.146	OSHA	Permit Required Confined Spaces
29 CFR 1910.147	OSHA	The Control of Hazardous Energy (Lockout/Tagout)
29 CFR 1910.151	OSHA	Medical Services and First Aid
29 CFR 1910.1001	OSHA	Toxic and Hazardous Substances -- Asbestos
29 CFR 1910.1200	OSHA	Hazard Communication
29 CFR 1926.35	OSHA	Employee Emergency Action Plans
29 CFR 1926.65	OSHA	Hazardous Waste Operations and Emergency Response
48 CFR 970.5204 2	USDOE	DOE Management and Operating Contracts. Integration of Environment, Safety, and Health into Work Planning and Execution
48 CFR 970.5204 78	DOE	DOE Management and Operating Contracts. Laws, Regulations, and DOE Directives
49 CFR 171.8	DOT	General Information, Regulations, and Definitions and Abbreviations
49 CFR 172.101	DOT	Hazardous Materials Table, Special Provisions, Hazardous Materials Communications, Emergency Response Information, and Training Requirements -- Purpose and Use of Hazardous Materials Table
40 CFR 68	EPA	Chemical Accident Prevention Provisions
40 CFR 261.3	EPA	Identification and Listing of Hazardous Waste. Definition of Hazardous Waste

6.2.2 Requirements of the DOE: Most DOE facilities are not covered by OSHA regulations. However, the DOE essentially adopts all of them and incorporates their requirements in its Directives (Orders). The DOE has an extensive system of regulations and orders covering OSHA topics that are binding on its employees, contractors, and subcontractors. Order 440.1A (DOE

2000b) applies to all decommissioning activities; its purpose is to establish a comprehensive worker protection program that ensures a level of safety and health that is at least equal to that required by the OSHA. Guide 440.1-1 is the corresponding guide providing general information and methodologies that the DOE finds acceptable in meeting the requirements defined in Order 440.1A. Most of the OSHA requirements are included in the Order and its Implementation Guide.

Figure 6.1 summarizes the taxonomy of worker protection documentation. Figure 6.2 shows the relationship of worker protection documents and other DOE Rules and Orders that are important to the overall worker protection program.

Regulations	29 CFR 1960 FEOSH			
DOE Requirements	DOE O 440.1A Worker Protection for DOE, Federal, and Contractor Employees			
Implementation Guides	DOE G 440.1-2 Construction	DOE G 440.1-3 Occupational Exposure Assessment	DOE G 440.1-4 Contractor Occupational Medical Program	DOE G 440.1-5 Fire Safety
Mandatory Manuals	DOE M 440.1-1 Explosives Safety			
Nonmandatory Technical Standards	DOE STD 1091-95 Firearms Safety			
Handbooks	Hoisting and Rigging	Electrical Safety	HAZWOPER	FEOSH

Figure 6.1 Worker Protection Management Taxonomy of Documents

6.3 WORK PLANNING

Incorporating worker protection features and requirements in the planning of decommissioning activities is the most cost-effective way to control hazards. OSH must be an integral part of planning and performing the work. A team of engineers, operations managers and employees, and appropriate worker protection professionals should conduct design reviews. OSH integration should start at the earliest design phase and continue throughout the decommissioning process to ensure that potential hazards are identified and evaluated and, to the extent feasible, eliminated or controlled through design changes. Though planning systems may vary, all of these systems should ensure the planning and budgeting of adequate resources to address worker protection. For activities performed at DOE sites, the resource implications associated with integrating OSH into all aspects of work planning, work execution, and performance monitoring must be considered in accordance with 48 CFR 970.5204-2.

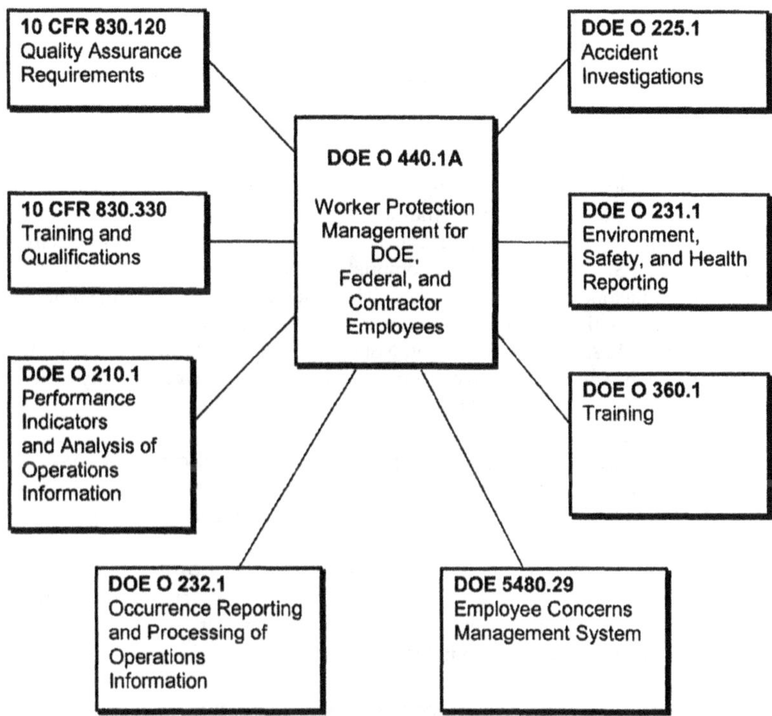

Figure 6.2 Major Interfaces between Order 440.1A and Other DOE Requirements

According to a DOE hazards analysis (DOE 1998c), workers find decommissioning more hazardous than facility operations because of potential involvement with degraded structures and equipment, and because tasks are typically unique, not repetitive. Specific noted concerns include:

- Unknown hazards
- Unanticipated and abandoned piping
- Unanticipated live utilities
- Unanticipated contaminated materials
- Falls (includes falling object impacts)
- Heavy equipment hazards
- Unplanned radiation exposure/contamination
- Unplanned asbestos and chemical exposures
- Structural hazards (sharp metal and debris)
- Biological hazards (e.g., rodents)

During planning, overall project management systems are developed, and goals and objectives are translated into tasks. A distinct project plan should be prepared for deactivation, S&M, and for actual decommissioning. These plans describe the work and methods to accomplish it. In general, these plans should include:

- Hazard identification, analysis, and control strategy
- OSHA requirements identification

- OSHA performance measures and progress metrics to be used
- Description of OSHA organizational responsibilities
- Waste management considerations, such as waste minimization and pollution prevention measures
- An approved facility safety basis, with its potential impacts

The OSHA requirements, listed in Table 6.1, appear as 29 CFR 1900-1990. Order 440.1A explicitly mentions the applicable OSHA requirements. Additionally, Appendix A of DOE-STD-1120-98, Vol. 2 (DOE 1998b) compiles potentially applicable OSH requirements for disposition activities as established by the DOE, OSHA, and EPA. This compilation may help project managers, contractors, and subcontractors identify applicable OSHA requirements to ensure the protection of workers during disposition activities. The DOE's mandatory and optional directives, and their intent, can be found in Table A-1 of the document. Not all directives apply to all disposition activities and situations; the specific directives applicable to a facility or work activity depend on the work to be performed and its associated hazards.

OSHA considerations associated with the use of subcontractors should also be considered, particularly in ensuring that subcontractor programs are in place, adequate, and monitored. The following program elements should be evaluated, where applicable, based on existing hazards.

6.3.1 Emergency Response: OSHA requires in 29 CFR 1926.35 that a written emergency action plan be developed to ensure employee safety from fire and other emergencies. The emergency plan should detail the types of evacuation to be used in emergency circumstances. The following elements, at a minimum, should be included in the plan:

- Emergency escape procedures and emergency escape route assignments
- Procedures to be followed by employees who remain to operate critical plant operations before they evacuate
- Procedures to account for all employees following emergency evacuation
- Rescue and medical duties for those employees who are to perform them
- The preferred means of reporting fires and other emergencies
- Names or regular job titles of persons or departments who can be contacted for further information or explanation of duties under the plan

The employer should review the plan with each employee when it is developed or changed, and whenever the employee's responsibilities change. Employees must know what to do in the event of an emergency, where the alarm is, and where written plans can be reviewed. Before implementing the plan, the employer should designate and train a sufficient number of people to assist in a safe and orderly emergency evacuation.

Guide 440.1-1 (DOE 1997b) explicitly includes through reference all the OSHA safety and health regulations for construction (OSHA 2003b). Specific aspects of its activities are covered by the Explosives Safety Manual (DOE 1998d) and the optional Firearms Safety Standard (DOE 1996b). Additional requirements are provided by the following documents:

1. Guide 440.1-3, Occupational Exposure Assessment (DOE 1998e), provides acceptable methodologies for assessing exposure to workers, including decommissioning.

2. Guide 440.1-5, Fire Safety (DOE 1995a), presents the interrelationship among the Orders, Contractor Requirements Documents, and other related requirements, that, along with site-specific criteria, define an acceptable fire protection program.

3. Order 210.1, Performance Indicators and Analysis of Operations Information (DOE 1997c), specifies the data required to demonstrate improving or deteriorating performance relative to identified goals. These data may suggest further improvement through the identification of good practices and lessons learned.

6.3.2 Injury/Illness Reporting and Recordkeeping: OSHA recording and reporting occupational injuries and illnesses (OSHA 2003c) are codified in 29 CFR 1904. Each employer is required to keep records of work-related fatalities, injuries, and illnesses. A case may not be new if it involves a pre-existing condition resulting solely from an unrelated event or exposure that occurred outside the work environment. A case is work-related if an event or exposure in the work environment either caused or contributed to the resulting condition or significantly aggravated a pre-existing injury or illness. OSHA defines the work environment as the establishment and other locations where one or more employees are working or are present as a condition of their employment. The work environment includes not only physical locations, but also the equipment or materials used by the employee during the course of work. Similar requirements exist for DOE facilities, as follows.

1. Order 225.1, Accident Investigations (DOE 1997d), presents the requirements for conducting investigations of certain accidents with the goal of improving performance and preventing the recurrence of accidents.

2. Order 231.1A, Environment, Safety, and Health Reporting (DOE 2003c), specifies the collection and reporting of information that is essential for evaluating operations and identifying opportunities for improvement.

3. Order 231.1A, Environmental Safety and Health Reporting (DOE 2003c) also insures that management is informed on a timely basis of events that could adversely affect safeguards and security interests, the health and safety of the public or workers, the environment, the intended purpose of the facilities, or the credibility of the agency.

6.3.3 Hazard Communication: A comprehensive hazard communication program is required by 29 CFR 1910.1200 (OSHA 2003a). The information concerning hazards and appropriate protective measures must be communicated to workers. This should include provisions for:

- A written hazard communication program, including lists of hazardous chemicals present
- Labeling the containers of chemicals
- Material safety data sheets
- A program to train workers regarding chemical hazards and protective measures

Worker training should include, as a minimum:

- Methods to detect the presence or release of a hazardous chemical in the work area (e.g., continuous monitoring, visual appearance or odor of hazardous chemicals)

- The physical and health hazards of chemicals in the work area
- Appropriate work practice, emergency procedure, and personal protective equipment

The standard requires a list of hazardous chemicals in the workplace as part of the written hazard communication program. The list may serve as an inventory of materials that will require the maintenance of Material Safety Data Sheets (MSDS). The most challenging aspect for decommissioning is the potential lack of an adequate inventory. The best way to prepare a comprehensive list is to survey the workplace. Purchasing records may also help. Chemicals in all physical forms (liquids, solids, gases, vapors, fumes, and mists) must be considered, rather than just liquids in containers. Special attention is required for the hazards associated with chemicals in unlabeled pipes. Moreover, it is necessary to consider chemicals generated during decommissioning operations (e.g., welding fumes, dusts, exhaust fumes).

Guide 440.1-1 explicitly includes through reference all the 29 CFR 1910 requirements. Training and Qualification (10 CFR 830.330) contains regulations pertaining to activities performed at DOE sites. Additionally, Order 360.1B, Federal Employee Training (DOE 2001b), establishes the requirements for employee training, education, and development, with the objective of improving the workforce performance related to its mission and strategic objectives.

6.3.4 Medical Services and Monitoring: The OSHA requires the availability of medical personnel for advice and consultation. Personnel must be adequately trained to render first aid if there is no infirmary, clinic, or hospital nearby. Where the eyes or body might be exposed to injurious corrosive materials, suitable facilities for quick drenching or flushing should be provided within the work area for immediate emergency use. First aid supplies are required to be readily available; those to be included are addressed in the Minimum Requirements for Industrial Unit-Type First-Aid Kits, as provided by the American National Standards Institute (ANSI 2003). When larger or multiple operations are being conducted at the same location, it is necessary determine the need for additional quantities and types of supplies and equipment. If it is reasonably anticipated that employees will be exposed to blood or other potentially infectious materials while using first aid supplies, employers are required to provide appropriate personal protective equipment in compliance with the provisions of the Occupational Exposure to Blood Borne Pathogens Standard (OSHA 1991).

A medical surveillance program is required for all employees who are or will be exposed to airborne concentrations of asbestos fibers at or above the time-weighted average (TWA) and/or excursion limit. A pre-placement medical examination should be made available before an employee is assigned to such a task. The employer also should provide or make available a termination of employment medical examination for any employee who has been exposed to airborne concentrations at or above the limits.

Annual examinations should be made available and the employer should keep an accurate record of all measurements taken to monitor employee exposure to asbestos. The employer should maintain this record for at least thirty years, in accordance with 29 CFR 1910.1001. Guide 440.1-1 explicitly includes through reference all the OSHA requirements. Additionally, Guide 440.1-4, Contractor Occupational Medical Program (DOE 1997e), provides the approaches and methods that it finds acceptable to meet the requirements for a contractor occupational medical program.

6.3.5 Fall Protection: The requirements for guarding floor and wall openings and holes are listed by OSHA (29 CFR 1910.23). Stairway floor openings should be guarded by a standard railing provided on all exposed sides (except at entrance to stairway). For infrequently used stairways where traffic across the opening prevents the use of fixed standard railing, the guard should consist of a hinged floor opening cover of standard strength and construction and removable standard railings on all exposed sides (except at entrance to stairway). Ladder-way floor openings or platforms should be guarded by a standard railing with standard toe board on all exposed sides (except at entrance to opening). Skylight floor openings and holes should be guarded by a standard skylight screen or a fixed standard railing on all exposed sides. Infrequently used pits and trapdoor floor openings should be guarded by a floor opening cover of standard strength and construction. Manhole floor openings should be guarded by a standard manhole cover that need not be hinged in place. Temporary floor openings should have standard railings, or should be constantly attended. Every chute or wall opening with a drop of more than four feet should be guarded by a rail, roller, picket fence, half door, or equivalent barrier. Alternatively, it is possible to use an extension platform for hoisting and handling materials; such platforms should have side rails or equivalent guards of standard specifications.

OSHA (29 CFR 1910.28) establishes safety requirements for the construction, operation, and maintenance of scaffolds used in the maintenance of buildings and structures. Scaffolds and their components should be capable of supporting without failure at least four times the maximum intended load. Scaffolds should be secured to permanent structures, through use of anchor bolts, reveal bolts, or other equivalent means. Unstable objects, such as barrels, boxes, loose brick, or concrete blocks should not be used to support scaffolds or planks. Employees should not work on scaffolds that are covered with ice or snow, or during storms or high winds. Tools, materials, and debris should not be allowed to accumulate in quantities to cause a hazard. Wooden pole scaffolds should not be erected beyond the reach of effective fire-fighting apparatus. No more than three men should be permitted to work at one time on a suspension scaffold, and each workman should be protected by a safety lifebelt attached to a lifeline. The lifeline should be securely attached to the structure (not scaffold) or to securely rigged lines, which will safely suspend the workman in case of a fall.

The minimum requirements for the construction, care, and use of the common types of portable wood ladders are listed by OSHA (29 CFR 1910.25). Stepladders longer than 20 feet should not be supplied, nor should two-section extension ladders longer than 60 feet. All ladders of this type should consist of two sections, one to fit within the side rails of the other, and arranged in such a manner that the upper section can be raised and lowered. Portable ladders should be so placed that the side rails have a secure footing. The top rest for portable rung and cleat ladders should be reasonably rigid and should have ample strength to support the applied load.

Similar OSHA requirements are listed for portable metal ladders in 29 CFR 1910.26. The minimum width between side rails of a straight ladder or any section of an extension ladder should be 12 inches. The length of single ladders or individual sections of ladders should not exceed 30 feet. Two-section ladders should not exceed 48 feet in length and over two-section ladders should not exceed 60 feet in length. Based on the nominal length of the ladder, each section of a multi-section ladder should overlap the adjacent section. Extension ladders should be equipped with positive stops to ensure the overlap. A metal spreader or locking device of sufficient size and strength to securely hold the front and back sections in the open position should be a component of each stepladder. The spreader should have all sharp points or edges covered or removed to protect the user.

Guide 440.1-1 explicitly includes through reference all the 29 CFR 1910 OSHA requirements. Guide 440.1-2, Construction Safety Management (DOE 1997f), provides project- and construction-specific guidelines to implement the proactive management of construction safety, with guidance on integrating the personnel and management techniques of OSH with other primary elements of project performance (quality, cost, and schedule).

6.3.6 Lock-Out/Tag-Out: The control of hazardous energy (lock-out/tag-out) covers the servicing and maintenance of machines and equipment that could cause injury to workers by any unexpected energizing or start up, or by the release of stored energy (see OSHA at 29 CFR 1910.147). An energy isolating device is capable of being locked out if it has a built-in locking mechanism or some means of attachment that will lock, such as a hasp. A lock-out/tag-out device ensures that the equipment being controlled cannot be operated until the device is removed. Lock-out and tag-out devices should be standardized in color, shape, and size within the facility. Additionally, in the case of tag-out devices, print and format should be standardized. The established procedures for the application of energy control (the lock-out/tag-out procedures) should cover the following elements and actions and should be done in the following sequence:

1. Preparation for shutdown: Before an authorized or affected worker turns off any equipment, the worker should have knowledge of the type and magnitude of the energy, the hazards to be controlled, and the method or means to control the energy

2. Equipment shutdown: The equipment should be shut down using the procedures established for that equipment. An orderly shutdown must be used to avoid any additional or increased hazard to workers as a result of the equipment stoppage

3. Equipment isolation: All energy isolating devices that are needed to control the energy to the equipment should be physically located and operated in such a manner as to isolate the equipment from the energy sources

4. Lock out or tag out device application: Lock-out or tag-out devices should be affixed to each energy isolating device by authorized personnel

Specific procedures should be used during shift or personnel changes to ensure the continuity of lock-out or tag-out protection. There should be provisions for the orderly transfer of lock-out or tag-out device protection between employees to minimize exposure to hazards from the unexpected energizing or startup of the machine or equipment, or the release of stored energy. The National Fire Protection Association (NFPA) provides more details in its Electrical Safety Requirements for Employee Workplaces (NFPA 2004). Guide 440.1-1 explicitly includes through reference all the 29 CFR 1910 OSHA requirements, while Order 440.1A includes the NFPA requirements. Additional details are provided in the DOE's optional handbook on electrical safety (DOE 1998f).

6.3.7 Confined Spaces: The OSHA 29 CFR 1910.146 requires a special program for confined spaces that must:

- Prevent unauthorized entry
- Identify and evaluate the hazards of confined spaces before employees enter them
- Develop and implement the means, procedures, and practices necessary for safe confined space entry operations

The confined space conditions should be evaluated to determine if acceptable entry conditions exist before entry is authorized. It is required to purge, inert, flush, or ventilate the permit space to eliminate or control atmospheric hazards. When testing for atmospheric hazards, it is necessary to test first for oxygen, then for combustible gases and vapors, and thereafter, for toxic gases and vapors. The workers should be provided with:

- Testing and monitoring equipment
- Ventilating equipment
- Communications equipment
- Personal protective equipment whenever feasible engineering and work practice controls do not adequately protect employees
- Lighting equipment
- Equipment, such as ladders for safe ingress and egress by authorized entrants
- Rescue and emergency equipment

At least one attendant must be present outside the confined space for the duration of operations. Guide 440.1-1 explicitly includes through reference all the 29 CFR 1910 OSHA requirements related to working in confined spaces.

6.3.8 Respiratory Protection: Respiratory protection should be accomplished as far as feasible by accepted engineering control measures (e.g., enclosure or confinement of the operational, general, and local ventilation, and substitution of less toxic materials). When effective engineering controls are not feasible, or while they are being instituted, appropriate respirators should be used. A written respiratory protection program and its site-specific procedures should include the following to satisfy the OSHA regulations in 29 CFR 1910.134:

- Procedures for selecting respirators certified for use in the workplace by the National Institute for Occupational Safety and Health (NIOSH)
- Medical evaluations of employees required to use respirators
- Fit testing procedures for tightly fitting respirators
- Procedures for proper use of respirators in routine and emergency situations
- Procedures and schedules for cleaning, disinfecting, storing, inspecting, repairing, discarding, and otherwise maintaining respirators
- Procedures to ensure adequate air quality, quantity, and flow of breathing air for atmosphere-supplying respirators
- Training in potential respiratory hazards during routine and emergency situations
- Training in the proper use of respirators, including putting on and removing them, any limitations on their use, and their maintenance
- Procedures for regularly evaluating the effectiveness of the program

More details regarding respiratory protection are provided by ANSI Z88.2, Practices for Respiratory Protection (ANSI 1992). Guide 440.1-1 explicitly includes through reference all the OSHA requirements, while Order 440.1A includes the ANSI requirements regarding respirator protection.

6.3.9 Hearing Protection: Protection against the effects of noise exposure should be provided when the sound levels exceed the 29 CFR 1910.95 limits. Feasible administrative or engineering controls should be used. If such controls fail to reduce sound levels, personal protective equipment should be used. A hearing conservation program should be administered whenever worker noise exposures equal or exceed an 8-hour TWA sound level of 85 decibels measured on the A scale (slow response). When any worker's exposure may equal or exceed this average, monitoring program must be developed and implemented. The sampling strategy should be designed to identify workers for inclusion in the hearing conservation program and to enable the proper selection of hearing protectors. Where circumstances make area monitoring generally inappropriate, such as high worker mobility, significant variations in sound level, or a significant component of impulse noise, it is recommended to use representative personal sampling. All continuous, intermittent and impulsive sound levels from 80 decibels to 130 decibels should be integrated into the noise measurements. Monitoring should be repeated whenever a change in process, equipment, or controls increases noise exposures to the extent that additional workers may be exposed at or above the action level, or the attenuation provided by hearing protectors may be rendered inadequate.

Audiometric testing should be available to all workers whose exposures equal or exceed the TWA. For workers who have experienced a significant threshold shift, hearing protector attenuation must be sufficient to reduce employee exposure to a TWA of 85 dB. The adequacy of hearing protector attenuation should be estimated using either the noise reduction rating developed by the EPA, or one of the three methods developed by the NIOSH (NIOSH 1975). Guide 440.1-1 explicitly includes through reference all the 29 CFR 1910 requirements regarding hearing protection.

6.4 HAZARD IDENTIFICATION

To develop a comprehensive HASP, the hazards of the project must be identified and analyzed to determine potential consequences to workers. Site characterization and a task-specific hazard analysis reveal the information needed to identify site/task hazards and to select worker protection methods. The more accurate, detailed, and comprehensive the information about a site and the tasks, the more the protective measures can be tailored to the hazards workers may actually encounter.

A hazard for this purpose is a source of danger (e.g., material, energy source, or operation) that can cause illness, injury, or death to personnel; harm to members of the public; or damage to an operation or to the environment. A hazardous substance or material includes any substance designated or reflected in 29 CFR 1910.120 with an exposure factor that may result in adverse affects to the worker, public, or environment. This includes:

- Any substance defined under section 101(14) of CERCLA
- Any biological agent or other disease causing agent that, after release into the environment and upon exposure, ingestion, inhalation, or assimilation into any person, either directly or indirectly, will or may reasonably be anticipated to cause death, disease, behavioral abnormalities, cancer, genetic mutation, physiological malfunctions (including malfunctions in reproduction), or physical deformations in such persons or their offspring

- Any substance listed by the DOT as hazardous materials under 49 CFR 172.101 and appendices (DOT 2003a)
- Hazardous waste, or combination of wastes as defined in 40 CFR 261.3 (EPA 2003a) or substances defined as hazardous waste in 49 CFR 171.8 (DOT 2003b)

Most accidents at nuclear facilities, even in decommissioning projects, are nonradiological. Decommissioning is more like construction than like facility operations. It is typical to note an increase in the number/amount of electrical incidents, as well as:

- Immediately dangerous atmospheres
- Welding/Grinding atmospheres
- Dust/Vapor nuisance atmospheres
- Scaffold uses
- Confined spaces
- Heavy lifts -- loading and rigging (e.g., a recent lift at a commercial site decommissioning was greater than 1000 tons).

The changes to the facility structures, systems, and components during decommissioning can also be factors in accidents and should be evaluated on an ongoing basis to ensure worker safety. Examples of facility alterations that affect safety include:

- Fire detection/suppression
- Service air or water
- Lighting
- Reductions in facility heating, ventilating, air conditioning (HVAC)
- Overhead cranes
- Communications
- Additions of temporary systems

The hazards should be identified and analyzed without regard for the likelihood or credibility of accident scenarios or consequence mitigation. Resulting information should be incorporated into the HASP and safety analysis documentation. The hazard analysis should rely on existing documented hazards analyses (if any) from earlier phases of the facility's life cycle. Documentation of the hazards analysis should formally record all identified hazards, including those that workers may encounter during decommissioning work activities. A formal record of controls used to support safe work execution in light of the identified hazards should indicate if actual corrective action differed from plan. The following elements should be included in the documentation for each hazard:

- Location
- Date found
- Description of hazard
- Referenced worker protection standard
- Planned corrective action
- Estimated cost of abatement
- Interim protective measures
- Abatement period (in calendar days)
- Scheduled abatement date
- Actual abatement date
- Risk level

The basic hazard analysis requirements of HAZWOPER primarily cover worker protection and emphasize an analysis of the impact to workers from hazardous substances and physical/biological hazards. Order 440.1A includes all OSHA requirements on identifying hazards. In addition to these, detailed requirements that may apply are listed in Table 3 of DOE STD-1120-98. Note that some of these requirements primarily concern facility safety rather than worker protection (i.e., assurance that facility structure and associated safety features are adequate to protect workers, the public, and the environment from hazardous substance inventories). The most common hazard analyses are listed herein.

6.4.1. Preliminary Hazard Analysis: A preliminary hazard analysis provides a broad hazard screening tool that includes a review of the types of operations that will be performed and identifies the hazards associated with these types of operations and facilities. The results of the preliminary analysis are used to: determine the need for further detailed analysis, serve as a precursor where further analysis is deemed necessary, and serve as a baseline hazard analysis where further analysis is not indicated. As required by OSHA, the preliminary analysis is most applicable in the conceptual design stage. It may also be useful for existing facilities and equipment that have not had an adequate baseline hazard analysis. The DOE, in addition to OSHA requirements, mandates the preliminary hazard analysis as support for certain nuclear facility SARs or BIOs.

6.4.2. Facility Hazard Analysis: A facility hazard analysis provides a baseline of anticipated facility hazards and their potential consequences based on available information from hazard identification and characterization activities, and from existing knowledge of the disposition work scope. Such as analysis requires a detailed study to identify and analyze potential hazards associated with each aspect of the facility and related equipment and operations. The analysis should include a systematic review of each facility component and task, and should consider:

- Facility design characteristics, such as electrical installations, platform heights, and egress concerns
- Types, interfaces, and location of the required equipment
- Proposed operations, including related hazardous substances and potential exposures, potential energy sources, locations of operations and required interfaces, resulting material and personnel traffic patterns
- Facility and equipment maintenance requirements, including confined space concerns, electrical hazards, inadvertent equipment startup or operation hazards

The facility hazard analysis may identify the need for other more specialized hazard analyses by functional experts, such as the health hazard analyses in Guide 440.1-3 and the process hazard analyses in 29 CFR 1910.119, Process Safety Management of Highly Hazardous Chemicals.

6.4.3. Process Hazard Analysis: A hazard analysis is required for any processes involving hazardous chemicals at or above the specified threshold quantities to satisfy the OSHA regulation 29 CFR 1910.119, Process Safety Management of Highly Hazardous Chemicals (OSHA 2003d). Such an analysis is also required by the EPA regulation 40 CFR 68, Chemical Accident Prevention Provisions (EPA 2003b), known as a risk management program. It should address:

- The hazards of the process
- The identification of previous incidents
- Engineering and administrative controls
- Facility siting
- Human factors, and the consequences of failure of engineering and administrative controls (including a qualitative evaluation of the range of the possible ES&H effects on employees in the workplace)

There has been confusion regarding various terms and acronyms for "hazard analysis" due to differences in terminology between guidelines of the DOE and the chemical industry. According to

DOE terminology, a preliminary hazard analysis is in accordance with its-STD-3009-94, Change Notice No. 2 (DOE 2002a). A process hazard analysis is in accordance with OSHA's Process Safety Management (OSHA 2003d) or the EPA's Risk Management Plan (EPA 2003a). OSHA's process analysis focuses on worker protection while the EPA's process analysis focuses on protection of the off-site public and environment.

6.4.4 Job Safety/Hazard Analysis: Operations and procedures should also be analyzed and reviewed to identify potential worker protection hazards and deficiencies. A job safety analysis (JSA) is the basic tool to identify hazards associated with work operations and procedures. A JSA can identify a large portion of the worker protection hazard requirements at most workplaces and is useful for both the dynamic work activities like construction and the relatively static activities involved in continuous operations. A JSA should be conducted:

- During the planning stage for new operations and procedures
- Before implementing changes to existing operations and procedures
- For existing operations and procedures that have resulted in injuries, illnesses, or near misses
- For existing operations and procedures that have not been adequately evaluated in the past or when there is no recent hazard analysis

JSAs should be updated periodically to ensure that new hazards have not been introduced since the last evaluation of the activity, and that procedures are still applicable to the operations. The principal elements of such an analysis are:

- Selection of operations and procedures to be analyzed
- Breakdown of operations and procedures to their component tasks
- Identification of hazards associated with each task and the controls necessary to protect workers against those hazards
- Development of operations procedures incorporating identified controls

Participation by affected employees and supervisors in the analysis process is essential. Their knowledge of the tasks and familiarity with the procedures actually used in performing the work, along with the associated hazards, will provide more complete information. In addition, these front line personnel can assist in determining the feasibility and effectiveness of proposed control measures. Detailed information on the conduct of such analyses is presented in OSHA documents (OSHA 2002).

6.4.5 Task Hazard Analyses: An analysis of individual facility disposition tasks or jobs (i.e., discrete units of work that when combined comprise a project) should be conducted to understand the impact from workers' interactions with hazards that may be introduced as a result of specific tasks. This analysis supports the development of work packages or other methods used in planning tasks. Task hazard analyses should be conducted throughout the life of a project as disposition tasks are planned and scheduled. The following guidelines should be used when conducting such an analysis.

1. The analysis should evaluate each step in the task's work instruction for hazards in the workplace and those introduced from chosen work methods. This process is

accomplished most effectively by performing a walkdown of the work area, as needed, feasible, and permissible, based on existing facility hazards (e.g., high radiation areas), using the workers who will perform the task. The analysis should review task steps and evaluate hazardous substances and physical hazards.

2. The analysis should involve project managers, engineering representatives, safety and health personnel, and workers.

3. The facility level analysis should be used as the basis and an input for performing a task hazard analysis.

4. Tasks should be screened against approved hazard baseline documentation (e.g., hazardous substance quantities, facility requirements, structural hazards) to ensure planned work is within the analyzed safety basis and to determine whether updates to documentation are necessary.

The extent of task hazard analysis will vary depending on experience and familiarity in conducting the task. For example, a work task, such as a previously conducted maintenance activity that is documented in current procedures and well understood, may rely on a review of task steps and a simple hazard checklist. Whereas a task that is new and unfamiliar to workers may warrant a more detailed task hazard analysis. Furthermore, for those activities that involve multiple tasks, the analysis should also include discussion of the sequencing of these tasks to minimize worker exposures, mitigate urgent risks or threats, and attain maximum reduction of radiological or chemical materials within the facility.

The following techniques are available to help perform hazard analyses:

1. *Safety Review:* A technique to provide a detailed evaluation of facility operations or processes. It is used to identify hazards associated with conditions, practices, maintenance, and other pertinent aspects of the facility or process.

2. *Change Analysis:* Is performed to ensure that design or proposed operational changes do not adversely affect the safety of the facility. The analysis identifies differences between the existing and proposed, identifies how the change will affect related features, and evaluates the effects of the differences and relationships on the overall safety of the facility. The change analysis is frequently used during design and construction.

3. *Energy Trace and Barrier Analysis:* Identifies potential energy sources and traces back to potential hazards, and helps determine if proper barriers (i.e., controls) are in place. It provides an effective tool to identify potential hazards for the preliminary hazard analysis.

4. *Failure Modes and Effects Analysis:* This is a critical review of the system (facility and operations), coupled with a systematic examination of all conceivable failures and an evaluation of the effects of these failures on the mission capability. It can help avoid costly facility modifications and should be initiated early in the design phase. Once performed, it provides valuable information if updated throughout the design process.

5. *Fault Tree Analysis:* This is a logic tree used to evaluate a specific undesired event. It is developed through deductive logic from an undesired event to all preceding events that must occur to cause the undesired event. The analysis can be applied at any point in the life of a facility and can be used to support the preliminary hazard analysis during design.

Detailed information on the selection and use of various hazard analysis methodologies and techniques is available in the American Institute of Chemical Engineers' Guidelines for Hazard Evaluation Procedures (AIChE 1992).

6.5 INTEGRATED HAZARDS ANALYSIS

All hazard analysis requirements share the same basic intent: to identify and analyze hazards so that a sound technical basis can be established for their control. Thus, an integrated hazard analysis is an opportunity to satisfy multiple requirements (i.e., facility safety, worker safety, and environmental protection). Typically, a hazard analysis is performed during project planning, when a general outline of the work is known but the details have yet to be determined. Thus, a task-specific analysis of hazards is needed during the planning of individual tasks using a job hazard analysis (JHA) or other suitable technique. The level of effort and techniques used to perform a facility hazard analysis will vary depending on the complexity of the project and its associated hazards. An analysis of each step in a job activity is undertaken before starting work. Detailed information for performing integrated hazards analysis is available from the System Safety Society (SSS 1999).

Controls necessary for the protection of workers should be based on a strategy that integrates various aspects of facility safety controls and is consistent with the hierarchy of controls:

1. *Hazard Elimination*: designing hazards out of chosen or alternative work methods. For example, substituting less hazardous substances, or using the smallest possible quantities in chemical decontamination of systems and building structures.

2. *Hardware Controls:* establishing engineered barriers to prevent unacceptable exposures to or contact with hazards, and to mitigate the consequence of accidents; to maintain safety systems, structures, and components for protection until the hazardous condition has been removed. Examples of include shoring for excavation, local exhaust ventilation systems, alarms, redundant control devices (e.g., valves), and barriers (e.g., temporary shielding).

3. *Administrative Controls:* limits on activities, occupational safety and health procedures, and work instructions to complement the above activities. These controls should also include inventory limits to prevent unauthorized consolidation of hazardous substances in a given facility area or the introduction of new hazardous substances into the facility.

4. *Personal Protective Equipment (PPE):* use may be necessary but should not be used without justification in lieu of the more reliable control strategies mentioned above. Use of PPE should be based on the hazard, used in accordance with established procedures and training, and periodically evaluated for effectiveness. Selection of PPE should take into account uncertainties in hazardous substance quantities and form (i.e., conservative assumptions to account for unknowns).

5. *Monitoring:* to verify adequate control of airborne hazards. Personnel exposure monitoring equipment, including for monitoring such physical agents as noise and radiological exposures should be used as part of an overall industrial hygiene program.

6. *Occupational Medical Program*: includes access for all workers to a board certified occupational physician. Workers should be qualified physically based on expected hazards and stresses associated with planned facility disposition tasks. Medical surveillance may be a necessary component of the occupational medical program to ensure the control of certain hazards, such as radiological, chemical, biological, and ergonomic hazards. If the disposition activity is performed at a Superfund site, the occupational medical program must comply with the requirements of 29 CFR 1910.120, Section F.

7. *Training:* helps define worker qualification associated with the planned task. Training and qualification should ensure that workers recognize any potential hazard that may be encountered. Specific training should be determined based on the type of activities to be performed and the hazards identified. Numerous OSHA regulations require training specific to hazard type (e.g., asbestos, lead, and radiation), work activities (e.g., construction and demolition), and job classification. HAZWOPER, for example, prescribes different levels of training depending on employee responsibility. Anticipated hazards and their controls should be clearly communicated to workers in HASPs (as required by HAZWOPER) or through other equivalent means, and in pre-job briefings. In addition to OSHA requirements, training programs are required at DOE sites to comply with Orders 440.1A (DOE 2000b) and 5480.20A (DOE 2001c). Additional guidance may be found in Guide 440.1-1.

For assurance of proper care of the workforce, it is prudent to evaluate each individual's physical condition prior to beginning work. This evaluation may include respiratory and cardiac evaluation to ensure that the worker can safely wear respiratory protection equipment. For individuals who may be exposed to lead, a baseline blood screening may be useful to ensure at the conclusion of work that the program effectively prevented lead intake by the worker (comparing blood levels pre-employment and post-employment). A similar baseline hearing evaluation is useful to ensure the potentially high noise environments during decommissioning do not have an adverse effect on an individual's hearing. As important as pre-screening a worker is to ensure their health, an effective on-site first aid capability is also essential. As noted above, decommissioning involves additional industrial safety and industrial hygiene challenges, and the facility project needs to ensure that it maintains trained personnel and equipment, commensurate with the additional risks. This may necessitate the addition of appropriately trained first aid personnel. Appropriate equipment (including personnel extrication equipment) is also needed. Equally important is that all personnel on site are trained and understand how to obtain emergency medical assistance at any time.

The application of some decommissioning techniques requires the implementation of special fire protection measures. The additional costs of these measures should be taken into consideration in the choice of a particular technology, (e.g., thermal cutting techniques or grinding). If the selected decommissioning strategy is long-term SAFSTOR, it may be necessary to install new fire detection and protection systems since the existing one may be too complicated for the envisaged requirement or it may have to be partially dismantled. The new system should be maintained during the entire SAFSTOR period and this may involve significant costs or additional commitments with the regulatory authorities.

The facility owner is ultimately responsible for the health and safety of the project workforce, even if personnel associated with external contracts perform much of the work. The contractor personnel should work under the facility programs, or the facility group should review and approve the contractor's programs. Audits and inspections should be incorporated into the facility programs, thereby including the contractors. If contractors work under their own programs, the facility owner should periodically audit performance and provide program oversight. Contractor safety programs and their performance record for similar work should be considered in the bid process.

Facility disposition activities require that safety controls change throughout the lifetime of the project, due to changing hazardous substance inventories, uncertainties, and discoveries. These changes can range from modifying or eliminating existing controls to implementing new, more restrictive, or modified safety controls. However, care should be taken to ensure that safety controls are not eliminated prematurely. Such a situation could compromise worker, public, and environmental protection (e.g., removal of fire protection systems when combustible materials are still present in the facility). It is expected that there will be less reliance on facility design and administrative features as the project progresses and as hazardous substances are removed. The following criteria should be used when determining if it is appropriate to eliminate a control:

- Hazardous condition being controlled is no longer present
- Hazardous substance's physical form has changed to a less dispersible form
- Hazardous substance quantities are no longer present or have been reduced to the point where the consequences of releases are no longer a concern

New safety controls should be developed for facilities entering a new disposition phase. Specifying controls for these types of events should emphasize mitigating vulnerabilities and should be implemented in the following order of priority (or combination thereof):

- Modifying operations and enhancing emergency planning and other contingencies rather than dedicating resources for enhancing facility structures, or
- Enhancing confinement integrity

For example, dispersible materials should be removed and contained, to the extent practical, and the containers physically separated (if possible) and secured to structures that provide enhanced stability or resistance to natural phenomena or fires. Finally, it may be useful to reevaluate the planned work scope and consider accelerated removal of releasable hazardous substances when the above controls cannot be practically achieved.

During integration of OSH into facility disposition activities, including transitions from one phase to another, hazard identification should incorporate the following activities:

- Review existing facility status by collecting and reviewing available facility operating records and existing hazard baseline documentation.
- Interview past and present employees, as necessary, to supplement information on past facility operations, including mishaps and incidents.
- Assess existing facility conditions and identify inherent hazards by performing a detailed facility walkdown, including radiological and toxicological surveys,

using a multi-disciplined team that includes the project manager, engineering representatives, OSH personnel, and workers.
- Review applicable lessons-learned for the facility and any similar facilities.
- Document the hazards associated with planned work activities.

To ensure that unanalyzed or uncontrolled hazards are not created, particular emphasis should be placed on identifying hazards that may be created or exacerbated due to the transition from one disposition phase to another or the time lapse between phases. For example, facility structures may deteriorate creating additional physical hazards to workers. Similarly, chemical hazards may increase during long-term S&M or any extended storage period. This is because chemicals left in process lines or storage tanks may be subject to radiolysis, container corrosion, concentration due to evaporation, decomposition reactions, or other hazardous conditions that may lead to decreasing the chemical's stability.

The need for intrusive characterization activities (e.g., sampling and analysis) should be evaluated based on the collection and evaluation of:

- Facility information
- Current status reports
- Operational history
- Deactivation records
- The remaining level of uncertainty regarding existing hazard (i.e., radiological materials, hazardous chemicals, or hazardous wastes)
- The existing facility condition

A characterization should be conducted if there is insufficient knowledge of hazards to understand the hazardous substance types, quantities, forms, potential exposures, and locations. Before any characterization activity, daily pre-job briefings should be conducted to discuss specific controls for the activity with workers, thus helping to increase hazard awareness.

Although typically not subject to HAZWOPER, deactivation and other disposition activities should still meet the intent of the hazard characterization requirements, including documentation and communication to workers of the following:

- Potential hazards that may be encountered during characterization (includes special hazardous substances, such as beryllium)
- Appropriate training and certification
- hazard controls, including engineering/administrative controls and PPE
- Work procedures
- An emergency response plan

In cases where characterization activities are conducted, it is mandated to provide an adequate level of protection to the workers performing these activities. For decommissioning activities subject to HAZWOPER, preparation of a HASP is required to ensure adequate controls for worker safety during the conduct of characterization activities. Furthermore, for DOE projects, these activities should be planned, analyzed, controlled, and executed in accordance with the requirements of the ISMS and Order 440.1A. Information resulting from planning and hazard identification activities should be used to determine the set of OSH directives applicable to the activity, as directed by P 450.2A, Identifying, Implementing and Complying with Environment, Safety and Health

Requirements (DOE 1996c), and 48 CFR 970.5204-78 (clause on laws, regulations, and DOE directives). Appendix A of DOE-STD-1120-98, Vol. 2 (DOE 1998b) may also assist in determining a candidate set of potential federal Applicable or Relevant and Appropriate Requirements (as used in CERCLA) for decommissioning projects; however, it should not be considered an exhaustive list of all potential requirements.

For the specific case of decommissioning activities involving only low-level residual fixed radioactivity that remains following removal of radioactive systems, components, and stored materials, alternative requirements may be applied in lieu of the safety management requirements contained within the orders applicable to nuclear safety. Such alternative Orders include 5480.23, Nuclear Safety Analysis Reports (DOE 1992), 5480.22, Technical Safety Requirements (DOE 1996d), 5480.21, Unreviewed Safety Questions (DOE 1991), 5480.19, Conduct of Operations (DOE 2001d), 5480.20A, Personnel Selection, Qualification, and Training Requirements for Nuclear Facilities (DOE 2001c), and 4330.4B, Maintenance Management Programs (DOE 1994). The conditions for when this is appropriate are discussed in a DOE memorandum prepared by the Office of Nuclear Safety Policy and Standards (DOE 1997g). As discussed in the memorandum, 29 CFR 1910.120 and 1926.65 (Hazardous Waste Operations and Emergency Response) may be applied in lieu of the above-mentioned nuclear safety order requirements. That is, provided that the QA requirements of 10 CFR 830.120 and the occurrence reporting requirements of Order 231.1A, Environment, Safety, and Health Reporting (DOE 2003c), are still applied.

6.6 HEALTH AND SAFETY PROGRAM

The HASP should ensure adequate protection for workers during the conduct of decommissioning activities. An HASP is required for decommissioning activities subject to HAZWOPER. Safety and Health Program Management Guidelines (OSHA 1989) publishes voluntary guidelines that outline what OSHA considers to represent the content and framework for safety and health program excellence. The guidelines stem from OSHA's experience in evaluating work sites through its state-run consultation projects and voluntary protection programs. The guidelines outline a management system to identify and control hazards on a proactive basis using four major program elements: management leadership and employee involvement, work site analysis, hazard prevention and control, and safety and health training. For DOE projects, Guide 440.1-1 explicitly includes through reference all the 29 CFR 1910 requirements.

The HASP should be developed using the information developed during the hazard analyses. The next source of information is to review the proposed tasks with the workers who will be performing these tasks to ensure that all task-related hazards are identified, along with the controls to eliminate or mitigate those hazards. The HASP should cover documentation and communication to workers of the following:

- Potential hazards that may be encountered during characterization (includes special hazardous substances, such as beryllium)
- Appropriate training and certification
- hazard controls, including engineering/administrative controls and PPE
- Work procedures
- An emergency response plan

Depending on the quantities and physical forms of radiological hazards, facilities containing such hazards may be subject to nuclear safety and radiation protection requirements that should be addressed by the HASP. While nuclear facilities undergoing decommissioning have nuclear safety and radiation protection programs in place, these are not separate from the overall HASP. Existing radiation protection programs must be incorporated into the HASP to ensure the overall program is coordinated and integrated. Therefore, the HASP should address the following regulations at NRC licensed facilities:

- 10 CFR 50 (NRC 2003d), Domestic Licensing of Production and Utilization Facilities
- 10 CFR 20 (NRC 2003e), Standards for Protection Against Radiation
- Regulatory Guide 1.70, Standard Format and Content of Safety Analysis Reports for Nuclear Power Plants (NRC 1978)

Similarly, for decommissioning activities at DOE facilities, the HASP should address the following DOE directives:

- Order 5480.21 (DOE 1991), Unreviewed Safety Questions
- Order 5480.23 (DOE 1992), Nuclear Safety Analysis Reports
- Order 5400.5 (DOE 1993), Radiation Protection of the Public and Environment
- Order 5480.20A (DOE 2001c), Personnel Selection, Qualification, and Training Requirements for Nuclear Facilities
- Order 4330.4B (DOE 1994), Maintenance Management Programs
- Order 5480.22 (DOE 1996d), Technical Safety Requirements
- Order 5480.19 (DOE 2001d), Conduct of Operations
- 10 CFR 835 (DOE 2003a), Occupational Radiation Protection

For the specific case of decommissioning activities involving only low-level residual fixed radioactivity that remains following removal of radioactive systems, components, and stored materials, alternative requirements may be applied in lieu of the safety management requirements contained within the DOE Orders applicable to nuclear safety (DOE 1997b).

It may be better for decommissioning to develop two OSH plans with differing scopes. The first plan should be developed early, to establish the base requirements and protocols needed to safely conduct work, and should be given to contractors to perform the actual decommissioning tasks and projects. An IAEA Safety Guide (IAEA 1999) and the DOE Standard 1120-98 specify the OSH planning information to be included:

- General description and timing of the work/project
- Known risks to workers (as far as known)
- Information for potential principal contractors (to demonstrate competency and adequacy of resources)
- Information required to prepare detailed plans for decommissioning

The second OSH plan should be highly detailed and task specific, and should be prepared jointly by the contracting workforce and facility OSH personnel. This plan should include, but not be limited to, the following sections:

- Introduction
- Facility policy statement for health and safety
- Roles and definitions under applicable regulatory authority
- Other relevant documentation
- Restrictions and constraints
- Specific nature of the work/project
- General health and safety objectives, principles, and standards
- Formal arrangements for monitoring health and safety performance
- Reporting and investigation of safety-related incidents
- Formal arrangements for consultation and communication with facility OSH personnel
- Selection and management of contractors
- Selection and training of personnel
- Emergency planning
- Project and site organization
- Documentation requirements
- Tables, figures, and appendices

6.7 ESTABLISHING CONTROLS FOR HAZARDS

Once the HASP has been completed, prevention and mitigation controls must be implemented. 29 CFR 1910.120 defines required engineering controls, work practices, and personal protective equipment to protect employees from exposure to hazardous substances and safety and health hazards.

6.7.1 Controls: Engineering controls may take the form of guardrails, confinement methods, filtered ventilation to control emission of materials, and temperature controls. Work practices include procedures, employee training, hazard communication, and job briefings. PPE consists of equipment worn by the worker to prevent harm or exposure; these include but are not limited to hard hats, safety glasses, respiratory protection, over-garments, gloves, and footwear. OSHA's policy is that engineering controls are the preferred option and every effort should be made to emphasize the use of engineering controls versus the use of PPE. Not every hazard can be removed or prevented by the use of engineering controls but every effort should to be made to develop and implement them in a cost-effective manner.

Worker training and the use of pre-job briefings play a critical role in controlling hazards at the work site. These pre-job briefings inform the worker of the hazards and how to prevent the hazard from harming them or others. Task-specific procedures are tools to ensure that the right way to work safely has been communicated to workers performing the task.

Overseeing programs in the field is a critical element of a successful decommissioning program. Early observation, documentation, and correction of "little things" can help prevent big things from happening. Event trend analysis is also a beneficial tool to help prevent OSH events in the field. It is necessary to maintain awareness of adverse trends and to take prompt action before an adverse trend becomes a significant injury or fatality. The scope of the OSH oversight should be tailored to the workforce. A purely contracted work force may require additional oversight than would a one with experience at the site. Despite the fact that safety programs and processes are managed by a small number of people, it is ultimately the responsibility of all individuals to ensure the OSH of the workforce. Table 6.2 lists some common cases of accidents and their potential consequences.

Table 6.2 Common Cases of Accidents and Their Potential Consequences

Personal Causes	Possible Accidents
Carelessness	Falls from heights
Inexperience	Being hit by a vehicle
Disobedience	Being hit by a falling object
Laziness	Tripping
Tiredness	Collapse of structures and machinery
Forgetfulness	Contact with moving objects or machinery
Practical jokers	Manual handling
Lack of safety awareness	Contact with electricity
Lack of knowledge	Ingestion of hazardous substance
Errors of judgment	Failure of pressurized systems
Lack of planning	Working in confined spaces
Unsafe attitudes	Exposure to excessive noise
Cleaning/Replacing/Drilling	Electrocution
Emergency batteries	Shorting and flash burns
Improper entry to confined space	Lack of oxygen

OSHA regulation 29 CFR 1910.120, Hazardous Waste Operations and Emergency Response, requires hazard controls to protect workers for any activities involving hazardous waste remediation or treatment, storage, and/or disposal facilities. Specifically, OSHA regulation 29 CFR 1910.119, Process Safety Management of Highly Hazardous Chemicals, contains requirements for preventing or minimizing the consequences of catastrophic releases of toxic, reactive, flammable, or explosive chemicals. Order 440.1A includes these OSHA requirements through reference. A listing of all hazard-specific regulations can be found in Volume 2, Appendix A of reference DOE 1998b. The DOE also requires the establishment of facility safety controls as specified in Orders 420.1A (DOE 2002b), 5480.23 (DOE 1992), and 5480.22 (DOE 1996d).

Although regulations provide expectations for operating facilities, the derivation and implementation of safety controls can be complicated because of a dynamic environment where hazards and work environment change frequently. Furthermore, a potential for uncertainty may exist in unknown hazardous substance forms and quantities. Most regulations fail to recognize that reliance on engineering safety controls is reduced as systems are removed from the facility. In addressing these issues, performance expectations of the program are:

- Establishment of worker safety controls
- Maintaining facility safety controls in a frequently changing work environment, including phasing out controls during the life of a facility disposition project
- Managing uncertainties in hazardous substance inventory or facility condition
- Documentation of hazards and their associated controls
- Assessing the adequacy of existing hazard baseline documentation

Uncertainties in material inventories or hazardous conditions should be reflected in safety controls. Such a situation can be encountered if intrusive characterization is needed to confirm material inventories (e.g., obtaining samples of materials in locations or vessels that are not readily accessible). When uncertainties exist, conservative assumptions may be made when specifying safety controls provided that:

- Hold points are established for conducting characterization or additional analysis to determine if the condition warrants establishing or changing a safety control.
- Assumptions are sufficiently conservative to ensure that safety is not compromised before or during characterization activities.

6.7.2 Technologies: The technologies available to improve the level of occupational safety and health can be classified as follows:

1. *Time Reduction Technologies*: Several tasks in a decommissioning project require substantial labor hours to accomplish. These tasks result in both a large cost to the overall project and a large potential exposure to OSH risks to the workforce. Much effort has been placed in developing technologies that reduce the labor hours in order to reduce the overall project costs. A secondary effect of many of these technologies is a reduced OSH risk to the workforce. Although there is a consensus that cost benefits come with the use of technologies designed exclusively to improve health and safety performance, these benefits are difficult to quantify. A key reason for the lack of quantification is the lack of worker safety and health exposures from the use of existing baseline technologies. Until a safety and health improvement can be demonstrated to be a prudent cost-beneficial task, such improvements will more likely come as secondary benefits. Those areas for which time reductions can positively affect OSH include technologies for characterization (addressed in Chapter 16), decontamination (Chapter 17), dismantling (Chapter 18), and waste management and treatment (Chapter 19). Examples of some of these technologies include:

   - Use of temporary plant services (air, water, cooling, and cleanup) that can result in time reduction, exposure reduction, and improved safety
   - In-situ characterization technologies that typically reduce the number of samples needed, thereby reducing sampling time
   - The use of scissor lifts and mechanical-assist scaffolding

   A number of relevant technologies are described in greater detail on the CD that accompanies this handbook.

2. *Increased Distance Technologies*: The most striking examples of increased distance technologies are in the use of robotics. Many examples of robotic technology are in use in dismantling, characterization, and size reduction. In contrast to the technologies developed to reduce decommissioning time, robotics were typically developed to reduce the hazards to workers or to allow access to areas inaccessible to humans. Robotic devices were developed to operate in environments viewed too hazardous to workers due to toxic or inert atmospheres, submersion, or other reasons. Another application of robotics is in locations too small for human habitation, for example small diameter piping inspection or cutting tools. Robotics is typically very effective when used for repetitive or similar tasks. This is

due to the typically higher capital cost of robotic technologies than solutions using a lower technology. When the capital costs can be amortized over multiple applications, their use becomes much more likely.

Another type of increased distance technology is in the use of manipulators or extension arms. These differ from robotics in that direct human manipulation of controls results in a direct manipulator/extension arm action, typically through mechanical or electro-mechanical interaction. These technologies generally provide for individualized use (e.g., a single manipulator arm is used for a single glove box). Extension arms or reach rods may be individually developed even down to individual valve manipulation. The benefit of extension arms is a lower cost than either manipulator arms or robotics.

3. *Increased Shielding Technologies*: These technologies can be viewed as those that place any type of additional barrier between the source of contamination and the worker. These can range from the simple application of lead blanket shielding to the use of gas permeable, moisture resistant, protective clothing. The use of PPE is an application of increased shielding technology, if the equipment provides a barrier between the contaminant and the affected worker. A challenge in the development of innovative PPE technologies is that their testing and evaluation may be viewed as experimentation on humans. As such, they often require longer time periods for development and incur higher developmental costs. These factors tend to reduce the number and type of technologies that enter the marketplace.

6.8 EVALUATING READINESS

Before work starts, it's crucial to complete a readiness evaluation to ensure that all hazards have been identified, appropriate OSH requirements have been met, and safety systems and controls are in place and able to perform their intended function. The scope and rigor of activities necessary to determine the readiness of a facility disposition activity will vary, depending on the type and magnitude of hazards present, the complexity of the work to be performed, and the extent of planned activities and hazards addressed in previous readiness evaluations. The scope of a readiness evaluation effort should focus on changes since the last detailed assessment (including external assessments that were performed on the facility). Potential changes could occur in operations, quantities, and physical forms of hazardous substances, facility structure or configuration, and personnel. The readiness evaluation should provide evidence that the following elements have been accomplished:

- All identified hazards have been adequately characterized to the extent practicable
- Appropriate OSH requirements have been identified and met
- An integrated JHA performed and controls established for worker protection
- Adequate safety and emergency response procedures and work instructions
- Personnel are knowledgeable of the work scope and the associated hazards
- Personnel have completed the training and qualifications for work to be performed
- Safety systems are operable and maintained according to design specifications
- Mechanisms for compliance with the provisions of all applicable environmental permits are in place

Simple checklists, as presented in Appendix J of DOE-STD-1120-98, Vol. 2, may be used to conduct the readiness evaluation. In all cases, readiness evaluations should be conducted by an organization not directly involved with day-to-day management of the activity. Requirements for performing readiness evaluations at DOE sites are found in Order 425.1C, Startup and Restart of Nuclear Facilities (DOE 2003b), and DOE-STD-3006-95, Planning and Conduct of Operational Readiness Reviews (DOE 1995b), respectively. An evaluation of readiness should be performed when there is a:

- Transition from the operations phase to facility disposition (e.g., to long-term S&M or deactivation)
- Transition between disposition phases (e.g., deactivation to decommissioning)
- Change in the contractor responsible for managing the disposition activity
- Significant change that affects the established safety basis or work scope (e.g., change in task)
- A positive Unreviewed Safety Question (USQ) is determined, as defined in Order 5480.21 for Hazard Category 2 or 3 nuclear facilities (DOE 1991)

6.9 MANAGEMENT OF CHANGE

Previously unknown hazardous substances or facility physical conditions may come to light during the removal of known hazards and the safety systems, structures, and components no longer needed for hazard control and confinement. As a result, work may be necessary that has not been planned for or included in existing safety documents. To ensure that the safety basis is current, adequate, and documented, it is important to develop a process for Management of Change (MOC). This process should evaluate all proposed activities, changes, and discoveries that may affect facility or worker safety. In addition, the process should address hazardous substance inventory maintenance to ensure the rigor of hazards analysis and associated safety controls are commensurate with the inventory changes.

The MOC process should be developed for all facility classifications and should establish a mechanism for evaluating the significance of any change, the need for additional analysis and safety controls, documentation affected or required, and the approval and training requirements for implementing it. An MOC screening and evaluation methodology should be developed for the following types of change:

- Minor changes that may impact job controls or instructions specified in work plans that should be implemented with minimal review (e.g., typographical errors, administrative details, or insignificant changes that have no potential ES&H impact)
- Changes that may impact the original work plans and may require safety evaluation, but do not require changes to existing safety documentation or work permits (e.g., hazardous substance in quantities or locations different than assumed)
- Changes that may impact the safety basis and require changes and approvals to the current facility safety documentation or work permits (e.g., unanalyzed hazards that require new analysis, safety controls, or changes to affect performance safety)

The evaluation of changing facility conditions or proposed disposition activities at DOE sites should be performed using a USQ process similar to the requirements outlined in DOE Order 5480.21

(DOE 1991). DOE O 5480.21 was superseded and been placed in website archives as a result of the publication of 10 CFR 830, Nuclear Safety Management in the Federal Register on January 10 , 2001. Subpart Part B, Safety Basis Requirements, and in particular Section 10 CFR 830.203, Unreviewed Safety Question Process, have replaced the formal requirements of DOE Order 5480.21 in regard to Unreviewed Safety Questions (USQ) and the process used for USQ review.

DOE Order 5480.21 is much more detailed than 10 CFR 830.21 and gives excellent informal guidance on possible approaches to an acceptable USQ process. 10 CFR 830 requires DOE contractors to use a DOE-approved USQ process. As the basis for most existing USQ processes, DOE O 5480.21 remains a most used detailed reference.

The USQ considerations establish a screening process to evaluate worker safety. For example, using the USQ process during a deactivation project allows the contractor to proceed expeditiously, without prior DOE approval, as long as the changes from planned disposition work do not affect the safety basis documented in the hazard baseline documentation. If a USQ exists, it does not necessarily mean that the activity is unsafe. Rather, identifying a USQ serves to alert facility management to potential conditions that could affect the facility's safety basis.

As stated in Order 5480.21, worker safety considerations are to be included in the MOC process and applied to each specific disposition work task. Screening and evaluation criteria should be developed and implemented to provide answers to the following questions:

1. Is there (a) an unanalyzed hazard, change, or increase in uncertainty in analyzed hazards, or (b) a change in hazardous substance type, form, or quantity as a result of the proposed activity, or (c) a discovery that could affect worker health and safety?
2. Are prescribed safety controls (including PPE) adequate to protect workers, as established by hazard baseline documentation, and are controls reviewed and approved?

A feedback and evaluation system should be implemented to review the safety of the project on a continuing basis. An unforeseen hazard or change to the work process may occur and require changes in prevention and mitigation controls. Feedback from workers should not only be encouraged, but required. It's an important safety and health consideration that helps identify unforeseen hazardous conditions and changes in the work environment. Those closest to the work will know if the safety measures are effective or interfering with the progress and a better method should be developed. DOE O 442.1A (DOE 2001e) addresses adequate feedback regarding employee concerns related to OSH.

Chapter 7

RISK ANALYSIS IN DECOMMISSIONING

7.1 INTRODUCTION

Decommissioning a nuclear facility requires a multitude of decisions that may have wide ranging effects. One critical aspect to consider in making decisions is the risk of potential adverse impacts on the facility workforce, the public, and the ecosystem. According to the American Association of Engineering Societies, risk analysis may be considered to include three essential modules: assessment, management, and communication. These modules are interrelated, as shown in Figure 7.1 (AAES 1998).

Figure 7.1 Overview of Risk Analysis

7.2 RISK ASSESSMENT

The scientific process of evaluating the adverse effects caused by exposure to a substance, activity, lifestyle, or natural phenomenon is called risk assessment. If risk assessment is performed in a probabilistic mode, it is called probabilistic risk assessment -- a process often used in conjunction with nuclear power operation. Risk assessment is often defined as:

$$\text{Risk} = \text{Probability of Occurrence} \times \text{its Adverse Consequence}$$

Potential adverse human effects, often evaluated in risk assessment, include premature death, increased incidence of birth defects, and increased incidence of cancer, including fatal (mortality) and nonfatal (morbidity). Not limited to human health, risk assessment includes an assessment of ecological risks. A typical example of a negative ecological effect evaluated by risk assessment is a reduction in the population of an endangered species. For decommissioning activities, however, assessed risks relate primarily to an adverse human health effect. For convenience, this chapter uses the increased incidence of fatal cancer as a general surrogate for all adverse human and environmental effects.

The Electric Power Research Institute has published guidance in its Decommissioning Economics and Risk Advisor (EPRI 1966). The National Academy of Sciences, as commissioned by several agencies, developed an approach for risk assessment in an evaluation of PCB-contaminated sediments (NAS 1990). According to that approach, once a source of release and the nature of contaminants are identified, risk assessment is divided into four major steps: hazard identification, dose response assessment, exposure assessment, and risk characterization. The relationship among these four steps is shown in Figure 7.2.

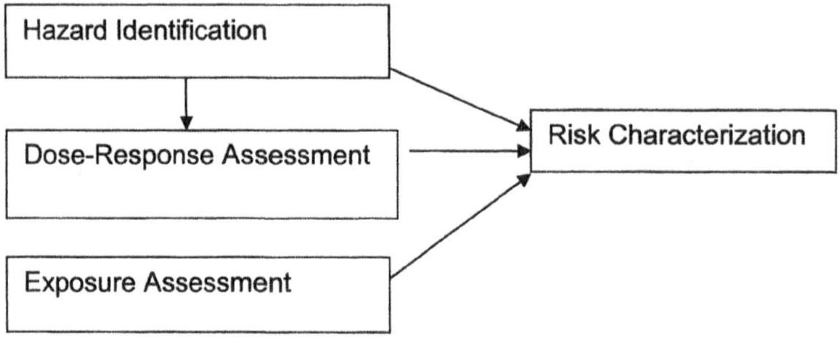

Figure 7.2 Attributes of Risk Assessment

The first step in a risk assessment is to identify the hazard. It includes identifying contaminants and determining whether the exposure to a particular contaminant can cause the adverse effect. The simple question to be answered in hazard identification is whether the contaminant expected to be released during the decommissioning activities can cause an adverse effect, such as cancer in humans. In hazard identification, all available information is evaluated about the effects of a contaminant to estimate the likelihood that it will harm people.

The next step is determining the response for a given exposure. This is known as a Dose B response assessment. U.S. regulatory agencies rely on the Linear No-Threshold (LNT) or stochastic hypothesis in assessing response to exposure of carcinogens, notably ionizing radiation. The LNT implies that exposure at any level results in some finite adverse effect. Another hypothesis, no longer advocated by major scientific organizations, would suggest that lower exposures result in proportionally larger adverse effects (the supra-linear response).

A third hypothesis suggests that for lower exposures the risk is less than the LNT would predict. A version of this is known as the linear-quadratic model. Finally, a hypothesis that continues to be seriously considered says that for lower exposures (a threshold) the risk is zero. The following Figure 7.3 graphically demonstrates these possible Dose B response models.

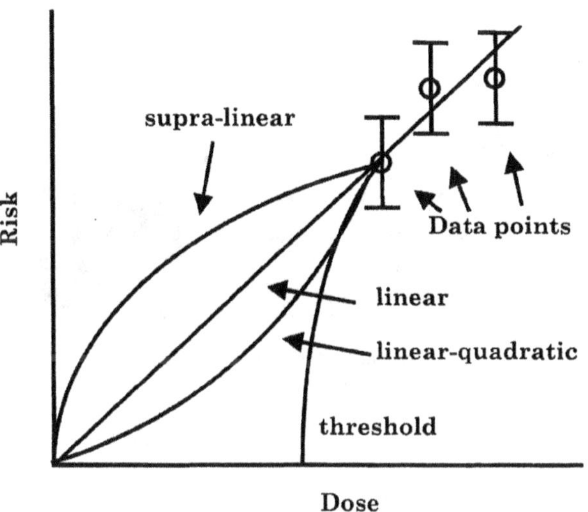

Figure 7.3 Dose B Response Curve

Similar to the challenges of hazard identification is the challenge posed by the lack of substantial human data on the specific response for a given exposure to an environmental contaminant. In the majority of cases, animal (mostly rodents) data are used to establish a dose response assessment; human exposure at high levels is also occasionally available. Despite the significant uncertainties about both, they are commonly used in dose response assessment. Fundamentally, the dose response function can be stochastic or not. "No stochastic effects" applies to those having a threshold implying that exposure below the threshold would not result in an adverse effect. One notable example of a threshold effect that has been determined to exist comes from radiation-induced cataracts. When the lens of the eye receives a radiation absorbed dose in excess of 0.6 to 1.5 Gy (60 to 150 rad), the possibility of cataract causation exists, while exposures below that threshold have been shown not to cause cataracts.

The impact of exposure is evaluated after establishing the dose response relationship. The exposure assessment step estimates how much of a substance a population inhales, ingests, or absorbs through the skin over a period of time; the duration of exposure; and whether the exposure was continuous or intermittent. The initial task is to identify the concentration or activity of the contaminant of concern at its source. Exposure assessment includes an evaluation of transport mechanisms from the contaminant to the individual (or receptor). As shown in Figure 7.4, there are typically several possible transport mechanisms to evaluate (ANL 2001).

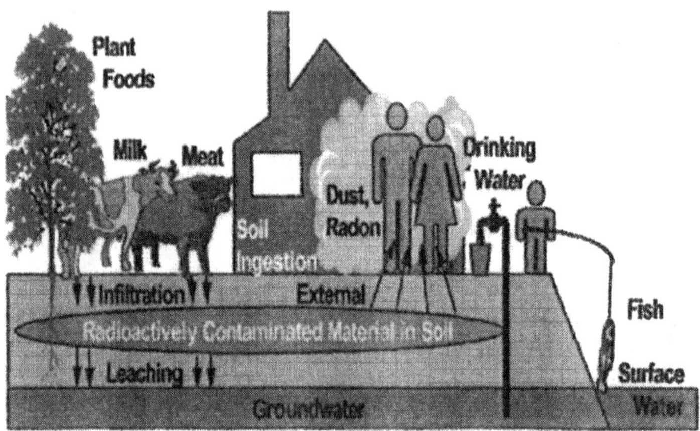

Figure 7.4 Exposure Pathways to Humans

As an example, transport mechanisms for surface contaminants can include:

- Air transport and receptor inhalation
- Air transport and receptor ingestion
- Air transport and surface deposition on receptor
- Air transport to a different surface and resuspension
- Direct contact with receptor

Note that if surface contaminants are washed away, they may enter liquid transport paths similar to the air transport paths.

The final step in risk assessment, risk characterization, synthesizes the three earlier steps to estimate the significance of the impact, combines the information on toxicity and exposure to describe what is likely to happen to people. It involves evaluating the respective uncertainties of each earlier step, determining potential data gaps or ambiguous results. It also involves choosing an appropriate dose response model. The risk characterization process tries to give an optimal answer, while acknowledging the potential range of answers and the uncertainties associated with each. Such an approach allows policy and decision makers to make the most informed decisions with respect to the risk to the public and environmental health.

By its very nature, risk assessment is characterized by uncertainty. Although much is known about environmental contaminants, limited data and knowledge still require assumptions throughout the risk assessment process. The NRC identified a process for the standardization of needed assumptions (NRC 1996b). This process risk assessment policy requires that regulatory agencies explicitly identify their assumptions, which are then used uniformly in their respective decisions.

In the hazard identification step, typically there are no human data. Consequently, animal exposure data must be used. An additional complication is that the animals used in the study are exposed to levels of contaminants that exceed by several orders of magnitude the exposure levels typically associated with decommissioning. Consequently, the hazard identification must not only make major assumptions of acceptability of extrapolating from test animals, typically rodents, but also on exposure from high levels to low levels. Problems associated with dose response assessment mirror levels of contaminants that exceed by several orders of magnitude the exposure levels typically associated with decommissioning. Consequently, the hazard identification must not only make major assumptions of acceptability of extrapolating from test animals, typically rodents, but also on exposure from high levels to low levels. Problems associated with dose response assessment mirror those of hazards identification. In the case of ionizing radiation, the several populations that have been exposed to ionizing radiation form the foundation of current risk assessment for radiation exposure and human exposure.

Like hazard and dose response assessments, exposure assessment must take various uncertainties into account to come up with the ultimate risk assessment. Since such variables as individual respiration or ingestion rates are not typically known, common practice is to use standardized models. For radiation exposures, the EPA has provided a number of guidance documents for risk assessment, including dose conversion factors for inhalation, submersion, and ingestion (EPA 1988); and external exposure to radionuclides in air, water and soil (EPA 1993). For cancer assessment, the documents include cancer risk coefficients for exposure to radionuclides (EPA 1994a, 1999a, and 2001). Other relevant documents include Technical Support Document for EPA regulations (EPA 1994b), and Risk Assessment Guidance for Superfund (EPA 1989).

7.3 RISK MANAGEMENT

Risk assessment scientifically characterizes a human or environmental impact. Risk management evaluates the potential costs of various risk minimization approaches and adds the societal and political factors to the overall risk analysis process. Although risk management may be thought to begin with the outcome of risk assessment, in reality both aspects typically work together throughout the risk analysis process.

Risk management comprises both objective and subjective aspects to help the risk manager make decisions. Objective aspects include characterized risk, comparative risk assessment, cost assessment, and cost-benefit assessment. Subjective aspects include risk perception, social and legal constraints, and intangible values. The impact of these critical elements of input for the risk manager should not be underestimated: because they allow affected stakeholders to interject opinions and interests into the risk decision-making process, the risk manager would be ill advised to ignore them.

7.3.1 Characterized Risk: The characterized risk is the outcome of the risk assessment process, giving the risk manager an overall evaluation of the human or environmental risks of a given task, exposure, or scenario, as well as the uncertainties associated with the values. It is important for the risk manager to understand both the values of the risk assessment and the bases for uncertainties. Understanding the uncertainties becomes increasingly important during the subjective aspects of risk management and during risk communication.

7.3.2 Comparative Risk Assessment: Comparative risk assessment comes into play most often when a manager must make decisions on resource allocation to minimize the risk among several similar contaminants. If the contaminants and effects are similar, it is reasonable that similar approaches would have been performed to characterize the individual contaminant risks and, more importantly, that the uncertainties in the assessments would be similar. The manager then has objective data to apply in optimizing any decisions. For example, if two chemical contaminants were both carcinogens and, for the same levels of exposure, one (A) produced an increased cancer risk of 1 x 10^{-4} and the second (B) produced an increased risk of 1 x 10^{-5}, then risk minimization resources may more appropriately be used to reduce the risk from the first (higher relative risk) contaminant.

7.3.3 Cost Assessment: Cost assessments also give the risk manager objective data. Each evaluated risk will typically have various alternatives to minimize the risk to humans. Typically, these alternatives relate to three approaches:

- Reduce the amount of contaminant, also known as source-term reduction
- Reduce the time of human exposure to the contaminant
- Reduce the amount of contaminant reaching the human (or increase distance)

The first approach, reducing the contaminant amount, is the basis for many environmental remediation techniques. Chemical or radiological contaminants in near-surface soil are classic conditions. A common method of remediation is to remove the top surface layer of soil and move it to an appropriate disposal facility, where the likelihood of its reaching human exposure pathways is substantially diminished. Restricting access to the area is a common risk reduction alternative when removal is not appropriate, . Given contaminated soil, fencing off the area to restrict personnel access will reduce exposure. The third approach, that of shielding humans, uses PPE during access to the affected area or by encases the affected area in some type of barrier. For soils, a common practice is to cover with a moisture resistant barrier and then additional uncontaminated soil. Both methods insert a barrier to reduce or limit the contaminant transport through one or more of the exposure pathways identified in Figure 7.4. Each risk reduction approach will have an associated cost of implementation. A clear understanding of the costs for each alternative, along with the outcome of the comparative risk assessment, gives the risk manager the objective information needed to perform the cost/benefit assessment.

7.3.4 Cost/Benefit Assessment: In cost/benefit assessment, the costs of each risk reduction alternative are viewed together with their overall impact on risk reduction. This combination allows the optimum approach to reduce risk on a purely objective basis. A hypothetical example follows to demonstrate this assessment.

Two contaminants were named in section 7.3.2. For a similar exposure level, Contaminant A produced an increased cancer risk of 1 x 10^{-4} and Contaminant B produced an increased cancer risk of 1 x 10^{-5}. If the probability of cancer from Contaminant A is denoted by r_A, from Contaminant B by r_B, and from A or B by r_{AB}, and assuming the two risks are independent, then the combined risk is as follows:

$$r_{AB} = r_A + r_B - (r_A \times r_B) = 1 \times 10^{-4} + 1 \times 10^{-5} - 1 \times 10^{-9} = 1.09999 \times 10^{-4}$$

Therefore, if these two contaminants were the only sources of concern, the total increased cancer risk would be approximately 1.1 x 10^{-4}.

Since the affected area is surface soil, three options were selected. These options, their costs, and respective risk reduction are as follows:

1. *Option 1* is to excavate the affected soil. For an expenditure of $1M, the remediation will reduce the concentration (source term) of Contaminant A by a factor of 10 and Contaminant B by a factor of 30.
2. *Option 2* is to erect a fence around the affected area and limit access. This option would cost $100,000 and would reduce the amount of Contaminant A that could reach humans by a factor of 20 but would not reduce the amount of Contaminant B.
3. *Option 3* is to encase the affected area with an impermeable barrier. This option would cost $800,000 and would reduce the amount of Contaminant A that could reach humans by a factor of 15 and Contaminant B by a factor of 10.

As listed in Table 7.1, Option 1 would reduce the total risk to a value of 1.03×10^{-5} or a reduction of total risk of ~ 10.6, with a cost of $1M. Option 2 would reduce the total risk to a value of 1.5×10^{-5} or a reduction of total risk of ~ 7.3, with a cost of $100,000. Option 3 would reduce the total risk to a value of 7.67×10^{-6} or a total reduction of risk of ~14.3, with a cost of $800,000.

Table 7.1 Risks and Costs Associated with Different Options

	Risk due to A	Risk due to B	Risk due to A and B	Risk Reduction Factor	Dollar Cost	Risk Reduction/ Cost
No Action	1×10^{-4}	1×10^{-5}	1.10×10^{-4}		0	
Option 1	1×10^{-5}	3.3×10^{-7}	1.03×10^{-5}	10.65	10^{6}	1.1×10^{-5}
Option 2	5×10^{-6}	1×10^{-5}	1.5×10^{-5}	7.33	10^{5}	7.3×10^{-5}
Option 3	6.67×10^{-6}	1×10^{-6}	7.67×10^{-6}	14.35	8×10^{5}	1.8×10^{-5}

On a purely cost/benefit approach, Option 2, the erection of fencing and access restriction, is clearly the best answer, conferring the greatest risk reduction per dollar expended. Yet this option may ultimately be the least likely to be implemented, for a variety of reasons. Nevertheless, regardless of the final decision, the sound assessment of comparative risks, costs, and cost-benefits are the crucial objective inputs needed for the subjective portion of the overall risk management process.

7.3.5 Perception of Risk: Perhaps the most challenging input to the risk manager is from the stakeholders' perception of risk for a particular activity. Although uncertainties to some extent cloud the risk assessment process, if effectively conducted it does provide a scientific basis for the risk and a comparison to similar or comparable risks. The public often does not share a common understanding and acceptance of either the scientific basis for risk assessment or the results obtained from the process. Rather, the public often perceives a substantially higher risk for a given project than the scientific assessment would suggest. In some cases, the perception of the level of risk is directly related to the level of media attention. The public can easily overestimate the risks of low probability but high consequence events, such as nuclear accidents, and underestimate a much higher probability but lower consequence event, such as an automobile traffic accident.

Social and legal constraints are noted as separate aspects, but most often are based on stakeholder risk perception. If a large enough group of stakeholders perceives a particular activity to be of sufficiently high risk (irrespective of the scientific basis of real risk) for a long enough time, laws may come to limit the perceived risky activity. The risk manager must be aware of such legal and/or social hurdles in order to deal with the challenges of stakeholder acceptance of the final risk decision. Inputs on intangible values also may come from stakeholders. An example would be a facility with surface soil contamination that was also heavily wooded and adjacent to a nature preserve. A reasonable concern for intangible values would suggest that stripping the top surface layer off along with trees and other foliage would not be a preferred remediation alternative.

The example on objective data in the previous section shows the potential impact of subjective elements of risk management, indicating how stakeholder inputs can alter the outcome for the overall risk management decision. The preferred option from a cost-benefit approach was erecting fencing around the facility. But stakeholders could see this option as the least preferable, raising the objection that fencing does not remove the contaminants. Typically, this approach will not win acceptance, even when contaminants may decay to background after a number of half-lives. And even if stakeholders agreed that this approach would successfully reduce the risk, the prospect of maintaining limited access to an area through some future date could also be a point of concern.

In many cases, stakeholders want changes to remediation options that could decrease their cost-effectiveness. In our example, the fencing option might become acceptable if the facility management agreed to add security provisions to the fencing and some form of ongoing environmental monitoring to assure that the contaminants were not migrating. This concern would also likely apply to Option 3, since there is no contaminant reduction for this option either. If the life-cycle cost for the added security and monitoring were $600,000, then Option 2 would still result in a risk reduction of a factor of 7, but the total cost would increase to $700,000. This is the same reduction in risk per dollar expended as in Option 1, soil excavation and disposal.

7.4 RISK COMMUNICATION

Many studies and substantial field experience show that although the risk assessment and management processes may well be done appropriately, substantial additional efforts will be required to successfully implement the decision if the information is not successfully communicated to the recipient. The communication of risk, beginning with risk assessment through the initial conclusions of the risk manager, requires care and deliberate attention to all affected parties and stakeholders. The subjective parts of risk management can inherently result in misunderstandings, with the potential for what is sometimes called public outrage -- an extreme level of public anger and fear about environmental risk. The result is a much greater effect on public reaction to a risk than the scientifically calculated risk warrants. As with other portions of the overall analysis, risk communication can be further defined by its individual aspects (the message, the source, the recipient, and the channel), each with its particular challenges.

7.4.1 The Message: The message consists of the outcome of risk management and the rationale for choosing a level of acceptable risk, based on a coherent and easily understandable form. The science and art of risk communication is continuing to evolve. Studies (AAES 1998) showing how the public perceives information indicate that public opinion may be adversely affected if the message included is seen to be:

- *Catastrophic*: A risk resulting from a large-scale, disastrous event (plane crash, nuclear reactor meltdown) is more dreaded than a risk affecting individuals singly (automobile accident);
- *Dreadful*: A risk that could cause a much feared or dreaded disease, like most cancers, is perceived to be more dangerous than a less feared disease;
- *Memorable*: A potential risk similar to a remarkable event embedded in the memory is viewed as much more dangerous than the risk of some unheard of or little known disease (NRC 1994c);
- *Uncertain*: A risk is communicated as being uncertain about its exact effect, severity, or prevalence;
- *Involuntary*: The level of acceptable risk is imposed to the recipient, and its prevention is the responsibility of another party (government, industry);
- *Unfair*: A population is exposed to a risk that people in a neighboring community or a different economic bracket are not (can be termed environmental justice);
- *Unfamiliar*: Exposure to an unfamiliar agent or a situation causing fear of the unknown.

The message may be viewed as the outcome of the risk assessment and management process. There should be one or more clearly articulated options, each assigned a numeric risk (typically values in the range of 1×10^{-8} to 1×10^{-2} or even wider), and a cost. Technical information should be translated into equivalent everyday terms easily understood by people not conversant with risk assessment. An example follows.

> As noted by the NRC, if an individual receives a Total Effective Dose Equivalent (TEDE) of 0.01 Sv (1 mrem), it would result in an increased cancer mortality risk of 4×10^{-4} (NRC 1996b). This would correspondingly mean that a TEDE of 0.001 Sv (100 mrem) would result in an increased cancer mortality of 4×10^{-5}. If an individual received this radiation exposure for each year for 70 years, it would result in an increased risk of 2.8×10^{-3}. The DOE notes that if an individual received an exposure of 100 mrem/y for 70 years, it would result in a projected reduction of 10 days from the overall 70-year lifespan (DOE 1998e). Obviously, this expression of risk for the same exposure is conveyed in a manner much easier to understand that simply stating premature death.

This approach should be used whenever possible; it can have a substantial positive impact on stakeholders' acceptance of analytical results.

7.4.2 Message Source: The message source relates to the perceived credibility of the organization conveying the message. Public confidence and trust in both governmental and large private organizations has diminished over the years. The prior relationship between the organization and stakeholders will have a direct impact on the success of risk communication. This is an issue that must be tended on an ongoing basis to help ensure that a key message will inspire trust. People become outraged if they lack confidence in the source of the risk and little trust in those who have a stake in the outcome. If a government agency or an industry wants to decommission a plant, the public thinks they'll try to cut costs by exposing the public to undue risk. But people are likely to accept risk information from those who have no stake in the outcome, such as personal physicians.

7.4.3 Message Recipient: The message recipient or the individual stakeholder receiving the message is perhaps the most important factor in risk communication. Even with a correct message, a trusted source, and a viable and objective message channel, the manner of receiving the message can vary widely and cannot always be ensured by the organization. This variability is affected by many attributes, such as education, socio-economic condition, potential receipt of other contradictory messages, or uncertainty in the risk analysis. The totality of these attributes provides the recipient's context for interpreting the message. Therefore, understanding the nature of an audience, especially any emotional triggers, is a key factor in helping to shape a message that will be accepted rather than refused.

7.4.4 Message Channel: The message channel is the mechanism or vehicle that conveys the message. Many channels are available, each with benefits and drawbacks. If the message will be viewed as of interest to a wide range of stakeholders, then local media will likely be a primary message channel. The benefit of such a channel is its wide, cost-free coverage and dispersal of the message. A potential drawback is that the media conveys the message as it chooses, which may or may not support the interests of the organization.

Those who start a decommissioning project often try to establish a committee representing various stakeholders. Such a group can provide a vital message channel and will likely be viewed as impartial, thereby supporting the organization's interests. Another key channel may be a local or national advocacy group that has taken on the role of concerned stakeholder. Such a group can use the information to support its own position, one that may support or oppose the interests of the organization.

Another channel is direct communication between the organization and the stakeholders who are personally and/or administratively affected. Such communications ensure that the intended organization message is conveyed to the appropriate recipients, but this channel is typically more time consuming and costly than others.

7.5 CASE STUDIES OF RISK ANALYSIS

There are several examples of successful risk analysis, starting with risk assessment, interacting with stakeholders, completing risk management, and performing risk communication. Two risk analysis cases highlight different aspects of the process. For a full discussion of all aspects, including stakeholder concerns, please see the referenced materials."

7.5.1 Corrective Measures Study at the Nevada Test Site: The first case study (DOE 1997h) describes certain potential closure options at the Nevada Test Site. Corrective Action Unit No. 92 (the Area 6 Decontamination Pond Facility) is a disposal unit located there. The site was required by the Nevada Division of Environmental Protection to characterize the facility under the RCRA. A corrective measures study was prepared to support the preparation of a closure plan. The purpose of the study was to identify and explain the selection of a recommended closure alternative based on:

- Level of risk
- Regulatory requirements
- Public/Regulatory acceptance
- Protecting human health and environment
- Compliance with waste management standards
- Long-term reliability and effectiveness
- Reduction of toxicity, mobility, and volume
- Long-term reliability and effectiveness

- Compliance with media cleanup standards
- Control of the release sources
- Feasibility
- Cost

Detailed characterization (including geotechnical and hydro-geological evaluations) resulted in specific contaminants of concern being evaluated for appropriate closure options, based on their relative risks to human health and the environment. The contaminants included arsenic, chlorobenzene, bis(2-ethylhexyl)phthalate, lead, total petroleum hydrocarbons, ^{60}Co, ^{90}Sr, ^{137}Cs, ^{210}Pb, ^{212}Pb, ^{226}Ra, ^{228}Ac, ^{238}Pu, ^{239}Pu, ^{240}Pu, and ^{241}Am. Three possible future land uses were established for the site: 1) a "reserved zone" for areas and facilities that provide support for short-term testing and experimentation, or short duration exercises and training; 2) a monitored and restricted zone to restrict public access to the land; and 3) a defense industrial zone, or land area designated for stockpile management of weapons. A possible future of returning the area to public access was not contemplated.

The results of characterizations conducted to identify the contaminants of concern were used as input to the risk assessments for the three pathways evaluated (ingestion, inhalation, and dermal contact) and ultimately to identify the risks to workers exposed to radiological and chemical carcinogens and other toxic materials. Projected exposures were also developed for various times after closure for both a no-action alternative and an engineered cap alternative. The results of these risk and dose assessments showed human and environmental risks to be acceptably low, even in the no-action alternative. Four alternatives were then explored and evaluated against the twelve evaluation criteria noted earlier:

- Alternative 1 No action
- Alternative 2 Closure in place by capping
- Alternative 3 Clean closure through excavation and disposal
- Alternative 4 Closure in place by partial excavation and capping

Although Alternatives 2 and 4 met the evaluation criteria, alternative 2 was selected since it provided the lowest overall risks. Alternatives 3 and 4 had overall increased risks primarily due to activities involved in removal, packaging, and transport of the contaminated soil. In addition, alternative 2 had the lowest life-cycle cost between it and alternative 4.

7.5.2 PCB Wastes at Ashtabula: This case study deals with options for the disposition of PCB-contaminated soil at the Ashtabula Environmental Management Project (Morris 2001). The site has 12,500 ft^3 (615 tons) of soil contaminated or suspected to be contaminated with PCBs in excess of 50 mg/kg (50 ppm) and uranium in excess of 30 pCi/g. Nine alternatives were evaluated against seven different risk criteria. The results of the evaluations are summarized in Table 7.2. Risks were categorized into five ranges:

- Lowest risk
- Low risk
- Acceptable risk
- Some concern
- Major concern

If each evaluation criterion were valued equally, the preferred option would be the one with the overall lowest score, reflecting the overall lowest risk. This would result in the selection of alternative 1, direct disposal at site A. In fact, this was the result of the analyses performed. In addition, it also resulted in the lowest overall life-cycle cost for the project.

Table 7.2 Cost and Risks for Options for Disposition of PCB-Contaminated Soil

Alternative	Schedule Risk	Permitting Risk	Technical Risk	Health and Safety Risk	Long Term Liability	Reindustri-alization Support
Direct disposal at waste site A $820 - 910K	1	1	1	1	3	3
Off-site treatment at B $9.3 - 10.5M	2	1	3	3	1	3
Off-site treatment at C and dispose at A $10.1 - 10.2M	3	2	4	3	1	3
Off-site treatment at D and dispose at A $3 - 3.1M	3	2	4	3	1	3
Off-site treatment at E and dispose at A $3.1 - 3.5M	3	2	4	3	1	3
On-site treatment via molten aluminum process $4.3 - 5.4M	4	3	4	3	1	1
On-site treatment using soil wash $2.6 - 3.4M	4	3	2	3	1	1
On-site storage for 5 yrs and disposal at A* $650 - 710K	2	5	3	4	3	2
Off-site storage for 5 yrs and disposal at A* $700 - 770K	2	5	3	4	3	2

Note: disposal vendors are identified by letters
* These costs were for storage only. They would be added to the eventual future costs of disposal.

Chapter 8

PUBLIC AND STAKEHOLDER PARTICIPATION

8.1 INTRODUCTION

The end of a nuclear facility's mission and its transition into decommissioning can create substantial hardship on nearby communities, particularly in the case of major facilities built at a distance from large metropolitan areas. A rural setting often promotes the establishment or growth of smaller communities around the nuclear facility that come to rely heavily on the facility for their economic well-being, either directly through taxes, or more indirectly, by services and support functions to the facility and its employees. Eliminating permanent staff during plant closing can result in significant negative economic effects, although a temporary local economic upturn is possible as specialized contractors, managers, and laborers move to the area for lengthy decommissioning activities. Because of these financial impacts, local community opinion could be anywhere from neutral to negative initially, even if shutdown were announced well in advance. The reaction can be more dramatic in the event of swift or unannounced plant closures.

This chapter gives a general overview of public and stakeholder participation. In the past, many agencies and organizations responsible for decommissioning made no distinction among various segments of the public. In effect, they identified stakeholders as anyone wishing to be involved. That approach fails to distinguish the general public from people clearly and unambiguously impacted by decommissioning decisions. Furthermore, such an approach makes it virtually impossible to reach the stakeholders of primary concern. The process described here is based on new developments in stakeholder participation and was used by the Peer Review Committee of the American Society of Mechanical Engineers and the Institute for Regulatory Science (ASME/RSI) and has since been applied in several peer reviews (ASME/RSI 2001 and 2002).

8.1.1 Regulated Public Participation: Organizations experienced in resolving contentious issues recognize the value of early public involvement. The NRC considers public involvement an important part of regulating the nuclear industry (NRC 2002a). The EPA acknowledges the need to integrate the knowledge and opinions of others into its decision-making processes. The EPA has developed detailed descriptions of public participation in compliance with legal requirements. The agency suggests that "Effective public involvement can both improve the content of the Agency's decisions and enhance the deliberative process. Public involvement also promotes democracy and civic engagement, and builds public trust in government" (EPA 2003c).

Virtually every major federal government activity requires some form of public participation. Briefly, a major proposed federal action is announced in the *Federal Register*, the official notification tool of the U. S. Government, and appropriate time must be provided for the public to comment. In many cases, the time and place of these meetings are also announced once public hearings are organized. Information may also be distributed through print and electronic media.

The National Environmental Policy Act, or NEPA as it is commonly called, generally applies to decommissioning activities involving the federal government. These include those regulated by the EPA, as well as facilities self-regulated by the DOE or licensed by the NRC. The NEPA process is also used in licensing and decommissioning nuclear materials facilities, such as those handling radioactive LLW. The Act can also apply to major activities managed by other agencies, such as those related to medical, agricultural, or industrial uses. However, NEPA specifically excludes certain activities, such as those that might have an impact on national security. (NEPA 1982)

As described in Chapter 5, the NEPA process starts with preparation of an environmental assessment (EA). Depending on the outcome of this assessment, an Environmental Impact Statement (EIS) may include mandatory public participation. The NEPA process includes detailed procedures for public participation, and the NRC makes opportunities for public involvement through roundtable meetings convened to enhance the success of the decommissioning process. Scoping meetings are also used to provide a forum for members of the public to supply information and opinion before NEPA actions (NRC 2002b). Certain other environmental regulations have specific requirements for public participation, including the following:

- 40 CFR Part 6: Procedures for Implementing NEPA by the Requirements of the Council on Environmental Quality
- 40 CFR Part 25: Public Participation in Programs under a Number of Environmental Laws

The regulations in these Parts suggest or require that agencies provide public notification, public comment periods, public meetings, public access to information, and/or other opportunities for the public to participate in the specified decision-making process. Requirements specify minimum required levels. Under certain circumstances, officials are urged to give the public opportunities for public involvement above and beyond the minimum. (EPA 2003c)

The EPA's regulations distinguish between *affected* and *interested* segments of the public. According to the EPA, affected individuals are those who may be directed or indirectly affected by the decision. They must be given an opportunity to participate in a decision about a proposed activity that will affect their environment and/or health. Interested organizations include affected individuals plus public interest groups, such as civic, social, environmental, labor, educational, professional societies, and many others. Note that EPA regulations are written for regulators, who are therefore not part of the public. Because EPA rules cover laws likely to apply to decommissioning, they are relevant to the public and stakeholder participation process at most projects.

8.1.2 Regulated Public and Stakeholder Participation: In contrast to public participation, with its aim of inclusiveness, stakeholder participation attempts to include a narrower, targeted group of people and organizations. Although the distinction between public participation and stakeholder participation is less than clear, there are some differences. Past decommissioning activities have identified several groups of stakeholders. Similarly, a number of organizations, agencies, and investigators have also tried to define stakeholders, as described herein.

Beyond sheer economics, concern from the local community can arise regarding the level of decommissioning and the selection of methods. Selection of technology and its relationship to level of cleanup can be a critical issue. In nearly all cases, an owner will need to openly discuss technology selection. A typical community reaction to a commercial nuclear power plant decommissioning is to push for selection of the thorough and prompt decontamination approach (DECON), since it moves the facility expeditiously toward free-release and the quickly assures people that the area is free from residual radioactivity. Local opinion might also favor prompt DECON because the additional staff typically required means increased revenues for local businesses.

8.1.3 Definition of Stakeholder: Many agencies and organizations have published their own definitions of stakeholders, as follows:

1. "Stakeholders" are representatives from organizations or interest groups who have a strong interest in the Agency's work and policies (EPA 2003c).
2. In the EPA's Common Sense Initiative, "multi-stakeholders" were defined to include "environmental organizations, environmental justice groups, labor unions, government regulators, and industry" (Coglianese 2004).
3. In its Project XL, the EPA divided stakeholders into direct participants, commentators, and the general public (EPA 1999b).
4. The World Bank identifies directly affected stakeholders, elected officials, line agency staff, and local government officials, as well as indirectly affected groups, such as nongovernmental organizations (WBank 2003).
5. The Inter-American Development Bank defines stakeholders as groups or categories of people who directly and demonstrably gain or lose rights and/or resources through development operations (Schwartz 1996).
6. Stakeholders have been defined as "Individuals who [are] actively involved in the project, or whose interest may be positively or negatively affected as a result of project execution or successful project completion." (Cleland 1998).
7. The U.S. Agency for International Development classifies stakeholders into primary, secondary, and key stakeholders (AID 1996).
8. The National Research Council prefers the alternative terminology of interested and affected parties rather than stakeholder (NAS 2000 and 2001).

8.1.4 Areas of Agreement: Regardless of the source, several commonalities emerge in the definitions of stakeholders and stakeholder participation:

- Stakeholder participation should be project-specific.
- People directly impacted by the project require special consideration.
- The intent of participation from all parties involved throughout the process is highly desirable and crucial to the success of a project.
- No specific exclusions of individuals or organization from a group can be made.
- There is no systematic or fool-proof process to identify who might be an interested or affected party.

8.2 STAKEHOLDER CATEGORIES

As revealed by the definitions discussed earlier, distinctions among concerned groups aid in understanding their needs and likely roles. The EPA indicates such in defining stakeholders as those who are or may be impacted by EPA decisions. Some facility owners starting a decommissioning project have also found it useful to distinguish among various segments of their stakeholders in order to tailor a successful approach to public participation. In effect, they divide stakeholders into segments based on the impacts of the project and location of the stakeholder. Such an approach can distinguish between the local organizations and the general public not residing in a defined impact area, such as that defined by a perimeter 50 miles from the facility (RSI 2003).

The process described here is a general methodology for structuring public participation by categorizing stakeholders, based on the results from studies of the collective impact from stakeholders involved in several peer reviews of DOE programs (ASME/RSI 2001and 2002). In this process, managing stakeholder participation is based on the clear identification of stakeholders and a well-defined approach for communication. Figure 8.1 depicts various classes of stakeholders created using the methodology.

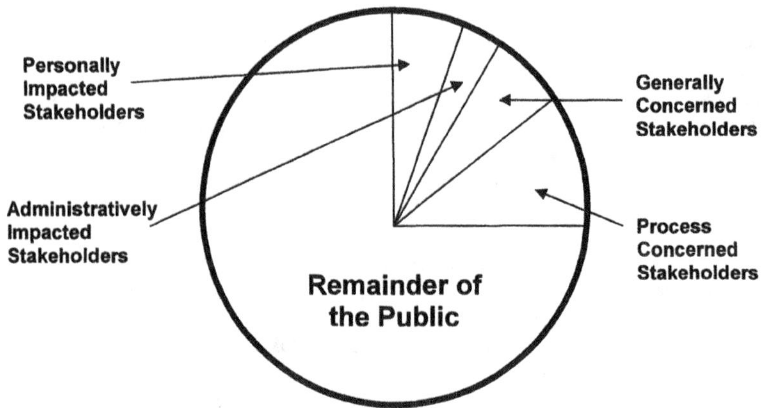

Figure 8.1 Categories of Stakeholders

8.2.1 Personally Impacted Stakeholders: These peoples' lives are directly impacted by the proposed action in terms of health, income, property value, or any other major aspect of life. Due to the nature of this category, impacts must be well defined, both scientifically and from a practical perspective. This category is identical to the EPA's definition of "affected individuals."

1. *Human Health Impact*: People at risk for potential human health impacts from any decommissioning action are a critical subset of stakeholders. The risks to these individuals or a combined nearby population must be calculated. Data should be based on:

   - Assumptions or studies regarding the future use of the facility
   - Contamination source
   - Potential technologies used to stabilize, remove, contain, and transport contaminated materials
   - Potential pathways for contamination to reach the public

To further support decision-making, potential risks can be calculated using models to estimate the hypothetical probability of human health impacts. There can be significant debate over the appropriate assumptions and models used in such calculations. Regulators generally have required guidelines and contamination thresholds for conducting most of these assessments and for their projected level(s) of remediation.

2. *Employment Impact*: A person whose job would be impacted by the proposed decommissioning project is also considered a personally impacted stakeholder, provided the job gain or loss is at the targeted facility or the proposed project can be reasonably postulated to cause the loss or gain. Examples include workers performing decontamination and/or its related work, such as waste packaging, transportation, and disposal.

3. *Financial Impact*: If the value of personal property could be financially affected, that individual would be personally impacted, as would someone whose business could be impacted by the proposed decommissioning. None are considered personally impacted if the project could impact the market price of a facility owner's corporate stock.

4. *Others*: Although not always necessary, some circumstances may warrant the creation of another group of directly impacted stakeholders. People impacted by foul smell, excessive noise, or similar environmental insults are also considered personally impacted.

8.2.2 Administratively Impacted Stakeholders: These are the elected, appointed, or employed people who must ensure that the proposed decommissioning is defined, reviewed, approved, and implemented in accordance with applicable laws, regulations, permits, licenses, or agreements. The participation of these stakeholders is usually required, by regulation or administrative protocol. Planning for their participation is more straightforward than for the personally impacted stakeholders. This category includes the following groups:

- Elected officials whose constituency represents, or is likely to speak on behalf of, a major portion of the personally impacted stakeholders
- Elected officials whose constituency includes a small portion of the personally impacted stakeholders, such as representing individuals from potentially impacted area (mayors, relevant state representatives, and members of the House and Senate)
- Individuals or groups responsible for preparing/implementing permits and licenses
- Members of regulatory agencies at the local, state, and federal level who are responsible for regulations, permits, licenses, and enforcement

In many cases, one or more regulatory groups have the authority to terminate a facility license upon satisfactory decontamination. Overlapping regulatory oversight can occur between various state and federal agencies, with jurisdiction over materials regulated under any or all of the laws described in Chapter 5. Because of the significant impact of regulatory agencies on decommissioning decisions, it is important to include this group of stakeholders in the process, particularly because state cleanup and exposure criteria may be more stringent than those of the NRC or EPA.

8.2.3 Generally Concerned Stakeholders: These people, by virtue of their personal philosophies, beliefs, or ideologies, are interested in or concerned about a proposed decommissioning. Obviously, personally impacted and administratively impacted stakeholders are also generally concerned.

However, this category is more inclusive; it's what the EPA defines as "interested organizations." All groups and organizations mentioned in the EPA regulations are included here. This group is most often represented by a specific entity, such as a public interest group with a well-defined purpose.

8.2.4 Process Concerned Stakeholders: A fraction of the public is concerned over the process used to manage and complete the proposed action. Although this group can be considered part of the EPA's interested organizations, its goals and objectives are somewhat different. Distinct groups within this category include those who:

- Consider the direct stakeholder participation in the decision process important to improve the quality of the decision and for stakeholder acceptance of the final decision
- Are concerned over potential undue influence that stakeholders may exert during decision-making and want to ensure influence commensurate with their vested interest in the outcome
- Work at similar facilities that could be impacted by decisions made by regulators
- Expect decision-making to be based on sound scientific principles and the best available data. Professional scientific and engineering organizations constitute an important core of this group. Among others, it would include the following:
    - American Nuclear Society (ANS)
    - American National Standards Institute (ANSI)
    - American Society of Mechanical Engineers (ASME)
    - American Society of Testing and Materials (ASTM)
    - National Academy of Public Administration (NAPA)
    - National Academy of Sciences (NAS)
    - Nuclear Energy Institute (NEI)

8.3 DEFINING THE PARTICIPATION APPROACH

The strategy for approaching stakeholders must consider their respective roles in the process. Each group should be identified and notified in accordance with how it may be impacted.

1. *Personally Impacted*: These people might be reluctant to participate directly in the decision process. Some are unwilling to devote sufficient time to study a subject or participate in meetings unless they perceive a significant impact on daily life. Since participation from this group is key, due to its potential impact, an affirmative outreach approach should be used to ensure the opportunity for participation. Organizations solely representing the local residents can be the primary focus, not only for identification but, more importantly, for notification of the stakeholders. Modes for reaching this class should include mailed invitations, personal contacts via phone, and other avenues considered likely to reward participation.

2. *Administratively Impacted*: Finding and notifying this group is easier than the personally impacted because it's a matter of known institutions. Elected officials work hard to get known within their constituency and are therefore readily identifiable. Regulatory agencies can also identify relevant stakeholders. Once identified, members of this category should be notified by phone and in writing.

3. *Generally Concerned and Process Concerned*: These groups are seldom obvious; typically, members must self identify. It is reasonable to attempt to reach these organizations via print and electronic mass media; in many cases the Internet has proven valuable for dispensing information and updates. Although there may be unaffiliated individuals within these two groups, they typically have access to the same information systems.

8.4 THE PUBLIC PARTICIPATION PROGRAM

Certain stakeholders may have no trust in the very agencies responsible for ensuring their participation -- they may be skeptical of the validity and fairness of any participation process. It is this skepticism that can raise demands for independent peer review of a project or its potential impact. Ideally, those responsible for managing participation related to a proposed action should be independent of those with a vested interest in its outcome. Neither the sponsoring agency nor the stakeholders should manage the participation process.

8.4.1 Timing of Involvement: All categories of stakeholders want to be involved, at a minimum, during the initial formulation of the decision process. Including them up front eliminates any complaint that decisions were already made and that any participation was merely "window dressing" to justify done deals.. Facility management must begin its project planning with the guarantee that all concerned parties will be involved immediately.

8.4.2 Required Information for Meetings: Concerns, especially from the personally impacted, are often expressed as fear -- perhaps because they feel incompetent to appreciate the technical intricacies of an issue. These fears can be substantially reduced or eliminated if the information provided addresses the issues of concern in a manner that is clear, concise, and easy to understand. Selection of a preferred decommissioning option that is both cost-effective and reduces the level of risk is more readily supported if the participation process supports and develops well-informed stakeholders.

Individual and group stakeholders deserve a level of relevant information sufficient for them to offer meaningful and informed input. Management should give them the same overview or detailed information given to other major participants, and with technical assistance when appropriate. This assistance can be come from project management, a regulatory agency, or an independent organization. Information should include:

- A summary of the need for the proposed decommissioning action
- A list of key issues being considered
- A description of the rules governing stakeholder participation
- A list of sources for further information, including a web site if established to provide information and/or to receive comments
- The names of trusted community leaders, such as doctors, pharmacists, and clergy, who have been included in the information dissemination.

It may be prudent to hire an Independent Inspector to serve as a technical consultant for the local stakeholders, particularly when they lack an understanding of basic radiation principles.

8.4.3 Participation in Decision-Making: To be credible, any stakeholder involvement process must create meaningful dialogue. A participation process should make every effort to ensure continual iterative involvement between decision-makers and stakeholders. If the potentially effective mechanism of the Internet is to be used for dialogue, its application must ensure an active effort to update information and respond to honest queries and comments timely (e.g., quarterly updates for projects requiring several years). In turn, stakeholders should be encouraged to be specific with questions and comments. But even generalities that may seem of little value as comments might offer insight about what to include as updates to information already given.

8.4.4 Reconciliation of Competing and Contradictory Interests: Personally impacted stakeholders might complain about being overshadowed by those who are not directly impacted. The generally concerned might complain that the decision process does not adequately consider their interests. Process concerned stakeholders might take the perspective that regulators, if overly influenced by the generally concerned, might pay insufficient attention to the best available scientific information. Because of the potential contradictory interests within and among the stakeholders, a properly managed stakeholder participation process must include relevant segments within each; for example, those whose jobs would be at stake and those who advocate elimination of an activity or service (and its relevant employees). The decision-maker must be able to get input from a diverse set of representatives and to ensure that the decision process considers their views.

8.4.5 Including Stakeholders in the Final Decision: Stakeholders will be frustrated if they think their participation has had no impact on the final decision. Once made, a full description of the final decision, made public as soon as possible, should include the following:

- How the final decision was determined
- A general outline of how the process progressed as a direct consequence of stakeholder participation
- Elements based on the desires of stakeholders
- Elements not based on the desires of stakeholders, including an explanation of why they were not accepted

Many contested issues can be resolved if decision-makers demonstrate that they functioned with a sincere consideration for the concerns from others. Stakeholders will be more likely to accept the results of a process if they perceive that process to be fair.

8.5 DEVELOPMENT OF GUIDELINES FOR PUBLIC AND STAKEHOLDER PARTICIPATION

Following are examples of successful public and stakeholder participation. Noteworthy experiences with stakeholder participation are discussed for the Maine Yankee and Big Rock commercial decommissioning projects in Part 5 of this handbook. Two involving the DOE are discussed below.

8.5.1 Restoration of the First Nuclear Reactor Site: This project dealt with decommissioning of the first nuclear reactor, known as Chicago Pile # 1 or CP-1 (Taboas 1997). The first successful nuclear chain reaction occurred on December 2, 1942 at 3:25 p.m., beneath the athletic field of the University of Chicago (Stagg Field). The initial construction site had been outside Chicago, but a strike by construction workers delayed its preparation and thus the University site was used.

Subsequently, CP-1 was moved and re-assembled (with the addition of concrete shielding and other appropriate modifications) as CP-2. The new location, known as Site A, was on land leased from the Cook County Forest Preserve District. Another reactor (CP-3) was added later, and the site became a major research facility.

Originally deactivated in 1956, it wasn't until 1990 that elevated levels of radiation were detected during routine monitoring. An investigation revealed that releases from portions of the disposal area were causing higher than normal levels of radiation, including tritium contamination. As a result, an immediate remediation effort began, with the goal of releasing the site for use as a public park. Restoration would require extensive characterization, risk assessment, and many related activities.

Numerous meetings over three years and at high administrative costs failed to yield any constructive outcome. Noting that more effort had been invested in discussion than in eliminating the contamination release, the new federal official in charge of environmental remediation revised the project's stakeholder involvement process. Two libraries were designated to hold all relevant documents as generated and to make them readily available to the general public. Stakeholder meetings were then convened to inform the public regarding the findings and project developments, and to earnestly seek stakeholder input in decision-making.

Three stakeholder groups were convened. The frequency of the meetings with each group increased in proportion to the schedule of remediation activities.

1. *Technical review group*: This group consisted of representatives of the DOE and its contractors, the Illinois Attorney General, the Department of Nuclear Safety, both federal and state EPAs, the U.S. Geological Survey, the Cook County Forest Preserve District, the County Board of Commissioners, and Willow Springs, the village closest to the site. A representative of Broken Arrow, a local environmental advocacy group, also participated.

2. *Community leaders*: This group was a roundtable consisting of civic leaders and associated federal and state legislators representing a radius of about 5 miles around the site. The roundtable was facilitated by a representative of the League of Women Voters.

3. *General public*: This group was made up of members from the broader community with interests in the proceedings.

Several widely advertised public meetings were held, and the public was encouraged to participate in discussing a number of site remediation options.

The base project was to remove physical hazards (concrete rubble, old fencing, pipes protruding from the ground, manholes, open tile drains, septic tank openings, and general debris associated with laboratory operations). The alternatives for the remaining restoration of Site A were narrowed to three, as described here.

1. *No action*: Leaving the site in the current condition after completing the site assessment. This alternative was defensible because comprehensive site risk assessments indicated a lifetime risk significantly lower than the EPA threshold of 10^{-4}. The advantages of this option were: no risk to workers involved in a potential remediation, no need for transfer of

contaminated soil, and no additional costs. There was solid legal foundation supporting this option. The disadvantages were: no change to the contaminated site condition, and opposition from the landowner, regulators, and the public.

2. *Limited site remediation*: Because extensive contamination was identified at two areas, removing soil from defined suspect areas would take care of a large fraction of the site contamination. Indeed, removing about the top 60 cm of soil would get virtually all radionuclides and most chemical contaminants. The advantages of this alternative were: the removal of the primary source of contamination, and therefore significant risk reduction, and acceptability to park users and the landowner. The disadvantages were: the need for transporting the contaminated soil, disposal at a different site, and a cost of about $650,000.

3. *Placement of a soil cover on segments of the site*: In this alternative, certain suspect areas would be covered with 60 cm of clean clay soil plus 15 cm of top soil to facilitate vegetation growth. The advantages of this option were: no intrusion into contaminated areas, no need for transfer of contaminated soil, and a reduction or elimination of risk associated with the transfer of contaminants. The disadvantages were: retention of slightly contaminated material in its current place, lack of full acceptance by the stakeholders, and the cost of about $600,000.

After candid discussions on potential risks and the availability of funds, there was a reasonable agreement on selecting Option 2 by both the DOE and local and state stakeholders. The Record of Decision for voluntary remediation included the removal of physical and tripping hazards (such as concrete rubble, fencing, and manholes). About 420 m^3 of soil were excavated and removed, stabilized with cement, and moved to a disposal facility. In addition, some graphite of considerable historic value was returned to Argonne National Laboratory. About 1700 m^3 of clean soil from a neighboring community were used to backfill the remediated area. The area was re-seeded with native vegetation and restored to its original condition. Final removal of debris was completed in March 1997. The project was completed for less than the projected cost and ahead of the agreed-upon schedule.

8.5.2 Groundwater Contamination at the Nevada Test Site: The Nevada Test Site (NTS) was established as the country's principal nuclear weapons testing facility. It is a remote, arid, and restricted area predominantly surrounded by tightly controlled federal lands and facilities. During operations, about 900 nuclear weapons were detonated at the site, primarily underground and either above or below groundwater levels, resulting in soil and groundwater contamination. The DOE developed its strategy for site remediation based on studies that included extensive monitoring data, development and application of hydrological models, and hydrogeological studies. The studies determined that tritium was of particular concern. Tritiated water moves with regular water, so its underground migration is identical to groundwater movement.

The proposed DOE strategy was to rely on radioactive decay. In effect, the strategy suggested that the migration of groundwater would occur at a rate slower than the radioactive decay of tritium (with a half-life of 12.3 years). A Community Advisory Board had studied the project since 1996 but found it a challenge to completely comprehend and provide meaningful citizen input (Dixon 2003). To help resolve the community issue, the DOE requested that a peer review panel (comprised of members of the Institute for Regulatory Science and ASME) coordinate the determination of a site remediation strategy (ASME/RSI 2001).

1. *The Process*: Before the peer review meeting, the DOE reviewed guidance on stakeholder definitions and participation under selected ASME procedures. In addition, the planning considered the tradition of all professional societies, indicating that all segments of the meeting (except executive sessions of the review panel) would be open to the public. The six panel members, having received extensive reading materials, met in June 2001 for presentations by the project managers and various stakeholders.

   During the introduction, the rules of participation were described. Members of the audience were told that they could ask questions of the speakers and could also make statements during the time designated for that purpose. In every case, stakeholders who wanted to ask a question or make a statement were required to indicate their name, affiliation, and their class of stakeholder. The audience was also asked to complete a questionnaire. The presentations, the discussion, and statements by the stakeholders during the meeting were recorded and transcribed.

   After the meeting, the questionnaires were collected and subsequently evaluated. Later, the panel met and wrote its report in accordance with the peer review procedures. The results were presented to the Community Advisory Board in a second meeting open to the public and stakeholders.

2. *Results*: The peer review gave invaluable and credible feedback to the DOE on the strategy to address the site's underground contamination issues (Dixon 2003). The stakeholder participation process was also peer-reviewed by the ASME, both before and during its implementation. The review supplied a level of feedback to the DOE far beyond what the committee could do on its own, thus providing the agency with the tools needed to communicate effectively and to focus subsequent public meetings.

8.5.3 Saxton Decommissioning: GPU Nuclear took the lessons learned from its experience with an advisory panel for the Three Mile Island remediation project (NRC 1994c) and applied it to the establishment of a Citizens Task Force for its decommissioning of the Saxton experimental reactor. The NRC recently recognized this strategy as a significant success in public involvement.

8.5.4 Lessons Learned: Both legal and moral imperatives encourage any decommissioning facility to begin stakeholder interactions. And, when conducted well, the process normally yields indisputable benefits. Both completed and ongoing projects have demonstrated that a properly tailored process that promptly involves the stakeholder, is thorough in its communication, and includes meaningful interaction, should result in better long-term decisions and prevent unnecessary delays. Lessons learned from these project experiences include the following:

1. The rules governing stakeholder participation must be clear and fair. They must be communicated to the stakeholders and must be applied fairly to everyone.

2. Managers must make an honest attempt to engage participation at the beginning of the planning process, and to focus on the balance of cost and risk.

3. Using a facilitator from an impartial entity is beneficial to the process, adding to the perception of equitable treatment for all concerned parties.

4. The chair or facilitator must ensure group focus by keeping the input actionable and constructive, reformulating where necessary.

5. Interaction with stakeholders should consider their categories.

An earnest effort must be made to listen and respond to the concerns of all stakeholders, thereby requiring informed and empowered facility representation. Many contested issues can be resolved if decision-makers demonstrate that they functioned with a sincere consideration for the concerns from others. It is a maxim in public participation that if stakeholders perceive the process to be fair, they will be more likely to accept the results.

PART 3

PLANNING AND IMPLEMENTATION

Nuclear facilities have a finite operating life. At some point the facility owners or managers must decide that it has reached the end of its useful life and must be shut down. Once the decision is made to decommission, various "end-state" options must be evaluated. The end-state may vary from unrestricted to a variety of restricted uses, respectively termed from "greenfield" to "brownfield." The end-state decision is the single most important decision in the planning process since it defines the extent of facility decommissioning and site remediation and directly implies the project cost, schedule, and scope.

Over the past several decades, considerable experience has resulted in valuable lessons learned that have been factored into the decommissioning planning and implementation process. Proper planning and management improves safety, reduces costs, and shortens schedules. Ideally, planning for decommissioning a facility starts during the planning for construction. At a minimum, planning must start years before shutdown, particularly in the case of power reactors. During planning, all parts of the decommissioning project are identified, often as formal plans to be approved by appropriate regulating agencies.

Decommissioning a nuclear facility often costs more than building it. The decommissioning should be managed at least to the same degree as any other major construction project, with the added complications of handling and disposing of nuclear wastes while meeting stringent regulatory requirements. The ultimate goal is to reach an end-state that has the approval of the owners, the regulators, and other stakeholders. A well-planned and implemented project assures that the goal will be met.

During the early planning phase, cost estimates and schedules must be developed, work scope identified, risks assessed, mitigation plans developed, regulatory compliance strategy developed, and management plans written and approved. During the actual decommissioning, managers must be aware of changing conditions and willing to consider them. Given how long an activity can take, regulatory requirements may change, stakeholders may ask for modifications, new technologies may become available, and numerous predictable and unpredictable events may take place. Rather than being hindered by them, a properly managed decommissioning uses them to advantage.

Chapter 9

INITIAL DECOMMISSIONING PLANNING AND PROJECT MANAGEMENT

Initial decommissioning planning occurs, whether intended or not, before the decision is made to permanently cease operations. The initial planning assessments completed by the facility owner are defined to generate specific and sufficient information, including costs related to continuing operation for specific time periods, ceasing operation by a specific date, and determining the type of decommissioning to pursue and a start date for preliminary activities. Key decisions finalized early in the planning process to make for a safe and cost effective project.

There are many examples of nuclear facilities where a determination was made to cease operations for an indeterminate time period, while assessments regarding the facility's future use were completed. Examples include the U.S. Department of Energy's (DOE) Fast Flux Test Facility remaining in hot standby until early 2003 (Lanais 1993), and Millstone Unit 1, a commercial nuclear reactor now in safe storage. In some cases, initial decommissioning studies are performed while a facility is in this shutdown pre-decision mode. A commercial power reactor facility in hot standby requires considerable expenditures to satisfy the U.S. Nuclear Regulatory Commission (NRC) requirements so that the facility might be returned to operation at a reasonable cost without re-testing equipment. This approach is costly since no operational benefits are derived but costs near operating level continue. The costs may be significantly less for small DOE reactors, and may be much less than normal operating costs for non-reactor facilities.

Though not required before starting decommissioning activities, determining the ultimate end-state of the facility and the site is a critical decision. Substantially different approaches may be implemented for a facility that will be remediated for potential residential use (greenfield) than for one that is to be partially demolished for potential industrial reuse (brownfield). In the greenfield approach, almost everything with discernible radioactive contamination would likely be removed and shipped off site for disposal. In the brownfield approach there are more options. For example, only some of the high radioactive content structures and components may be dismantled for off site disposal. Only a portion of the above- and below-grade structures might also be removed, along with some of the contaminated soil beneath. While above-grade portions of buildings on a brownfield site may be demolished, their foundations may remain in place with little or no decontamination. Structures intended for reuse might have their internal radioactive components removed and the remaining interior surfaces decontaminated to sufficiently suitable limits to allow unrestricted reuse of the remaining space.

9.1 PROJECT SCOPE

Several factors must be evaluated to determine the scope of the eventual decommissioning project. These include:

- Initial site and facility configuration and characterization
- Initial project bounds and contents
- Final site configuration

- Final site boundary
- Expected release criteria

It is essential to have a clear understanding early in the decision process about what structures, systems, components, buildings, and grounds are to be permanently decommissioned or decontaminated and modified. The systems and buildings housing the primary radioactive materials (e.g., reactor containment or laboratory hot cells) would clearly fall within the decommissioning project. The facility and site characterization for radiological and hazardous materials will be extremely valuable when making that decision. It may be obvious that certain components and soil areas will not be removed and shipped off site due to their high radioactivity levels and the risk they pose to the public and the environment. But there could be substantial disagreement about whether a groundwater sampling station 500 meters from the primary radiation source should be included in the overall project.

To reach such decisions, the owner must characterize the entire site and its structures and determine the cleanup limits to apply to the characterization information. The characterization surveys should be sufficiently definitive to support final decisions as to the soil, structures, and components to be removed and what may remain on site. Characterization continues throughout the project to secure the additional detail necessary for use at any given project stage.

Note that performing highly detailed surveys at the beginning of a project is often extremely wasteful of both money and time since much of the information may never be needed and alternate approaches used later may render much of the original data useless. A thorough review of the current and past history of the site and its use, and that of its buildings and structures, is essential to determining the appropriate extent of characterization. Once an adequate initial characterization survey is accomplished, then interaction with facility management and key stakeholders will pave the way for agreement on project scope.

9.2 PLANNING THE DECOMMISSIONING OPTION

The objective of decommissioning is to remove a nuclear or radioactive facility from service and restore it to such a condition that there is no unreasonable risk from residual radioactive material to public health or safety or the environment, either from occupational exposures or public exposures from related activities (e.g., transportation of radioactive wastes). Although decommissioning can also include nonradiological contaminants requiring remediation even at significant distances from the facility (such as water and soil) this chapter will consider activities primarily from the perspective of radiological sources. The three decommissioning alternatives described in Chapter 3 -- DECON, SAFSTOR, and ENTOMB, or a combination of these, e.g., partial decommissioning followed by a storage period before completion of decommissioning -- are detailed in NUREG-0586, Supplement 1 (NRC 2002). Actions taken for any of these alternatives or combinations are assumed to reduce risk to workers, the public, and to the environment from the radiological and environmental contaminants remaining on site.

9.2.1 DECON: This is the only alternative that may result in termination of the license, and may take from fairly short time periods for small facilities to up to several years for a large one. The nuclear utility generally has a strong interest in applying to the NRC to change the its operating licensee to a Possession Only License (POL) since the change enables the owner to achieve some savings by downgrading operating reactor requirements. Hence, the activities to achieve a POL are generally conducted soon after plant operations permanently cease. However, a utility may delay the start of full scale decommissioning for a prolonged time period (subject to the 60-year constraint on overall time allowed for decommissioning in NRC regulations). DECON time frames for some facilities are noted in Table 9.1 (Boing 2001).

The end-state of the DECON approach may lead to termination of the license and unrestricted use; that is, a greenfield state, which allows site/facility reuse for any purpose, including residential and agricultural. DECON may also lead to an end-state of restricted use. In this case, the facility would require administrative or engineering controls to maintain an acceptable low risk to the public and the environment. This brownfield state is an appropriate approach for facilities that will continue to be used by the facility owner for the extended future. Restricted-use DECON may be a viable alternative to the SAFSTOR or ENTOMB options in the cases where facilities have substantial amounts of radioactive material, waste transportation issues, and/or lack of disposal options.

Table 9.1 Time and Cost Estimates for Facilities in Decommissioning

Facility Name	Type	Decommissioning Period	Cost (US$)
EBWR	Research reactor	1986-1996	19.6M
Janus	Research reactor	1995-1997	2.1M
CP-5	Research reactor	1991-2000	29.5M
Yale University HILAC	Accelerator	Completed 1975	~ 100k
ANL – Bldg 212 Plutonium Gloveboxes	Gloveboxes	1992-1996	6.9M
Carnegie-Mellon Synchrocyclotron	Synchrocyclotron	Completed mid 1970's	~ 500k
Cambridge Electron Accelerator	Accelerator	Completed mid 1970's	~735k
NUREG-0586 Reference Research Reactor	1 MW_t pool reactor	2 year estimate	1.25M estimate (1986 $)
NUREG-0586 Reference Test Reactor	60 MW_t reactor	4 year estimate	24.2M estimate (1986 $)

Because all DECON work is completed within a few months or years following permanent shutdown, personnel radiation exposures are generally higher than for other decommissioning alternatives that enable the owner to extend the work over longer periods, thereby allowing for additional radioactive decay. For example, if the major radionuclide on site were ^{60}Co, with a half-life of 5.27 years, the radiation exposures using the DECON option would be approximately four times higher than they would be if the facility had been placed into SAFSTOR for ten years. Prompt DECON is accomplished at the expense of larger initial commitments of funding, personnel radiation exposure, and potential waste disposal volumes than for the other alternatives.

A great advantage of the DECON alternative is the availability of an on-site work force highly familiar with the facility. Another advantage is that prompt DECON eliminates the need for long-term security, maintenance, and surveillance that would be required when decommissioning is conducted for longer periods of time.

Some of the factors related to the DECON option differ between commercial utility plants and government facilities. The beneficial reuse of real estate and the time value of money are important factors for commercial utilities, while the costs for maintaining technical specifications and facility requirements to ensure operational safety have a high cost impact until the terms of the license are changed. In contrast, the more important factor for government projects is the extensive cost of ongoing surveillance and maintenance (S&M). Advantages and disadvantages to the DECON option are presented in Table 9.2.

Table 9.2 DECON: Advantages and Disadvantages

Advantages	Disadvantages
Facility and site becomes available for earlier reuse	Higher radiation exposure to workers and to the public
Knowledgeable facility workforce available to support project	Larger initial commitment of funds
Reduction in security, surveillance, and maintenance costs due to reduced project duration, which would be required for other alternatives.	Potentially larger volumes of radioactive wastes generated
Greater certainty about availability of low-level radioactive waste burial space	Project complications if site must continue to store high-activity radioactive wastes on-site.
Lower project costs due to non-escalation and less risk of uncertainty	May require use of new (not yet developed) technology
Known regulatory climate	

As noted earlier for a brownfield DECON, nonradioactive equipment and structures may not need to be dismantled or removed. Once decontaminated to levels permitting reuse, they may either be used or demolished at the owner's option. This is often the case in government projects.

9.2.2 SAFSTOR: SAFSTOR (safe storage) is the alternative for a nuclear facility to be placed in (or prepared for) safe storage, and maintained until subsequently reused or decontaminated to levels that permit release for reuse (deferred DECON). Under NRC regulations for a power reactor, the facility may remain in SAFSTOR followed by DECON for up to 60 years. As noted earlier, a key beneficial factor of the SAFSTOR option is the reduction in personnel radiation exposure from radioactive decay during the storage period. Overall reductions in hazards, and site hazard classification, also typically occur during safe storage, allowing reductions in S&M costs. Several subcategories of SAFSTOR are possible:

1. *Hot/Cold Standby*: an interim state where nuclear facility operations have stopped for an indeterminate period. Evaluations as to the facility's operational future and potential end-state may be determined during this period. Other facilities may be kept in this mode if the facility mission is currently not needed but may be required in the future (e.g., Nevada Test Site). Operations in hot/cold standby are similar to those conducted during long S&M outages. Protection of the workers, public, and environment is maintained through compliance with operational technical specifications and license requirements.

2. *Custodial SAFSTOR*: requires a minimum initial cleanup and decontamination effort, followed by a period of continuing care with the active protection systems (principally ventilation, waste management, and radiation monitoring) kept in service throughout the storage period. Full-time on-site surveillance by operating and security forces is required for radiation monitoring, equipment maintenance, and prevent accidental or deliberate intrusion into the facility. Continued environmental monitoring is also required to ensure radiological

contaminants remain within the confines of the facility. Protection of the workers, public, and environment is maintained through compliance with a reduced set of operating technical specifications and license requirements, active systems, and some passive barriers.

3. *Passive SAFSTOR*: requires a more comprehensive initial decontamination effort, enough to permit deactivation of the active protective systems during the continuing care period. The structures are completely secured and electronic surveillance is provided to detect accidental or deliberate intrusion. Periodic monitoring and maintenance is required to ensure both the appropriate confinement of radioactive contaminants and that structures maintain physical integrity over time. More emphasis is placed on engineered controls to protect the worker, public, and environment during passive SAFSTOR compared to custodial.

4. *Hardened SAFSTOR*: requires comprehensive cleanup and decontamination, and effective passive barriers around areas with significant quantities of radioactivity. These barriers are of sufficient strength to make accidental intrusion virtually impossible and deliberate intrusion extremely difficult. Surveillance requirements are limited to detection of intrusion at the barriers, maintenance of the integrity of the structures, and periodic monitoring. This mode requires substantial emphasis on both engineered and institutional controls to ensure the protection of the workers, the public, and the environment.

All categories of SAFSTOR require some positive action at the conclusion of the period of passive continuing care to allow for the release of the property for reuse and to terminate the license(s) for radioactive materials. Depending on the nature of the nuclear facility and its operating history, the necessary action can range from a radiation survey that shows that the radioactivity has decayed and the property is releasable, to the dismantling and removal of residual radioactive materials. These latter actions, whatever their scale, are generically identified as deferred DECON.

SAFSTOR is used to satisfy ES&H requirements while minimizing the initial commitments of time, money, occupational radiation exposure, and waste disposal. SAFSTOR may have some advantage where there are other operational nuclear facilities at the same site, and may become necessary if there are obstacles to radioactive waste disposal. Modifications to the facilities are typically limited to those that ensure the security of the buildings against intruder, and to those required to ensure containment of radioactive, toxic, or hazardous materials.

The reduced initial effort (and cost) of preparing for safe storage is tempered by the need for continued surveillance and physical security. Electronic monitoring devices are used by local law enforcement or private security agencies as continued surveillance for intruders, fire, and increase in radiation levels. These devices require periodic checks and maintenance, as do the site structures themselves. The duration of the storage/surveillance and dismantling periods can vary from a few years to up to several decades, depending on the type of facility and the applicable regulatory requirements.

In the SAFSTOR option, the facility owner, with the regulator's approval, decides the length of the safe storage period. Limited to 60 years, the duration of safe storage is based on consideration of several factors, including desirability of terminating the license, radiation dose, and potential waste volume reductions, availability to waste disposal capacity, and other nuclear facilities at the site. SAFSTOR may be an attractive option for keeping an operating license active on the site if the end-state use may be a differing form of nuclear facility. Selecting this option keeps future use options for modification, new construction, or facility upgrades a viable possibility. Similarly, the decision on the extent of decontamination during the period of preparation for safe storage, and the resultant subcategory of SAFSTOR to be used, depend on safety considerations and the planned duration of the S&M period. For example, if ^{60}Co, with a half-life of 5.27 years, is the controlling source of occupational exposure, a chemical decontamination campaign achieving a decontamination factor (DF) of 10 (i.e., radioactivity levels reduced to 1/10 of original) will result in approximately the same dose reduction as a decay period of 17 years.

Deferred DECON would be used at the end of SAFSTOR to remove residual radioactivity and terminate the license. If the safe storage period is sufficiently long, radioactive materials may have decayed to levels low enough to permit release for reuse without additional decontamination. This would not apply in the case of a reactor that had been operated long enough to produce significant amounts of long-lived isotopes (^{59}Ni, ^{94}Nb, or dispersed transuranics). Following DECON, the remaining nonradioactive systems and structures would be demolished at the owner's discretion. Such a greenfield end-state is the plan for both the Big Rock and Maine Yankee nuclear power plants (see Part 5). This will entail the complete demolition of all buildings identified in the facility operating license.

A disadvantage of SAFSTOR is the reduced availability of personnel familiar with the facility operations at the time of deferred decontamination. Good historical records, including pictorial records, will reduce the additional training time. Other disadvantages include the inability to reuse the facility or site for an extended period of time, potential future regulatory uncertainty, and the ongoing costs for maintenance, security, and surveillance. Advantages and disadvantages to the SAFSTOR option are presented in Table 9.3. A discussion of the SAFSTOR option for a research reactor may be found in the article, "Decommissioning the World's Premier Facility for Radiological Research: The Janus Reactor" (Taboas 2000).

Table 9.3 SAFSTOR: Advantages and Disadvantages:

Advantages	Disadvantages
Substantial reduction in radioactivity as a result of the decay that occurs during the storage period.	Shortage of personnel familiar with the facility at the time of deferred decontamination and dismantling
A reduction in worker radiation exposure (as compared to DECON option)	Limited availability for site reuse during SAFSTOR period
A reduction in public radiation exposure (due to fewer radioactive waste shipments as compared to DECON option)	Uncertainties of availability and costs of low-level radioactive waste storage at the time of decontamination and dismantling
A reduction in radioactive waste volumes (compared to DECON option)	Continued need for security, surveillance and maintenance during the SAFSTOR period
Lower annual costs following cessation of operations	Probable higher total project costs due to price escalation during the SAFSTOR period and the intermediate S&M costs.
SAFSTOR period allows more time for development of permanent disposal options such as a high level waste repository	

9.2.3 ENTOMB: ENTOMB may be selected as an option when the residual radioactivity can decay to levels permitting release of the facility within a reasonable time (i.e., while confidence remains in both the continued structural integrity of the entombing structure and in the reliability of radioactivity containment and access restriction). An example would be a facility contaminated primarily with ^{60}Co or tritium, where relatively short half-lives would allow essentially all the residual material to decay over a ~100-year control period.

ENTOMB is also a viable decommissioning option for a nuclear facility with long-term barriers available to prevent exposure to residual radioactivity. A candidate for this would be radioactive contamination emitting only low energy alpha and/or beta particles. However, a few radioactive isotopes found in fuel reprocessing plants, nuclear reactors, fuel storage facilities, and mixed oxide facilities have half-lives well in excess of 100 years. Thus, the basic requirement of continued structural integrity of the entombment cannot be ensured for these facilities and ENTOMB would not be a viable alternative for these cases.

The ENTOMB option for a complex structure, such as a reactor, requires the radioactive materials remaining in the entombed structure to be characterized well enough to ensure their decay to acceptable levels by the end of the surveillance period. If this cannot be satisfied, deferred DECON would become necessary, making ENTOMB more difficult and costly than the options of prompt DECON or SAFSTOR. The ENTOMB option, when implemented by owners, increases the number of sites dedicated to containment for very long periods of time. In some cases, the specific activity of residual TRU may require entombment due to the lack of alternate disposal. (SAIC 1982).

Though the ENTOMB option does not actually exist in an approved form at this time, its advantages and disadvantages are presented in Table 9.4 . Cost, engineering, and regulatory factors will have to be addressed for it to be a viable option in the future. It is not considered practical for commercial nuclear plants, and may be used by the DOE only after careful reviews and approvals have been obtained. This option could be used for modern DOE projects using such engineered safety features as long-term isotope retention barriers or other similar technologies. Modifications of the ENTOMB approach, such as 75-year storage rather than the SAFSTOR's 60, have been taken by the DOE. For example, the Hanford C-Reactor is in a modified ENTOMB state, maintained with robust institutional controls, and review of the facility status and decommissioning approach at specified periods (e.g., 25 years). As long as the entombed facility remains structurally sound, with engineering and institutional controls intact, the facility may be safety retained in this mode for many years. Further discussion of the C-Reactor is provided in Chapter 29 of this handbook.

Table 9.4 ENTOMB: Advantages and Disadvantages

Advantages	Disadvantages
Reduced cost (as compared to DECON OPTION)	Facility unavailable for reuse
Reduced personnel and public radiation exposure (as compared to DECON OPTION)	Maintains radioactive materials at distributed locations rather than centralized repositories

9.3 ECONOMIC INPUTS TO THE DECISION PROCESS

Optimum decommissioning decisions can only be made after an owner has obtained and compared a variety of financial and economic data. Key variables include:

- Cost estimates for each potential decommissioning option considered
- Decommissioning funding approaches, whether a fully funded external fund or annual budget allocation
- The effects of project delays (both for timing to begin the project and delays while the project is in progress)
- Projections on growth rates for decommissioning funds
- Projections on interest rates (determination of future cost of money)
- Evaluation of the variability in these parameters

The preferred decommissioning option will likely be the result of several iterations of the economic estimates. Before an economic analysis of any option can be completed, it is imperative to have prepared a baseline estimate for cost comparisons during evaluations. Estimates prepared by in-house facility staff should be independently validated, and vice versa. Project tasks/activities with high individual costs, or high potential variability in costs (e.g., radioactive waste processing and disposal, or stakeholder programs and support) should receive particular focus. Chapter 12 describes the process of developing cost and schedule estimates; these provide the bases for comparing alternatives, benefits, and risks.

9.4 PROJECT MANAGEMENT AND PLANNING

Managing a decommissioning project involves technical, financial, administrative, legal, regulatory, and operational resources and disciplines. Decommissioning is indeed a *project*, not a continuing series of activities associated with operations. In fact, the Project Management Institute (PMI) defines project management as:

> "A temporary endeavor undertaken to create a unique product or service. Temporary means that every project has a definite beginning and definite end. Unique means that the product or service is different in some distinguishing ways from all similar products or services. Project management is the application of knowledge, skills, tools, and techniques to project activities in order to meet or exceed stakeholder needs and expectations from a project."(PMI 2000)

This working definition does not apply to repetitive corrective maintenance, routine preventive maintenance, facility maintenance (e.g., grass cutting or janitorial services), or surveillance and repetitive programs (e.g., routine pump maintenance). The objective of project management, from initiation through closeout, is high quality projects that meet customer needs and are implemented consistently across the organization, with the highest efficiency and lowest cost.

Planning for decommissioning is an integral part of managing the project. While initial planning sets the path forward for the project, in-process planning is a continuing function as the project unfolds, site conditions change, regulations evolve, and owner/licensee objectives change. This chapter will provide an overview of the planning process and identify key elements that must be considered, regardless of the size of the project.

Attached to this handbook is a compact disk (CD) that, along with other information, contains a detailed listing of decommissioning activities from planning through closeout. The CD includes a document proposing a standardized list of decommissioning cost items prepared by internationally recognized teams of experts from the European Commission (EC), International Atomic Energy Agency (IAEA), and the Organization for Economic Cooperation and Development/Nuclear Energy Agency (OECD/NEA). While the development of this program was focused on preparing a standard chart of accounts, the resulting document provides a comprehensive listing of activities associated with decommissioning of virtually any type of facility. The reader is encouraged to review this document in the CD's section on additional resources prior to developing the elements of a site-specific project management process and system.

In general, PMI recommends dividing a project into six phases. These phases and their decommissioning equivalent, listed in Table 9.5, form the organizational structure for this chapter.

Table 9.5: The Standard Phases of a Project

PMI Phases	Decommissioning Phases
Project Initiation	Pre-Decommissioning
Planning - Preliminary Studies	Post-Shutdown Pre-Planning
Planning - Detailed Design	Post-Shutdown Detailed Planning
Execution	Decommissioning Activities and License
Termination	
Controls	Project Management
Closeout	Site Restoration

The phases applied to decommissioning are typically further divided into several more detailed elements for improved control of activities, costs, and schedule.

9.5 PROJECT INITIATION – PRE-DECOMMISSIONING

Project initiation begins when the decision is made to permanently shut down a facility and proceed to decommissioning. The owner/licensee organization will conduct scoping analyses to define the end-point objectives for the facility. The radiological release level to be achieved for the facility or site is an important early step in this process since it will establish the level of residual radioactivity. Similarly, the clearance level is the level of radioactivity permissible to release equipment from the site for reuse in the community. The end points of the facility (there may be more than one -- for the facility and for the site) are highly dependent on these two criteria.

The endpoint objectives typically involve preliminary cost studies of decommissioning alternatives to evaluate their relative cost and schedule advantages and disadvantages. The preliminary studies, typically used to secure funding authorization to proceed, take cost bases from previous estimates, modified where necessary to account for site-specific changes in assumptions, site conditions, or funding constraints imposed by management or regulatory agencies.

The more formidable tasks of project initiation include:

- Planning the transition of the existing staff from operations to decommissioning
- Setting up a management organization

- Evaluating whether to use an outside Decommissioning Operations Contractor (DOC) or to self-perform the project
- Identifying resources and how to get them
- Performing preliminary site characterizations to bound the scope of the work
- Preparing baseline cost and schedule estimates to monitor and control expenditures

These tasks will be described in greater detail in the following sections.

9.5.1 Release and Clearance Criteria

1. *Release Criteria - Radiological*: Many applicable radiological release criteria are possible, since, depending on the project, the criteria may be set by the DOE, the Environmental Protection Agency (EPA), the NRC, various state and local agencies, or all of these. In regard to license termination for a commercial reactor site, the NRC states (NRC 2003a):

    "A site will be considered acceptable for unrestricted use if the residual radioactivity that is distinguishable from background radiation results in a Total Effective Dose Equivalent (TEDE) to an average member of the critical group that does not exceed 25 mrem per year, including that from groundwater sources of drinking water, and that the residual radioactivity has been reduced to levels that are as low as reasonably achievable (ALARA)."

    Various NRC documents tabulate the specific allowable contamination levels per 100 square centimeters of surface area for unrestricted work areas and the estimates of picocuries per gram of soil for a given radioisotope that will result in 25 mrem per year by particular pathways.

    The DOE document most often used for site release criteria and surface cleanup limits for unrestricted reuse is DOE Order 5400.5 (DOE 1993). This Order, stipulating cleanup limits for surface contamination, soil, water, direct radiation, and the site in general, identifies 100 mrem/yr as a maximum radiological pathways limit for an individual. Certain DOE projects, however, have adopted the NRC 25 mrem/yr pathways dose as their criterion.

    The EPA often prefers to use a somewhat lower limit of 15 mrem per year to an individual by all pathways for a site, non-inclusive for the water pathway. The EPA's limits for drinking water are stated as, "The average annual concentration of beta particle and photon radioactivity from man-made radionuclides in drinking water shall not produce an annual dose equivalent to the total body or any internal organ of greater than 4 mrem/year" (EPA 2003a). The agency also lists additional standards for ^{226}Ra and ^{228}Ra, and gross alpha activity in drinking water in community water systems (EPA 2003b). Various states may have adopted even more stringent radiological release limits. Maine, for example, adopted a 10 mrem/yr pathways dose to an individual.

2. *Release Criteria - Nonradiological*: Although hazardous chemical cleanup criteria are generally of lesser importance in decommissioning most commercial nuclear power plants, cleaning up hazardous chemicals may constitute the majority of the work in many DOE projects where asbestos and polychlorinated biphenyls (PCBs) are routinely encountered.

3. *Environmental and Related Requirements*: The detailed environmental cleanup criteria implicit in the federal laws are extensive; they are primarily contained in the EPA's 40 CFR publications and in state and local environmental cleanup regulations.

4. *Stakeholder Input to Decommissioning Decision Process*: Broadly defined, a stakeholder is "any group or individual who is affected by or who can affect the future of the facility decommissioning program." A variety of stakeholder groups may have interacted with the owner/licensee throughout the operating life of a nuclear facility. The relationship with these stakeholders can substantially affect the approach and timing selected for decommissioning. Some will be supportive, some adversarial, and others open to interaction. It is important for the owner/licensee to interact and gather information as to their interests and intents with respect to the decommissioning process. The owner then should evaluate the potential impact each stakeholder may have on the decision-making process. Additional discussion on stakeholder interaction is provided in Chapter 8.

9.5.2 Transition Planning, Organization, and Resources:

9.5.2.1 Transition Planning: The first major task in project initiation is establishing the project management organization. The existing site management organization usually includes functional personnel representing critical skills necessary to implement the decommissioning project. A major element of planning includes the transition of personnel from an operating to a decommissioning perspective. This change in culture can be a traumatic experience for facility workers who may have been working on site since the construction and startup of the facility. Resource planning involves decisions regarding those functional groups needed to support decommissioning, and the key individuals within those functional groups. Key employees are usually offered incentives to remain with the project until specific milestones are completed. The remaining employees may be offered positions in one of the necessary functional groups, positions in other units of the company or organization, continuing positions with perspective subcontractors, early retirement, or severance.

Since the facility itself must also undergo the transition from operations to permanent shutdown, most existing personnel will be needed initially to prepare for decommissioning. However, the number and duration of their services must be clearly defined and communicated to them to ensure a clear understanding of the beginning and endpoints of their services. Specific management staff is tasked to address the changes to be made to the facility, how they are to be accomplished, and the endpoint of the defined scope.

The size of the staff will depend on the type of facility, level of difficulty, and degree of control mandated by management to oversee the work. Small, single mission facilities may have only a handful of professionals as its entire staff. Nuclear power reactors typically have operating staffing levels of approximately 250 - 350 per reactor, including professionals, managers, and technicians (maintenance, operations, and environmental safety and health (ES&H). Large, multi-mission national laboratories have large staff -- also with a wide range of technical, professional, and managerial positions. Although the numbers and types of personnel required vary widely based on facility size and project scope, the staffing needs for decommissioning will usually require either significant retraining or new skill development for the incumbent staff or significant changes in the number and types of personnel.

As discussed in Chapter 5, the Transition Implementation Guide (DOE 2001a) provides guidance at DOE sites for transition activities that occur between operations and decommissioning. Transition includes:

- Declaring the facility excess
- Establishing an integrated transition team
- Performing site characterization
- Planning for facility stabilization
- Integration of safety management
- Verification of satisfactory stabilization
- Transfer of the facility to the disposition program.

Commercial decommissioning projects have produced reports valuable to any licensee approaching the shutdown decision. Two such articles have been published by the Electric Power Research Institute (EPRI 2000 and 2001). Chapter 15 addresses transition issues in detail; it is a key element for a successful decommissioning project. The following subsections provide an overview of this important activity.

9.5.2.2 Owner-Licensee Staff: The advantage of using existing employees to retain site specific plant knowledge is coupled with the benefit of not having to implement employee termination and severance issues. Yet, that staff will ultimately recognize that it will be performing activities to eventually terminate specific responsibilities. Therefore, delaying the decision for purely social needs is only a temporary resolution. The most important issue is identifying key personnel with the plant knowledge needed to safely and efficiently complete the project. Retaining on-board resources may be problematic if there are not personal incentives for the decommissioning activities to be done expeditiously. Decommissioning experience has shown that work incentives keyed to specific milestones is an effective mechanism to motivate the workforce.

In many cases, the decommissioning organization is a combination of the existing plant staff and subcontracted resources. The owner/licensee usually will retain the license, and therefore will be responsible for the public health and safety, security, and financial considerations. The existing facility personnel are probably best suited to perform these functions because they are familiar with the plant systems and structures, and are well-trained in organizational procedures and financial reporting mechanisms. (The accounting for decommissioning activities can and probably should be rather rigorous to enable the owner to track project and task performance, especially when subcontractor payments, milestones, and incentives are tied directly to the project.)

The functional responsibilities to be performed by individual organizational units must be defined early; and the owner must establish a clear description of functional scope and position for each organizational unit, including the essential detailed understanding of activities required to achieve cost-effective work performance. For example, the integrated approach to decommissioning by the workforce requires positive integration within and between functional groups. A good example regarding the importance of positive interaction related to work activities would be the manner by which one unit removes waste material, followed sequentially by other units completing packaging, transportation, and final disposal. This integration and coordination is to be expected and planned for in detail during the development of the organization.

9.5.2.3 DOC Staff: The decision whether to employ a DOC or to self-perform the management function will set the stage for the number and composition of the management staff. The use of a DOC has the benefit of an experienced contractor completing the project. The lessons learned at other decommissioning projects could be a major asset for the project since the knowledge/skills of specific methods and procedures implemented successfully elsewhere could rapidly accelerate the initial planning and mobilization activities. The DOC is usually provided cost incentives by the type and form of contract proposed. Fixed-price contracts or cost-plus-incentive-fee contracts have defined project performance motivations built into the agreement. These involve either a rapid completion date accelerating the amount of profit a contractor can earn or definitive incentive payments tied to performance goals for safety, schedule, person-rem exposure, and waste minimization. Selecting the most efficient mode for retaining or contracting management and technical workers, and the optimum time for switching from the current mode to the new approach to workforce management, have been major factors in the level of success experienced by owners on recent decommissioning projects.

The CD attached to this Handbook contains a listing of project management and support service positions starting at 08.0200, "Project Management and Engineering Services," and continuing through 08.0503, "Industrial Safety." The division of responsibilities between the owner/licensee and the DOC is a management decision and contractual arrangement.

9.5.3 Pre-Planning Fuel Disposition, Facility Endpoints, and Funding

9.5.3.1 Fuel Disposition: During the pre-decommissioning phase the owner/licensee will conduct cost-benefit studies to evaluate whether spent fuel should be stored in the fuel storage pool or transferred to on-site dry storage, such as an Independent Spent Fuel Storage Installation (ISFSI). Considerations would include:

- The length of wet storage required for decay heat removal
- Estimated schedule for ultimate transfer to a federal waste disposal repository
- Dry cask design
- Licensing
- Construction duration
- Schedule impact on the desired facility decommissioning schedule

Evaluations of the available cask vendors and construction companies (to construct the storage pad) would also be performed during this period.

Technical Specifications (NRC-licensed facilities), or Authorization Basis documents (DOE facilities) identify the technical conditions and limitations under which the facility was licensed to operate. These rigorous requirements may be relaxed as fuel is permanently removed from the reactor or the level of transuranic materials is reduced at the facility, thereby reducing the potential for a design basis accident. The associated reduction in maintenance, surveillance, and security activities and costs are a significant savings in personnel resources and project budget.

Following permanent shutdown and transition, both NRC and DOE facilities follow a program of facility deactivation to achieve a safe shutdown that is economical to monitor and maintain for any extended period. Systems that are not essential to decommissioning may be drained and de-energized in preparation for safe storage until they can be removed for disposal. The DOE's Deactivation Implementation Guide provides additional guidance (DOE 1999a).

At NRC-licensed plants, the common practice is to deactivate the facility to a "cold and dark status" to minimize operating and maintenance costs. In some cases that is a misnomer, such as for facilities in potentially frigid regions where system freezing can present a safety problem and condensation in unheated buildings can cause the icing of floors. In such cases, appropriate heating must be maintained.

9.5.3.2 Facility Endpoints: During this phase the owner/licensee will identify preliminary endpoints for each building, structure, or site area. The age and condition of structures are usually a major factor in this decision process. The projected condition after decommissioning (following potentially destructive demolition techniques used to extract radioactivity and components) is usually the driving force in those decisions.

The owner/licensee must evaluate if it has any proposed plans for reuse of the buildings/structures or sites. In some cases, these facilities may be reused for new power production facilities, or for temporary warehouse/storage areas. Such facilities are often at a distance from other communities and the potential for leasing the property to an outside party may be minimal. However, at the DOE Oak Ridge site, the federal government successfully established the East Tennessee Technology Park (at an attractive financial arrangement) specifically to attract local companies to reuse the decommissioned facilities.

Though not required before starting decommissioning activities, determining the ultimate end-state of the facility and the site is a critical decision. Substantially different approaches may be implemented for a facility that will be remediated for potential residential use (greenfield) than for one that is to be partially demolished for potential industrial reuse (brownfield). In the greenfield approach, almost everything with discernible radioactive contamination would likely be removed and shipped off site for disposal. In the brownfield approach there are more options. For example, only some of the high radioactive content structures and components may be dismantled for off site disposal. Only a portion of the above- and below-grade structures might also be removed, along with some of the contaminated soil beneath. While above-grade portions of buildings on a brownfield site may be demolished, their foundations may remain in place with little or no decontamination. Structures intended for reuse might have their internal radioactive components removed and the remaining interior surfaces decontaminated to sufficiently suitable limits to allow unrestricted reuse of the remaining space.

9.5.3.3 Evaluations for Funding Authorization: Pre-shutdown evaluations are usually performed to determine the level of funding required and the rate of withdrawal from the Decommissioning Trust Fund (DTF). In some states (such as California), the state retains control as to when to release any decommissioning funds. The state public utility commission usually requires a determination of the total decommissioning cost (perhaps for several options or alternatives), the rate of withdrawal, and sometimes a risk analysis as to whether the planned program will accomplish the objectives safely and cost-effectively. The federal government also requires that funding requirements for DOE sites are realistic, and may retain an independent cost estimate team to review the preliminary estimates for validity. Upon satisfactory validation, the government submits its request to Congress for authorization to proceed.

9.6 PRELIMINARY STUDIES – POST-SHUTDOWN PRE-PLANNING

Following shutdown, the owner/licensee staff will begin preparing the facility for decommissioning. Elements of this period are:

1. *Site Characterization*: A site radiological and hazardous/toxic materials characterization should be performed to bound the project scope. As noted earlier in section 9.1, it is essential to have a clear understanding early in the decommissioning decision process about what structures, systems, components, buildings, and grounds are to be permanently decommissioned or decontaminated and modified. The systems and buildings housing the primary radioactive materials (e.g., reactor containment or laboratory hot cells) would clearly fall within the decommissioning project. The facility and site characterization for radiological and hazardous materials will be extremely valuable when making that decision. It may be obvious that certain components and soil areas will not be removed and shipped off site due to their high radioactivity levels and the risk they pose to the public and the environment. But there could be substantial disagreement about whether a groundwater sampling station 500 meters from the primary radiation source should be included in the overall project.

2. *Regulatory Notifications and Public Meeting*: As described in Chapter 3, licensed power reactors must notify the NRC by written certification within 30 days of permanent cessation of operations. A separate certification, with no time limit, is required when fuel has been permanently removed. Within two years after permanent cessation, the staff must prepare a Post-Shutdown Decommissioning Activities Report (PSDAR) as the first step in gaining regulatory approval to begin. The NRC requires that a public meeting be held in the vicinity of the facility (within 90 days of PSDAR submittal) to allow the public the opportunity to voice its opinion on the proposed approach. Two years before the requested license termination date identified in the PSDAR, the licensee must submit a License Termination Plan (LTP) to identify the specific activities required to terminate the site license.

3. *Implementation of Revised Technical Specifications and Authorization Bases*: The proposed changes to the technical specifications and authorization bases would be implemented at this time. Individual changes may have to be phased in as specific activities are performed, for example when fuel is removed from the reactor to the spent fuel pool and then to the dry cask storage facility. For DOE facilities involving transuranics (TRU), a preliminary decontamination campaign may have to be carried out to reduce TRU to levels sufficient to permit relaxed authorization bases.

4. *Staff Reorganization into Transition Staff Positions*: The transition staffing plan is typically implemented at this time. Key functional staff positions are filled by existing personnel, and sometimes by outside personnel on an as-needed basis. Excess personnel may be offered positions in other parts of the company or offered severance packages at this time.

5. *Bid Specifications Preparation*: Bid specifications would be prepared during this period to invite competitive proposals and quotes for the DOC for long-lead procurement items (special under-water cutting equipment, system decontamination equipment/vendors, etc.), spent fuel dry cask vendors, ISFSI pad construction companies, and independent consultants

to advise and/or oversee engineering work. Existing owner/licensee procurement procedures are usually adequate for these activities, although contracting a low bidder is not always the best course of action. Extensive contractor reference checks, interviews, and sample documentation should be reviewed prior to awarding contracts for these project elements.

9.7 DETAILED DESIGN STUDIES – POST-SHUTDOWN DESIGN PLANNING

During this phase, the owner/licensee and the DOC (if used) identify functional responsibilities and start to work together more closely to ready the facility for decommissioning. It also includes the award of major contracts and subcontracts, contractor mobilization, the preparation of detailed procedures, and facility changes to ready the site for decommissioning activities. Typically, the owner/licensee retains responsibility for the following:

- Overall project management
- The site license
- Worker/Public health and safety
- Environmental protection
- Surveillance and maintenance
- Security
- Quality assurance
- Public relations
- Cost estimates and schedules
- Procurement

The staff would include a project manager, assistant project manager, engineers, health physics technicians and radiation protection manager, licensing specialists, environmental manager, plant manager, quality assurance manager, and financial managers, all with associated support personnel. This staff operates in an oversight and advisory role during the project.

A DOC, taking responsibility for the physical work to be performed, would assign a project manager, assistant project manager, project engineers, superintendents for dismantling crews, waste management (packaging and handling) crews, radiological and health physics technicians, and various craft labor crews to perform the work. A DOC would also prepare detailed procedures to control the work and submit them to the owner/licensee for review and approval.

This phase also includes contracts with vendors for major elements of the project, including dry storage casks, ISFSI construction, modifications of perimeters and fencing to accommodate increased truck traffic, temporary crew trailers and support facilities, phone systems, radio systems, and temporary warehousing trailers. Long-lead items and service contracts would include such items as remote underwater cutting equipment, mockup facilities, specialized decontamination equipment, waste disposal facility allotments, transportation contracts (truck, rail, and barge, as appropriate). These services and equipment will be part of the critical path for the project.

9.8 PROJECT EXECUTION

This is the dismantling phase of the project, where the radioactive, hazardous, and toxic materials are removed to acceptable residual levels so the license may be terminated. This phase involves the bulk of overall costs and must be closely monitored during execution. The project management function is critical for rigorous control of safety, conformance to approved procedures, in-process engineering evaluations, quality assurance, and security. Activities include:

- The draining and deactivation of systems

- Decontamination of systems and structures
- Removal and disposal of piping and components
- Segmentation and/or removal and disposal of the reactor vessel and internals
- Demolition of biological shielding
- Scarification (surface removal) of contaminated concrete floors and walls
- Removal of hazardous materials (asbestos, lead, mercury, PCBs)
- Final site survey for license termination
- Waste processing, packaging, transportation, and disposal

As discussed in Chapter 5, the DOE's Decommissioning Implementation Guide (DOE 1999b) covers project execution in decontamination and dismantling of facilities.

9.8.1 Decommissioning Activities:

1. *Draining and Deactivation*: Although some of the non-essential (to decommissioning) systems may be deactivated and drained immediately following shutdown (including engineered safeguard systems), others are needed to support wet storage in the reactor vessel and the fuel storage pool. The reactor vessel inventory of water is needed to provide shielding for the highly activated vessel internals during segmentation and removal. In addition, if full or partial primary system decontamination is required to reduce occupational exposure before removal, the reactor water cleanup system, primary coolant system, and associated waste management must be maintained operational until decontamination is complete. These latter systems would remain active, and some maintenance and surveillance would be required to ensure their availability. All systems must be deactivated and drained prior to equipment removal.

2. *Decontamination*: Full or partial system decontamination is sometimes required to reduce occupational exposure during removal. Chemical processes (as described in Chapter 17) may be used to reduce the internal surface levels of contamination. These processes rarely clean the interior of piping and components to free-release levels, but have been highly effective at reducing dose to workers. The potential for contaminated chemical spills during these activities is great, and keen oversight control must be maintained during treatment.

3. *Spent Fuel Transfer and Storage*: Reactor facilities with on-site spent nuclear fuel must transfer the fuel to some form of wet or dry storage until the federal government is able to take possession and place it in a long term repository. As discussed in Chapter 14, an extensive cost-benefit analysis is required to demonstrate the most appropriate option for each facility. The owner/licensee has the option to wet store fuel under the existing 10 CFR 50 license, or apply for a 10 CFR Part 72 license specifically for the dry storage ISFSI. In any event, decontamination and removal of components will not be final until all such fuel is removed from the building(s) to be decommissioned. The owner/licensee is responsible for the license until all the fuel is removed, and the ISFSI (if constructed) is decommissioned.

4. *Removal*: Piping and component removal follow system decontamination. Components must be size-reduced to fit into transportable and disposable containers. Contamination control during cutting is an important factor to avoid cross-contamination of (known) clean components or facility areas. Pre-planning for these size determinations should be a function

of the management team to ensure that secondary cutting prior to packaging will not be required. Larger components, such as steam generators and pressurizers, are usually removed intact, with the steam generator or pressurizer shell being used as the transport container. This also requires the coordination of the management team to ensure these large components are removed safely, and cost-effectively prepared for shipment, loaded onto transport vehicles, and transported to the disposal facility.

5. *Waste Management, Packaging, Transport, and Disposal*: The project management organization must include a dedicated waste management team. The team must make arrangements with the disposal facility to determine the waste acceptance criteria, and the size and weight restrictions imposed by the disposal facility. Contracts must be awarded to the disposal facility to guarantee availability of sufficient disposal space to handle the total waste stream. Multiple disposal facilities are often used for different levels and types of radioactive materials.

   This waste management group is also responsible for determining the number and type of shipping and disposal containers required to accommodate the waste streams generated from decommissioning. Container size and weight limitations often determine the size reduction requirements for removal of components and piping. Vendor contracts must be awarded for timely supply of necessary containers. Each package for transport must be inventoried prior to shipment to determine the number of curies, exterior dose rate, and weight.

   The management group must also award contracts to transport companies for moving the waste packages for disposal. Transport company vehicles must be inspected to ensure they are road worthy and safe for transport. Accurate records must be maintained throughout the project to document what was shipped and where it was sent for disposal.

6. *Surveillance and Maintenance*: Certain systems and structures must be maintained during the implementation phase of decommissioning since they support a safety function. As noted earlier, spent fuel cooling and storage pools must be maintained until the fuel is transferred to dry storage or its final disposition. Cranes, lighting, ventilation systems, and certain waste processing equipment are needed to support surveillance and maintenance during decommissioning. The DOE has provided guidance for these activities in its Surveillance and Maintenance During Facility Transition and Disposition. (DOE 1999c)

9.8.2 License Termination and Final Surveys:

As mentioned earlier, the project team is required to prepare and submit an LTP to the NRC within two years before the expected license termination date, as requested in the PSDAR. The LTP describes the methods and activities to be performed to meet the residual radioactivity limits for license termination. The LTP must include:

- A site characterization
- Identification of remaining dismantling activities
- Plans for site remediation
- Plans for the final site survey
- End use of the site

- Estimate of remaining decommissioning costs
- A supplement to the environmental report, describing any new environmental change

DOE facilities prepare a decommissioning project plan that usually includes how the project will be managed. The project plan describes what must be done to the facility during the various phases to protect ES&H, and to dismantle it or place it in long-term storage. The management plan describes how the project will be managed, including its integrated safety management program.

Health physics staff performs final site surveys to determine that the criteria for license termination have been achieved. The staff identifies the final survey to be performed, the methodology for determining the number of sampling and survey points, and the appropriate instrumentation for such determinations. Usually it identifies thousands of such points, and must follow a formal protocol for performing the measurements.

9.8.3 Confirmatory Surveys: The NRC frequently retains the Oak Ridge Institute for Science and Education (ORISE) to perform a confirmatory survey to independently validate termination data. Therefore, the project management organization must ensure the residual systems and structures are ready for verification. The NRC will terminate the license upon satisfactory validation of the data.

9.9 CONTROLS – PROJECT MANAGEMENT

Project management is a continuum of activities to control the project and to accomplish the project objectives. Changes to the project scope, direction, and progress occur repeatedly, and re-planning is necessary "on the run" virtually every day. The management organization must be flexible to cope with these changes and to anticipate their eventual impact on project cost, schedule, and other parameters. Quality Assurance (QA) is another part of project control, whereby the objectives of safety management are identified, and quality control monitoring and auditing is performed under a rigorous program that includes a QA program plan and manual, implementing procedures, design reviews, audits, and closeout requirements.

Some companies prepare a technical baseline to identify the technical objectives and requirements at the outset of a project. They may also prepare a scope baseline to identify all activities, the labor and material resources required, and expected end products. These two baselines should be agreed to by both the management organization and the owner/licensee.

From a controls standpoint, the management team should prepare a baseline cost estimate, baseline schedule, and often a baseline occupational exposure estimate. These baselines then establish the milestones against which the project can be measured for performance. Variations from the baselines are measured as variances against pre-assigned limits of reporting and justification. When the parameters of performance are exceeded, a written justification should be required to explain why the variance occurred, and what measures will be taken to correct the process and recover the overrun. This self-evaluation process is a valuable control tool to monitor performance of a project and to make the organization accountable for its actions.

9.9.1 Quality Assurance: The QA program is an integral part of the management process. The objective is to ensure the safety of all elements of the project through the establishment of policy and procedures that are incorporated into management, planning, execution, controls, and closeout.

A QA program does not replace the project management program, but rather monitors and audits the satisfactory performance of the project.

A formal QA program should be prepared for decommissioning since its activities are substantially different than operations. It should reflect the needs for health and safety, cost and schedule, design, control of data, documentation, and project closeout. In most cases, the existing program can be used as a starting point so as not to necessitate a complete change in processes and procedures. QA requirements should be clearly described. Recent programs have adopted a graded approach so as not to unnecessarily burden the project with requirements that are no longer applicable.

For commercial reactors, the revised QA program should address the NRC's 18 criteria for nuclear power and fuel reprocessing plants (NRC 1971). The DOE's projects should be in accord with its

- QA requirements under nuclear safety management (DOE 2000a)
- DOE changes to QA in Order 414.1A (DOE 1999d) and 10 CFR 830.120-122 (DOE 2000b)
- The DOE's Quality Assurance Management System Guide (DOE 1999d)

As noted earlier, a graded approach to QA should be adopted early in a project to avoid unnecessary requirements. In each case, the specific requirements (e.g., the NRC's 18 criteria) should be addressed as applicable or nonapplicable to the project.

1. *QA Procedures*: Procedures should be prepared to implement the graded approach for the project. Routine administrative procedures (format and content of procedures, preparation of design calculation notebooks, auditing procedures, etc.) may be adopted from the existing program. Newly required procedures should be prepared and approved in accordance with QA requirements.

2. *QA Project Plan*: The QA Project Plan (QAPP) should identify the specific requirements of the QA program that apply to the project. A QAPP should be prepared for the overall project, and for each major task involving safety-related or quality-related functions. A brief summary of typical QAPP topics and requirements is as follows:

    - Identification of project team members and responsibilities
    - Points of contact
    - Schedule of design (project) reviews -- start, mid-point, end of project or task
    - Control of design calculation notebooks
    - Identification of drawings, specifications, and procedures to be developed
    - Document control system for issue and revision of documents
    - Schedule of project audits
    - Project closeout requirements

9.9.2 Project Baselines: Project baselines are established during the initiation phase of the project to monitor and control the processes and activities throughout the program. They give feedback to management on how the project is being managed, and whether intervention at higher levels is necessary to prevent a safety incident, or cost and schedule overruns (or other parameter).

9.9.2.1 Baseline Cost Estimate: A baseline cost estimate is key to detailed project planning. This estimate is usually prepared after a detailed facility characterization, wherein the essentially static conditions of radioactive contamination can be determined from historical assessments and actual field measurements. The scope of planned work can be identified for the conditions existing at the facility at a fixed point in time. Estimates of the planned activities can be made based on the inventory of equipment and structures to be decontaminated, removed, packaged and shipped for disposal. The size of the management organization can be established and the duration of the project determined. Usually a Work Breakdown Structure (WBS) is developed to identify specific milestones for completion. Costs to accomplish each element of the WBS are used as the control measurement parameter. As noted earlier, the CD attached to this handbook provides guidance on the activities and chart of accounts that should be included in the baseline cost estimate.

9.9.2.2 Baseline Schedule: In like manner, the schedule for performance of these WBS elements is identified and used to measure performance and progress. The schedule is developed in concert with the costs so as to establish the size of the field labor resources required. A precedence network is established to show the inter-relationship of WBS elements and associated costs.

9.9.2.3 Program Evaluation and Review Technique (PERT)**:** PERT is a process to monitor project performance based on baseline cost and schedule estimates. Also called a cost and schedule control system, it relies on several parameters developed for specific activities in the WBS. The Association for the Advancement of Cost Engineering International (AACE) defines project management performance measurements and explains how to apply them. (AACE 2003)

Cost and schedule variance are usually given an action range of acceptability wherein an overrun or under-run is explained in a statement of the problem, cause, and corrective actions planned or taken to bring the specific activity into conformance. The variance analysis is self-correcting in that a correction is still accountable within the next reporting period -- if the correction is insufficient, the following reporting period will show a continuing variance, and require additional correction.

1. Cost and Schedule Parameters

BCWS	=	Budgeted Cost of Work Scheduled	What was planned to be done
BCWP	=	Budgeted Cost of Work Performed	What was done at budget rates
ACWP	=	Actual Cost of Work Performed	What was paid for the work
BAC	=	Budget at Completion	Original budget plus changes
EAC	=	Estimate at Completion	Actual cost plus estimate to complete
	=	(ACWP) + (BAC – BCWP)	
CV	=	Cost Variance	Earned work minus actual cost
	=	(BCWP) – (ACWP)	
SV	=	Schedule Variance	Earned work minus work planned
	=	(BCWP) – (BCWS)	
SPI	=	Schedule Performance Index	Measure of schedule performance
	=	(BCWP) ÷ (BCWS)	
CPI	=	Cost Performance Index	Measure of cost performance
	=	(BCWP) ÷ (ACWP)	

(The relationship between BCWS, BCWP, and ACWP is shown in Figure 9.1.)

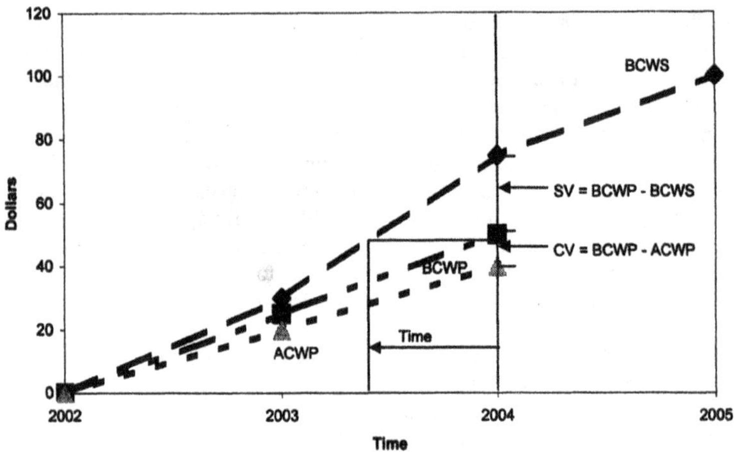

Figure 9.1 Relationships of BCWS, BCWP, and ACWP

A positive variance and an index of 1.0 or greater indicate favorable performance. The project manager will usually establish dollar, man-hour, or percentage thresholds requiring variance analyses. The thresholds may be based on the year-to-date BCWS, as follows:

BCWS < $250,000	%CV =	(BCWP – ACWP) x 100 ÷ BCWP	= ± 10%	and
	$CV =	(BCWP – ACWP)	= ± $15,000	
	%SV =	(BCWP – BCWS) x 100 ÷BCWP	= ± 10%	and
	$SV =	(BCWP – BCWS)	= ± $8,000	
BCWS ≥ $250,000	$CV =	(BCWP – ACWP)	= ± $25,000	
	$SV =	(BCWP – BCWS)	= ±$15,000	

(Note: The values are for illustration and do not represent recommendations. Management should establish the thresholds and ranges of variability for each project.)

2. Earned Value Parameters

EV = Earned Value (% complete) x (BAC for that account)

Overall = Overall Progress $\frac{\text{BCWP for all accounts}}{\text{BCWS for all accounts}}$

Performance measurements using the SPI and the Critical Path (CP) Total Float (TF) is another way to monitor project performance.

TF > 0

SPI > 1.0 Ahead of schedule on CP; more work being done than planned

SPI = 1.0 Ahead of schedule on CP; some shortfall in work on non-critical activities

SPI < 1.0 Ahead of schedule on CP; significant shortfall in work on non-critical activities

TF = 0

SPI > 1.0 CP on schedule; more work being done on non-critical activities than planned
SPI = 1.0 CP on schedule; total work volume as planned
SPI < 1.0 CP on schedule; shortfall in work on non-critical activities

TF < 0

SPI > 1.0 CP activities behind schedule; total work more than planned; indicating excess attention to non-critical activities
SPI = 1.0 CP activities behind schedule; total work volume as planned; indicating too much attention to non-critical activities
SPI < 1.0 CP activities behind schedule: total work less than planned; need more overall effort

3. Forecasting Parameters: There are three general methods for forecasting:

a. Forecast Costs EAC = (ACWP) + (BAC – BCWP)
b. Forecast Rate of Progress EAC = BAC ÷ CPI
c. Forecast Using Curves by extrapolation

9.9.2.4 Baseline Exposure Estimate: In a similar manner, a baseline occupational exposure estimate is prepared for each major activity at the beginning of the project, and actual exposure is recorded and compared against the baseline value for that activity. Variances against this baseline must be explained and corrective actions identified to bring the activities into conformance.

9.9.2.5 Other Control Parameters: Additional parameters may be established for monitoring and control of the project, depending on the activities to be performed. These might include generated waste volumes or weights, number of shipments of waste per reporting period, and the number of reportable safety accidents or incidents. The number of such additional parameters should be evaluated relative to the anticipated risk associated with a specific activity and the overall effect on completing the objective in a timely and safe manner.

9.10 CLOSEOUT – SITE RESTORATION

Removal of residual radioactivity to site clearance levels, followed by an independent final site survey and verification, qualifies a facility for license termination from a radiological standpoint. However, the remaining nonradioactive systems and structures still constitute a potential safety hazard and thus a liability for the owner/licensee. The destructive methods by which components are removed to facilitate decommissioning typically leave structures in an unsafe condition for long term maintenance and surveillance. Many components, such as large tanks, feedwater heaters, moisture separator/reheaters, steam generators, and pressurizers are removed in one piece to simplify handling and transport for disposal. Building walls and roofs are often cut open to permit heavy crane access for a direct lift from the building and placement onto a transport vehicle. If the owner/licensee has no intended reuse of these buildings, it may not be economical to repair or replace their walls and roofs. Maintaining the structures for an indefinite period of time can be an expensive option, and the potential for break-in by vandals and possible injury can lead to lawsuits and negative public image. A less costly option may be the immediate demolition of these structures to make the site safe to the public.

9.10.1 Regulatory Guidance: There is limited regulatory guidance with respect to demolition of structures following removal of radioactivity. Two regulatory provisions provide guidance for the owner/licensee to evaluate building demolition and site restoration. One is the Uniform Building Code; the other is the collection of the state's individual liability regulations for property transfers.

1. *Uniform Building Code, or BOCA Code*: The Uniform Building Code, often taken from the BOCA Code (BOCA 1996), is adopted or modified by each state and includes provisions for unsafe structures. According to the BOCA Code, there is a right to deem unsafe as:

   > All buildings or structures that are or hereafter shall become unsafe, unsanitary or deficient in adequate *means of egress* facilities, or which constitute a fire hazard, or are otherwise dangerous to human life or the public welfare, or which involve illegal or improper use, occupancy or maintenance, shall be deemed unsafe buildings or structures. All unsafe structures shall be taken down and removed or made safe and secure, as the code official deems necessary and as provided for in this section. A vacant building, unguarded or open at door or window, shall be deemed a fire hazard and unsafe within the meaning of this code.

   Based on these considerations, the owner/licensee would either retain a DOC or hire demolition contractors directly to demolish and remove the remaining buildings and structures so that the site can be placed in a safe configuration until a new use is identified.

2. *Property Transfer Laws*: A number of states require industrial property owners to determine the potential hazards that may exist in buildings, or on land being offered for sale to another party. In general, regulations require that an existing owner notify the state Department of Environmental Protection (or equivalent agency) of the intended transfer of property, and to identify the hazardous substances, underground storage tanks, or spills prior to the transfer. New Jersey, for example has enacted the Industrial Site Recovery Act to regulate the transfer or closure of industrial property. The seller must prove there has been no discharge of hazardous substances, that any has been cleaned up, or, if the seller has no proof, obtain approval for a cleanup plan and remediation of the site. Noncompliance can result in civil penalties to the seller of $25,000 for each offense. If the violation continues, each day is considered an additional and separate offense. The seller is required to enter into a remediation agreement with the state, and should it fail to fulfill its obligations, the state will hire contractors to perform the work and require the seller to reimburse the state.

   An owner/licensee who suspects there are hazardous substances may determine that it is less costly to remove such substances, demolish the structures, and remediate the soils while staffing is still available. This evaluation should be part of the cost-benefit analyses performed to determine the end point of the facility.

3. *Facility Demolition and Site Restoration*: If there is no hazardous/toxic (or radiological) material on site, the owner/licensee (or DOC) may contract a conventional demolition contractor for the structures. This may result in lower cost since equipment would not become contaminated and there would be no need for radiological training, protective clothing, or respiratory protection (accept for dust control). These decisions should be documented in appropriate cost-benefit analyses based on fixed-price bids from contractors.

9.10.2 Facility Closeout and Documentation: Following final necessary building demolition and site restoration, the owner/licensee should document the overall site decommissioning, license termination, and restoration process followed. This documentation should be recorded in the local town or state agency, and in the land deed. A final report of the project serves as a record of what work was performed and how the facility meets all current regulations for license termination.

9.10.3 Disposition of the ISFSI: If the owner/licensee constructs an ISFSI, it will be responsible (and liable) for its maintenance and surveillance until all the fuel is transferred to a federal repository. Based on the government's queue for acceptance of fuel from each specific installation, this period can extend for twenty years or more, thereby essentially prohibiting sale of reuse during this period. Proof of financial ability must be provided to the satisfaction of regulators. The ultimate ISFSI decommissioning will require the decontamination of any activated or contaminated materials and the demolition and removal of storage containers and the concrete storage pad.

Chapter 10

CHARACTERIZATION PLANNING AND SURVEYS

10.1 INTRODUCTION

This chapter describes the collection of information needed to plan the decommissioning of a nuclear facility. Characterization is frequently thought of as synonymous to radiation measurement, but the term has a broader meaning. Characterization is the collection of all information needed to describe, in adequate detail, the following:

- The hazards present at/in the facility
- The condition of the facility structure as it may affect worker health and safety
- The extent, nature, and concentration of radiological and hazardous chemical contamination
- The institutional, legal, and technical restraints on decommissioning alternatives

Characterization is an iterative process that precedes and parallels actual decommissioning. The characterization program discussed here is in support of decontaminating structures to protect workers and future site occupants, but this effort is intimately related to the consideration of ground contaminants, beneath and around the structure, and their potential travel to off-site environs.

10.1.1 Overview: This chapter begins with a discussion of the process of gathering information. The primary focus is on identifying the types of information important to planning, and how to obtain that information. This is followed by a discussion on characterization. Next is an overview of measurement techniques for quantifying contaminants, including field measurements and laboratory analysis of samples. The final section addresses release criteria, how to establish a standard for determining when the facility has been sufficiently cleaned, verification that the standard has been met, and how to clear materials and structures for unconditional release.

10.1.2 Definitions: The following technical or unusual terms are used elsewhere in this chapter.

1. *Aliquot* -- A specific fraction of a sampled material
2. *Analyte* -- The substance that the analysis is seeking to identify or quantify
3. *Bulk contamination* -- Contamination distributed within the volume of the host material, as distinguished from surface contamination. It frequently arises either from neutron activation or from physical transport process, such as diffusion, adsorption, or capillary action.
4. *In-Situ* -- In the field; in place; at the location of interest. Used to distinguish measurements made at the facility from laboratory measurements on samples.
5. *Isopleth* -- A graphical plot map or physical diagram of lines of constant value. Common ones are the isobars (lines of constant pressure) shown on weather maps; herein it refers to a plot of concentration of contaminates or dose rate.
6. *Planchet* -- A flat metal disk that holds samples for radioactive measurements.
7. *Source term* -- The amount of radioactive or hazardous chemicals available.

10.1.3 Characterization: Characterization is integral to the decontamination process. The aspects are, in the order performed, safety, scoping, preliminary, detailed, in process, and confirmatory.

1. *Safety*: a characterization encompassing precautionary investigations to determine the magnitude and nature of residual hazard. In addition to contamination, the structural condition of the facility is investigated, the hazard of occupancy for cleanup work is characterized, and appropriate information is gathered to plan structural work. This is a necessary first step, particularly for old, abandoned facilities.

2. *Scoping*: This characterization constitutes an historical site assessment (HSA) to gather all available information describing the extent of contamination. Process knowledge helps estimate the source term. Records of routine monitoring and special surveys help identify problem areas, and contaminant levels. Much of this information can be drawn from interviews with supervisors, workers, and protection staff who know about contamination during actual facility operation, particularly off-normal events.

3. *Preliminary*: The characterization goal is to discriminate among options for corrective action. Preliminary characterization and scoping are input to preparing a detailed plan. It usually includes brief surveys to confirm and fill gaps in historical information. Previously unmeasured locations, materials, or contaminants may have to be characterized. Measurements may have to be more sensitive than that of scoping studies, particularly as related to differentiating bulk from surface contamination.

4. *Detailed*: Detailed characterization helps define decommissioning activities in enough detail to prepare a definitive baseline in terms of scope description, and estimated schedule and cost. Detailed radiological and hazardous chemical characterization is a primary element of the project plan and schedule. Confirmatory characterization designed in conjunction with the detailed characterization can be cost-effective, such as by the use of similar sampling grid and protocol. Contaminants are measured at concentrations to ensure meeting project requirements. Surface contamination is measured thoroughly so as to not overlook localized elevated levels. Bulk contamination is determined for the structure, storage, and equipment, so as to meet dose limits on applied scenarios. Monitoring challenges include ventilation, sanitation, and power systems, particularly in ducts, chases, vents, drains, and pipes.

5. *In-Process*: A significant amount of characterization is necessary, such as comparing before-and-after measurements to determine if the decontamination goals are being achieved. It serves to assess progress, to determine whether the decontamination processes being used are working, and to ascertain when the regulatory release criteria are being reached.

6. *Confirmatory*: A final characterization demonstrates compliance with acceptance criteria. It must be sufficiently thorough and documented to ensure reliable data, and to provide convincing evidence of compliance with regulations.

10.1.4 Sources of Information: Discussions on various aspects of characterization planning are presented throughout this handbook. Various DOE sources expand on some of the topics presented here, particularly chapter 7 of the DOE's 1994 Decommissioning Handbook (DOE 1994), and sections 3.8 to 3.11 of its 2000 Handbook (DOE 2000c). DOE's Decommissioning Implementation

Guide (DOE 1999b) presents basic information on performing preliminary characterization and hazards analysis, developing a characterization plan, and the actual characterization plan.

The IAEA maintains a series of technical reports on decommissioning. A 1992 report addresses monitoring programs (IAEA 1992). Another addresses various aspects of measurement, including a detailed list of literature references (IAEA 1998). The ASTM also addresses preparation of characterization plans (ASTM 1997a), and a recent standard guide (ASTM 1997b) addresses the selection of portable radiological survey instruments, including such factors as counting efficiency, geometry, and radiation attenuation. A similar approach is taken in NUREG-1507 (NRC 1998a), which also addresses detection limits. This report, however, evaluates various types of detection instruments for both laboratory and field monitoring. Decommissioning of structures not necessarily associated with radioactivity is also examined (Briggs 1997).

Several federal agencies cooperated in a Multi-Agency Radioactive Site Survey and Investigation Manual (MARSSIM) involving radiation surveys for free release of sites and structures (NRC 2000a). This guide, discussed in detail in Chapter 11, emphasizes the data quality objectives (DQO) approach, and calculation techniques for selecting sampling patterns. The NRC has issued a manual for radiological surveys (NRC 1992a) and a related guide for decommissioning plans (NRC 2000b). The EPA has also prepared guidance for the DQO Process (EPA 2000a).

The National Council on Radiation Protection and Measurements (NCRP) has issued pertinent reports, including one on the calibration of survey instruments for radiation fields and surface contamination (NCRP 1991), and recommended screening limits for contaminated soil (NCRP 1999) that is generally applicable to confirmatory characterization beneath and around structures.

10.2 GATHERING INFORMATION

The purpose of characterization is to help plan actions and make better decisions. This includes:

- Identifying and understanding the hazards present
- Defining the required work scope
- Selecting techniques for reducing or eliminating the hazards
- Monitoring progress towards meeting hazard reduction goals
- Documenting the nature and quantity of materials and waste generated to satisfy treatment, storage, and disposal facility acceptance criteria
- Verifying that cleanup standards have been met

As noted in Chapter 4, most of the references cited emphasize survey and measurement methods for characterizing contamination. While measuring contamination is a major part of characterization, there may also be is a wealth of other equally important information available. The history of a facility, for example, is frequently written in its operating records. In many cases, when a facility ceases operation, the records are discarded. If still available, they should be analyzed and preserved. The following describes different types of records that might be of value as part of an HSA.

10.2.1 Construction Records: If these are available, they can be valuable sources of information.

1. *Structural records* can help locate construction joints, expansion joints, and temporary construction openings. This helps plan demolition and removal of large equipment.
2. *Descriptions and photographs of large component installations* could be helpful in selecting rigging, letdown areas, and pathways for equipment removal.
3. *Photographs showing construction progress* could be helpful in planning demolition sequences and may indicate stable configurations of partial assembly. They also help determine hidden features, such as embedded objects, drain routing, and rebar density.

10.2.2 Design Records: A primary source of information about the building and how its contents are interrelated is the documentation produced during the design phase.

1. It is essential to seek available *engineering drawings*. If they have been lost or destroyed, they may need to be redone. The engineering design needed to support decommissioning is frequently centered on a small but essential set of drawings.
2. *System descriptions and functional specifications* indicate the intended use of operational components. They are particularly important if decontamination is to be achieved by using existing equipment or circulating chemicals in process equipment and lines.
3. Certain aspects of *purchase specifications, equipment drawings, and vendor documentation* serve a number of purposes (specially if for use during decommissioning, or salvaged):
    - Internal configurations, if the equipment needs to be decontaminated
    - Weight, lifting points, and foundation details during dismantling
    - Operating manuals, spare parts lists, and maintenance requirements
    - Details of construction, including materials of manufacture, if internal components will be dealt with separately (e.g., disposed of as hazardous waste).

10.2.3 Operating Records: Facility contamination normally occurs because of routine or abnormal events. These records help provide information about the nature and extent of contamination.

1. *Operating procedures* generally have been refined over the years to accommodate upset conditions that may not have been anticipated during the design. For example, process or contaminated fluids may have been dumped to a sump or pumped into another tank or system if the original facility design did not make adequate provision.
2. The primary purpose of keeping an *operating log* is to establish a record of all operational occurrences. This may reveal sources of contamination, spill duration, and cleanup measures taken (or not taken). A log will also contain names of personnel who may still be contacted.
3. *Survey reports* are the best available starting point for mapping the extent of contamination. Exposure records, ALARA studies, and other such documentation generated by the health physics (HP) department will provide valuable information about exposure levels, detection, decontamination techniques, and other details for dealing with contamination.
4. Records generated and maintained to satisfy the waste disposal site, such as *shipping logs and manifests*, will give an indication of waste generation and cleanup activities.
5. *Chemical inventory logs* are valuable for identifying leftover material in containers and for determining the presence of nonradiological contamination.
6. *Weekly, monthly, and annual reports* may summarize operational occurrences and highlight problems that occurred and what was done about them.
7. For NRC licensees, *10 CFR 50.75(g) record requirements* are important to the safe and effective decommissioning, such as spills and other unusual occurrences.

10.2.4 Safety Documentation: a quick way to estimate a facility's residual hazards.

1. *Fire hazards analysis* provide information about flammable loadings, fire barriers, detection and suppression provisions, egress routes, and details of construction that might be valuable for planning decommissioning. This information can be used to select or preserve personnel protection features, and to identify the presence of material (for example, asbestos fire barriers) that may require special provision.
2. *Emergency plans* contain information about the hazards that might result from upset conditions. It may be a starting lead for selecting specific actions to remove or abate hazards, such as draining tanks, removing sources of hazardous energy, or disposing of reactive chemicals. These plans may also describe notification requirements in the event of accident, and sources of medical, fire-fighting, or other emergency assistance.
3. *Accident investigation reports* document significant hazards that caused an accident or upset condition. This also helps provide additional insight into residual hazards. The root cause analysis and lessons learned typically reported in such documents may suggest ways of dealing with such hazards.

10.2.5 Licensing Documents: Regulatory agencies require documents showing that hazards to workers, the public, and the environment have been recognized and that appropriate provisions are in place to control them and mitigate the effects of an accident.

1. *Safety analysis reports*: Licensees must describe facility design in detail, systematically address possible accidents and upset conditions, and present design features that control the hazards or mitigate their effect. The Preliminary Safety Analysis (PSAR) describes the intended design and is the basis for granting a construction permit. It is revised and reissued as a Final Safety Analysis Report (FSAR) to describe the "as-built" condition. The FSAR and technical specifications are the basis for granting an operating license.

2. *Authorization basis documents*: The DOE established an authorization basis to regulate its own facilities. It places each facility into one of seven facility hazard categories to determine safety analysis and documentation requirements. Safety analysis is divided into hazard identification, hazard analysis, and accident analysis. Based on these analyses, documentation becomes part of the safety basis; once approved, the distribution and changes are controlled. The following safety documents might be used, depending on the facility hazard category and the age of the facility (when some listed were never used):

   - Process Hazards Analysis
   - Preliminary Hazards Analysis
   - Operational Safety Requirements
   - Technical Justification for Continuous Operation
   - Technical Standards
   - Test Authorizations
   - Unreviewed Safety Question Evaluations
   - Hazard Assessment Document
   - Health and Safety Plan
   - Basis for Interim Operation
   - Process Hazards Review
   - Safety Analysis Report
   - Auditable Safety Analysis
   - Technical Safety Requirements
   - Technical Specifications
   - Administrative Limits

3. *Regulatory agency permits*: The EPA and state and local authorities control a facility's releases and emissions by means of permits. The permits and supporting documentation are sources of information for planning characterization.

Some facility documentation at government facilities (and some equipment) may have been classified. For those, it is important to understand the limits that classification places on freedom of action. It will be necessary to have at least some personnel cleared to a level sufficient for appropriate review. Although the ideal option is to destroy the classified features of the material, or declassification, frequently, the easiest alternative is to find an authorized user and transfer custody. Classified records need proper review, storage, handling, and disposal by an authorized official.

10.2.6 Personal Testimony: Much information exists only in the minds of workers and managers. The most obvious source is in present and past facility staff, but others may have undocumented valuable knowledge. Information may not have been recorded because it was overlooked or considered unimportant, too routine to document, or a source of embarrassment.

This type of information may be difficult to gather but can be very valuable. Individuals with the desired knowledge may have moved on, retired, died, or simply forgotten. The more time passes, the more difficult it will be to collect and the more incomplete or inaccurate it will be.

1. *Faculty/Staff*: Major sources of anecdotal information are those who worked at a facility. Ideally, persons who know the facility well will help plan decommissioning. The operating organization may be consulted for present or former key workers. Frequently, former staff keeps in touch with each other and provides leads for contacting colleagues.

2. *Regulators*: Regulators may be a significant source of information. State regulators and the regional offices of federal agencies may have active files and staff familiar with the facility and its hazards. Records of oversight inspections and enforcement actions may provide insight into facility conditions that elevated regulatory visibility.

3. *Public*: Members of the public may have participated in advisory committees, or have an express detailed interest. Their credibility depends on how direct the information might be. If the information is correct, their insights may represent a different viewpoint. Public feedback also may illuminate weaknesses or gaps in the direct body of information that should be an integral part of the cleanup process. In particular, old newspaper articles and broadcast-video libraries may provide some details or insights into unusual occurrences.

10.2.7 Walkdown Inspection: A fundamental element of any initial site assessment is a walkdown inspection, during which a team of trained observers carefully notes the condition of the facility. A visual inspection may have to be repeated if observations require clarification during estimating, planning, or design. A systematic walkdown inspection should include the following elements:

1. Participants should be trained observers with detailed knowledge of requirements in their areas of expertise (fire protection, structural, industrial safety, etc.) and enough experience to recognize abnormal conditions.

2. Inspections should be planned in advance. Any possible hazards facing the inspectors should be recognized and corrected. An inspection checklist, as shown in Table 10.1, is a good tool for this purpose. Each item should be checked off for:
   - Obvious deficiency - Requiring correction
   - Questionable, further investigation required
   - No obvious deficiency - Remarks/Observations

3. The inspection should be monitored and supported by persons not on the walkdown team. In the event of an accident, immediate response and assistance should be available. Some hazards may not be identified until the walkdown is completed.

4. The results should be documented thoroughly. Data summaries, pictures, maps, and diagrams can make the information more useful.

Videos and photographs are good supplements to visual observations in a walkdown. Hazards generally are higher in a facility that has been idle for some time than in a similar facility that is still staffed and operating. Consequently, the number of entries should be minimized. A photographic record can be a satisfactory substitute for a second walkdown. Photographs and videos may show hazardous conditions in a more clearly and easily understood manner than mere verbal descriptions. In addition to their usefulness as a characterization tool, they can document achievements and results of the decommissioning effort.

A radiation protection technician should be the first person in the walkdown team to enter suspected radiation areas and perform a survey to ensure safe entry. The technician may also perform field measurements to verify previous records and postings, to map out the extent of radiological contamination, and to determine whether it has spread since the last time it was measured.

10.2.8 Sampling and Analysis: The walkdown inspection described in the previous section relies on visual observation and portable instrument measurements. This is not sufficient to fully characterize hazards. It will be necessary to take samples of materials in the facility and have them analyzed to determine constituents and concentrations of contaminants. Some of the types of samples that may be appropriate are as follows:

1. *Process Vessel Contents*: To clean out tanks and piping, it may be necessary to first sample the contents. Many processes involve bulk materials that include chemicals in unusual form. These may change drastically when the operational process is ended. For example, suspended solids form sludge, components "plate out" and become bonded to surfaces, and chemical changes occur with temperature changes or exposure to air.

2. *Stored Material*: Identifiable material must be characterized according to its hazard potential. If the label is illegible or lost, the material must be identified. Confirmatory sampling may be needed to correct mislabeling.

3. *Components of Unknown Composition*: In old facilities, the materials of construction and component parts may differ significantly from what is accepted today. Examples are lead paints, electrical insulation containing PCBs, mercury switches, and asbestos insulation. Sampling may be the only way to determine whether such materials are present.

Table 10.1 Walkdown Inspection Checklist

MATERIAL CONDITION OF STRUCTURE

1 Footings and Slabs on Grade
- a Uneven settling
- b Surface cracking

2 Foundation Walls, Retaining Walls
- a Leaks
- b Cracks, holes, deformation
- c Paint or coatings
- d Drainage system

3 Excavation/Backfill
- a Erosion
- b Improper grading
- c Settlement
- d Drainage ditches, swales
- e Drainage collection, dispersion, runoff

4 Beams, Columns, Bracing (Steel, Wood, Concrete)
- a Cracking
- b Misalignment, excessive deflection, buckling
- c Impact damage
- d Protective coating/corrosion
- e Loose or missing connections or damaged welds
- f Hoist, crane, elevator support structure and connections

5 Interior Floors
- a Cracking
- b Deflection
- c Protective coatings
- d General damage
- e Floor drainage

6 Roof Structure
- a Corrosion or deterioration
- b Leaks
- c Protective coatings
- d Cracking, buckling, or damage
- e Loose or missing connectors
- f Damaged welds
- g Parapets
- h Fire barriers
- i Bracing and bracing connectors
- j Secure lightning protection system

7 Roofing
- a Membrane or surface cracked, buckled, punctured
- b Leaking
- c Roof penetrations damaged or leaking
- d Drainage (scuppers, drains, standing water, etc.)
- e Plant/algae growth

8 Access Ways/Stairs
- a Corrosion or deterioration
- b Cracking
- c Deformation, buckling
- d Damaged welds or connections
- e Protective/Anti-Slip coatings
- f Handrails
- 6 Leakage or contamination in elevator pits

9 Walls-Exterior
- a Cracking, buckling, deformation
- b Leaks
- c Protective coatings
- d Siding - corroded or damaged
- e Doors damaged or not functional
- f Windows
- g Connection to building structure or foundation
- h Material (e.g., transite)

10 Interior
- a Walls - damage, cracking, buckling
- b Protective coatings
- c Doors, windows, other penetrations
- d Sub-Ceilings
- e Floor finishing or coverings
- f Fixtures (electrical, mechanical)

11 Mezzanine or Platform Structures
- a Corrosion, decay
- b Loose, missing, or damaged connectors
- c Protective coatings
- d Cracking, buckling, or deformation
- e Stairs, ladders, handrails
- f Beams, columns, bracing
- g Equipment supports and connections

ADMINISTRATIVE CONTROLS

12 Security Gates, Entrances, and Fences
- a Corrosion
- b Sagging, torn fencing, splits
- c Missing or broken hardware or components
- d Impact damage
- e Entry points locked
- f Facility custodian identified and posted
- g Appropriate radiological postings
- h Entries controlled and logged; entrants briefed on hazards and tracked

Table 10.1 Walkdown Inspection Checklist (Continued)

13 Surveillance and Maintenance
- a Surveillance program in place
- b Surveillances up to date and documented
- c Essential equipment identified
- d Maintenance program for essential equipment in place
- e Equipment maintenance current and documented

FIRE PROTECTION

14 Fire Detection Systems
- a Detector system intact
- b Detectors still operable
- c Detector testing up-to-date
- d General condition of components
- e Supports and connections

15 Fire Alarms
- a Alarm system intact
- b Alarms still operable
- c Alarm testing up-to-date
- d General condition of components
- e Supports and connections
- f Alarms notify remote response station

16 Fire Suppression Systems
- a Sprinkler system intact
- b Sprinklers operable and maintained
- c Fire water pressure provided to sprinkler system
- d Standpipes in good condition and valves operable
- e Halon/CO_2 discharge systems intact
  - Intact, operable, and maintained
  - Charged and pressurized
  - Alarms/safety delays operable, tested, maintained, and documented
- f Portable extinguishers
  - Present, charged, and with current inspections
  - Appropriate to type of fire hazard (Class A, B, etc.)
  - Locations clearly marked, unobstructed, and visible
  - Mounting brackets sound

17 Fire Safety
- a Two means of egress provided from all spaces
- b Egress paths unobstructed and clearly marked
- c No accumulations of combustible material
- d Emergency lighting present and operable

PERSONNEL SAFETY HAZARDS

18 Slips, Trips, Impacts and Falls
- a Slippery floors
- b Loose or missing stair treads, ladder rungs, or handrails
- c Irregular walking surfaces
- d Unguarded floor openings
- e Bent, unsecured, or missing deck plates
- f Low overhead clearance ("head knockers")
- g Low floor obstructions, pipe, or other components in walkways ("ankle biters")
- h Moving machinery without guards or cover

19 Respiratory Hazards
- a Confined space (suffocation hazard) not properly identified
- b Unsatisfactory air quality due to inadequate ventilation
- c Airborne pathogens (mold, fungus, legionella, spores)
- d Toxic fumes from stored material.
- e Polychlorinated biphenyls (in paint, electrical insulation, plastics)
- f Other RCRA hazardous chemicals present
- g Chemicals stored properly (labeled, sound container, inventoried)
- h MSDS available for all chemicals
- i Service gases (acetylene, argon, nitrogen, etc.)

20 Living Organisms
- a Evidence of vermin (insects, rodents)
- b Evidence of animal infestation (snakes, birds, raccoons)
- c Toxic plants (poison ivy)

21 Hazardous Materials
- a Asbestos (siding, bulk insulation, gaskets)
- b Mercury (switches, ballasts, fluorescent bulbs)

22 Hazardous Energy
- a Electric circuits energized
- b Steam present under pressure
- c Steam leaks
- d Compressed air or other gases
- e Stored mechanical energy (springs, counterweights, flywheels)

SYSTEM FUNCTIONALITY

23 Heating, Ventilating, and Air Conditioning
- a Air handling equipment
  - Operability and condition
  - Supports and connections

Table 10.1 Walkdown Inspection Checklist (Continued)

- b Steam System
  - Operability and condition
  - Leaks
  - Supports and connections
  - Insulation
- c Chilled Water
  - Operability and condition
  - Leaks
  - Supports and connections
  - Insulation
- d Ductwork and Accessories
  - Integrity, damage, corrosion
  - Supports and connections

24 Electrical
- a Corrosion of enclosures
- b Electrical equipment exposed to moisture or water
- c Electrical equipment not secured, loose, or damaged
- d Wiring, conduits, or raceways damaged
- e Switches or controllers damaged or not functional

25 Lighting (Exterior, Interior)
- a Damaged or non-functional fixtures
- b Inadequate light for facility
- c Switches damaged or not functional

CONTROL OF PROCESS MATERIAL

26 Distribution of Remaining Process Material
- a Bulk quantities stored in facility
- b Residual material in tanks and piping
- c Spilled material (locations)
- d Material held up in ventilation

27 Evidence of Contamination
- a Discoloration on floors, walls, or other surfaces
- b Vapors detected by sniffers; unusual odors
- c Postings and other signs and labels
- d Evident corrosion or deterioration of pipes, vessels, glove boxes, etc.
- e Open or broken containers

4. *Core Bores*: A common problem during decommissioning is concrete that has become contaminated within its volume rather than just on the surface. This can occur by neutron activation or diffusion from the surface. Core drill samples are used to determine the ultimate depth of contamination. Although less common, core drill samples can also be taken on other structural materials, such as earthen floors or composition walls.

5. *Smear Samples*: The removable portion of surface contamination is transferred to a wipe (smear sample) that is then analyzed in a laboratory environment. This is one of the primary techniques of radiation characterization through all phases of decommissioning. Samples also may be collected by chipping or scraping corrosion material, paint, or dirt.

6. *Air, Soil, and Water Samples*: Airborne particles are collected on filters to measure material resuspended from surfaces, e.g., dust. Gases are collected to measure volatile contaminants. Water samples are collected to measure dissolved and suspended contaminants that remained in process vessels and piping, or entered the sanitary system. The immediate surroundings should be investigated to determine whether contaminants have escaped from the facility.

10.3 CHARACTERIZATION PLANNING AND IMPLEMENTATION

10.3.1 Planning Approach: Decommissioning can encompass a variety of actions, ranging from an uncomplicated cleanup of a simple facility, through a complex deactivation, to a complete dismantling and site restoration. Each project will be different, but in each, sufficient information must be obtained to plan the work scope, select a remedy, commit the resources, monitor progress, and verify that the work has been satisfactorily completed. Since characterization can be costly, one of the challenges is to perform it appropriately for proceeding to the next step, without collecting excess information. A systematic approach to characterization should be followed for all projects.

The first step is to gather all existing pertinent facility information. The information is used to describe conditions germane to the potential activities. For instance, a project that will alter the structure (such as by partial dismantling) will need a summary of the details of construction, layout, and overall condition. Decontamination will require knowledge of the composition of contaminants, their physical distribution throughout the facility, and the underlying materials.

The next step is to define the nature of the proposed facility transformation. The preliminary plan will not describe completely what needs to be done or the options to select for alternative actions. Existing information may be sufficient to choose the best method for removing a component or decontaminating a room, but frequently the information to make that decision is unavailable. For example, the depth of contaminant penetration must be determined to decide whether to clean surface of a concrete floor or to plan complete excavation and removal.

After comparing the available information to the planned actions and possible alternatives, it is necessary to fill the information gaps required for choosing a course of action, designing a solution, and preparing definitive work plans and cost estimates. A systematic definition of such missing information is fundamental to the characterization effort. For each gap in information, identify:

- The exact parameters that are not known
- The anticipated application
- The effort to obtain minimal sufficient characterization

After the historical data search, visual inspections, and preliminary characterization of the facilities are completed, the available data are evaluated to select the project scope, and to inform the regulators and stakeholders. The data should be analyzed and presented to include addressing the alternative impact of doing nothing instead of proceeding with the work scope.

The preliminary baseline risk should be evaluated to support these initial communications. Risk assessment evaluates factors that influence the health risk, in order to select actions that protect public health. How to conduct to risk assessment in the pre-decommissioning phase is described in more detail in Chapter 7. In general, a risk assessment involves the following steps:

1. *Hazard identification*: Defining chemical or radiological contaminants and the potential detrimental effects on workers, the public, or the environment.

2. *Source characterization*: Identifying the nature and extent of contamination.

3. *Exposure assessment*: Determining the pathways the contaminant could follow through the surroundings to result in exposure, such as ingesting drinking water contaminants or inhaling air contaminants.

4. *Dose response assessment*: Identifying the effects of contaminant concentrations in the exposures to maximally exposed individuals, a critical group of individuals, and the body's most sensitive organs.

5. *Risk characterization and uncertainty analysis*: Evaluating the predicted doses, taking into account the probability of exposure for each decommissioning work scenario (including the no-action option).

This preliminary evaluation should also identify any gaps in knowledge that would control the choice of actions to reduce the hazards. Any assumptions that were made during these preliminary evaluations should be identified and validated by additional characterization.

The next major step in planning is to determine the best methods to use for field measurements and surveys, and the type of sample analyses needed. In selecting measurement methods, it is important to choose the appropriate number of measurements. Using the DQO process, defined by the EPA and further detailed in Section 10.3.4, is suggested as a method of determining sapling sites.

Once the characterization plan is prepared, it's time to set the organizational structure, schedule worker training, and to designate points of contact with the owner, and the safety and security organizations. Security classification may have to be considered for those entering certain facilities and structures. Briefing sessions for regulatory and public stakeholders should be scheduled before and during decommissioning to communicate the process and define the end-point.

For the safety and scoping characterizations, it's usually appropriate at this point to go into the facility and obtain missing information. A more formal process will be needed for the preliminary, detailed, and confirmatory characterizations. This will involve preparing a written plan and submitting the plan for review by others, such as the regulatory authority and/or the waste disposal facility. The regulator wants to ensure that the data are collected in a sufficiently rigorous manner to support planning remedial actions, and to support conclusions that the work has been completed satisfactorily. The waste disposal facility needs to receive material of known composition in a manner that is compatible with the facility's license and its own operating parameters.

10.3.2 Characterization Implementation: Various different characterization surveys are used to determine the amount and location of radiological or hazardous chemical materials present at any given point in time. The types of surveys and their purpose are presented in Section 10.1.3. Two of the more significant surveys are the initial characterization to detail the magnitude and nature of the hazards that the task force will be subject to, and the final (confirmatory) survey that ensures the cleanup criteria specified for the project were actually met.

Detailed characterization is generally a major effort, carefully planned in advance of most facility work. A written plan is advisable, and is frequently required by regulation. It may be contracted out to another organization, but remains a key document that may be required to demonstrate that the resulting waste meets acceptance criteria. Each facility is unique but the written characterization plans will cover similar ground. Table 10.2 presents the outline of an adequate characterization plan. MARSSIM provides extensive guidance on:

Table 10.2 Suggested Characterization Plan Outline

Introduction and Executive Summary

Part A. Characterization Needs

1. Facility Description
2. History of Operations
3. Previous Characterization and Other Available Information Summary
4. Evaluation of Existing Conditions
5. Missing Information (and why it is needed)
6. Important Factors Influencing Characterization
    - a. Radionuclides of Interest and Associated Limits
    - b. Hazardous Materials of Interest and Associated Limits
    - c. Waste Disposal Plans and Disposal Site Acceptance Criteria
    - d. Measured, Estimated, or Calculated Inventory of Waste Constituents
    - e. Scaling Factors or other Deductive Methods Used When Direct Measurement is not Possible

Part B. Implementation of Characterization

For each of the distinct methods of characterization activities listed in B.1, include a discussion of:

- a. Locations in the facility where that method will be used
- b. Specific instruments or other equipment that will be used
- c. Planned number of measurements in that location
- d. Disposal path for secondary wastes, if any, arising from characterization activities

1. Characterization Methods
    - Visual Inspections and Photographs
    - Gamma Radiation Level Measurements
    - Beta-Gamma Scan Surveys
    - Alpha Surveys
    - Smear Surveys
    - Water/Sludge Samples
    - Surface/Subsurface Soil Samples
    - Embedded Piping/Component Samples
    - Concrete Samples
    - Metal Samples
    - Paint Samples
    - Ultrasonic or Infrared Liquid Level Determination
    - Samples of Materials of Unknown or Suspect Composition
2. Assurance Samples and Surveys Are Representative
3. Sequence and Schedule of Survey Activities
4. Characterization Organization and Management
5. Health and Safety Considerations for Persons Engaged In Characterization

Part C. Quality Assurance Plan

1. Planning and Control
    - a. Operational Chain of Authority
    - b. QA Procedures
    - c. Documentation and Reporting Requirements

Table 10.2 (Cont'd)
Suggested Characterization Plan Outline

2. Data Quality Objectives
    a. Precision
    b. Accuracy
    c. Representativeness
    d. Completeness
    e. Comparability
3. Storage, Handling, and Shipment of Samples
    a. Sample Identification and Labeling System
    b. Preservation and Holding Time
    c. Storage and Handling
    d. Shipment
4. Chain of Custody
5. General Controls
    a. Standards, Controls, and Blank Samples
    b. Relating Sample ID to Collection Location
    c. Preservation (and Archiving) Samples
6. Calibration Procedures and Frequency
7. Analytical Procedures
8. Data Reduction, Validation, and Reporting
9. Performance Oversight and Audits

- Radiological characterization, notably the development of derived concentration guideline levels (DCGL) for comparison with measurements
- Defining the characterization program in terms of DQOs
- Identifying contaminant release pathways
- Maintaining QA for all measurements

If the acceptance criteria for residual contamination have not been met, further removal of contamination will be necessary, possibly using a different method. In-process surveys are iterative, and help to decide whether to continue or change techniques. For volumetric contamination, where removal of a portion of the surface (e.g., by scabbling) has been planned, surface measurements may indicate places where the contamination has penetrated deeper than originally estimated, allowing removal of additional material to continue. Chapter 11 provides further discussion on in-process (adaptive) sampling and analysis, and specific computer software to aid in the process.

The technologies incorporated into any aspect of the project depend on the radiological and/or chemical status of the source term. Selection will be strongly influenced by source characterization data. The project's selected technology must support the project's ALARA goals by limiting the exposure to the workers while executing the decommissioning work. Generally speaking, technologies are selected to achieve multiple project goals, including:

- Performing the work in the most cost-effective manner
- Completing the project as rapidly as possible
- Minimizing any negative safety and health impacts to the workers

- Minimizing off-site environmental impacts
- Minimizing the generated volume of waste
- Minimizing the effects of any residual source term on future occupants

After the work has been completed, final measurements must demonstrate that the goal has been achieved and that the remaining contamination levels meet the project's acceptance criteria. The confirmatory surveys, done to the same protocols and grids as were used for the detailed survey, confirm the final results. MARSSIM details how to perform such surveys.

These surveys are frequently performed by the organization responsible for decommissioning, and a separate, parallel, and independent set of samples and surveying is performed by the regulator. For an NRC-licensed decommissioning, the licensee performs the license termination survey and the NRC typically performs a confirmatory survey. This practice may be adjusted by the agency for smaller sites or projects. NRC rules for termination surveys are found in 10 CFR, Parts 30.36, 40.42, 50.82, 70.38, and 72.54.

The DOC performs the final surveys in accordance with the DOE's Decommissioning Implementation Guide, and the DOE or its independent contractor performs an independent verification survey.

10.3.3 Quality Assurance: The QA program depends on the correlation between measuring actual performance in relation to specified standards, and the relevance of the standards to the outcome. All QA-related activities are described in a formal QAPP. QA for characterization by survey requires periodic calibration and maintenance of monitoring instruments, checking data entries (measurement and location) by selective replication, and tracking records with regard to instrument, operator, location, and date. A reliable background pattern must be established for radiation, radionuclides, or other contaminants, based on earlier monitoring at the contaminated location or monitoring at similar locations. Application of calibrated monitoring instrument response must be evaluated, whether to radiations from the actual radionuclides at the site or airborne effluents from the actual chemical contaminant. The quality and frequency of data related to the extent of contaminants at each location must be checked for thoroughness and documented.

The major contaminants in a structure are readily identified at most facilities. Following a guide such as MARSSIM can provide a simple path to unambiguous characterization of areas acceptable for release and those that need decontamination. In contrast, difficulties may arise in circumstances where the contaminant:

- Limit is near the detection limit
- Limit is near the ambient background
- Levels are wildly variable
- Is difficult to find (e.g., bulk contamination)
- Is difficult to monitor (e.g., emits only alpha or weak beta particles)
- Is mobile.(e.g., polonium, tritium)

Each of these circumstances requires special characterization efforts based on meticulous QA to distinguish between areas suitable for release and those that require cleanup.

The sample collection plan must be evaluated for its ability to represent the extent of the contaminant with respect to area and concentration. Techniques for sample collection must meet project specifications, and sample preservation must meet criteria for maintaining radionuclides and other contaminants in their original chemical and physical forms. A chain of custody tracks the sample from collection through analysis and data reporting.

A major concern for in-situ measurements and sampling is the degree to which they represent the entire area being characterized. The monitoring instrument may respond to various nearby and distant radiations and record a value that is not necessarily an average radiation flux or dose rate. A surface sample may not match the area that it represents. MARSSIM presents calculations for selecting the number of measurement or sampling sites that would provide a proposed degree of reliability in reporting measurement results. More or less rigorous patterns may be applied to various regions to obtain reliable values at levels near the limiting values or less reliable values far below such limits. As part of preliminary characterization, brief studies should identify and locate areas with distinctive differences in contamination, and estimate, based on measurement or sampling, the extent of exposure or concentration fluctuations within relatively small distances.

Chemical and radiochemical analyses must use written procedures and be performed by qualified analysts under acceptable laboratory conditions. Method reliability must be tested periodically by analyzing replicates, blanks, samples with known contents, and inter-comparison samples provided by independent laboratories. The analyst must maintain a notebook for all analytical activities, including reagent preparation, sample analysis, QC actions, and training.

Instrumental processes, including radiometric, and their operators are subject to the same QA requirements as are chemists and their analyses. In addition, the detection instruments must be checked at frequent intervals for background and operating reliability. Any background fluctuations due to airborne contaminants (such as radon progeny) or facility operation must be considered, as well as occasional cross-contamination due to other samples. For radionuclides, the relation of count rate to activity (the counting efficiency) depends strongly on photon or particle energy, amount of sample, and sample density. Therefore, such calibrations must be performed over the range of cited factors. And, since sample concentration and purity are also major factors in instrumental response for chemical analyses, calibrations also must be performed over the range of encountered sample characteristics. The reported data usually are qualified in terms of uncertainty (e.g., standard deviation) in measurement or minimum detectable levels. These levels are estimated from counting statistics.

10.3.4 Data Quality Objectives: The DQO process is a planning tool for designing data collection efforts. The basic idea is to identify the question to be answered by the data, then design the data collection and analysis process to provide an answer to that question. The process, presented in detail in an EPA document (EPA 2000a), is a seven-step planning approach to develop sampling designs for data collection activities that support decision-making. This process uses systematic planning and statistical hypothesis testing to differentiate between two or more clearly defined alternatives. A summary of the seven steps is presented in Figure 10.1.

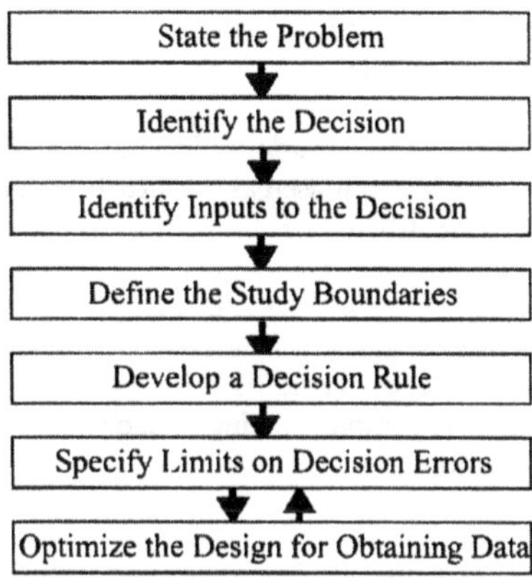

Figure 10.1 DQO Process

The DQO process is iterative and allows incorporation of new information and modification of output from previous steps as input for a subsequent step. Although the principle of systematic planning and the DQO process are applicable to all scientific studies, the process is particularly designed to address problems that require a decision between two clear alternatives. The final outcome is optimizing the design for collecting data (e.g., the number of samples to collect, and when, where, and how to collect), together with limiting the probability of making decision errors.

10.4 SURVEYING, SAMPLING, AND MEASURING TECHNIQUES

In every case, characterization combines surveys and samples. Surveys are measurements made at the location of the contamination with portable instruments chosen for their ability to detect the expected contaminants. Samples are taken and sent to a laboratory facility when the contaminant is suspected but unknown, when portable instruments are inadequate, or cannot differentiate between multiple constituents. Sampling is used for bulk contamination because most portable survey instruments are only capable of surface measurements.

10.4.1 Hazardous Materials: Nonradiological hazardous materials are generally categorized as:

1. RCRA Characteristic: Materials that are hazardous by virtue of characteristic properties (flammability, corrosivity, reactivity, toxicity), as defined in 40 CFR 261 Subpart C

2. RCRA Listed: Specific substances listed as hazardous materials in Subpart D

3. TSCA Regulated: Substances that are regulated by the Toxic Substances Control Act (TSCA 1976), including asbestos, lead paint, radon, and PCBs

The handling, packaging, treatment, storage, and disposal of these materials are covered by complex regulation. They are described in more detail in Part 2 of this handbook. This section discusses the methods for detecting and quantifying those materials, both by field surveys with portable equipment, and by laboratory analysis of samples.

Hundreds of substances have been identified as hazardous. The key for characterization is to recognize or anticipate those that could have been used as construction materials or as process chemicals. Hazardous materials can be identified by examination of construction and equipment records, review of process history and records, and visual inspections. This identification effort will result in three classes of material: those that are known to be hazardous, those that are known not to be hazardous, and suspect materials. It is the suspect materials that will require further investigation to determine whether or not they are hazardous. Table 10.3 is a listing of commonly found hazardous materials at nuclear facilities. A complete listing of such materials is not feasible because there are so many different types of facilities.

Table 10.3 Common Hazardous Materials

Materials	Category	Occurrence
Asbestos	TSCA	Thermal insulation, gaskets, fire barriers, floor tiles, shingles, composition siding
Cadmium	RCRA Characteristic	Batteries, control rods
Lead	TSCA	Shielding, counterweights, batteries, paint
Mercury	RCRA Characteristic	Switches, illuminating lamps, vacuum systems
Polychlorinated Biphenyls (PCB)	TSCA	Oils (lubricating, hydraulic), ballasts, plastics (electrical insulation), paint
Process Chemicals	(depends on specific substance)	Chemicals originally used in facility processes that have not yet been shipped off-site for disposal; any other stored process chemicals that must be disposed prior to project completion
Solvents	CERCLA/RCRA	Cleaners, degreasers, chemical reagents
Petroleum products	CWA & Oil Pollution Act of 1980	Lubricants, packing, some plastics, fuels
Poisons	FIFRA/RCRA	Insecticides, fungicides, herbicides, rodent poisons, chemical reagents

There are a very large number of possible hazardous contaminants, a vast array of analytical techniques for identifying and quantifying them, and a very complex set of laws and regulations that apply to them. The EPA, which has enforcement authority for most related regulations, has set the standard for characterization of hazardous materials. Publication SW-846 (EPA 2000b) is a manual of about 3500 pages that contains an official compendium of analytical and sampling methods that have been evaluated and approved for use in complying with RCRA regulations.

10.4.2: In-Situ Monitoring: Most analytical techniques used to identify hazardous materials cannot be used outside a laboratory because of the size of the equipment or the nature of the operation.

Much research has occurred to develop field-deployable survey instruments, and a limited number of such devices are available. Almost all field instrumentation is very useful for determining the presence of a suspected contaminant. However, field deployable, portable instruments do not usually have sufficiently low detection limits to demonstrate the absence of a suspected material. Therefore, in-depth characterization will rely mostly on samples, and on analysis by an EPA and state certified laboratory.

A good resource for selecting in-situ monitoring instruments is located on the Internet at an EPA web site called the Field Analytic Technologies Encyclopedia. The site contains a wealth of detail on technologies for field characterization; some of these are listed in Table 10.4. These sound practices and well-established methods notwithstanding, research and development of innovative technology continues. Some newly developed characterization technologies are listed in Table 10.5. The CD provided with this handbook contains further details on such innovative technologies for which DOE reports have been published.

10.4.3: Sampling and Laboratory Analysis: The large number of different substances that are considered hazardous contaminants means that characterization plans must choose from a large number of different analytical methods to detect and quantify them. The methods run the gamut of physical and analytical chemistry and physics. The number of such methods is too great to include practical descriptions in this manual. The EPA Publication SW-846 is a good starting point for choosing analytical characterization techniques. Another source is the highly useful matrix, Field Sampling and Analysis Guide (Appendix E) from the Federal Remediation Technologies Roundtable. The technologies listed in Table 10.6 are based on a federal review of commonly used field techniques. In addition, there are several web sites that give an overview and explanation of available methods at an introductory level. The web sites may be found by their title or key subjects:

- SCIMEDIA: Encyclopedia of Analytical Instrumentation
- CLU-IN: Site Characterization and Monitoring
- Field Analytic Technologies Encyclopedia

In almost all cases, the certified analytical laboratory will have technical support personnel to advice on detection limits required for meeting the various EPA regulations and those technique best suited for a suspected contaminant. The learning curve for characterization is enormous. Consultants or staff members with significant experience in this area are needed since an inexperienced person will almost always have to resample and reanalyze.

10.4.4 Radiation and Radionuclides: An important prelude to performing measurements is to gather the available information on the potential and actual radionuclide contamination and external radiation fields. The selection of survey instruments, sampling sites, and radiochemical analytical procedures to measure the appropriate radionuclides at existing levels and sites is guided by this information. Potential contamination is identified by considering the source term and records of earlier monitoring. Ideally, a synthesis of available information will permit mapping the surface distribution of each radionuclide in terms of concentration isopleths. These values are compared to concentration limits specified for final use and are used to identify the radionuclides subject to decontamination. More commonly, the available information indicates the presence of the radionuclides that are candidates for cleanup and the locations that must be cleaned.

Table 10.4 Technologies for In-Situ Characterization

Technology	Principle of Operation	Use
Colorimetric Indicator Tests	Reagents that change color when exposed to specific contaminants	Air samples, PCB in fuel, chlorine in oil and soil, PCP and TCE in water
Fiber Optic Chemical Sensors	A portion of the optical fiber cladding is removed and replaced with a chemically selective layer. The sensor is then placed directly into the media to be analyzed. Interaction of the analyte with the chemically selective layer creates a change in absorbance, reflectance, fluorescence, or light polarization. The optical change is then detected by measuring the change in light characteristic carried by the optical fiber.	Volatile petroleum constituents such as BTEX and chlorinated VOCs such as TCE, PCE, and carbon tetrachloride in water, air, or soil gas
Gas chromatography	The sample is transported via an inert gas called the mobile phase through a coiled tubular column filled with a material called the stationary phase. As the mobile phase moves through the column, the components of the sample move with different speeds because of differences in affinity with the stationary phase material. This causes a physical separation of the sample components.	Phenols, phthalates, amines, chlorinated pesticides, PCBs, PAHs, chlorinated hydrocarbons, volatile organic chemicals (VOC), organophosphorus compounds, chlorinated herbicides, MTBE, halogenated VOC, SVOCs/base neutral acids (BNA), dioxin
Graphite Furnace Atomic Absorption Spectroscopy	The sample is vaporized in a graphite furnace to free atoms at ground state. Ground-state atoms then absorb energy, in the form of light, and are elevated to an excited state. This happens at specific wavelengths for each type of atom, resulting in an absorption spectrum. The absorbed wavelengths are used to identify the analytes.	Aluminum, arsenic, barium, boron, cadmium, calcium, chromium, cobalt, copper, iron, lead, lithium, magnesium, manganese, mercury, molybdenum, nickel, potassium, selenium, silicon, silver, sodium, titanium, terbium, vanadium, ytterbium, and zinc in soil/water.
Immunoassay	Antibodies that have been developed to bind with a target compound or class of compounds	PCP, 2,4-D in Water and Soil, PCBs in Soil, TPH in Soil, PAHs, Toxaphene, Chlordane, DDT, TNT Explosives in Water and Soil, RDX Explosives in Water and Soil
Infrared Spectroscopy	A collimated beam of infrared light is passed through the sample. The contaminant (analyte) absorbs the infrared light at frequencies specific to the functional groups present in the molecule. Thus, each contaminant has characteristic infrared spectra, which can be used to identify the compound. Also, the amplitude of absorbance is proportional to the analyte concentration.	Air monitoring for volatile organic chemicals (VOCs); fence-line or site perimeter monitoring, worker exposure monitoring, emission rate assessment, air impact measurement during emergency removals, air impact evaluation during remedial actions, vapor suppression technique evaluation, accidental release early warning systems, industrial facility monitoring.
Laser-induced Fluorescence	Laser light excites electrons in target material. When they drop back to ground state, light of different characteristic frequencies (fluorescence) is given off by the analytes.	Gasoline, diesel fuel, jet fuels, fuel oil, motor oil, grease, and coal tar
Mass Spectrometry	The analyte is vaporized, ionized, and accelerated through a magnetic field. Differences in mass-to-charge ratio between the components cause physical separation due to the influence of magnetic field on motion of the component.	Semi-volatile organic chemicals (SVOC), base-neutral acids, dioxins, Volatile Organic Chemicals (VOC), halogenated VOC, pesticides and PCBs.
X-Ray Fluorescence	A radioactive source generates X-rays that excite inner shell electrons in target material. When the atoms drop back to ground state, x-rays of different characteristic frequencies (fluorescence) are given off by the analytes.	Arsenic, barium, cadmium, chromium, copper, lead, mercury, selenium, silver, and zinc. Field-portable fluorescence units run on battery power and use a radioactive source were developed for analysis of lead-based paint and now accepted as stand-alone technique for lead analysis.

Table 10.5 Innovative Technologies for Characterization

Technology	Summary	Tech
Chemical Analysis		
Alloy Analyzer	Uses x-ray Fluorescence spectrum analysis to identify and quantify elements in metal	2397
Field Raman Spectrograph	Screens, monitors, and identifies a wide variety of contaminants from more than 50 meters away	873
Lead Paint Analyzer	Identifies and quantifies metals and elements in lead based paint	2317
Portable x-ray Fluorescence Spectrometer	Provides nondestructive, real-time elemental analysis	1790
x-ray Fluorescence Metal Analyzer	Determines common alloy elements in the field with lab-like accuracy	2001
Internal Duct and Piping		
Internal Duct Characterization System	Remotely operated system inspects and characterizes contaminated ventilation ductwork	42
Pipe Crawler Internal Piping Characterization System	Characterizes radiological contamination inside piping systems	1810
Pipe Explorer Surveying System	Characterizes radiological contamination inside piping systems	74
Piping Inspection (Remote Using Specialty Cameras)	Performs and records visual radiological inspections of pipes	1811
Small Pipe Characterization System	Characterizes contamination in pipes 2-3 inches in internal diameter	43
Mapping		
Gamma Cam Radiation Imaging System	Displays relative strength and location of gamma radiation	1840
Gamma Ray Imaging System	Displays and plots location of radioactive contamination	1793
Indoor Radiation Mapping Using Laser-Assisted Ranging and Data System	Performs and reports radiological surveys	1946
Interactive, Computer Enhanced, Remote Viewing System	Provides reliable geometric description of a remote environment	33
Radiation Monitoring and Characterization		
Cogema 3-D Gamma Imaging	Provides precise information on the quantity and location of gamma radiation-emitting sources	2302
Decommissioning In-Situ Plutonium Inventory Monitor	Performs in-situ assay of plutonium contaminated equipment with passive neutron counting and 3-D imaging	2241
Electret Ion Chambers	Measures Alpha Radiation	2315
In-Situ Object Counting System	Identifies specific nuclei and corresponding radioactive inventory	2098
Long Range Alpha Detection for Component Monitoring	Measures alpha contamination on surfaces of items and pipes	2382
Wireless Remote Radiation Operating	Remotely monitors personnel radiation dose and area exposures	2104

Table 10.5 Cont'd
Innovative Technologies for Characterization

Technology	Summary	Tech #
Robotic & Remote		
2-D Linear Motion System	Semi-robotically operates decontamination tools and characterization instruments on walls	1476
Geophex Unmanned Airborne Unmanned Survey System	Robotic helicopter sweeps areas for metallic and nonmetallic anomalies	295
Position Sensitive Radiation Monitoring	Motorized characterization and data analysis system surveys contaminated floor and wall surfaces	1942
Rosie: Mobile Teleoperated Robot System	Deploys tools for equipment removal and other Decommissioning tasks	1799
Soil		
Compact Subsurface Investigation System	Retrieves soil samples in congested areas where standard sampling equipment cannot fit	2153

As a practical matter, external radiation monitoring is the logical first step in radiological characterization. It is a nondestructive procedure for establishing an overview of large spaces with relatively small effort. The gathered information, however, is limited by the sensitivity of the instrument with regard to type of radiation and amount detected. Radiation detectors are selected with these limitations in mind, and the survey design must be adjusted to the characteristics of the radiations emitted by the radionuclides of interest. Table 10.6 (DOE 1994) provides a listing of survey design techniques.

Because gamma rays are highly penetrating, they can be detected at large distances and within solids, but with low counting efficiency. Beta particles are less penetrating and more efficiently detected. Weakly penetrating alpha particles are detected only on surfaces, are readily absorbed by thin layers of paint and dust, and require the equipment to be placed within less than 10 cm of the contamination. The detected energies of these radiations differ from those of emitted radiations because their lower energies have lower penetrating power. Indications of detection sensitivity are given in NUREG-1507 (NRC 1998a).

Fissionable materials, notably ^{239}Pu, ^{238}U, ^{235}U, and ^{233}U, emit neutrons that are measured with specialized detectors. Their detection sensitivity depends on neutron energy, and results may have to be corrected for other radiations measured at the same time. Additional neutrons can be produced in these contaminants by exposing them to radiation sources, or (more conveniently) measured radiations can be stimulated with a neutron source.

Table 10.6 Radiological Measurement Techniques for Use on a Decommissioning Site

	APPLICABILITY		
TECHNIQUE	***In-Situ***	***Onsite Facility***	***Laboratory***
Alpha survey	Surface contamination		Surface contamination
Beta survey	Surface contamination		
Gamma survey	Surface contamination, Bulk contamination	Bulk contamination	Bulk contamination
Gamma spectrometry	Surface contamination, Bulk contamination	Surface contamination, Bulk contamination	Surface contamination, Bulk contamination
Neutron	Bulk contamination	Bulk contamination	
Alpha smearable	Surface contamination		Surface contamination
Beta smearable	Surface contamination		Surface contamination
Gamma smearable	Surface contamination	Surface contamination	Surface contamination
Dose rate	Surface contamination, Bulk contamination	Bulk contamination	
Alpha spectrometry			Surface contamination
Beta spectrometry			Surface contamination
Liquid scintillation			Surface contamination, Bulk contamination

Conventional survey instruments and their applications, extracted from the MARSSIM are described in Tables 10.7, 10.8, and 10.9. They are usually limited in application to identifying the presence of radionuclides and indicating the magnitude of the resulting radiation dose. Any calculation of radionuclide amounts on a surface or concentration in a material requires assumptions that may not be valid unless the radionuclides and their radiations are known, the number of radionuclides is limited, and the geometric relation of source to detector is well defined.

Some radiations can be distinguished from others by varying the shielding of the portable detector or using multiple detectors. Some detectors are selected for relatively uniform response over a wide range of photon or particle energy. Others provide radiation dose rates applicable to humans (i.e., tissue equivalent) instead of the detector material. Portable intrinsic germanium (Ge) gamma-ray detectors with spectrometers are available for distinguishing photons by energy with good resolution.

To locate contaminants in a small room, the surveyor moves the monitoring instrument across floors and walls in response to visible or audible signals. The surveyor fixes exact locations of elevated contaminant regions by the highest readings. For monitoring large expanses, an optimal program will specify instrument path, path intervals, and rate of instrument travel to attain readings that are significant at specified contaminant levels and avoid overlooking localized contaminants at such levels. The manual recording of measurements and grid locations is being replaced by automated measurements and remote determination of locations. Robots can replace human surveyors with hand-held detectors if useful for relatively inaccessible locations, as in ducts.

Table 10.7 Radiation Detectors for Alpha Surveys

Detector Type	Description	Application	Remarks
Gas Proportional	<1 mg/cm^2 window; probe area 50 to 1000 cm^2	Surface scanning; surface contamination measurement	Requires a supply of appropriate fill gas
Air proportional	<1 mg/cm^2 window; probe area ~50 cm^2	Useful in low humidity conditions	
Scintillation	ZnS(Ag) scintillator; probe area 50 to 100 cm^2	Surface contamination measurements, smears	
Solid state	Silicon surface barrier detector	Laboratory analysis by alpha spectrometry	
Passive, integrating electret ion chamber	<0.8 mg/cm^2 window, also windowless, window area 50 to 180 cm^2, chamber volume 50 to 1000 ml	Contamination on surfaces, in pipes and in soils	Useable in high humidity and temperature

Table 10.8 Radiation Detectors for Beta Surveys

Detector Type	Description	Application	Remarks
Gas Proportional	<1 mg/cm^2 window; probe area 50 to 1000 cm^2	Surface scanning; surface contamination measurement	Requires a supply of appropriate fill gas
Ionization (non-pressurized)	1 to 7 mg/cm^2 window	Contamination measurements; skin dose rate estimates	
Geiger-Mueller	<2 mg/cm^2 window; probe area 10 to 100 cm^2	Surface scanning; contamination measurements; laboratory analyses	
	Various window thickness; few cm^2 probe face	Special scanning applications	
Scintillation	Plastic scintillator	Contamination measurements	
Solid state	Silicon surface barrier detector	Laboratory analysis by beta spectrometry	
Passive, integrating electret ion chamber	<0.8 mg/cm^2 window, also windowless, window area 50 to 180 cm^2, chamber volume 50 to 1000 ml	Low energy beta including H^3 contamination on surfaces and in pipes	Useable in high humidity and temperature

Table 10.9 Radiation Detectors for Gamma Surveys

Detector Type	Description	Application	Remarks
Gas Ionization	Pressurized ionization chamber; non-pressurized ionization chamber	Exposure rate measurements	
Geiger-Mueller	Pancake (<2 mg/cm^2 window) or side window (~30 mg/cm^2)	Surface scanning; exposure rate correlation (side window in closed position)	Low relative sensitivity to gamma radiation
Scintillation	NaI(Tl) scintillator; up to 5 cm by 5 cm	Surface scanning; exposure rate correlation	High sensitivity. Cross calibrate with PIC (or equivalent) or for specific site gamma energy mixture for exposure rate measurements.
	CsI or NaI(Tl) scintillator; thin crystal	Scanning; low energy gamma and x-rays	
	Organic tissue equivalent (plastics)	Dose equivalent rate measurements	Detection of low energy radiation
Solid state	Germanium semiconductor	Field gamma spectrometry	
Passive, integrating electret ion chamber	<7 mg/cm^2 window, also windowless, window area 50 to 180 cm^2, chamber volume 50 to 1000 ml		Useable in high humidity and temperature

Placing small and relative inexpensive radiation dosimeters at numerous locations can delineate radiation fields. Thermo-luminescent dosimeters (TLD) and photographic films are typically used as dosimeters. They are collected after a period of time sufficient to accumulate a radiation dose that is detectable above the background dose, and are measured in a dosimeter reader.

Wiping an area with a smear and then counting this smear at a fixed location near a radiation detector will determine removable surface contamination. Conventionally, a 5-cm diameter filter-like paper is wiped by hand over an area of 100 cm^2 with reasonably consistent surface movement and pressure. Larger areas can be smeared mop-like with cloth wipes. The wipes may be measured on location or at a laboratory. The latter permits more sensitive and discriminating radiation detection, and also radiochemical analysis to identify radionuclides and eliminate interfering solids and radionuclides. Smears are a useful indicator of radionuclides removable under the specified conditions, but not of fixed surface contamination that may become removable under changed conditions, such as washing, scraping, or heating.

Airborne radioactive particles, usually due to re-suspension from surfaces and ducts, are monitored by collection on filters in air samplers. Filters are made of paper, glass fibers, or plastics, in 5-cm diameter circles or larger. Flow rates may be as low as 1 m^3 per day or higher than 1,000 m^3 per day. Vapors emitted from surfaces may be collected on charcoal, silica gel, or molecular sieve, depending on the chemical form. Air with the gases to be detected can also be collected directly in plastic bags or in pressurized containers for counting. Radionuclide measurements may be made in real time, or the filters and gases may be measured and analyzed in the laboratory.

Portable monitors incorporate radiation detectors facing filters that are fixed, move continuously, or move periodically. The output can be recorded continuously and be connected to visible or auditory alarms to notify of excessive airborne levels. Some of these monitors are also used to measure particulate radon progeny in air. If the intention is to measure other radionuclides, the radon progeny background must be subtracted.

In most instances, an in-situ radionuclide measurement with a monitoring instrument provides a count rate (count per minute) that is not directly related to the radionuclide concentration on a surface or in a solid. The two quantities can be related by calibration if the radionuclide source and the geometric relation between source and detector are specified. These requirements can be met if only a few radionuclides are in the area and their distribution approximates a simple form, i.e., a uniform plane or a point. Typically, the measurements distinguish elevated levels from lower contaminant or background levels, suggest locations for collecting samples for radiochemical analysis, and may indicate a radiation exposure rate for personnel protection.

10.4.5 Sampling and Radiochemical Analysis: Samples to be measured for radionuclide content are brought to the laboratory for sensitive measurement and analysis. Samples may be scrapings, cores, or portions of solid or liquid material. They may be measured at a temporary or conventional on-site facility or at a contracted laboratory. Measurements range from gross radioactivity to spectral and/or radiochemical analyses.

Detection sensitivity can be enhanced by reducing and stabilizing the radiation background electronically, by using massive shielding, or by counting for long times. Larger detectors can be used to achieve higher detection efficiency, although the radiation background also may be higher. Processing large samples and removing interfering material before the measurement can also increase sensitivity.

Gamma-ray spectral analysis is a relatively simple technique for identifying and measuring radionuclides that emit photons. Relatively large samples, typically of the order of 1 kg, are measured without processing other than weighing them and placing them in a standard container for counting. A high purity (intrinsic) Ge detector doped with lithium (Li) provides high resolution but requires liquid nitrogen or other refrigeration for stable measurements. If there is no need for high resolution because only a few radionuclides are present, thallium-activated sodium iodide detectors [NaI(Tl)] may provide appropriate counting efficiency without cooling. Plastic scintillation detectors can be made much larger, but they have poor energy resolution.

Certain samples and radionuclides require special treatment. For example, samples measured for ^{226}Ra by counting photons from the ^{214}Pb and ^{214}Bi progeny of gaseous ^{222}Rn must be placed in sealed containers and held until the in-growing progeny approach equilibrium with ^{226}Ra. Detectors for radionuclides that emit low energy (well below 50 Kev) photons and characteristic X-rays require thin windows because the radiations are significantly attenuated by passage through material. Thinner GeLi and NaI(Tl) detectors can be used to reduce interference from more energetic photons. Mixing is required when radionuclides are distributed non-uniformly in solids or may become so in liquids.

Gross alpha and beta particle measurements are performed with proportional and Geiger-Mueller ionization detectors to screen the magnitude of the radionuclide concentration in a sample. Combined alpha or beta activity concentrations are estimated through an assumed counting efficiency. Concentrations can be measured accurately only when a single radionuclide is known to be in the sample. Thin samples are required for counting beta particles and very thin samples for alpha particles, unless only surface activity is to be measured, i.e., count rate per cm^2 instead of per kg. In general, the amount of a liquid sample to be evaporated on a planchet is limited to yield 100 mg or less dried weight. A similar mass limit applies to a solid sample that is slurried on the planchet and dried. The sample should be spread as uniformly as possible to match the counting efficiency of the sample to that of a standard.

The liquid scintillation counter is another detector for radionuclides that emits alpha or beta particles. Small volumes of liquids (typically 5 - 10 ml) are mixed with an organic solvent that contains an organic scintillator. Other components may be added to improve solubility and to match the energy of the emitted light to the optimum response of the photomultiplier system. In some cases, radionuclides in liquids are adsorbed on scintillating spheres, or particles are suspended in a gel. Liquid scintillation is particularly useful for measuring radionuclides, such as ^3H and ^{14}C, since these emit low-energy beta particles that would be mostly or entirely absorbed in thick samples and ionization detector windows. Another benefit of this detector is the option for spectral analysis, or at least multiple windows, to distinguish among beta particle groups of significantly different energies. Some systems also permit differentiation by pulse shape between alpha particles and beta particles.

Individual radionuclides are chemically separated to extract small amounts from bulky samples and to remove all non-isotopic radioactive impurities. Isotope dilution analysis, the process of adding an isotopic carrier or tracer, is usually applied to permit calculating chemical recovery without requiring quantitative analysis. The radiochemical analyses that are commonly performed in contract laboratories are listed in Table 10.10 (DOE 1994). The extent of chemical separation effort depends on the needed purification.

The major radionuclide in a sample may be purified in a few steps, while a radionuclide that constitutes a small fraction of the total radioactivity or that is chemically similar to important impurities may require many steps. The sensitivity and associated minimum aliquot values given in the table presumably pertain to routine analyses to attain the usually required detection limits. Processing larger samples, concentrating the radionuclide of interest, and counting for much longer periods in detectors at a low and stable background can reach much greater sensitivity. On the other hand, lesser sensitivity may be required to delineate the extent of contamination. In general, greater analytical sensitivity costs more.

After chemical purification, the sample is counted with the detectors described above. The difference is that now only the specified element remains, with a single or possibly two or three radioisotopes, and the mass is usually reduced greatly. The identification of these radionuclides by element is confirmed by observing the types and relative amounts of their radiations, and by repeated counting to determine the half-life.

If the final sample has been prepared sufficiently thin, it can be analyzed by a spectrometer with a silicon diode detector to identify radioisotopes by alpha-particle energy. Such spectral analysis also permits addition of isotopic radioactive tracer for the actinide radionuclides that have no stable isotope for isotope dilution analysis.

Table 10.10 Radiochemical Analyses Available from Off-Site Laboratories

ANALYSIS	LIQUID MATRIX *Sensitivity (pCi/L)*	LIQUID MATRIX *Minimum Aliquot (L)*	SOLID MATRIX *Sensitivity (pCi/g)*	SOLID MATRIX *Minimum Aliquot (g)*
Gross alpha	1.0	0.2	10.0	0.1
Americium 241	-	-	0.2 dry	100.0
Americium isotopic *	1.0	1.0	0.6 dry	2.0
Curium isotopic	1.0	1.0	0.6 dry	2.0
Neptunium 237	1.0	1.0	0.6 dry	2.0
Plutonium isotopic *	1.0	1.0	0.6 dry	2.0
Plutonium 241	20.0	1.0	10.0 dry	2.0
Polonium 210	1.0	1.0	0.6 wet	2.0
Radium 226	1.0	1.0	0.2 dry	100.0
Radium 228	3.0	2.0	0.2 dry	100.0
Thorium isotopic *	1.0	1.0	0.6 dry	2.0
Uranium	1.0 µg/L	0.005	0.01 µg/g dry	0.5
Uranium isotopic *	1.0	1.0	0.6 dry	2.0
Carbon 14	100.0	0.05	10.0 wet	1.0
Calcium 45	6.0	1.0	-	-
Chlorine 36	10.0	1.0	-	-
Iodine 125	20.0	0.1	-	-
Iodine 129	10.0	0.1	1.0	50.0
Nickel 63	100.0	0.05	10.0	1.0
Lead 210	1.0	0.5	2.0	100.0
Promethium 147	50.0	0.15	-	-
Hydrogen 3	20.0	0.1	500.0 pCi/L	Enough sample to extract 5 ml water
Ruthenium 106	20.0	1.0	-	-
Strontium 90	5.0	1.0	1.0	2.0
Technetium 99	30.0	0.1	15.0	1.0
Iron 55	100.0	1.0	-	-
Gross gamma	4.0	0.2	10.0	0.2
Gamma isotopic *	20.0	1.0	0.2	100.0

* Mixture of various isotopes of the element

10.4.6 Recent Developments in Radiometric Measurements: Some alternatives are available to the radiochemical analyses described in Section 10.4.2. Recent developments in chemistry provide the ultra-sensitive technique of inductively coupled plasma mass spectrometer analysis. This method attains approximately the same sensitivity as radiochemical analysis for long-lived radionuclides (those with half-lives of thousands of years and longer). Liquid samples can be made volatile without extensive prior treatment. The problem of interference has a different aspect than for radiochemical analysis because it is caused by the presence of ions in the plasma that have the same mass as the radioisotope being measured.

Mass accelerators are used to separate ^{3}H and ^{14}C for measurement at very low levels. Neutron activation analysis has been used for measuring long-lived radionuclides, such as ^{129}I, producing short-lived ^{130}I for counting and also ^{128}I for comparison with stable iodine (^{127}I).

The long-range alpha particle detector listed in Table 10.5 is an example of a new approach. This technique measures the ionization produced by alpha particles in air rather than the alpha particles themselves. The air above a radionuclide that emits alpha particles is swept to the location of the ionization detector, so that radionuclides may be measured in locations that are inaccessible to a detector. This system is convenient for the inside of a pipe or duct, in narrow chases, and in corners.

The Cogema system, also listed in Table 10.5, obtains the gamma-ray distribution pattern in two dimensions for relatively high radiation levels and matches the pattern to a visual image. Multiple images may then be combined to present a 3-dimensional model of the source. The concept has been applied to neutron measurements.

Electret chambers (the third item) are ionization detectors that can be used in the field without power source, because the amount of ionization is measured by the decrease in voltage of the electret in the form of a charged Teflon plate.

The in-situ object counting system is a portable germanium detector with multi-channel analyzer and computer. The collected data are analyzed with a program that includes the option for Monte Carlo simulation of the geometric relation of source and detector.

The wireless, remote, radiation operating system has electronic radiation detectors that transmit data to a central system and then reset themselves.

The radiation detection systems listed under the internal duct and piping category generally are radiation detectors that crawl through pipes under remote control while transmitting data. Prevention of contaminant accumulation is a significant consideration. The mapping category describes techniques for scanning an area both radiologically and visually to locate elevated gamma radiation levels.

10.5 RELEASE DETERMINATION

The confirmatory characterization plan is based on the cleanup specifications of the decommissioning plan. Regulations are important sources of these specifications for the end-state conditions of structures and associated land, and for wastes buried on site or shipped to repositories. The EPA, the NRC, and state health and environmental protection agencies enforce the regulatory requirements that must be met by confirmatory characterization. The values must be compared to the acceptable values calculated by models structured for scenarios of human exposure. Default calculation models and tables of limiting concentrations are provided by the regulatory agencies, but site-specific alternatives can be proposed in the decommissioning plan. Levels of direct external radiation must be compared to the criteria for free release for the site, and to Department of Transportation (DOT) requirements for shipping wastes.

Guidance for the confirmatory characterization radiological survey, available in Section 4.3 of the MARSSIM, enables the planner to estimate the number of measurements needed to attain the specified degree of confidence in the monitoring results. The main concerns are that measurements meet the concentration and dose rate criteria, and that no unmeasured locations and materials exceed the criteria. Nearby locations that may have been contaminated by the facility must also be shown to meet health and environmental protection criteria. These locations include surfaces and structures beyond the fence line, stream beds, and subsurface structures, soil, and water.

Wastes to be transported from the site must be characterized for shipping and for disposal. Bills of lading must have the information to meet DOT regulations, as indicated in Chapter 6 of this handbook. For disposal, the wastes must be described to match NRC and DOE radiological waste criteria, and to identify RCRA and mixed wastes, as described throughout Part 2.

Thorough documentation of characterization results is necessary to demonstrate that regulatory limits are met and that reasonable potential data needs can be met in the future. Stakeholders may consider a non- regulatory exposure scenario as equally important, or may have different proposals for site use. Moreover, changes in contaminant or operational knowledge may change the risk assigned to certain contaminants and may require re-evaluation of characterization data.

10.5.1 Defining Standards for Free Release or Continued Control: The purpose of a decommissioning project is to permanently remove the facility as a source of significant hazard, so that the site can be freed from continued restrictions on access or use. The amount of residual contamination is reduced to below a specific level representing an acceptable hazard. When that level of contamination is achieved, the facility, its contents, or what remains after it is demolished can be released. The residual contamination levels that represent free release criteria are generally referred to as authorized limits. In the U.S., the regulatory approach to setting authorized limits varies and depends on which government agency has jurisdiction. Brief overviews of standards and regulations by the NRC, the EPA, and the DOE are presented herein.

10.5.1.1 NRC: NRC regulations in 10 CFR Part 20 include explicit radiological criteria for decommissioning (62 CFR 139, pp. 39057-39092). Subpart E contains dose-based radiological criteria for restricted and unrestricted release, consisting of a TEDE limit for residual radioactivity above background. The criteria include a TEDE limit of 25 mrem/y that should not be exceeded by an average individual among those who could potentially receive the greatest exposure from any residual activity within a facility or on a site. The NRC uses essentially the same release criteria for contamination on surfaces as does the DOE (except for tritium), but published as guidance rather than as regulations in Regulatory Guide 1.86 (NRC 1974).

Regulations require that the entire applicable source term and all credible dose pathways be considered when determining whether residual radioactivity is indistinguishable from background or when calculating TEDE. The criteria also require a licensee to reduce any residual radioactivity to ALARA levels. The evaluation of ALARA should be based on a multi-variant analysis that considers both on- and off-site radiological and nonradiological risks, and evaluates individual and collective dose for both public and worker populations. Although every reasonable effort should be made to reduce residual radioactivity to levels that allow unrestricted release of the site, the regulations accept the possibility that restrictions may be required in the site's future use. If such restricted use is proposed, 10 CFR 20.1405 defines additional criteria that the licensee must meet.

10.5.1.2 EPA: The EPA has draft site cleanup regulations in 40 CFR 196. Case-specific site release criteria are derived for concentration limits on the basis of a TEDE limit of 15 mrem/y to members of the public. There is no requirement for using the ALARA process (EPA 1994). Under these rules, the cleanup will be performed so that radionuclide concentrations in groundwater will not exceed the current maximum contaminant levels specified under the SDWA. In general, ingestion of any radionuclide in water at these concentrations will not cause a dose in excess of 4 mrem/y. Note that the EPA also recommends that structures meet the indoor radon guidelines and that homeowners apply mitigation procedures if ^{222}Rn exceeds 4 pCi/L in air.

10.5.1.3 DOE: DOE Order 5400.5 provides limits specific to decommissioning. However, if the DOE sites are remediated under CERCLA, the EPA and state regulatory agencies must approve the release conditions. The DOE Order imposes requirements on the release of residual radioactive materials on DOE property. It requires that release comply with a public dose limit of 100 mrem/y from all sources and pathways, except medical and background. Compliance with the all-sources dose standard is typically ensured by calculating DCGL using both the all-sources standard and a single-source dose constraint that is a fraction of the all-source standard (e.g., 30 mrem/y).

The DOE also requires its sites to complete four actions to protect the public and the environment before release of property containing residual radioactive materials:

- The property is appropriately surveyed to characterize its radiological condition.
- Residual radioactive material is determined to meet applicable release limits.
- Required documentation is completed.
- The owner or recipient of the released property is appropriately notified of the radiological status of the property and the availability of required documentation.

For isotopes where potential public exposure pathways are well defined, specific regulatory release criteria (expressed in radioactivity concentrations) have been adopted. For example, the DOE sets release criteria for radium and thorium in soils, for radon in occupied or habitable buildings, and for contamination on surfaces of structures, building components, equipment, and other materials. Release criteria are established in 40 CFR 192 for radium in soils; for radium, uranium, and gross alpha particle activity in groundwater; and for radon in occupied or habitable structures.

Case-specific release criteria are derived using the ALARA process when DOE Orders or applicable regulations have not established the release criteria. The ALARA process involves, among other things, use of computer codes or other computational methods to make dose-to-concentration conversions. Considerable professional judgment is typically involved when property is released. Because of the complexity of radiation transport, it is important to establish early and effective communication with decision-makers and the public. Detailed knowledge of the source term and exposure pathways is essential when interacting with regulators and stakeholders.

1. *Grounds and Structures - Radiological*: The final radiation survey plan will provide the basis for verifying that the facility, site, and the contiguous adjacent environs meet radioactivity levels that permit release for unrestricted use. The criteria will vary, depending on the nature of the matrix (e.g., surface or bulk contaminated) and the type of contaminant (e.g., ^{60}Co).

Criteria for release for unrestricted use of some contaminants can be found in the NRC's Regulatory Guide 1.86 and in the DOE's Order 5400.5. For other contaminants, the criteria must be developed to the satisfaction of the regulatory agency, using such methodology as pathway analyses. Volume 4 of NUREG/CR-5512 (NRC 1999a) gives a pathway methodology for developing radiological release criteria for structures, and RESRAD is a computer program for calculating allowable residual radioactive materials based on pathway analysis. These are discussed further in Chapter 11.

For NRC-regulated facilities and soil, the release criterion is stated in subpart E of 10 CFR 20.1402, Radiological Criteria for Unrestricted Use. The criterion is based on residual radioactivity distinguishable from background radiation resulting in a TEDE below 25 mrem. This guide applies facilities and soil and does not apply to equipment.

The Surface and Volume Radioactivity Standards for Clearance (ANSI 1999) relates primary radiation dose to a derived screening criterion for the release of both surface- and volume-contaminated materials. The standard is not a substitute for regulatory release criteria, but is protective of the public and the environment. It is generally the responsibility of the decommissioning contractor or the facility owner to propose final acceptance criteria for unrestricted release and to obtain agreement from regulators.

2. *Ground and Structures - Hazardous and Mixed Waste*: The criteria for release of areas previously contaminated with hazardous waste are specified in the environmental laws and regulations. Chapter 6 discusses release criteria for various contaminants. Release criteria for mixed waste invoke both hazardous and radiological criteria. The waste must frequently meet requirements for unconditional release.

3. *Materials and Equipment for Reuse*: The decommissioning process allows the reuse or salvage of excess materials and equipment to help defray some of the costs. Reusable items may be released either conditionally or unconditionally. If the item is to be reused in another contaminated facility, it is not necessary to clean and survey the equipment for unconditional release. It only needs to meet criteria for its intended use.

10.5.2 Survey Units and Dose Assessments: Detailed characterization is needed for determining whether a particular unit meets the release criteria. A survey unit, an area of specified size and shape for which a final status survey is designed and conducted, results in a release decision. Surveys should obtain all data required for making a decision, but avoid the collection and analysis of superfluous samples. Usually there are two conditions that lead to the determination that a particular survey unit requires further remediation before unrestricted release, the average level or a small area within the survey unit exceeds the regulatory limit.

Sampling at discrete points within the survey unit is a simple method for determining if the first conditions exist. The term sampling is used here in its statistical sense, namely obtaining data from a subset of a population. Sampling in this sense includes both direct in-situ measurements and the collection of physical samples for laboratory analysis. On the other hand, sampling at discrete points within a survey unit is not an efficient method of determining if the second condition exists. Scanning is a much better method for detecting isolated areas with elevated activity, but generally is not as sensitive as sampling. An efficient characterization design is the use of both scanning and sampling at distinct locations to accurately determine the final status of a survey unit.

Source term knowledge is the key to calculating the anticipated internal and external dose to decommissioning workers and members of the public. Computer codes and models can assist in converting residual radioactivity concentrations to estimated radiation doses, or vice versa. The codes and models vary in terms of intent and the algorithmic approach to calculating dose. The following listing of computer models is intended to provide a basis for further research into the selection of the proper software program for the specific decommissioning initiative at hand.

10.5.2.1 CAP88-PC: The CAP88-PC (EPA 1992) is a set of computer programs, databases, and associated utility programs for estimating dose and risk from radionuclide emissions to air. Derived from versions of AIRDOS-EPA (EPA 1979) and DARTAB (ORNL 1981), this package allows users to perform full-featured dose and risk assessments in a personal computer to demonstrate compliance with 40 CFR 61.93(a). It differs from the dose assessment software AIRDOS-PC in that it estimates both risk and dose, offers a wider selection of radionuclide and meteorological data, provides the capability for collective population assessments, and allows users greater freedom to alter values of environmental transport variables.

CAP88-PC uses a modified Gaussian plume equation to estimate the average dispersion of releases from up to six sources. The program computes radionuclide concentrations in air, rate of deposition on ground surfaces, concentrations in food, and intake rates by ingestion of food produced in the assessment area. Dose and risk are estimated by combining the inhalation and ingestion intake rates with the air and ground surface concentrations (with dose and risk conversion factors provided).

10.5.2.2 COMPLY: The COMPLY code (EPA 1989a) is a computer program that may be used to demonstrate compliance with the national emission standards for hazardous air pollutants in 40 CFR Part 61, Subpart I. It has various levels of complexity, the simplest being a computerized version of the tables of concentration and possession limits in compliance guidance (EPA 1989b). The program has three levels beyond the possession or concentration limit tables. Level 2 is the lowest, requiring a minimum of input, and level 4 requires the most input. (Levels 2 and 3 correspond to the NCRP screening levels 2 and 3; its screening level 1 is not used (NCRP 1989). The higher the input, the less conservatism is built into the dose estimate. The most complicated is an air dispersion calculation using a wind rose. At all levels, the program helps determine compliance.

10.5.2.3 DECOM: The DECOM code (Till 1988) allows the user to calculate doses from radionuclides with different concentrations in multiple soil layers (in 15 cm increments). The code calculates doses to a hypothetical receptor from the following pathways and exposure routes:

- External gamma ray dose from radionuclides in soil
- Inhalation dose from contaminated dust
- Ingestion dose from intake of contaminated plants, meat, milk, and water

10.5.2.4 GENII-S: The GENII-S code (Leigh 1993) is a comprehensive package of models that address routine and accidental releases of radionuclides to air or water, as well as residual contamination from spills or decommissioning operations. Both population and individual doses can be calculated. The addition of the sensitivity and uncertainty analysis shell enables the user to perform both stochastic and deterministic analysis. The code allows the user to model up to two layers of contaminated soil: the top 15 cm, and deep soil. It calculates doses to a hypothetical receptor from the following pathways and exposure routes:

- External gamma dose from radionuclides in soil and air
- Inhalation dose from contaminated dust
- Ingestion dose from intake of contaminated soil, plants, meat, milk, fish, and water

10.5.2.5 NUREG/CR-5512 Methodology: The methodology described in NUREG/CR-5512 (NRC 1999a) incorporates a set of generic screening models to calculate radiation doses received by an individual from residual contamination in buildings and soil, and from potential groundwater contamination. These models are codified in the DandD software. The methodology encompasses building occupancy and renovation scenarios and a residential scenario. For the residential scenario, the contaminants are limited to the top 15 cm of soil. The potential pathways and exposure routes considered are similar to those considered in the above Section 15.5.2.4.

10.5.2.6 PATHRAE-EPA: The PATHRAE-EPA code (EPA 1987) was designed to calculate average annual and maximum annual effective doses and cancer risks to an on-site critical population group and an off-site population at risk. The approach used in PATHRAE is similar to that used in PRESTO, as described next.

10.5.2.7 PRESTO-EPA-CPG: The PRESTO-EPA-CPG code (EPA 2000c) is designed to estimate radiation doses to individuals and critical population groups over 1,000 year from the disposal of LLW. On-site doses resulting from farming and intrusion and off-site doses from exposure to contaminated air, surface water, and groundwater are considered. The following pathways and exposure routes are considered for hypothetical individual receptors, both on and off site:

- External gamma dose from radionuclides in soil, water, and air
- Inhalation dose from contaminated dust
- Ingestion dose from intake of contaminated plants, meat, milk, and water

10.5.2.8 RESRAD: The original DOE RESRAD code (ANL 1993a) is a pathway analysis code that calculates radiation doses to a hypothetical individual living on a contaminated site. RESRAD allows the user to define up to nine pathways and three exposure routes:

- External gamma radiation from radionuclides in soil
- Inhalation of contaminated dust and radon gas
- Ingestion of contaminated plants, meat, milk, aquatic foods, water, and soil

Several scenarios, including residential, industrial, and recreational, can be modeled by adding or suppressing pathways and entering appropriate values for occupancy and consumption rates. Several additions to the basic RESRAD code have also been developed and are described in greater detail in Section 11.2.

10.5.2.9 RESRAD-BUILD: The RESRAD-BUILD code (ANL 1994) is a pathway analysis model designed to evaluate the potential radiological dose (normally external) incurred by an individual who works or lives in a building contaminated with radioactive material that resides on building

surfaces and its contents, or within the structure itself.. The material within the structure can be released to the indoor air by mechanisms, such as diffusion (radon gas), mechanical removal (decontamination activities), or erosion (removable surface contamination). The transport of radioactive material from one building compartment to another is calculated with an indoor air quality model. The model considers the transport of radioactive dust particulates and radon progeny due to air exchange, deposition and resuspension, and radioactive decay and in-growth.

The design of RESRAD-BUILD is similar to that of the RESRAD code: the user can construct the exposure scenarios by adjusting the input parameters. Typical building exposure scenarios include long-term occupancy (resident and office worker) and short-term occupancy (remediation worker and visitor). The RESRAD-BUILD code considers six exposure pathways:

- External exposure directly from the source
- External exposure to materials deposited on the floor
- External exposure due to air submersion
- Inhalation of airborne radioactive particulates
- Inhalation of aerosol indoor radon progeny (in the case of the presence of radon predecessors)
- Inadvertent ingestion of radioactive material, either directly from the sources or from materials deposited on the surfaces of the building compartments

RESRAD-BUILD considers multiple sources of contamination:

- Residing on the surface of the floor, walls, or ceiling
- Within the building material, such as drywall, concrete floors, steel I-beam, metal pipes and wires
- Accumulated inside the building, such as in the air exchange filter or in the drain

It will provide calculated doses for external exposure (deep effective dose), internal exposure (committed effective dose equivalent - CEDE), and TEDE.

Source term information input into RESRAD-BUILD includes the following:

- Source type (point, line, area, volume) and the coordinates of its center
- The radionuclides and their activity
- Exposure scenario (office worker, resident, industrial worker, remediation worker, building visitor, or any individual spending time inside the contaminated building)
- The location (coordinates) of the receptors, for up to 10 receptors

A site-specific model may be developed if the available generic software models are inadequate for a site, but the selected model and associated parameters must be thoroughly justified. Generally speaking, regulators and stakeholders are skeptical of dose calculations derived from site-specific dose models without documented facts supporting the model. Source term knowledge is at the heart of every dose calculation.

10.5.3 Performing Final Surveys: The purpose of the final survey is to demonstrate that the facility has been properly decontaminated to meet the release criteria previously specified by the regulatory agency. Final surveys are to be conducted according to the criteria of the regulatory agency. In the case of the DOE, the final survey must satisfy the unrestricted release guidelines in DOE Order 5400.5. Decommissioning projects regulated by the NRC must meet the requirements of the approved DP. The MARSSIM guidance is applicable to those projects that use established dose- or risk-based release criteria. The approach to performing the final survey is described in detail in NUREG/CR-5849 (NRC 1992a). This document gives guidance and recommendations on proper measurement techniques. The NRC's Draft Guide DG-4006 provides guidance suitable for releasing a facility or site unconditionally (NRC 1998b).

10.5.4 Classifying Waste Streams:

1. *Identification for Disposal*: Proper characterization of waste streams is of paramount importance to a decommissioning project. Waste should be segregated into major categories, typically including LLW, mixed, hazardous, and sanitary waste.

   Each disposal site specifies the waste acceptance criteria the generator must meet. They require isotopic identification and quantification of activity. The disposal site operator may also establish additional criteria. Waste generators must certify to the disposal site that the waste they propose to send will meet all acceptance criteria. A conservative estimate of the total curie content within the volume of the material must be based on certifiable characterization data. Typical methods may include swipe-to-curie ratios based on statistical volumetric analysis.

   Mixed waste contains both radioactive and hazardous materials under the RCRA or the TSCA. Typical mixed waste could include radiological contaminated PCBs, mercury, lead, lead paint, and cadmium. They require proper characterization, segregation, and certification based on the designated disposal site.

   Nonradiological hazardous wastes include RCRA- and TSCA-regulated materials without the radiological component. These must also be properly characterized, segregated, and certified per the requirements of the receiving disposal site.

   Sanitary waste consists of material containing no radiological or hazardous waste. Several subsets exist within this category of wastes and each may have different disposal criteria invoked by the receiving disposal site. Typical wastes generated are construction rubble, asbestos-containing materials, and general debris.

2. *Packaging, Certification, and Manifesting*: The characterization required for shipping, packaging, and certification must conform to DOT regulations and the requirements of the receiving waste disposal site.

10.5.5 Considering Ancillary Contamination: Any nearby facility or surface area that has been contaminated by the facility being decommissioned must be incorporated as part of the original scope, or as a separate remediation project. This involves including the facilities nearby to all characterization, decontamination, and unconditional release characterization efforts. Ancillary facilities should be identified during the initial characterization efforts described in Section 10.2.

Examples of these areas are pipes, contaminated ground and surface water, and paved parking lots and roads where contaminants may have been paved over.

10.5.6 Project Acceptance by Management, Regulators, and Stakeholders: Decommissioning at DOE facilities must comply with the community relations and public participation requirements established by the CERCLA and DOE policy (DOE 1995a). Similar public forum policies apply to NRC-regulated facilities. Guidance on these policies is provided in Chapter 8.

Full documentation that is traceable to the original surveys, instrument calibration standards, or other primary sources, is of paramount importance for final acceptance for unconditional release. The documentation package should be of sufficient detail to enable a third party to evaluate the results of the final survey and make an independent determination as to whether the release criteria have been met. It should be anticipated that the regulatory agency would perform an independent confirmation of the final survey prior to releasing the site unconditionally.

Chapter 11

REGULATORY CHARACTERIZATION AND RISK ANALYSIS MODELING

11.1 INTRODUCTION

Performing site and facility characterization assessments is often complex. Chapter 10 presented the needs for a facility undergoing decommissioning. As discussed in Chapter 7, a key aspect to risk assessment is the evaluation of radiological exposures to hypothetically maximum exposed individuals. Computer-based tools have been developed to help address these challenges. This chapter specifically discusses three computer-based aids developed by government agencies for the conduct of decommissioning, namely RESRAD, DandD, and Compass, together with MARSSIM, and the adaptive characterization approach they support. RESRAD and DandD are used to conservatively calculate exposures to the maximum exposed individual through multiple pathways, based on measured values of residual radioactive contamination. These two codes are the primary dose modeling approaches considered acceptable by the NRC and DOE. Compass is a code used to automate some processes from the MARSSIM characterization protocols.

The MARSSIM methodology provides information on planning, conducting, evaluating, and documenting the final status radiological surveys of building surface and surface soils for demonstrating compliance with dose- or risk-based regulations or standards. MARSSIM is a multi-agency consensus document that was developed collaboratively by four federal agencies having authority and control over radioactive materials: the Department of Defense (DOD), the DOE, the EPA, and the NRC. The manual was originally issued in December 1997 as NUREG-1575 (NRC 1997a) and then as NUREG 1575, Rev.1 (NRC 2000a) in August 2000 to incorporate comments received on the 1997 version. Errata and modifications that were included in Revision 1 are shown on page xxviii of that document. In general, the two versions are quite similar; the former is part of the documentation of completed projects and the latter is used for future and on-going projects.

This chapter also addresses two approaches to characterization in decommissioning. MARSSIM is widely used to optimize a characterization program while helping ensure statistical validity of the results. The program is focused toward field radiological characterization. The second approach is an adaptive sampling and analysis program that has been demonstrated to improve cost and time performance objectives for characterization (DOE 2001c).

11.2 RESRAD

The original RESRAD code, from the term RESidual RADioactivity, was developed at ANL in 1989 (Gilbert 1989). The RESRAD Family of Codes, developed over a number of years, is described in detail on the Internet at the ANL's web site (ANL 2001). The Code Family consists of the basic RESRAD dose pathways code and six specialized pathways codes: RESRAD-Offsite, RESRAD-Build, RESRAD-Recycle, RESRAD-Biota, RESRAD-Ecorisk, and RESRAD-Chem. The early development of this code and some of its additional features are described in Section 11.2.1.

The more modern version of the base RESRAD code (RESRAD Version 5.0) was introduced in 1993 (ANL 1993a). It is a pathway analysis code to calculate radiation doses to a hypothetical individual from a contaminated site, allowing the user to define up to nine pathways and three exposure routes: external gamma radiation, inhalation, and ingestion.

Several scenarios, including residential, industrial, and recreational, can be modeled by adding or suppressing pathways and entering appropriate values for occupancy and consumption rates. One of the major decisions is the future use of the land and buildings after decommissioning. The decision will help define the contamination cleanup levels required and will help in defining the scenarios, pathways, and critical groups that are important for the site dose assessment. This decision will allow for identification of:

- Potential human activities on or near the site that could result in exposure (scenarios)
- Migration and exposure pathways of the radionuclides (pathways)
- Critical receptors (the critical group)

Scenarios are defined as reasonable and plausible sets of human activities and of future uses of the site. As such, scenarios provide a description of future land uses, human activities, and behavior of the natural system. With an understanding of the potential human activities and physical system, models of the site can be developed and translated into mathematical models. These models may be implemented through, and solved by, the application of corresponding analytical or numerical models and computer codes. The objective is to calculate a dose that is compared to dose criteria for assessing whether the site complies with applicable requirements. The modeling code RESRAD was developed to estimate these dose rates and has been used extensively in decommissioning.

The selection of an appropriate scenario, identification of its pathways critical group, and the dose assessment is site-specific. There are two generic critical groups used for identifying pathways, as described in NUREG/CR-5512.

1. *Building occupant for reuse of structures*: This scenario accounts for exposure to a fixed and removable thin layer or surface contamination sources within a structure. The building occupant is defined as a person who works in a commercial building following decommissioning. The pathways that apply to the building occupant include:

    - External exposure to penetrating radiation from surface sources
    - Inhalation of re-suspended surface contamination
    - Inadvertent ingestion of surface contamination

2. *Resident farmer for contaminated soil sites*: This scenario accounts for potential exposure to residual radioactive contamination in the surface layer of soil. The resident farmer is defined as a person who lives on a site following its remediation, grows some portion of food products, and drinks water from an on-site well. The pathways that apply to the resident farmer include:

    - External exposure to penetrating radiation from volume soil sources while outdoors
    - External exposure to penetrating radiation from volume soil sources while indoors
    - Inhalation exposure to re-suspended soil while outdoors

- Inhalation exposure to re-suspended soil while indoors
- Inhalation exposure to re-suspended surface sources of soil tracked indoors
- Direct ingestion of soil
- Inadvertent ingestion of soil tracked indoors
- Ingestion of drinking water from a groundwater source
- Ingestion of plant products grown in contaminated soil
- Ingestion of animal products grown on the site (i.e., after animals ingest contaminated drinking water, plant products, and soil)
- Ingestion of fish from a water source with a contaminated surface

This information could describe a critical group, referred to here as a site-specific critical group, which is different from the generic group. Use of a site-specific critical group would occur in cases where, for example:

- Major pathways (e.g., the groundwater pathway or agricultural pathways) associated with the generic group could be eliminated, because of physical or site use reasons
- There was a specific sensitive group on the site
- Restricted use was proposed for a site

The critical group is defined as an individual or relatively homogenous group of individuals expected to receive the highest exposure under the assumptions of the particular scenario considered. The average member of the critical group is an individual who is assumed to represent the most likely exposure situation on the basis of prudently conservative exposure assumptions and parameter values within the model calculations.

In the building occupancy scenario, a person who works in a commercial building following remediation represents an average member of the critical group. A person who lives on the site represents the average member of the critical group in the resident farmer scenario. Thus, RESRAD has a resident farmer scenario built in to compute this dose. A special purpose branch of RESRAD, RESRAD-BUILD (ANL 1994), is widely used to compute external doses to personnel from contamination on the internal surfaces of buildings and has been previously described.

Modifying scenarios and developing site-specific critical groups requires information regarding plausible uses of the site and demographic information. Such information might include 1) considerations of the prevailing (and future) uses of the land and site-specific issues, such as historical and planned future land use, and 2) physical characteristics that constrain site use. It may be necessary to evaluate several potential critical groups to determine the group receiving the highest exposure, based on different combinations of site-specific scenarios developed from expected pathways and demographics. It is especially important to evaluate the homogeneity of specific groups to determine if what appears to be one group are actually multiple groups.

Similar considerations apply for restricted release. However, now the nature of the critical group is likely to change due to site restrictions and institutional controls that can restrict certain kinds of activities or land or water uses. The detailed definition of the scenarios considered must include the impact of the control provisions on the location and behavior of the average member of the appropriate critical group. The following should be evaluated when developing site-specific scenarios, critical groups, and pathways for restricted release:

- Whether the generic scenarios defined above are applicable to the site and, if so, for each scenario
- Whether major exposure pathways can be modified or eliminated from further consideration based on site-specific conditions

Pathways may be added or eliminated, as appropriate, using site-specific data, and it is possible that different scenarios and associated pathways may be necessary for complex site-specific scenarios. This evaluation should include adequate justification, based on site-specific data, for eliminating scenarios and/or pathways from the analysis.

An analysis of exposure pathways should begin with at least the pathways described above. After considering those pathways, a more thorough pathway analysis may be needed. The objective of proceeding from generic to more site-specific pathways is to focus resources on the pathways and their associated models that have the highest likelihood of significant exposure to the critical group. Applying this pathway analysis process will result in a set of the dominant pathways for the site-specific scenarios that can be further revised using site-specific conditions.

11.2.1 Code Development: The RESRAD code was initially developed in support of residual radioactive material guidelines established by DOE Order 5400.5 (DOE 1993). ANL developed the computer program using the models that were initially developed as part of a DOE-wide effort that began in the early 1980s and involved most of the national laboratories and DOE program offices. It has been used to evaluate risk at more than 300 sites. The NRC has also approved use of RESRAD to evaluate dose at licensed nuclear facilities. The EPA uses the code in evaluating cleanups.

RESRAD determines site-specific residual radioactive material cleanup guidelines based on calculation of the radiation dose to hypothetical residents or workers on the site. The pathway estimates radiation doses and cancer risks to a critical population group. The nine environmental pathways considered in the code are direct exposure, dust inhalation, radon, and ingestion of plant foods, meat, milk, aquatic foods, water, and soil. Figure 11.1 shows these dose pathways.

The special purpose code, RESRAD-BUILD evaluates potential health impacts in buildings contaminated with radioactive materials. A variety of building compartments and contamination forms are modeled, along with corresponding human exposure pathways.

11.2.2 Code Evolution: Risk assessment, the likelihood for an adverse effect to occur, has become an important tool in the determination of environmental quality objectives. It can be applied to the evaluation of a wide range of technological and environmental issues in a variety of environmental and demographic settings. To do so requires a set of analytical methods to quantify the hazards or sources of the risk – the pathways or mechanisms resulting in exposure to sensitive human or ecosystem populations, and the nature/extent of the impacts resulting from exposure. Developing a thorough understanding of the perceived (as compared to actual) risk and the resulting social and economic impacts can be an equally important analytical requirement in the evaluation of options for limiting or managing risk.

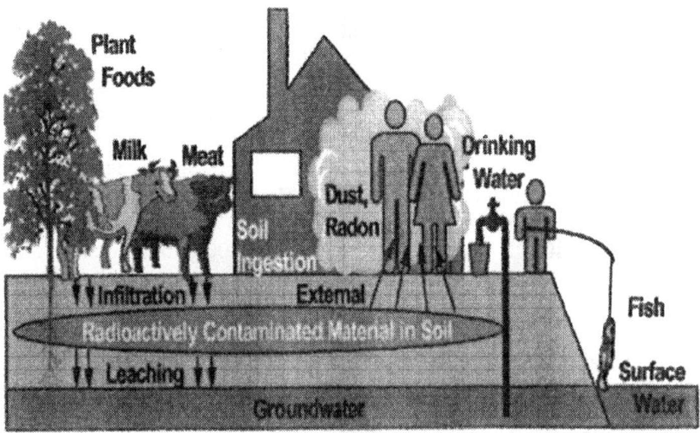

Figure 11.1 Pathways Considered in RESRAD

Risks to off-site receptors from contamination migrating from on-site sources are assessed using a variety of computational methods that evaluate contaminant exposure through multiple pathways. Contaminants and/or pathways contributing the majority of risk are identified, and any ensuing risk mitigation may be allocated to deliver the greatest reduction in risk to the identified receptors.

In 1999, the NRC tasked the adapting of existing RESRAD and RESRAD-BUILD codes for use in site-specific dose modeling and analysis. Subsequently the code was revised and customized to be consistent with the current NRC guidance for both deterministic and probabilistic dose modeling being developed in the Standard Review Plan for Decommissioning (NRC 2000b).

The objective of dose modeling is to assess TEDE to an average member of the critical group from residual contamination, including any contamination that has reached ground sources of drinking water. The assessment offers a reasonable translation of residual contamination through probabilistic dose analysis and ranking parameters according to relative importance. Compliance with the NRC-prescribed dose criteria may then be assessed by the modeling results.

A composite scoring system was developed to rank the parameters. The rankings were divided into three levels – high, medium, and low priority. Then they were ranked on the basis of four criteria:

- Relevance of the parameter in dose calculations
- Variability of the radiation dose as a result of changes in the parameter value
- Parameter type (physical, behavioral, or metabolic)
- Availability of data on the parameter in the literature

Overall, 14 parameters were ranked as high priority. The data were obtained from a variety of published sources representative of a national distribution. Because they are based on national averages, it may not be fully appropriate for a site-specific assessment. Potential correlation among parameters was also studied. The data obtained and used as the input are described in the Data

Collection Handbook (ANL 1993b), which provides the data definitions, typical ranges, default values used in RESRAD, relation to other parameters, and most measurement techniques.

11.2.3 Using the Code: The RESRAD code is a user-friendly, multiple pathway analysis code designed to be run on personal computers. Hydro-geological, meteorological, geochemical, geometrical (size, area, depth), and material-related (soil, concrete) parameters are used. RESRAD has a built-in parameter sensitivity analysis capability. Users are referred to the User's Manual for RESRAD Version 6 (ANL 2001) for use of this feature and to Parameter Distributions for Use in RESRAD and RESRAD-BUILD Computer Codes (Kamboj 2000) for guidance.

11.3 DandD

Dose assessment typically involves mathematical models to represent the contaminated site. The codes used in dose assessment may be for generic screening or be site-specific. A screening analysis usually uses simple sites with building surface and/or with surface soil contamination. Simple and conservative models/codes and parameters, under generic scenarios and default site conditions, are usually employed to define the screening DCGL equivalent to the dose criteria. Because of the conservative nature of the screening analysis approach, the resultant levels obtained are typically more restrictive than the site-specific levels. Currently, the only generic screening code acceptable to the NRC is DandD, Version 2.1. The source code and user manual for DandD Version 2.1 are documented in NUREG/CR-5512, Volume 2 (NRC 2001a). This section presents a generic description of the code and describes the generic issues associated with this screening model.

Several different types of dose modeling scenarios may be used. The most common are:

- Building occupancy (generic screening – based on NUREG/CR-5512)
- Residential farmer (generic screening – based on NUREG/CR-5512)
- Urban construction (contaminated soil, no suburban or agricultural uses). This scenario is meant for small urban sites cleared of all original buildings; only contaminated land and/or buried waste remains
- Residential (a restricted subset of the residential farmer scenario, for those urban or suburban sites where farming is not a realistic projected future use of the site)
- Recreational (where the site is preserved for recreational uses only)
- Hybrid industrial building occupancy (adds contaminated soil; building may or may not be contaminated)
- Drinking water (no on-site use of groundwater; off-site impacts from the contaminated plume)

The building occupancy and the residential scenario can be evaluated with DandD. The building occupancy scenario accounts for exposure to fixed and removable residual radioactivity on the walls, floor, and ceiling, and presumes that the building will be used for commercial or light industrial activities (e.g., an office building or warehouse). The resident farmer scenario accounts for exposure involving residual radioactivity that is initially in the surface soil (the top 6 cm). The DandD software assists in determining the extent of decommissioning required for allowing unrestricted release of property. DandD automates the scenarios, models, mathematical formulations, and assumptions documented in NUREG/CR-5512, Volume 1 (NRC 1992b).

11.3.1 Code Development: NRC staff tested and evaluated the code's performance and conducted code/model comparisons with the deterministic RESRAD and RESRAD-BUILD codes (NRC 1999a). Licensees and stakeholders also evaluated the performance of DandD. These evaluations identified several areas where DandD may be overly conservative.

A probability density function was established for the range of values for each parameter. A single set of default parameters was selected by sampling the probability density functions, to maintain a 90 percent confidence level that doses would not exceed the dose limit for a combination of all radionuclides. A detailed discussion of the way the default parameters were selected is contained in NUREG/CR-5512, Volume 3 (NRC 1999b).

11.3.2 Limitations: An acceptable dose assessment analysis need not incorporate all the physical, chemical, and biological processes at a site. The scope of the analysis and level of sophistication in the conceptual model should be based on the overall objective of the analysis. A conceptual model can be simple if it still provides satisfactory confidence of performance. For an initial screening analysis, little may be known about the site from which to develop a conceptual model. Computer codes used for screening analyses are generally intended to provide a generic and conservative representation of processes and conditions expected for a wide array of sites. Accordingly, the generic conceptual model in such codes may not provide a close representation of conditions and processes at a specific site. Such a generic representation is still acceptable, as long as it provides a conservative assessment of the performance of the site.

11.3.3 Issues in Performing Screening Analysis: Major issues with screening analysis include:

- The definition of screening and the transition to a site-specific analysis.
- Qualification of the site for screening, in terms of site physical conditions and compatibility with the modeling code's assumptions and default parameters.
- The acceptable screening tools (e.g., code, look-up tables), approaches, and parameters that can be used to translate the dose into equivalent screening concentration levels.

A screening analysis, for demonstrating compliance with the dose criteria in Part 20, Subpart E, may not be applicable for certain sites because of the status of contaminants (e.g., location and distribution of radionuclides) or site-specific physical conditions. Therefore, the site source term characteristics should be evaluated to ensure consistency with the source term assumptions in the screening model/code used. In addition, specific physical conditions at the site should be reviewed to ensure that they are valid for the model and code assumptions. The selected screening parameters and pathways should be reviewed to ensure that they are consistent with the parameters and pathways of the code. Lastly, a review should ensure that the site can be evaluated within the envelope for a simple site. If the determination is made that the site is complex, then DandD may be inappropriate and the RESRAD code would be more suitable.

When using the screening approach for demonstrating compliance with the dose criteria in 10 CFR Part 20, Subpart E, the facility assessment must demonstrate that the particular site conditions (e.g., physical and source term conditions) are compatible and consistent with the DandD model assumptions. In addition, the default parameters and default scenarios/pathways must also be used in the screening dose analysis. Therefore, reviewers should examine the site conceptual model and the generic source term characteristics.

11.3.4 Site-Specific Analysis: For site-specific analyses, the intent is to provide a realistic assessment of doses based on more site-specific information and/or data since more is known from which to develop a conceptual model. The licensee should provide a schematic or verbal description of the problem to be analyzed. Even when using a computer code that has a predefined conceptual model, it is important for the licensee to identify any site features or conditions that may differ from those assumed in the code. In developing a site-specific conceptual model or identifying potential limitations with a predefined one, the following issues should be considered:

- Assimilation of existing data and information
- Scenario definition and pathway identification
- System conceptualization
- Dose assessment

The following list identifies site features and conditions that may be incompatible with code assumptions:

- Highly heterogeneous radioactivity
- Wastes other than soils (e.g., slags and equipment)
- Multiple source areas or with radionuclides that may generate gases
- Contaminated zones thicker than 15 cm (6 in)
- Chemicals or a chemical environment that could facilitate radionuclide releases (e.g., colloids)
- Soils that have preferential flow conditions that could lead to enhanced infiltration
- A perched water table, surface ponding, or no unsaturated zone
- Groundwater discharging to springs or surface seeps
- Existing groundwater contamination
- Potential groundwater use is not expected to be located immediately below the contaminated zone
- Significant transient flow conditions
- Significant heterogeneity in subsurface properties
- Fractured formations
- Groundwater dilution that would be less than 2000 m^3 (70,000 ft^3)
- Overland transport of contaminants is of potential concern
- Radionuclides that may generate gases
- Stacks or other features that could transport radionuclides off the site at a higher concentration than exists on site

11.4 MARSSIM

The MARSSIM, in both its 1997 and 2000 revisions (NRC 1997a and 2000a), is a nationally consistent consensus approach to conducting radiation surveys and investigations at potentially contaminated sites. This approach has enough should be both scientifically rigorous and flexible enough to be applied to a diversity of site cleanup conditions. The title includes the term survey because it provides information on planning and conducting surveys. It also includes the term investigation because the process outlined in the manual allows one to begin by investigating any site (i.e., by gathering data or information) that may involve radioactive contamination.

A final site survey typically requires a demonstration to the responsible federal or state agency that a cleanup effort was successful and that the release criterion (a specific regulatory limit) was met. The demonstration of compliance with respect to conducting surveys is comprised of three interrelated parts:

1. *Translate*: Translating the cleanup/release criterion (dose per year, specific risk) into a corresponding DCGL level (surface or soil contamination levels) through the use of environmental pathway modeling (the screening code DandD, and the more detailed RESRAD).

2. *Measure*: Acquiring scientifically sound and defensible site-specific data on the levels and distribution of residual contamination, as well as levels and distribution of radionuclides present as background, by employing suitable field and/or laboratory measurement techniques.

3. *Decide*: Determining that the data obtained from sampling does support the assertion that the site meets the release criterion, within an acceptable degree of uncertainty, through application of a statistically based decision rule.

The MARSSIM furnishes guidance on the *Measure* and *Decide* modules for contaminated soil and buildings. This guidance describes a performance-based approach for demonstrating compliance with a dose- or risk-based regulation. This approach includes processes that identify data quality needs and may reveal limitations that enter into conducting a survey. The data quality needs, stated as Data Quality Objectives (DQOs), include performance measures and goals in relation to a specific intended use of the data (EPA 2000a and 2002a).

The DQO must be developed on a site-specific basis. However, because of the large variability in the types of radiation sites, it is impossible to provide criteria that apply to every situation. As an example, MARSSIM presents a method for planning, implementing, assessing, and making decisions about regulatory compliance at sites with radioactive contaminants in surface soil and on building surfaces.

In particular, it describes generally acceptable approaches for:

- Planning and design work scopes, characterization, remediation support, and final status surveys for sites with surface soil and building surface contamination
- Historical site assessment
- QA/QC in data acquisition and analysis
- Conducting surveys
- Field and laboratory methods and instrumentation, and interfacing with radiation laboratories
- Statistical hypothesis testing and the interpretation of statistical data
- Documentation

Thus, MARSSIM incorporates standardized and consistent approaches for planning, conducting, evaluating, and documenting environmental radiological surveys, with a specific focus on the final status surveys carried out to demonstrate compliance with cleanup regulations. Since these approaches may not meet the DQO at every site, other methods may be used as long as an equivalent level of performance can be demonstrated.

Several issues related to releasing sites are beyond the scope of MARSSIM. These include translation of dose or risk standards into radionuclide specific concentrations, or demonstrating compliance with ground or surface water regulations. MARSSIM can be applied to surveys performed at vicinity properties, but the decision to apply it is outside the scope of the manual. Other contaminated media (e.g., subsurface soil, building materials, groundwater) and the release of contaminated components and equipment are also not addressed by the manual. With the main focus on final status surveys, it continues a process of following remediation activities that are intended to remove subsurface contaminants. Therefore, some of the reasons for limiting the scope of the guidance to contaminated surface soils and building surfaces include the following:

1. Contamination is limited to these media for many sites following remediation.
2. Since many sites have surface soil and building surfaces as the leading source of contamination, computer models that calculate concentrations based on dose or risk generally consider only these as source terms.
3. MARSSIM also recognizes that there may be other factors, such as cost and stakeholder concerns, that have an impact on design surveys but are outside its scope. Unique site-specific cases may arise that require a modified approach beyond what is presently described in the manual. Such cases may be associated with:
   - The release of sites contaminated with naturally occurring radionuclides in which the concentrations corresponding to the release criteria are close to background variability.
   - Sites where a reference background cannot be established. (The process of planning, implementing, assessing, and making decisions about a site described in the manual is applicable to all sites, even if its examples do not meet a site's specific objectives.)

MARSSIM thus becomes a formal framework for planning data acquisition surveys so that the data sought will yield the kind of information actually needed for making key decisions, such as whether or not to release a particular site following remediation.

11.4.1 Structure of the Manual: MARSSIM is presented in a modular format, with each module containing guidance on conducting specific aspects of, or activities related to, the survey process. Followed in order, each module leads to the generation and implementation of a complete survey plan. Although this approach may involve some overlap and redundancy in information, it also allows many users to concentrate only on those portions that apply to their own particular needs or responsibilities. The procedures within each module are listed in order of performance, and options are provided to guide a user past portions that may not be applicable. Where appropriate, checklists condense and summarize major points in the process. The checklists may be used to verify that every suggested step is followed, or to flag a condition requiring specific documentation to explain why a step was not needed.

Based on a graded approach, MARSSIM also contains a simplified procedure that many users of radioactive materials may, with regulatory approval, be able to use to show compliance with the release criterion. Sites that may qualify for simplified release procedures are those where radioactive materials were:

- Of relatively short half-life and have since decayed to insignificant quantities
- Kept only in small enough quantities so as to be exempted or not require a specific regulatory license
- Used or stored only in the form of leak-proof, sealed sources
- Combinations of the above

11.4.2 Roadmap: The manual includes a roadmap design as a quick reference for users already familiar with the process of planning and performing radiation surveys. It offers basic guidance, references to further guidance, and rules of thumb. Detailed guidance includes the planning, implementing, and evaluating of environmental and facility radiological surveys to demonstrate compliance with a dose- or risk-based regulation. MARSSIM focuses on demonstration of compliance during the final status survey following scoping, characterization, and remedial action.

11.5 COMPASS

COMPASS is the Computerization of the MARSSIM for Planning and Assessing Site Surveys. Developed by Oak Ridge Associated Universities (ORISE 2002), it was designed as a tool for facilitating the statistical calculations used in planning the MARSSIM final status surveys. It should be operated under the same premise as the manual in that it is not prescriptive and provides flexibility for some site-specific situations. The use of COMPASS does not guarantee compliance with applicable regulations; its use is simply to automate some of the processes and protocols.

Two useful modules of COMPASS are practice in using MARSSIM and site planning. The practice module allows for exploration of the development of statistical testing and prospective power for a hypothetical survey. It also provides for practice in the use of surrogates and DCGL modification for soil surveys. The site planning module allows for planning the final status surveys using DQO self-help "wizards." When the self-help is complete, COMPASS can generate a survey plan report. When the surveys are complete, the statistical validity of the data is evaluated to determine if the release criteria have been satisfied. Self-help wizards are also provided for this process.

COMPASS also provides listings of MARSSIM-recommended survey unit sizes and listings of recommended scan minimal detectable concentrations for NaI detectors obtained from NUREG-1507 (NRC 1998a). It also extracts a listing of reference background count rates for various instruments. Lastly, it provides a discussion on the use of surrogate measurements. COMPASS can be downloaded from the Oak Ridge web site on the Internet (ORNL 2002a).

11.6 ADAPTIVE SAMPLING AND ANALYSIS PROGRAM

Typical site characterization methods rely on pre-planned sampling programs and off-site analysis of samples to determine the extent, specific composition, and quantity of hazardous and radioactive contamination. This process is usually very labor intensive, hence, costly and time-consuming. Static work plans specify the numbers and locations of samples to be collected and the analyses to be performed on those samples. Sampling crews are mobilized, samples collected, and the crews demobilized before final results become available. The drawbacks of this traditional approach are the high costs per sample, pressure to over-sample to ensure that all needed samples are taken in one campaign, and the additional sampling programs often required for resolving uncertainties raised by the initial analysis results.

A key step in the characterization of hazardous and radioactive wastes is determining the extent of contamination. The proper number and placement of sampling locations is required to minimize characterization costs and ensure reasonable confidence in the estimated extent of contamination. Because soft information (i.e., historical records, computer modeling results, past experience, etc.) for a site is usually just as important as hard laboratory results, the approach taken must include a quantitative way of accounting for both soft and hard site data.

An alternative to traditional sampling programs is an adaptive sampling and analysis program, which relies on field analytical methods to generate sample results quickly enough to have an impact on the course of the sampling program. Rather than a static work plan, an adaptive program is based on dynamic work plans that specify the logic for how sampling numbers, locations, and analyses will be determined as the project proceeds. It also relies on rapid, field-level decision-making to ensure that the sampling stays on track. Such a program requires field analytical methods that are appropriate for the types of expected contaminants and media present at the site, and a way of supporting decision-making in the field that is appropriate for the goals of the project.

Adaptive sampling and analysis programs use a dual approach to the sampling strategy problem. First, they use a Geographical Information System (GIS) specifically designed for site assessment work to integrate, manage, and display site characterization data as it is being generated. Such a GIS is Site Planner™, developed by ConSolve, Inc. as a graphical, object-oriented database designed to provide qualitative support of environmental site assessments. Its purpose is to give technical characterization staff a sound understanding of their site data in near real time. Coupled with the GIS is Plume™, an ANL interactive software package. Plume™ was developed as a separate technology and adds quantitative support to adaptive sampling and analysis programs. The software merges soft site data with hard sample results to form images of contamination location, provide quantitative measures of the potential benefits to be gained from additional sampling, and indicate the location of the best new sampling locations. Plume™ uses advanced Bayesian and geostatistical procedures to complete these tasks. In this approach, using site maps and all available information, the analyst develops prior probabilities of the exceeding of threshold concentration levels for constituents of concern. Output generated includes:

- Graphics, such as maps, fence diagrams, and boring logs that provide the characterization staff with a qualitative picture of the extent of contamination and its environmental context
- Measures of contaminant extent and its uncertainty
- Estimates of the benefits to be gained by obtaining additional samples
- Locations of the best new sampling locations

The most recent developments have focused on integrating these techniques into soil remedial actions to make those actions more precise.

The adaptive sampling and analysis approach, supported through use of SitePlanner™ and Plume™, is designed specifically for characterizing the presence and extent of contamination in groundwater, surface soils, and subsurface soils. The baseline method for characterizing the extent of contamination includes rigid pre-planned field sampling events, with selective sampling based on best engineering judgment. Specimens are sent off site for analysis and can require days to months for turnaround of sampling results depending on the sampling and analysis protocols. Multiple sampling events are typically required to reach a high level of confidence that characterization is complete. Conversely, the adaptive program provides optimization of the sampling locations' on-site analysis using lower cost field analytical methods, and single-stage sampling to arrive at the same levels of confidence (DOE 2001c). Four distinct advantages of the adaptive program over the baseline method are as follows:

1. *Better than traditional approaches*: Plume™ can estimate the value that additional sampling data may provide, allowing stakeholders to weigh benefit/costs of collection. SitePlanner™ allows rapid site visualization as the data are generated. Both tools provide for better characterization.
2. *Faster than traditional approaches*: fewer samples are collected and additional field sampling events can be eliminated, which results in expedited site characterization.
3. *Safer than traditional approaches*: worker exposure is reduced with fewer samples collected and fewer field-sampling events. Many field measurements can be performed in-situ, which reduces or even eliminates wastes generated during the sampling process.
4. *More economical than traditional approaches*: fewer samples are collected and overall project costs are reduced.

When adaptive sampling and analysis is used to support soil remediation work, a more precise excavation plan can be developed. This reduces overall remediation costs by focusing the work on those soils that fail restoration goals.

11.6.1 Technology Description: The adaptive approach to sampling and analysis uses real-time data collection techniques and in-field decision-making to guide the progress of data collection at hazardous waste sites. An adaptive approach to site characterization and/or remediation is based on a dynamic work plan that specifies how data collection decisions will be made in the field; it does not, however, specify the exact locations and numbers of samples to be collected. In a typical data collection program, the off-site laboratory analysis of soil samples is the primary QA/QC check for the real-time data; it is not used as the principal data source for decision-making. During an adaptive program, the course of data collection work is driven by the results as they are obtained. In its extreme form, the next sampling location might be determined by all previous results. More commonly, data planning and acquisition take place in sequential segments. For example, results from one day's work might be used to plan the activities scheduled for the next day. Figure 11.2 graphically illustrates this process.

An adaptive program requires three key components to be effective:

- Real-time data collection techniques appropriate for the contaminants of concern and their cleanup guidelines

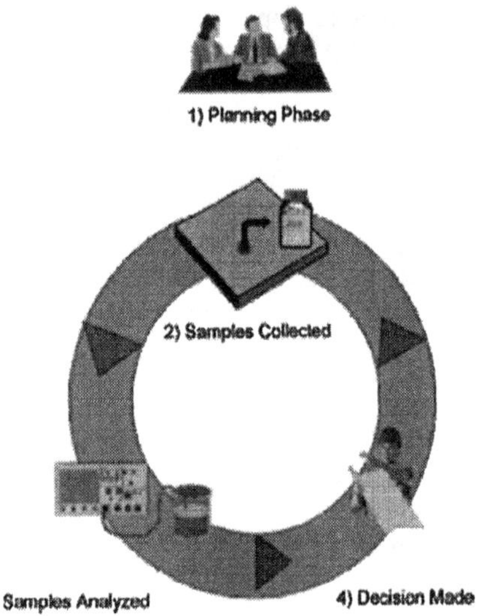

Figure 11.2 The Adaptive Sampling and Analysis Process

- An in-field decision-making methodology for determining the course of data collection in response to real-time data streams
- Rapid in-field decision-making for qualitative and quantitative decision support (DOE 2001c)

Qualitative decision support is defined as providing on-site technical staff with an accurate understanding of the progress concerning the sampling program as quickly as possible. Since large adaptive programs produce hundreds of samples per day, managing, integrating, and displaying the results pose a serious logistical problem that might interfere with program progress if not adequately addressed.

A typical configuration includes some type of in-field database system, along with some form of GIS for data display. Good qualitative support is the prerequisite for quantitative decision-making. SitePlanner™ was developed for environmental management and display needs. Several other competitive software packages are available, including ArcView® GIS and SiteView.

Quantitative decision support requires the ability to estimate contaminant extent based on sampling results, determine the uncertainty associated with those estimates, measure the expectations from additional sampling (i.e., reductions in uncertainty), and find new sampling locations that yield the most value. Such support must take into account two general characteristics of contamination at hazardous waste sites. The first is that, while there may be initially few, if any, discrete sample results to use as a base for a sampling program, there typically is a wealth of pertinent information. The second is that spatial auto-correlation is usually present at hazardous waste sites and must be accounted for when drawing conclusions from discrete sample results. As noted earlier, Plume™ furnishes quantitative support for adaptive sampling programs using a combination of Bayesian analysis with geostatistics to guide the program design and implementation.

Geostatistics is used to interpolate sampling results from locations where hard data exist (results from the analysis of collected samples) to other locations that lack sampling data. Geostatistics is grounded in the presence of spatial auto-correlation -- the fact that two samples collected very close to each other will have similar results, but samples separated by a large distance may have results that are totally unrelated. When sample results are correlated and the level of correlation is a function of the distance separating the samples, spatial auto-correlation exists. For the purposes of contaminant extent delineation, the primary issue is not the absolute value of a contaminant observed but whether that value exceeds some action level or cleanup goal. In this context, sample results can be reduced to being either above or below an action level. A specialized form of statistics, called indicator kriging, may be used to determine the spatial distribution of contamination above and below action levels. The design and implementation process for contamination delineation follows these steps:

1. *A set of decision points forming a regular grid is laid across the site.* Decision points are so named because at each point a decision will have to be made, based on the available information. Will this point be considered clean (i.e., the probability of contamination above the prescribed action level is acceptably low) or contaminated (i.e., the probability is unacceptably high)? The acceptable level of uncertainty serves as the criterion for differentiating between decision points that may be considered clean and points that must be treated as contaminated. (This value must be selected with agreement from the regulators and stakeholders involved.) For example, if the acceptable level of uncertainty is set at 0.2, a decision point with probability of contamination greater than 0.2 will be considered contaminated, while decision points with less than 0.2 will be considered clean. This treatment of uncertainty is consistent with the Type I and Type II error analyses advocated by the EPA's DQO approach to environmental restoration decision-making (EPA 2000a).

2. Based on the soft information available for the site (e.g., site maps, historical, past experience with similar sites, etc.), *a probability is assigned to each decision point* that captures the initial assumptions for the probability of contamination above action levels at that location. Some cases will yield absolute certainty of its presence or absence, while others may not be sufficiently definitive to draw any conclusion. The result is the initial conceptual model for the site.

3. *The probabilities at each of the decision points are updated with hard data*, if sample results are initially available. Bayesian analysis can be combined with indicator geostatistics to accomplish the required updating (ASTM 1996). The site is then broken into three regions:
   - Where the probability of contamination associated with decision points is below the predefined acceptable level of uncertainty (accepted as clean, with perhaps only minimal confirmatory sampling)

- Where the probability of contamination is so high that there is no need for sampling to confirm the presence of contamination
- Where the probability of contamination above action levels is neither very low nor very high (gray regions that represent areas of uncertainty in the context of the presence or absence of contamination above prescribed action levels)

4. *The final step is actual sampling*. There are alternative decision rules to "drive" data collection. The most common is to focus on maximizing the area defined as clean. This rule tends to produce an adaptive sampling program that starts at the fringe of known contamination and works its way in. As data are collected, the underlying probability model is updated using Plume™, the value of collecting additional information evaluated, and additional sampling locations selected that maximize the area classified as clean. Sampling stops when the additional value of sampling no longer warrants the investment. This becomes a simple cost calculation that weighs program costs with the expected volume of clean soil and thereby avoids remediation costs if sampling moves forward. Other potential decision rules include maximizing the area known to be contaminated or minimizing the area categorized as uncertain.

The process is the same regardless of the decision rule used. Sampling locations are selected to provide the most benefit in the context of the selected rule. These would be sampled, their results analyzed, the probabilities of contamination associated with the decision point grid updated with the sample results, the extent of contamination determined again (along with the number of "uncertain" decision points remaining), and a decision made whether additional sampling locations are justified. If so, the next best set of locations would be selected and the process carried through iteration. The program stops when the expected information gain from additional sampling does not warrant the cost of collecting and analyzing additional samples. SitePlanner™ and Plume™ are used to optimize sampling strategies. This is shown graphically in Figure 11.3.

11.6.2 Baseline Technologies and Alternatives: The baseline method for characterizing the extent of contamination includes pre-planned field sampling events with selective sampling based on best engineering judgment. A phased approach is typically used for data collection, using a statistical basis or regular grid to develop a conceptual model. Samples are sent for lab analysis and require days to months for turnaround, depending on the protocols required. Multiple sampling events may be required to reach assurance that characterization is complete. A variety of baseline approaches for site characterization are discussed herein.

11.6.2.1 The Observational Method: This is an investigative process developed and used in the 1920s to 1950s for geotechnical characterization of soils and engineering design prior to construction. Characterization, design, and construction proceed hand in hand. As construction proceeds, observed changes in the soil system are used to modify the design. A critical element of the method is an early assessment of most probable conditions. Application to characterization or remedial investigation would mean focusing on determining general site conditions and identifying the most probable conditions and reasonable deviations as the basis for a flexible approach to remediation. The observational method does not provide explicit sampling design guidance.

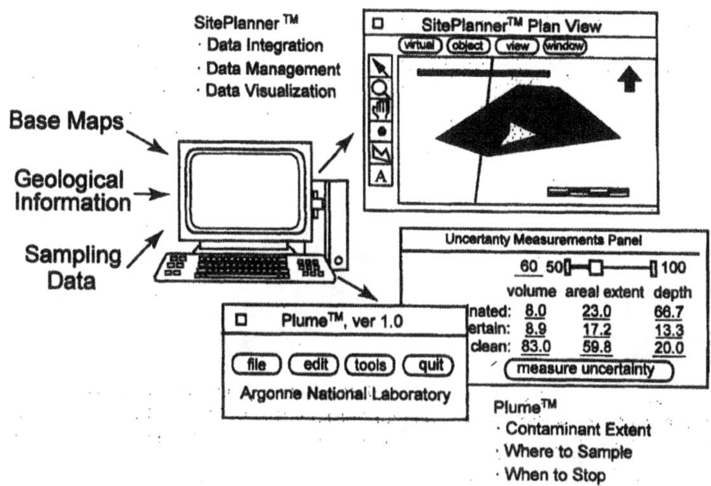

Figure 11.3 Approaches to Optimize Sampling Strategies

11.6.2.2 Expedited Site Characterization: This method was developed as a time-saving, cost-effective approach for hazardous waste site investigations (DOE 1998a). The expedited approach effectively shortens the length of the assessment period and may significantly reduce costs at many sites. It is not a specific technology or system but a methodology for effectively and efficiently conducting a site characterization. The principal elements include:

- A field investigation conducted by an integrated team of experienced professionals working together
- Analysis, integration, and validation of the characterization data as they are obtained
- A dynamic work plan that lets the team take advantage of new insights from recent data to adjust the work plan

This methodology emphasizes the delineation of the hydro-geological framework of the potentially contaminated site, followed by the delineation of the contaminant pathways and distribution, and the selection of the most effective measurement technologies. Adaptive sampling and analysis programs (ASAPs) can be integrated into an expedited approach for site characterization.

11.6.2.3 SACM and SAFER: The Superfund Accelerated Cleanup Model (SACM) was developed by the EPA; the Streamlined Approach for Environmental Restoration (SAFER) was developed by the DOE. These are complementary approaches for speeding up CERCLA remedial investigation/ feasibility study projects. SAFER integrates DQO with the observational method. An adaptive sampling and analysis program approach can be easily incorporated into its characterization portion to quickly yield a conceptual model of the site.

Data collection proceeds until the characterization goals are met. Consequently, the need for additional site characterization efforts is greatly reduced. In contrast, traditional characterization programs that rely on off-site laboratory analyses for information often require repeated mobilizations to clarify sample results.

Chapter 12

COST ESTIMATES AND SCHEDULE

12.1 INTRODUCTION

The decommissioning cost estimate and schedule do not stand-alone; they are an integral part of the planning for a project from the concept to the final implementation. The cost estimate and schedule are linked inseparably, as changes to the cost affect the schedule, and vice versa. Only with an accurate cost estimate and schedule can management usefully track costs and project trends.

Reliable cost estimating is one of the most important elements of decommissioning planning. Alternative technologies may be evaluated and compared based on efficiency and effectiveness, and measured against a baseline cost as to the feasibility and benefits derived. When the plan is complete, those cost considerations ensure that it is economically sound and practical for funding.

Estimates of decommissioning costs have been performed and published by many organizations. The results of an estimate may differ because of different work scopes, different labor force costs, different money values because of inflation, different oversight costs, the specific contaminated materials involved, the waste stream and peripheral costs associated with that type of waste, or applicable environmental compliance requirements. Some of the divergence in costs, however, cannot be easily explained, and this lack of consistency prohibits direct estimating by measuring standard quantities, such as initial capital cost, facility size (megawatts), square footage of facilities, or volumes of waste streams. At some point, it may be possible to multiply one or more of these measurements by some predetermined number to arrive at a cost estimate. But until such a system is proven to be reliable, a reasonable degree of reliability and accuracy can only be achieved by developing decommissioning cost estimates on a case-by-case basis.

There is no universally accepted standard for developing cost estimates, or for that matter a clear unambiguous reference for terminology used in decommissioning. However, AACE International, founded as a resource for cost estimating methods, has established an education and certification program for cost estimators to lend consistency to the process, and has published a book to guide estimators in the new and evolving practices from all facets of industry (AACE 1984).

12.2 TYPES OF COST ESTIMATES

Not all cost estimates represent actual costs. To maintain computational precision when working with computers, it is often necessary to carry a calculation to the nearest whole dollar (or even worse, to the nearest cent). The reason for that apparent precision is that otherwise, spreadsheets will not balance, series calculations won't close, and sums won't match the inputted values. But the concomitant tendency to believe that such precision exists in the real world, and to expect such estimates to be cast in stone, is obviously troublesome to the estimator. Similarly, the accuracy of available input information will also affect the developed cost estimate; it will generally improve as the project implementation grows closer. Nevertheless, it remains true that when two estimators have developed costs for the same project, comparisons are nearly impossible to correlate, and raise doubt in the mind of the reviewer as to the veracity of either estimate.

To supply some help here, the AACE identifies the types of cost estimates and corresponding levels of expected accuracy. These cost estimate types are summarized in the following paragraphs:

1. *Order-of-Magnitude Estimate*: One without detailed engineering data, where an estimate is prepared using scale-up or -down factors, and approximate ratios. It is likely that the overall scope of the project has not been well defined. The level of accuracy expected is -30% to +50%.

2. *Budgetary Estimate*: One based on the use of flow sheets, layouts, and equipment details, where the scope has been defined but the detailed engineering has not been performed. The level of accuracy expected is -15% to +30%.

3. *Definitive Estimate*: One where the details of the project have been prepared and its scope and depth are well defined. Engineering data would include plot plans and elevations, piping and instrumentation diagrams, one-line electrical diagrams, and structural drawings. The level of accuracy expected is -5% to +15%.

It is apparent from these estimate types and levels of accuracy that even in the most accurate case, a definitive estimate is only accurate to -5% to +15 %. The cost estimator exercises judgment on the level that the data will support. In developing a funding basis for a project, the estimator must include enough margin to account for a potential budget overrun in line with the uncertainty.

12.3 DEVELOPING THE COST ESTIMATE

This section describes cost-estimating methods and their appropriate application, and recommends a specific approach for decommissioning planning and cost estimating.

12.3.1 Approaches to Cost Estimating: Costs may be estimated in a number of ways. Recorded experience from other decommissioning projects, estimating handbooks, and equipment catalogs performance data, among other sources, may be used to develop cost data. The techniques used for preparing cost estimates will necessarily vary with the project's degree of definition; the state-of-the-art of the project; the availability of databases, cost estimating techniques, time, and cost estimators; and the level of engineering data available. A study of the item or task, in light of the degree of estimating difficulty, should indicate the method or combination of methods to be used in estimating the cost of that particular item or task. Some common estimating techniques are described herein.

1. *Bottom-up Technique*: Generally, a work statement and specifications or a set of drawings are used to "take off" (extract) material quantities required for executing each discrete task performed in accomplishing a given operation. From these quantities, direct labor, equipment, and overhead costs can be derived.

2. *Specific Analogy Technique*: Specific analogies depend upon the known cost of an item used in prior systems as the basis for the cost of a similar item in a new system.

Adjustments are made to known costs to account for differences in relative complexities of performance, design, and operational characteristics.

3. *Parametric Technique*: Parametric estimating requires historical databases on similar systems or subsystems. Statistical analysis is performed on the data to find correlations between cost drivers and other system parameters, such as design or performance parameters. The analysis produces cost equations or cost estimating relationships that may be used individually or grouped into more complex models.

4. *Cost Review and Update Technique*: An estimate may be constructed by examining previous estimates of the same or a similar project for internal logic, completeness of scope, assumptions, and estimating methodology.

5. *Expert Opinion Technique*: This may be used when other techniques or data are not available. Several specialists may be consulted iteratively until a consensus cost estimate is established.

The method recommended in this handbook is the bottoms-up technique, based on a building block approach known in DOE nomenclature as the work breakdown structure (WBS). The building block approach follows the same logic whether the estimate is being generated to support a demolition or a construction scenario. In this approach, a decommissioning project is divided into discrete, measurable work activities, ideally at a level of detail sufficient for the estimate to apply to all occurrences of the activity. This estimating approach was developed and presented in the AIF/NESP-036 report (LaGuardia 1986).

12.3.2 Cost Element Definitions: It is constructive and helpful to group elements of costs into categories to better determine how they affect the overall cost estimate. To that end, the cost elements are broken down into activity-dependent, period-dependent, and collateral costs as defined in the following paragraphs. Contingency, another element of cost, is applied to each of these elements on a line-item basis (as will be described separately) because of the unique nature of this element of cost.

1. *Activity-Dependent Costs*: Activity-dependent costs are those costs associated with performing decommissioning activities. Examples of such activities include removal, decontamination, packaging, shipping, and burial. These activities lend themselves to the use of unit cost and work productivity factors (or work difficulty factors) applied against the plant and structures inventories to develop the decommissioning cost and schedule.

2. *Period-Dependent Costs*: Period-dependent costs include those activities associated primarily with the project duration: engineering, project management, dismantling management, licensing, health and safety, security, energy, and quality assurance. These are primarily management staffing level costs, developed by estimating the manpower loading and associated overhead costs based on the scope of work to be accomplished during individual phases within each period of the project.

3. *Collateral and Special Item Costs*: In addition to activity and period-dependent costs, there are costs for special items, such as construction or dismantling equipment, site

preparation, insurance, property taxes, health physics supplies, processing liquid radwaste, and independent verification surveys. Such items do not fall in either of the other categories. Development of some of these costs, such as insurance and property taxes, is obtained from owner-supplied data.

4. *Contingency*: Contingency is defined by the AACE as "a specific provision for unforeseeable elements of cost within the defined project scope, particularly important where previous experience relating estimates and actual costs has shown that unforeseeable events that increase costs are likely to occur."

The cost elements in a cost estimate are based upon ideal conditions. All activities are performed within the defined project scope, without delay or interruption, inclement weather, tool/equipment breakdown, craft labor strikes, waste shipment problems, or burial facility waste acceptance criteria changes, and changes in the anticipated plant shutdown conditions, etc. Such is not the real world of any human endeavor. As with any major project, events occur that are not accounted for in the base estimate. Therefore, a contingency factor is applied.

Early decommissioning cost estimates included a contingency of 25% that was applied to the total project cost. More recent and accurate approaches apply contingencies on a line-item basis, yielding a weighted average contingency for the cost estimate. An AIF/NESP study discusses the types of unforeseeable events that are likely to occur in decommissioning, and provides line-item contingency guidelines to apply for various activities. Table 12.1 summarizes these contingency values.

Table 12.1 AIF/NESP-036 Contingency Values

Activity Category	Contingency, %
Engineering	15
Utility and DOC Costs	15
Decontamination	50
Contaminated Component Removal	25
Contaminated Concrete Removal	25
Steam Generator/Pressurizer/Circ. Pump Removal	25
Reactor Removal	75
Reactor Packaging	25
Reactor Shipping	25
Reactor Burial	50
Conventional Radwaste Packaging	10
Conventional Radwaste Shipping	15
Conventional Radwaste Burial	25
Clean Component Removal	15
Supplies/Consumables	25

5. *Scrap and Salvage*: The cost estimate should include an evaluation of the scrap and/or salvage values from materials determined to be clean, or that were never exposed to

radioactive or hazardous material contamination. The evaluation should be based on recent cost data obtained from one or more of the references included this chapter.

For purposes of this chapter, salvage will be defined as removed materials that have an identified market for resale or reuse at a specific facility. Accordingly, pumps, motors, tanks, valves, heat exchangers, fans, diesel engines and generators, etc., are candidates for salvage. Scrap is defined as removed materials certified to be clean, and that may be sold to a scrap dealer for ultimate recycling as a raw material. Examples of scrap materials are copper wire and bus bars, stainless steel plates and structural members, carbon steel and stainless pipe, carbon steel structural shapes, beams, plates, etc.

The market for salvageable materials from nuclear facilities is limited, owing to the very specific purpose for which they were intended. Market prices fluctuate depending on the needs of the economy, the availability of new replacement parts, and the buyer's expense to remove the component intact and to package it and transport it to its new application in reusable condition. Each of these factors will tend to reduce the resale value of salvaged materials. Salvage material companies tend to underbid for these items in an effort to return a high profit on the resale. For steel, scrap material is sold on an as-is, where-is basis. There are no warrantees or representations as to the reusability of the item.

The market or scrap metals is also affected by the market conditions, the amount of virgin materials available, and what the cost would be to remove it from its installed location. Market prices are usually posted daily in the Wall Street Journal. These prices are for what a buyer would be willing to pay for scrap. For carbon steel, scrap material is sold on an as-is, where-is basis. The buyer assumes the costs of preparing it for resale to steel mills (cutting it to firebox size – 18" x 60").

For nuclear plant estimating purposes, it is reasonable to assume that the DOC will remove the scrap materials from their installed locations. The certified clean scrap will be placed on the facility loading dock for a local scrap dealer to haul away at not cost or credit to the nuclear decommissioning account. The scrap dealer will resell the scrap to recover his costs to load, haul, and resell the materials.

6. *Work Breakdown Structure*: The WBS is used to categorize cost elements and work activities into logical groupings that have a direct or indirect relationship to each other. The work groupings are usually related to the accounting system, or chart of accounts used for budgeting and tracking major elements of the decommissioning costs. The DOE adopted the WBS approach for decommissioning its facilities (as well as for other projects) as a means of tracking costs consistently from one project to another. While this hasn't always worked out perfectly in every case because of facility differences and differing decommissioning approaches, the general objectives have been maintained.

7. *WBS Levels*: The WBS elements are generally arranged in a hierarchal format similar to a company's organization chart. The topmost member or level of the WBS would be the overall project. The second level would be the major cost groupings under which project costs would be gathered. The next level would be the principal component parts

of each direct or indirect cost category for that cost grouping. Subsequent levels are often used to track details of the component parts of the grouping so that a clear understanding of all the cost bases can be made. From a cost perspective, WBS element levels rarely exceed the sixth level; as such, detail is usually unnecessary for cost reporting purposes. In most cases the costs are "rolled up" to Level 3 or Level 2 summary costs for management information.

8. *Chart of Accounts*: The project management or accounting software used on major projects usually identifies categories of costs in terms of a chart of accounts. The chart of accounts is where the individual cost items of labor, equipment, consumables, capital expenditures, recycle services, transportation or disposal services are budgeted and cost-controlled on a rigorous basis. The input data to the chart of accounts would include labor hours and costs, procured equipment, procured consumables, major capital equipment procurement, recycle invoices, transportation waybills and invoices, and waste disposal invoices. Most project management and accounting software packages are capable of relating the chart of account information directly into the WBS format for project reporting purposes.

12.3.3 Cost Estimating Process: A thorough cost estimating process flows from an overview of the project to the scenarios evaluated or selected, to the assumptions critical to the approach, to the details of the cost elements and the work schedule, and then to a summary of the principal cost drivers. While there are no hard and fast rules for formatting the process, there are logical guidelines to follow so that cost estimates can be easily tracked and compared. The guidelines provided in this section have been proven through extensive use in the industry and can be adapted for virtually any decommissioning project.

12.3.3.1 Scope of Work: The scope of work for the project must be clearly stated at the outset of the estimate to ensure the estimator and reader understands what is included in the estimate, and the extent of effort required. The scope should identify the building(s) to be decommissioned, the systems and structures to be removed and dismantled, the below-grade structures to be abandoned in place or excavated, and the site restoration required, including backfill of subgrade voids, and grading of the surface to adjacent natural or existing topographical profiles. Any special landscaping requirements should be included so as to complete the site restoration work.

Establishing the framework for an estimate sets the requirements for data acquisition. Format should be established "up front," such that subsequent data acquisition is not compromised. Once the project has been defined, the estimating model chosen, and the format specified, limiting assumptions and project constraints must be identified. Defining the estimate's bases allows for the future assessment of perturbations in these different factors. It also permits manipulation of input variables to evaluate the sensitivity of the estimate to the underlying bases.

12.3.3.2 Decommissioning Options: The decommissioning options to be evaluated, or the selected scenario to be estimated should be identified. Typical such options include SAFSTOR, ENTOMB, and DECON, as defined in Chapter 9 of this handbook. In many cases the options may be composites of the foregoing scenarios to accommodate special site-specific objectives or constraints. The special conditions of each option should be identified, as they will have a direct affect on the assumptions that have to be made and the technical approaches that can be implemented during decommissioning.

12.3.3.3 Collection of Information: The planning for the data collection phase of the cost estimate is as important as the actual collection process. Careful attention must be directed at the outset to clearly identify what information is needed, how it is to be obtained, and how it is to be used. Often valuable time and labor is expended collecting detailed data that is either unnecessary or particularly difficult to access. Unless there is an identified need, personnel and time resources should not be expended on gathering volumes of data.

A site-specific estimate uses defined engineering data, including site and plot plans, general arrangement and architectural drawings, piping and instrument diagrams, one-line electrical diagrams, equipment specifications, and reference manuals, etc., to provide a basis for the plant systems and structures requiring decontamination and dismantling.

1. *Site and Facility Characterization Data*: Site and facility characterization data usually involves identifying the radiological and hazardous materials contained in the facility, or that has been deposited on the site on exterior building surfaces, in soils, and groundwater. Obvious characterization data includes radiological contamination, asbestos, lead (solids or paints), PCBs, mercury (in switches), chromates, and potassium compounds.

   Typically, a Site Characterization Plan is prepared to identify the scope of the characterization program, the extent of survey, surveying protocols, expected results, and reporting format. The actual characterization program should not automatically be put out to bid to the lowest responsible bidder on a fixed-price basis. The primary objective is to ensure that the characterization program has identified all the important contaminants that will affect costs and the extent of remediation required. If the characterization requirements are not carefully scoped out in the bid request, a fixed-price contractor might be inclined to stop or reduce investigative work when the contractor runs out of money. Under such circumstances, a time-and-materials, or cost plus incentive fee approach may more closely satisfy the needs and objectives of the characterization program.

   Some characterization programs attempt to go to extensive details to identify specific levels of contamination within systems, structures, and soils. For the purpose of a cost estimate, all that is needed is the determination of clean versus contaminated materials or hazardous materials, and the identification of the principal radioactive isotopes. The radioactive contamination in virtually all commercial nuclear facilities is classified as LLW, with the exception of special nuclear material (uranium, plutonium, etc.). Most of this low-level waste is Class A waste as defined in 10 CFR 61 (NRC 2003b), and is suitable for disposal in near surface disposal facilities, or readily processed by commercial waste processing companies (recyclers). The only areas that would require more extensive investigation are those where the components in a system are potentially clean, or very highly contaminated. In these cases, a more thorough investigation might determine that potentially clean components can be free-released, and highly contaminated components might have to be decontaminated to reduce occupational exposure prior to removal.

For reactor vessels, internals, and surrounding concrete biological shields, an estimate of neutron activation is needed to determine the disposition of each component. The reactor vessel and internals will contain Class A, B, C, and greater than class C (GTCC) wastes as defined in 10 CFR 61 (NRC 1983). The disposition of these wastes and associated costs represent the major element of the waste disposal costs. Computer codes, such as ANISN (ORNL 1995) and ORIGEN (ORNL 2002b), are one-dimensional codes for determination of the radionuclide inventory, and are usually adequate for estimation purposes. More detailed three- and two-dimensional codes, such as TORT-DORT (ORNL 1998), may be used as part of detailed engineering and planning.

2. *Inventory of Systems and Structures*: An inventory plan should be established at the outset of the data collection program. The plan should identify the scope of systems and structures to be included in any estimate. It should list the specific information to be gathered, including component identification, location, and physical size (overall dimensions, volume, and weight). Other information that may be valuable might include pump flow capacity, motor horsepower, or steam turbine flow rate, etc., to assist in estimating the physical size of the components.

Piping inventory is perhaps the most difficult and time-consuming data to collect. To physically walk down and log the actual diameters and feet of pipe in each system would take several months of effort, and at high cost. Similarly, to work from the facility isometric drawings that identify every diameter and foot of type for each system would also be an extensive effort. An alternative approach that gives satisfactory information for purposes of cost estimating is to obtain from plant construction records the actual purchased quantities of piping of each size for each system. Often these quantities are not broken down into the individual systems for which there were purchased. In that case an empirical approximation can be made using the inventory of valves obtained from the piping and instrumentation diagrams (P&IDs). Reviews of numerous piping inventories have indicated there are approximately seven feet of pipe for every valve. Using this ratio and the valve inventory, the construction record inventory can be assigned to each system in the plant.

Ventilation and air conditioning ductwork should be included in the inventory. This information can be obtained from P&IDs, ventilation system drawings, or all from a walk down of the plant. Electrical cable and conduit inventory can also be included in the inventory. This inventory can be obtained from the owner-licensee construction records, and assigned to each building based on an allocation of the number of electrical systems in that building.

Component inventory can be obtained directly from P&IDs, or from the general arrangement drawings, or by a walkdown of each area in the plant. Some owner-licensees have performed a detailed inventory of all components and equipment at the site. This information may be available in electronic format (compact disks, computer tapes, etc.). In this case, the data can be incorporated directly into the estimator's inventory code as an ASCII file. Caution is advised, however, in that this computerized inventory may be too detailed, including small instrumentation valves and fittings that might be housed in a single cabinet. For decommissioning purposes, the cabinet would be removed in one piece, including the individual components contained within. The estimator must review this component inventory for such situations.

Structures inventory can be obtained from the structural drawings of the facility. The type of structure is important to the method of removal to be employed. Steel structures, such as columns, beams, pipe-whip restraints, seismic restraints, stairways, and gratings, may not be contaminated and can be removed as clean scrap. Concrete structures, including floors, walls, ceilings, and roofs, may be either clean or contaminated depending on their location and use. The type of construction might include poured concrete floors and walls, concrete blocks, pre-cast concrete slabs, nonreinforced and reinforced poured concrete, and heavily reinforced massive concrete. It is important to distinguish each of these structures because it will affect the method of removal, and the associated costs. There are computer programs available, such as WinScale™ (Roctek 2000), that can digitize drawings directly into the database. Such a program provides the ability to determine the volume and weight of structure segments within each building, and provide retrievable documentation of the estimating basis.

The contaminated soil inventory should be included in the estimate. This item can represent many millions of dollars of remediation costs. The affected area and depth of soil contamination should be identified as part of the site characterization program. If necessary, test borings should be made to quantify the total inventory of contaminated soil.

3. *Local Area Labor, Materials, and Equipment Costs*: For a truly site-specific estimate, the estimator needs to use local area information on labor rates, materials, and equipment costs. Local labor rates may be obtained from the owner-licensee using existing union bargaining agreements, or on-site subcontractor labor rates. The labor rates should be broken down into direct labor plus fringes, union dues, training fees, workers compensation, subcontractor overhead, general and administrative costs, and profit.

Owner-licensee staff costs should be obtained for each of the major organizational functions that the owner-licensee is to perform as part of the decommissioning project oversight. These staff labor costs should be broken down into direct salary, benefits, overhead, and general and administrative costs. All on-site and home office overhead expenses necessary to maintain this staff on site should be included. This information is usually available from the company human relations department.

Similarly, the DOC staff labor costs need to be included. If the DOC has not been selected for the site, a reasonable approximation would be to use the owner licensee staff costs as a proxy for the DOC costs. In his case, an appropriate adjustment would be made for a DOC overhead, general and administrative cost, and profit.

Local material costs can be identified for such items as torch cutting gases, small tools, radiation protection equipment (coveralls, shoe covers, gloves, tape, respirators and filters, etc.), and other consumables. The cost for fuel oil for heating, diesel fuel, gasoline, and electricity should be included. Reference books for equipment rental often include the cost of fuel associated with the operation of equipment. Electricity is more difficult to estimate as it depends upon the amount of electrical equipment in use at any time. One approach is to obtain electricity consumption rates from the owner-licensee during periods of shutdown of the plant. Discussions with the owner-licensee might identify the types of equipment that would be operable during these periods.

Local equipment costs should include purchase or rental of heavy equipment to support the decommissioning activities. This would include forklift trucks, front-end loaders, mobile cranes, and pickup trucks. Special equipment, such as remote cutting tools for reactor vessel and internal segmentation, should be included. Sometimes the specialized items are identified and cost estimated for their use within a specific activity.

4. *Insurance, Taxes, Permits, Engineering, and Regulatory Fees*: Collateral cost items include insurance, taxes, permits, engineering, and regulatory fees. Insurance costs may be obtained from the owner-licensee based on previous premium experience, or from insurance companies directly. Insurance should include property damage, and liability coverage for the duration of decommissioning project. Property damage insurance and liability insurance usually have decreasing premiums as the amount of radioactivity is reduced on site, and the potential risk is correspondingly reduced.

   Information on local taxes may be obtained from the owner-licensee based on previous experience. However the owner-licensee should discuss potential reductions in taxes with the local authority as the value of the property is reduced during the decommissioning process.

   The cost to obtain permits to decommission should be included in the estimate. This would include local building permits and permits from the National Pollutant Discharge Elimination System (EPA 2002b).

   The cost for engineering services to perform specialized engineering analyses, detailed planning and scheduling, and site preparation activities should be included. The average hourly rate for such services may be obtained from the owner-licensee from previous experience on site.

   Regulatory fees include the NRC fees associated with reviewing licensing documentation, performing on-site inspections, and providing a resident inspector. The NRC publishes the hourly cost of labor to perform these services in the Federal Register.

12.3.4 Identification of Decommissioning Activities: The decommissioning process can be broken down into a series of repetitive or similar tasks: decontamination, removal, waste processing, packaging, shipping, disposal, and site remediation and restoration.

1. *Decontamination*: On-site decontamination of systems and structures is often performed to reduce radiation exposure and local area contamination. The cost-benefit of performing decontamination should be determined prior to its inclusion in the cost estimate. The estimator should compare the cost and occupational exposure of direct removal and disposal, versus the cost for decontamination, secondary waste generation and disposal, occupational exposure, and ultimate disposition. Except for very high exposure levels, or high contamination levels, decontamination may not be cost-effective. The estimator should be wary of undocumented vendor claims of decontamination to achieve free-releasable conditions.

   Decontamination techniques include chemical and mechanical methods, and vendor information is readily available on productivity rates, decontamination factors, and secondary waste generation. The selection of any particular method for cost estimation should be made with the concurrence of the decommissioning engineering organization.

2. *Removal*: Removal of systems components is generally accomplished using destructive measures. This would include mechanical cutting, torch cutting, or in some cases tearing or ripping of small components from their installed locations. In some cases it is more cost-effective to mechanically disassemble the components by reversing the manner of their installation. The estimator needs to exercise his judgment as to whether mechanical disassembly is more effective than erecting contamination controlled tents to prevent airborne contamination from torch cutting.

3. *Waste Processing*: Several vendors offer waste processing services for decommissioning, including recycling and volume reduction. These companies generally package contaminated components on site and ship them to their off-site facilities for segregation, decontamination, and disposal. The advantage of using waste processors is the generally lower cost labor available at the off-site processing facility, and the ability to achieve waste volume reduction. The cost of transportation is usually a small component of this service.

4. *Packaging*: Packaging activities include the cost to provide the transport and disposal containers, and the cost to prepare the containers to receive the waste. Each container must be moved to the loading site, covered in plastic to prevent surface contamination, and loaded with waste. The container must be identified for the transport bill-of-lading documentation.

5. *Shipping*: Transport of the waste containers may be by truck or rail, or by barge. In most cases truck transport is used because is less expensive. Rail and barge shipments are usually cost-effective for individual large components. Truck transport cost schedules are readily available from trucking companies. Rail companies usually require detailed information on the size and weight of the components, and the route that must be taken to bring the component from the nuclear facility to the disposal site. Railroad bed weight and bridge width limitations are highly restricted in the Northeast corridor, and may require a circuitous route to avoid these obstructions. There are fewer limitations west of the Mississippi River, and rail transport may be more cost-effective.

6. *Disposal*: Disposal by near-surface burial is possible at several facilities in the United States. Currently these include Barnwell, SC, Clive, UT, and Hanford, WA. These facilities publish disposal rates for each class of waste material. The estimator should clearly understand the disposal rate schedule and limitations imposed by the burial facility operator.

7. *Site Remediation and Restoration*: The extent of site remediation and restoration should be clearly stated in the estimation basis and assumptions. Generally, site structures are removed to a depth of three feet below grade, and the remaining subgrade voids backfilled with clean rubble and soil fill. Typically, grass is planted to prevent erosion of the soil.

12.3.5 Development of Unit Cost Factors: The methodology used to develop cost estimates follows the basic approach presented in the DOE's original Decommissioning Handbook (DOE 1980) and the AIF/NESP-036 study report (LaGuardia 1986). Battelle Pacific Northwest Laboratory later updated its reference estimates for the NRC (NRC 1995 and 1996) to adopt this unit cost factor (UCF) methodology. This USF method simplifies the calculations for estimating activity costs.

The unit cost factor method provides a demonstrable basis for establishing reliable cost estimates. The detail of activities provided in unit cost factors for activity time, labor costs (by craft), and equipment and consumable costs provides assurance that cost elements have not been omitted. These detailed unit cost factors, coupled with the plant-specific inventory of piping, components, and structures, provide a high degree of confidence in the reliability of the cost estimates.

1. *System Removal UCFs*: Removal of system components, such as piping, valves, pumps, heat exchangers, tanks, filters, electrical equipment, conduit, and cable, can be estimated using the unit cost factor methodology. Separate unit cost factors are used for cleaning components as opposed to removing contaminated components. In general, the unit cost factors identify the scope of work for each type of equipment, and the duration (in minutes) for the individual activities to remove the component. The methods for removal are described, including the expected productivity rate for the estimated duration. Activities that can be performed in parallel are identified, and the appropriate critical path duration used as the basis for the overall removal activity duration.

   Work difficulty factors are applied to account for site-specific conditions that would extend the duration of removal. These factors, described in more detail in the following section, include respiratory protection, ALARA, accessibility, protective clothing, work breaks, and worker productivity. The overall effect of these factors is to increase the time required to remove each component.

   The appropriate crew composition and size to perform these activities is identified, and the associated fully burdened labor rate is applied for each worker. The number of crew and the burdened later rate are multiplied by the component removal duration. The subcontractor's overhead and profit is added to this labor cost.

   Material and equipment costs are included to account for consumable materials, and equipment needed to remove the component. This would include the consumption of cutting gases, rental of torches, contamination control plastic for tents, absorbent material to collect spills, etc. The costs for these materials are summed, and a contractor's markup added to the total.

The total cost for labor, and materials, and equipment is summed and the cost divided by the appropriate unit value (foot of pipe length, size of valve, weight of heat exchanger, etc.). Unit cost factors are developed in this manner for each different type of equipment in the plant.

2. *Structure Removal UCFs*: Removal of structures follows the same format as the system removal UCFs. Separate UCFs are developed for each type of structure, including structural steel, concrete block, reinforced and non-reinforced poured concrete, and heavily reinforced massive concrete. The methods for removal are described, including the expected productivity rate for the estimated duration. Activities that can be done in parallel are identified, and the appropriate critical path duration used as the basis for the overall removal activity duration. In the same manner as for systems removal, work difficulty factors are applied to develop the overall duration for the activity.

   The appropriate crew composition and size to perform these activities is identified, and the associated fully burdened labor rate applied for each worker. The number of crew and the burdened later rate are multiplied by the component removal duration. The subcontractor's overhead and profit is added to this labor cost.

   Material and equipment costs are included to account for consumable materials, and equipment needed to remove the component. This would include the consumption of cutting gases, rental of torches, contamination control plastic for tents, large concrete drilling machines for blast holes, hydraulic ram-hoes, front-end loaders, and hauling trucks, etc. Where appropriate, multiple crews and equipment may be used to shorten the overall duration. The cost for these materials is summed and a contractor's markup added to the total.

   The total cost for labor, and materials, and equipment is summed and the cost divided by the appropriate unit value (ton of steel, yd^3 of concrete, or ft^2 of area, etc.). Unit cost factors are developed in this manner for each different type of structure in the plant.

3. *Site Remediation UCFs*: Site remediation activities include excavation of subgrade structures, excavation of buried pipe, backfill of subgrade voids, grading to adjacent terrain, and site landscaping. In each case, the specific activity is described in the UCFs, the appropriate duration for each activity is included, parallel paths are identified, and a critical path determined.

   In a similar manner to systems and structure removal, crew labor costs, material and equipment costs are determined, and the total cost for removal divided by an appropriate unit value (cubic yard of soil excavated, or square feet of area seeded).

12.3.6 Development of Work Difficulty Factors: The application of work difficulty factors is intended to account for the productivity losses associated with working in a difficult or hazardous environment. The approach is widely used at operating power plants to account for difficulty in performing maintenance activities during outages. The application of this methodology to decommissioning activities is a natural and reasonable extension of this work adjustment factor.

1. *Respiratory Protection Factor*: The respiratory protection factor is intended to account for the difficulty of a worker performing activities while wearing a full-face respirator, or supplied-air mask. The respirator impedes breathing, obscures vision due to the mask window and fogging, and adds stress from the straps around the head. The respiratory protection factor can have a value of from 10 % to 50 %.

2. *ALARA Factor*: The ALARA factor is intended to account for the time spent preparing for an entry into a high radiation or high contamination area. This time is used to alert the crew to the potential hazards in the area, the specific activities to be accomplished while in the area, and emergency procedures to be implemented for immediate evacuation. This factor also accounts for the periodic training the crew would receive to maintain their radiation training and certification. The ALARA factor can have a value of from 10 % to 15 %.

3. *Accessibility Factor*: The accessibility factor is intended to account for difficulty of working on scaffolding, on ladders, in pipe tunnels, or in other confined spaces. The limited degree of motion possible under these working conditions reduces worker productivity. The accessibility factor can have a value of from 10 % to 20 %.

4. *Protective Clothing Factor*: The protective clothing factor is intended to account for the time and worker needs to put on protective clothing for each entry and exit from a radiation control area. Typically, this represents four changes per day, assuming suiting up in the morning, a morning break, a lunch break, an afternoon break, and end of the shift. The protective clothing factor can have a value of from 10 % to 30 %.

5. *Work Break Factor*: The work break factor is intended to account for the time a worker needs to take a morning break, a lunch break, and an afternoon break. Experience has shown worker productivity under stressful conditions improves when workers are allowed a morning and afternoon break. The work break factor can have a value of from 5 % to 10% (nominally taken at 8.33%).

6. *Work Productivity Factor*: The work productivity factor is intended to account for site-specific productivity differences in the workforce. These differences may arise through union bargaining agreements, severe weather factors (heat or cold), or other limitations. The work productivity factor adjustment is at the discretion of the estimator.

12.3.7 Assembling the Activity Cost Estimate: Assembling the activity cost estimates involves applying the cost factors to each item of inventory in the systems, structures, and the site. The process may be done manually, on a spreadsheet, or programmed to read the inventory and UCF files. Usually the cost estimate is formatted on a master spreadsheet to vertically list each system and structure and site area by name to help identify the scope included in the estimate. Individual activities of decontamination, removal, waste processing, packaging, shipping, and burial are listed horizontally in columns. Contingency is added to each activity on a line item basis and summed in a separate column on the cost estimate.

1. *Application of UCFs to Systems, Structures, and Site*: To assemble the cost estimate, the estimator should prepare intermediate activity spreadsheets that apply the individual UCFs to each item of inventory within a specific system, structure, or site area. For each activity (decontamination, removal, waste processing, packaging, shipping, and burial), an appropriate contingency percentage should be applied. The results of each line item would be entered onto the master spreadsheet. The process should be repeated for each system, structure, and site area. The master spreadsheet should also list undistributed costs, including the owner-licensee period-dependent costs (staff costs and DOC costs), and other undistributed costs (collateral costs).

2. *Evaluation of Crew Utilization*: The estimator should review the results of the estimate for practicality. For example, from UCFs the estimator can track how many hours of work are expended in a given work area. From the expected duration to complete that area, the number of workers required should be reasonable. Typically, a crew of five requires approximately 1600 ft^2 of work area, which includes the area of a component being worked on, toolbox space, and laydown space. Using this guideline and the overall floor area within a building, an estimate of the crew size that can be efficiently employed in that building on that floor can be developed. Similar checks should be made when there is a large amount of floor space but very few components to be removed.

12.4 DEVELOPING THE PROJECT SCHEDULE

The project schedule is an integral part of a detailed cost estimate. The duration of a project affects the costs through the period-dependent costs, and the selected technology for the activity-dependent work and associated costs affects the schedule. The estimator must keep these two interrelated elements in balance when preparing an overall cost and schedule estimate.

12.4.1 Activity-Dependent Schedule: The activity-dependent schedule draws from the cost estimate database to establish durations for each activity in the schedule. Each of the UCFs provides a duration estimate to perform the activity. The activity duration multiplied by the quantity of an item in the inventory provides an estimate of the overall duration to perform that activity. The UCFs also provide a manpower estimate to perform that activity. The number of labor hours multiplied by the quantity of an item in inventory provides an estimate of the overall manpower resources to perform that activity. These two elements, activity duration and activity manpower, are the input factors to the project schedule.

12.4.1.1 Work Process Flow Chart: The cornerstone of project planning and schedule development is a formal documented scope. The written scope statement should include:

- Project charter
- Assignment of key personnel
- Project constraints
- Assumptions
- Scope statement
- WBS

Activity sequencing requires the determination and documentation of the relationship between and among activities. The precedence diagramming method is typically used to structure the relationship between activities. Sequencing usually begins with a chronological ordering of activities, based on a logical progression of events. For convenience, the estimator may choose to divide the project into individual periods to track similar kinds of activities. For example, Period 1 would include all preparatory work, Period 2 would involve decontamination, dismantling, and license termination, and Period 3 would be site restoration. Within each period the estimator would sequence the activities consistent with known schedule drivers. In Period 1, preparation of the de-fueled technical specifications and PSDAR would be accomplished prior to site mobilization. Individual durations for these activities would come from the estimator's experience, or from other project experience.

Activity definition requires using the scope document and the WBS together to develop discrete activities that are unique and can be associated with a deliverable. Typical elements of the project may include the following:

- Preliminary site characterization
- Safety documentation
- NEPA documentation (for federal projects)
- Permits and approvals
- Site characterization
- Conceptual design
- Preparation of decommissioning plan
- Preliminary design
- Detailed design
- Procurement
- Dismantling and demolition
- Waste treatment
- Transportation and disposal
- Inspection
- Final site survey
- Independent verification
- Site restoration

Each activity in the work process flow chart has a predecessor and corresponding successor activity. A complex project would involve multiple parallel paths, to reduce the overall schedule.

Resources other than people can also be planned and analyzed as part of the schedule development, and are routinely included in project schedules. Other resources could include radiation exposure, critical pieces of equipment, use of stationary cranes, or utilities. Including these resources in the schedule will allow the estimator to determine whether there are critical resource restraints during particular periods of time. Once all resources are loaded into the schedule, critical analysis of the resource constraints and resource profiles can be developed to show the time-phased consumption of resources.

Schedule development is an iterative process involving the analysis of start and finish dates, activity relationships, activity durations, resource availability, and work calendars to optimize the overall schedule and project goals. Summary activities may also be developed as part of the WBS hierarchy. This will help with preparation of higher level reporting to stakeholders or management.

12.4.1.2 Determining the Level of Detail: This is a critical decision. If the schedule is prepared at too fine a level, the project runs the risk of being overwhelmed with data that inevitably the project control staff is unable to maintain. On the other hand, a schedule with too little detail is insufficient to use in tracking progress, anticipating problems, or developing risk strategies. As a general rule, the estimator should schedule activities only at the level needed to control the work. This may be somewhat judgmental and depends on the skill of the project team, its past experience, the complexities of the activities, and the risk involved in each activity.

12.4.1.3 Evaluation and Optimization of Critical Path: The critical path is the longest sequence of activities in the work process flowchart and controls the overall length of the project. Incremental change to any critical path activity will result in a corresponding change in the overall schedule.

The estimator and the project management team should evaluate the critical path to determine what technological changes, parallel path changes, or duration estimate changes can be made to shorten it. The overall schedule duration is one of the major cost drivers in a decommissioning project. Once adopted by the estimator and project management team, the schedule becomes the baseline schedule for the project. It is against this schedule that project performance will be measured. It requires owner-licensee buy-in and a commitment of management resources to support it.

12.4.2 Development of Management Staff: The owner-licensee management staff is one of the major cost drivers in the estimate. Management costs are period-dependent; that is, the costs are a function of the duration of the overall project. The owner-licensee management organization to oversee the project must reflect the level of activities being performed during each period. Similarly, the DOC staff is also a major cost driver in the estimate. It is also a period-dependent cost and must reflect the little of activities being performed during each period.

12.4.3 Owner-Licensee Staff: The project management team usually develops the owner-licensee management staff. Each management position is determined for the specific function needed to be performed to support the decommissioning project. It should include force account labor and all team members from the project manager through supervisors. Typically, a straw man organization chart is developed for review by the project management team. The team should review the specific project positions, and the number of personnel in each position for the duration of each period.

12.4.4 Decommissioning Operations Contractor Staff: The DOC staff must also be estimated to develop the overall cost estimate and budget for the project. In a similar manner to the owner-licensee staff, the DOC staff positions are identified for each function and for each period. A draft organization chart is also developed for review by the project management team. Once the DOC positions are selected, the cost estimate can be modified to reflect the actual staffing levels and associated costs for each period.

12.4.5 Software and Flexibility: The several project management software systems and schedule systems on the market today are relatively inexpensive. Items to consider in selecting scheduling software include knowledge of the project team and scheduling personnel, compatibility with subcontractors systems, ability to integrate with the timekeeping and cost reporting systems, and the ability to communicate with other team members.

12.4.6 Developing Overall Cost and Schedule Estimate: The last phase in developing the cost estimate and schedule is assigning the cost elements to the work breakdown structure. Typically, the WBS is used to collect and monitor costs of the project. The WBS cost estimate is usually established as the baseline estimate for the project. At the owner/licensee's discretion, the baseline may be adjusted periodically to account for changes in scope of work, funding constraints, or schedule changes as a function of acceleration or delays in the project. Each element of the WBS structure may be separated to establish a budget for that work scope. The owner-licensee functional managers can be held accountable for each of the budgets within their jurisdiction.

12.5 USING THE PROJECT SCHEDULE

Effective management of a decommissioning project depends on a rigorous project schedule. It's the road map that outlines how the decommissioning will move from conceptual engineering to completion and site restoration. As a tool, it will support different purposes for different members of the project team. It shows the engineer or the field supervisor what needs to be done and when, and what other activities may affect the start and completion of the work. It gives the project manager an overview of where the project is going, a means of assessing project performance, understanding of differences between planned and actual progress, a means of allocating resources and adjusting resources, and a framework for assessing, managing and controlling a project risk. It informs stakeholders on the scope of the work, the status of the decommissioning, and how it might affect their interests. Finally, it gives senior management a measure of project performance.

The project schedule is also a universal communication tool. When used properly it is an excellent vehicle to communicate expectations, responsibilities, discuss problem areas, anticipate future events, and initiate contingency plans. It is primarily up to the project manager and the project sponsors to establish the expectations regarding use of the schedule.

12.5.1 Interrelationship with Overall Project Management System: The project schedule cannot be developed or implemented in a vacuum, and must be fully integrated with the other elements of the decommissioning project management system.

The project schedule must be developed and maintained consistent with the overall project scope documents and the project cost estimate. Inconsistencies between these documents will undermine the level of commitment and understanding of the overall project goals. If, for example, the schedule does not include some relatively insignificant activity elements of the decommissioning that nevertheless are required for the project to continue, the risk exists that this activity may be completely forgotten, or viewed as something relatively unimportant thereby resulting in a schedule delay. The schedule also needs to be consistent with the project communication and procurement functions. The schedule must consider the timing and duration of important procurement functions.

12.5.2 Management Responsibilities: Management is responsible for implementing the overall project and for managing key stakeholder issues. Without management expectation that all project participants will use and maintain the schedule, it can very easily be forgotten. Yet, strict and timely resource management is what preserves original project milestones and regulatory commitments. And the schedule will always dictate the costs.

12.5.3 Stakeholder Involvement: Decommissioning projects often will include stakeholders that are removed organizationally since they are outside the normal communication chain of the project organization. Although they are often invisible to the majority of the project team, they may have a profound influence over the success of the project. The stakeholders should be identified as part of the overall planning process, and they should be considered and consulted as part of the preparation of the overall decommissioning schedule.

12.5.4 Responsibility Matrix: Clear lines and statements of responsibility must be established before developing the schedule. These jurisdictional issues should be reinforced during schedule review and, at the completion of schedule development, each and every activity in the schedule must have an owner-licensee that understands and fully acknowledges project responsibilities.

1. *Assigning Resources for Project Scheduling*: To use the schedule as a tool for management of the decommissioning, management must commit sufficient resources to prepare, optimize, and maintain the schedule. Assigning the preparation and maintenance of the project schedule as a collateral duty does not work. Management must assign the development of the schedule to someone as his or her number one priority. This is usually assigned to a project controls position where one or more persons are responsible for schedule and cost management functions.

2. *Relationship of Owner-Licensee Schedule with Subcontractor Schedule*: The owner-licensee and the DOC need to work as a team. Close coordination of the respective schedules must be maintained so that there is no misunderstanding and there are no surprises.

3. *Management and Stakeholder Involvement and Expectations*: It is critically important that senior management routinely enforce and support the continued use of the schedule. Senior personnel should attend project status meetings, actively ask questions, and challenge personnel on performance of work according to the schedule. If management either has its own schedule, or is perceived to have its own milestone dates that are in conflict with the original schedule, the entire project team will quickly loose commitment and interest in support of the plan. Management should also refrain from dictating the schedule.

   Stakeholders must continue to be involved during the implementation phase of the decommissioning project. External stakeholders, such as the regulatory agencies and the external community, have strong interests, responsibilities, and implicit authority over the activities, and retain authority over the final product. The project team must keep them informed and involved.

12.5.5 Forward-Looking Philosophy: During project execution, the schedule must be used as a forward-looking tool and not just a mechanism for reporting what happened, and when. There should be routine meetings with all project personnel to review the current project status and look at those subsequent activities needed to support the project goals. Team members should discuss problems meeting the schedule, resource requirements to support the schedule, significant handoff issues (to make sure personnel know when one activity ends and another starts), risk evaluation issues, and whether additional planning or contingency schedules should be prepared.

12.5.6 Communications: During project execution, the current schedule must be distributed to all affected parties. The schedule must be clear about what is going to occur, who is responsible, and what the deliverables are. Schedule updates need to be shared with all parties, including project team members, stakeholders, contractors/subcontractors, and those responsible for resources. Given email and the Internet, there are many options available for communicating schedule updates.

12.5.7 Schedule Detail and Frequency of Updates: The project manager must decide on the level of detail at which to track schedule progress and the frequency of schedule updating. The level of detail may also vary depending on the phase of the project. During long and routine activities, periodic progress may be more readily measured and tracked by commodity reporting rather than by breaking the schedule down into further detail. The schedule detail will have to be adjusted based upon the overall schedule knowledge and familiarity of the project team. Some training of team members may be needed to make sure that everyone is able to read and understand the schedule as a tool. The frequency of schedule updates is based on the significance and speed of ongoing activities, and the resources available to maintain the schedule.

12.5.8 Resource Planning: During project execution, personnel with significant or potentially project-limiting resource needs should receive periodic resource requirement reports that identify levels of resources during future time periods to a level that the manager or team member can anticipate bottlenecks and take appropriate action to secure additional resources, revise the activity plan, or change the work calendar.

Chapter 13

DECONTAMINATION DECISIONS AND PROCESS

13.1 INTRODUCTION

The decisions and process for decommissioning nuclear facilities play a key role in controlling radiation exposures. The immediate goals of such projects are to reduce or eliminate occupational and public radiation exposures, primarily by isolating contaminants from surfaces and subsurfaces, and to enable cost effective radioactive waste management and disposal. Decisions that define the applicability of various processes and techniques are at the heart of decommissioning. Potential techniques include washing, heating, chemical or electrochemical action, mechanical cleaning, and others. For large facilities and equipment, applicable methods are mostly chemical or mechanical. For components that have a role in a facility with a continuing operation mission, dilute chemical agents are typically used to nondestructively dissolve just the outermost surface film bearing most of the contamination. Conversely, if there is no potential for reuse, the base metal can also be dissolved. The process for a small contaminated metal shack may be brief, as compaction may be obviously the most viable option, while a large highly radioactive facility may require more complex processes, including options, such as the wide use of robotics. Intermediate project objectives on the path to restoring the facility to a defined end state include:

- Removing loose radioactive contaminants and fixing the remaining contamination
- Minimizing residual radioactive contamination requiring protective storage
- Maximizing recycle and salvage of equipment and materials
- Minimizing volume requiring disposal as radioactive waste
- Segregating waste for shipping and disposal in the least restrictive acceptable form (e.g., unrestricted, clean rubble, sanitary, special class, surface contamination only, hazardous, radioactive low-level, mixed hazardous and radioactive, remote handled, and transuranic waste)
- Minimizing any period requiring protective storage or long-term monitoring

When chemical decontamination is indicated, it may be necessary to temporarily dedicate appropriate facilities for precipitation, filtration, evaporation, demineralization, stabilization, recycling, and or reclamation. Residual concentrates may represent a significant source of radiation before transportation for treatment and/or disposal. Each step is an expense to the budgeted occupational dose and increases the risk of an unplanned release, such as the uptake of radioactive material that results in higher doses than those from handling the contaminated system without extensive decontamination. Planning is therefore the process of selecting options based on a continual and iterative cost-benefit analysis.

13.2 DECISION PROCESS

A decision about decontaminating a structure, system or component, and about the specific approach to use, is not only site-specific but also typically task-specific. It is often an iterative process involving characterization and remediation. Before deciding on decontamination, it's wise to conduct at least a semi-quantitative cost-benefit analysis; and if the decision is yes, then the question turns to the particular decontamination process and procedures to be employed.

The decision processes used at federal facilities are more diverse than in the licensed sector. This is due in part to the wide range of activities, and in part to inherent flexibility of vertical integration of all major elements, from greater control of the facility land use, to control of disposal sites. An example at the Rocky Flats Environmental Technology Site involved a major challenge -- the disposal of extremely large pieces of equipment contaminated with radioactive or hazardous materials. Past practice had been size reduction to fit equipment into standard waste containers previously approved for transportation and disposal. Size reduction of durable large equipment is typically extremely labor-intensive, exposing workers to industrial, chemical, and radiological hazards. Instead of decontamination, application of a well formulated spray plastic proved to be effective as a strong-tight industrial package. This allowed the intact equipment to be effectively transported to a low-level waste receiver site on a flat bed trailer. Conscious application of a clear decision process resulted in the use of polyurea spray coating (instead of decontamination), and had extremely positive cost, schedule, and safety impacts to the decontamination project.

13.2.1 Release of a Structure for Re-Use: If the project's purpose is to free-release an existing contaminated portion of a building or structure (e.g., for office space), then extensive decontamination is likely to be required. Typical techniques will provide for the bulk physical decontamination of surfaces to meet very low-level surface contamination limits before release.

Building surfaces may be decontaminated to readily allow for reuse, without radiological restrictions on inhabitants, even if contamination remains in confined areas, such as cracks in structural concrete and in soil underneath the building. Blocking the pathway of contamination in these confined areas makes it not accessible to potential receptors; hence, a facility can be made available without restriction for specific use, such as for office space. The remaining radioactivity may still have to be eventually addressed in the future, such as in structural repair or demolition. For example, the Brookhaven Graphite Research Reactor, in Long Island, NY, served as a museum for several years after decommissioning but before entering the process for final disposition.

Achieving a completely unrestricted release condition, or a "greenfield" end-state, often requires complete removal of above- and below-grade structures, and any contamination underneath. Any dose above background normally precipitates an appropriate risk assessment documenting the acceptability of potential risks. Thus, the degree of decontamination and the methods and processes used are often intimately connected with the risk assessment process and supporting documentation.

13.2.2 Limiting Occupational Dose: Chemical decontamination of highly radioactive components often reduces potential radiological dose from equipment removal. When considering such a purpose, decision-makers should conduct a cost-benefit analysis to see if the desired dose reduction might best be achieved some other way. As previously noted, chemical decontamination can generate secondary waste that is difficult to dispose of, including any equipment and structures used in the process. Obvious options to consider include allowing time for significant radiological decay, in-situ shielding and immobilization, avoidance of segmentation, and the use of robotics and other remote handing techniques.

Robotics, such as for remote cutting, dismantling, scabbling, and packaging should be carefully considered where high external dose rate presents an exposure problem. Very high radiation fields frequently mandate the use of robotics and other remote handling and dismantling techniques.

A period of time delay to allow for radioactive decay is a reasonable alternative to chemical decontamination as a method to limit occupational dose. This is particularly applicable when the dose is generally due to external radiation from relatively large quantities of gamma emitting radioisotopes in the area to be decommissioned.

In order for time to result in a significant dose reduction, the half-life of the contaminant must be short when compared with the period of delay before the start of decommissioning. Typical gamma emitters ^{60}Co and ^{137}Cs cause high external doses and have a respective half-life of roughly 5 and 30 years. For practical purposes, the period of time that a facility might reasonably be in SAFSTOR status before decommissioning would be around 60 years. During this period, the dose from ^{60}Co would decrease by a factor of over a thousand. However, the dose from ^{137}Cs would only decrease by roughly a factor of four. A thousand-fold reduction from ^{137}Cs decay would take about 300 years. Thus, a time delay is highly useful in the case of such activation products as cobalt, and not as applicable for such relatively long-lived fission product as cesium.

13.2.3 Nine Factors to Remember: The following factors all support one critical goal -- to determine the actual sources of contamination. Experience indicates that sources are not always self-evident. For example, for a pipe reading 500 mrem/hr, it cannot be assumed that the source is internal. One project discovered that the source term was removed along with the insulation. Management at the TMI remediation project assumed that much of the dose in the reactor building was due to highly contaminated water in the basement. However, rates barely changed when the water was removed, resulting in another large effort to pinpoint and shield the source.

1. A careful evaluation must be made in the comparison between the Decontamination Factor (DF) obtained as a result of actual decontamination, versus the Occupational Dose Equivalent (ODE) reduction factor. As an example, in a cubicle where the area dose equivalent rate is due to a contaminated line in the room, the DF will be close to the ODE factor. However, if there are several other sources of radioactivity in the cubicle and contributing to the area dose equivalent, decontamination of only one system in the room (e.g., the contaminated line) may not be cost-effective. In this scenario, the DF for the system decontaminated is significantly larger than the ODE reduction factor for the area as a whole.

2. Decontamination can be performed on multiple systems, individual systems, or portions of individual systems. The selected scope and sequence of decontamination should maximize the benefits.

3. The dismantling of piping and components insulated with asbestos or otherwise containing hazardous material requires significant time due to required respiratory protection and other personnel protective equipment. For these, the internal decontamination of components may provide significant dose equivalent savings.

4. Segmentation of components at the work location may produce airborne radioactivity. Personal protection requirements may be altered if is successful decontamination allows workers to avoid using previously required respiratory protection.

5. Whether an area is decontaminated or not, other dose reduction techniques, such as shielding and hot spot flushing, should be considered during dismantling.

6. The ability of existing systems to provide flow paths and flow rates required during decommissioning must be addressed. A hazard assessment may trigger modification of existing piping, pumps, and other equipment, and possibly developing new controls.

7. The sequence of decommissioning work should be arranged so that a pathway is first cleared with a relatively low external dose rate, through which components and waste bins collected from other areas can be readily and rapidly removed to the shipping area.

8. In regard to surface and potential airborne contamination, decontamination activities should be carefully planned so that an area already cleaned is not re-contaminated by subsequent work in an adjacent area.

9. Commercial available software is available (with various levels of complexity) to model the physical and radiological layout of a facility, which can be reasonably effective in sequencing work and scheduling workers to minimize radiation exposure.

13.3 TIMING

Some of the effects of timing in minimizing occupational exposure have already been briefly discussed. There are instances when extensive decontamination is not part of a decommissioning project, even though moderate surface contamination is widespread. For example, in the case of complete demolition of a small, slightly contaminated structure where the off-site disposal is essentially all low-level waste, the added cost of extensive decontamination of scrap would not be justified since essentially all materials would require disposal as radioactive waste. However, some decontamination will likely occur at almost all nuclear facilities after their primary mission ceases. A key driver for this timing variability results from radioactive decay.

To evaluate the impact of radioactive decay, samples should be taken, as reasonable, to determine the radionuclide mix and the total source term. A table of estimated decommissioning activity radiation dose equivalent levels should be assembled for both decontamination and non-decontamination scenarios for each major work area of interest. These estimated dose equivalent levels should be adjusted for anticipated radioactive decay prior to dismantling activities. If the composition of the mixture is not known, then use the half-life of the operational nuclide that resulted in the greatest portion of occupational radiation exposure. Typical nuclides in this category include ^{60}Co and ^{137}Cs.

The ultimate decay factor developed should be applied to the radiological source term affecting personnel radiation exposures. Additionally, if the facility generated waste products that consist solely of short-lived radionuclides, then the application of the associated decay factor will also indicate the reduced future radioactive waste liability.

13.4 ELEMENTS OF A COST-BENEFIT ANALYSIS

A cost-benefit analysis is a practical approach to determine if decontamination will be appropriate and, if so, which particular technology should be used. The analysis should consider all relevant factors of the specific project. Relevant parameters include the direct cost of the decontamination, the potential savings in personnel radiation exposure, the dollar value of savings in reduction of time in critical path work, radioactive waste processing, and in radioactive waste disposal costs. Figure 13.1 summarizes specific elements that should be addressed in a decontamination cost-benefit analysis (Demmer 2000).

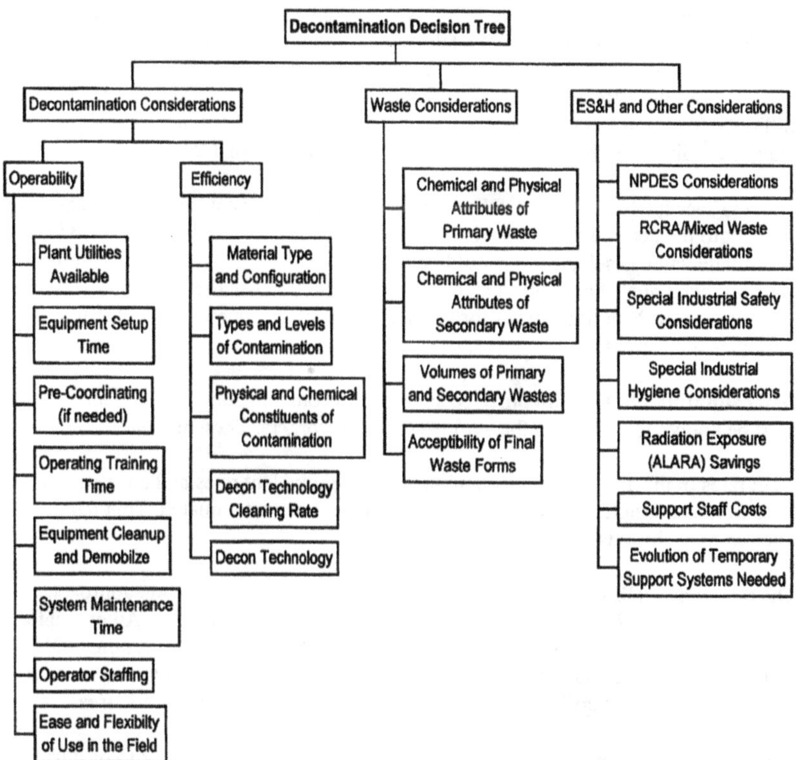

Figure13.1 Specific Elements in a Decontamination Cost-Benefit Model

13.4.1 Assumed Costs per ODE: A necessary assumption for performing an economic evaluation of a given decontamination approach is the assignment of a monetary value to the radiation dose that was avoided for decommissioning personnel as a result of the decontamination conducted. This value is the avoided cost of the occupational dose equivalent expressed in dollars per person-rem.

This value is used in facility economic analyses to identify the incremental cost that a project is willing to incur in order to avoid a given level of personnel radiation exposure. For example, if a facility uses a value of \$2,000 per person-rem, and it was proposed to spend \$5,000 to avoid an exposure of 300 person mrem, then the proposal should be approved, as it is less than the value of \$2,000 person-rem.

- The IAEA uses used a value of \$20,000/PSv, with a range of \$10,000/PSv to \$40,000/PSv depending on the specific country's socioeconomic conditions (IAEA 1994)
- The NRC uses a value of \$200,000/PSv (NRC 1998a)
- The ANL derives a value of \$100,000/PSv, based on 10 CFR 50 Appendix I, and NUREG-0880. ANL also provides a table summarizing examples of expenditures to limit cancer risk
- Derived from the NRC and EPA criteria and guidelines. The values range from \$1 million to \$200 million per cancer averted, equivalent to \$12,500 to \$2,500,000/PSv using the radiation-induced probability of cancer causation of 1.24 E-2/PSv.

Historically, the "acceptable" annual ODE at nuclear facilities has continued to decrease. An additional input to the determination of this site-specific value may come from facility stakeholders. Whatever ultimate value is selected, it should be used for all site analyses and communicated to facility planning personnel. This communication is essential because the value of \$/PSv should be used for additional cost-benefit analyses for tasks and projects conducted in the facility's radiological controlled areas.

13.4.2 Determination of ODE Savings: To support a decontamination decision analysis, several factors must be known to determine the avoided ODE to workers. These include:

1. ODE rates in the area, including "ambient" dose equivalent rates and dose equivalent rates from the structure, system, or component to be decontaminated. This factor is referred to as the existing ODE rate.

2. The estimated ODE reduction factor for the proposed decontamination approach. Although related, the ODE reduction factor is not the same as the DF. An estimated DF is typically available from the decontamination technology provider. As the ODE factors are work area specific, an estimate of this value should be obtained from health physics technicians familiar with the radiological conditions in the work area.

3. The estimate of exposure labor hours worked in the affected area prior to or during decontamination, but not including the time of performing.

4. An estimate of labor hours worked in the affected area actually performing the decontamination.

5. The estimate of labor hours worked in the affected area after decontamination is complete. If the contamination levels in the work area without decontamination were sufficiently high that respiratory protection was required for the workers, then

decontamination to levels where no respiratory protection is required may result in substantial time savings for this period.

6. The estimated dose in PSv received due to additional waste management activities from decommissioning operations.

7. The estimate of labor hours worked in the area if decontamination were not performed. This typically is the same as the sum of pre- and post-decontamination labor hours, although not in all cases. This represents the no-decontamination labor hour estimate.

Using the above factors, one can compute the total ODE savings (in person-rem) resulting from the reduction in dose from the proposed decontamination process. The resulting savings for personnel dose reduction are then compared with the appropriate costs and savings achieved by the process under evaluation to determine the relative costs and benefits of the proposed action.

13.4.3 Waste Management: As with personnel radiation exposure, a potential benefit to decontamination is a reduction in the costs for processing and disposal of radioactive wastes at the facility undergoing decommissioning. The determination of radioactive waste management costs is a key decommissioning decision. Typically a facility must manage, process, and dispose of various categories of radioactive wastes, each with different cost models. Each variety of waste streams should be evaluated since they reflect different volumes, processing, and disposal cost.

The general approach for evaluating the value of specific radioactive waste management practices is consistent with that used to evaluate personnel exposure. The probable waste costs must be evaluated and compared with and without decontamination. Decontamination generates waste by removing radioactive material. In addition, the media supporting decontamination, ranging from paper towels to large volumes of contaminated liquids, constitute a secondary waste stream, also requiring disposal. Decontamination may also result in additional costs from the use of large volumes of clean water, which subsequently becomes a low-level liquid radioactive waste with its own processing and disposal costs. Planning and attention can keep the volume of secondary waste from being over an order of magnitude larger than the volume of primary decontamination waste. Decontamination also generates non-radioactive wastes. The additional cost should also be captured as a different waste stream with its own associated unit costs.

13.4.4 Project Time: Completing decontamination in a work area that results in eliminating respiratory protection requirements, will typically improve worker productivity. Likewise, decontamination to levels such that worker protective clothing or other equipment can be reduced will also improve productivity – by at least 25%. In a long duration projects, these productivity improvements may result in accelerated completion, hence reduced cost.

To estimate overall values for the potential critical path savings, ask how many expected hours or days of project time the proposed decontamination activity will save. The cost-benefit of time saved may vary based on the significance of the work area task in relation to the overall project schedule. Tasks that directly affect the overall project duration (critical path tasks) and are completed early can result in substantial financial benefit. The project management group should be consulted to obtain a value for the financial benefit of time saved.

13.5 ESTIMATING DIRECT DECONTAMINATION COSTS

In addition to the task-related costs identified in previous sections, the following costs should be considered in the overall cost-benefit evaluation:

- Capital equipment if purchased, or vendor cost if leased
- Mobilizing and demobilizing equipment, including craft labor (riggers, pipe fitters, electricians) and equipment (including the use of cranes and forklifts)
- Utilities and other consumable goods
- Maintenance costs for decontamination equipment
- Direct labor and equipment maintenance
- Contract and subcontract work scopes

Unless a particular technology is used repeatedly throughout a facility, leasing high-cost equipment is typically preferable to buying. This may be problematic for the technology provider, as it is reasonable to expect that the equipment used may itself become contaminated. Purchased equipment should be amortized over the entire decontamination project. In some cases, decontamination technology providers will also supply people to operate and maintain the leased equipment, with maintenance costs included as part of the overall service or lease.

In addition to evaluating the vendor-provided production rates, preventative and corrective maintenance should be included in estimating how long a particular technology solution will be used. It is also good to apply a contingency factor, as vendor-specified production rates typically are optimum values that are not realized in the field. The overall direct cost for decontamination is simply a summation of all of the above costs, which are directly attributable to the conduct of the decontamination task or project. This total cost is then applied as:

Application cost = [(Cost per Person-rem Avoided) x (Person-rem Avoided)]
+ [(Radioactive Waste Processing Savings) + (Project Time Savings)]
- [(Direct Cost of Decontamination) + (Indirect Cost of Decontamination)]

13.6 ESTIMATING INDIRECT DECONTAMINATION COSTS

During decontamination, indirect resources may be required for general support, such as:

- Support staff (e.g., ES&H support, management oversight, QA/QC reviews)
- Developing site-specific procedures (and associated training)
- Engineering support to allow temporary systems to hook up to facility systems
- Site-specific operations and ES&H training (for contracted personnel)

13.6.1 Permanent vs. Contracted Workforce: If decontamination is expected to occur soon after a facility ceases operation, then the existing site labor force may be used to cost-effectively support the effort. For many mechanical decontamination approaches, minimal or no training will be required for site personnel. In some instances, however, it is more economical to use a contracted workforce, such as when a facility has been in safe storage for a long time and when limited permanent staff remains on site. Typically, a contracted work force is required for more specialized equipment or decontamination processes (e.g., large fluid system chemical decontamination).

Whenever the use of contracted labor is anticipated for decontamination, training in site operations, and ES&H protocols at a minimum are necessary. The costs for the development and conduct of training should be included as an indirect cost associated with decontamination.

13.6.2 Status of Needed Plant Systems: If decontamination occurs shortly after cessation of operations, then it is reasonable to expect that most of the facility utility and support systems will be operable. Key systems include:

- Radioactive and nonradioactive waste management
- Electrical supply
- Compressed air and/or other gases
- Ventilation
- Radiation monitoring
- Water and other utilities

If existing utility and support systems are no longer available, then temporary stand-ins may be required. The costs for providing these temporary utility and support systems can be substantial since they frequently include engineering design (to ensure operation within site ES&H envelope), component procurement, installation, and maintenance. These costs also should be included as indirect decontamination costs in the equation given earlier in section 13.5

13.7 ADDITIONAL DECONTAMINATION ISSUES

Not all factors influencing the ultimate decision process are cost-quantifiable. Some approaches will remove portions of the structure, system or component. For example, concrete scabbling may remove ~ 2 cm per pass; if additional passes are required to meet the cleanliness requirements, then ultimately a substantial fraction of the thickness may be removed. Similarly, chemical decontamination may result in the removal of interior surface layers of metal within a pipe. Successive passes can substantially thin the wall thickness. These are just two examples of decontamination that can result in such a sufficiently degraded structure that its structural integrity may become questionable. In these cases, it is useful to gain the insights of qualified engineers to ensure that any structures, systems, or components required after decontamination is completed will be able to serve their intended function safely.

Although typically not a concern in decommissioning, another item to evaluate is how the chosen method fits within the overall project schedule. If the most cost effective decontamination approach is shown to impair the overall project schedule, then a higher-cost but faster approach may be warranted.

Chapter 14

SPENT FUEL AND DECOMMISSIONING

14.1 INTRODUCTION

This chapter presents information related to spent fuel acceptance, management, storage options, and the operational and financial considerations faced before and during decommissioning. The generation of electricity by nuclear fission creates highly radioactive waste in the form of spent fuel. As each reactor is refueled, one-fourth to one-third of the fuel is removed and placed in the plant's storage pool, which shields and cools the spent assemblies.

Since these fuel pools were initially designed for provisional storage, most plants cannot store all the assemblies generated over the expected reactor operating life. In the late 1960s and early 1970s, assemblies could be transferred off site for reprocessing, freeing up additional space in the storage pools. It was anticipated that the DOE would provide interim storage at centralized sites. In the meantime, sites took measures to extend their own storage capacity, often by replacing the storage racks in the pools with denser ones to extend capacity as much as practical. When that was no longer sufficient, supplemental storage was acquired, typically as an independent facility. With additional delays in the development of the government's waste management system, plants shut down over the next ten years will have a full compliment of fuel in on-site storage and will need to address its disposition in planning for decommissioning.

The NRC gives no specific guidance on managing or funding such storage. It does, however, refer a licensee to the provisions of 10 CFR 50.54(bb) (NRC 2001b), which require the licensee to submit for NRC approval the program by which it intends to manage and fund the management of all irradiated nuclear fuel at the reactor until possession transfers to the DOE. Whether funding is available as government reimbursements, damage claims, or through rate relief, the owner must determine the magnitude of potential financial liability.

The Nuclear Waste Policy Act assigned the responsibility for disposal of spent nuclear fuel to the DOE and created a Nuclear Waste Fund (Section 302) for collecting money from the consumers of the electricity generated by commercial nuclear power plants. The date targeted for startup of the waste management system was 1998.

After pursuing a national site selection process, the Act was amended in 1987 to designate Yucca Mountain, Nevada, as the only site to be evaluated for geologic disposal of HLW. Also in 1987, the DOE announced a five-year delay in the opening date for the repository, from 1998 to 2003. In 1989, an additional 7-year delay was announced, primarily due to problems in obtaining the required permits from the state of Nevada to perform the required characterization of the site. Based on the anticipated delays, it is unlikely that spent nuclear will be disposed of in the Yucca Mountain repository before 2010.

14.2 SPENT FUEL ACCEPTANCE ASSESSMENT

On November 14, 1997, the U.S. Court of Appeals for the District of Columbia Circuit issued a decision in Northern States Power Company, et al., v. U.S. Department of Energy. In the decision, the Court reaffirmed its earlier Indiana-Michigan ruling that the DOE had an unconditional obligation to begin disposal of the utilities' spent nuclear fuel by January 31, 1998 (NSP 1997). Since the agency was not in default when the decision was issued, the court declined to prescribe "remedies" in the likely event the DOE failed to uphold its obligation. More recently, the Court of Appeals for the Federal Circuit ruled in favor of Yankee Atomic Power Company (MY 2000) in its pursuit of damages. However, even with the ruling, the DOE has no plans to receive spent fuel much before 2010.

The DOE's inability to accept spent fuel has resulted in utilities' need to expand storage capabilities or build new interim storage facilities, and commercial utilities will likely remain responsible for storing spent fuel for the foreseeable future. The methodology used to assess the requirements for interim storage is based on comparing commercial spent fuel generation with the repository's planned acceptance rate. Contributions from those plants that have ceased production are subtracted from the cumulative projection of commercial spent fuel generation in the U.S., resulting in a listing, by year, of the cumulative metric tons of heavy metal of commercial spent fuel generated and eligible for repository placement (DOE 1995b). Correspondingly, the DOE's acceptance rate for spent fuel at the repository is also summed, using the repository's design capacity values. The DOE's stated premise is "first generated, first accepted." The working formula uses the date when the cumulative acceptance of spent fuel at the repository equals the accumulation generated by the industry (in the year that a particular unit ceases operation) to define the date when all spent fuel will have been removed from that unit (DOE 2001d).

14.3 LONG-TERM SPENT FUEL MANAGEMENT

Developing an appropriate scenario for long-term spent fuel management at a particular reactor or site is based on qualitative assessments of the two prevailing industry-standard options -- wet storage using the existing spent fuel pool and dry storage using an on-site ISFSI. Both options are currently in use at shutdown and operating sites, and either method is suitable.

Under the prompt decommissioning scenario, remediation of the radiological contamination and associated waste from the site must occur as soon as possible. With wet storage, the fuel pool and associated systems, structures, and components must remain operational until the DOE removes the last fuel assembly from the site. Since this process may extend over 20 or more years, wet storage is generally incompatible with prompt decommissioning. Dry storage, however, allows the spent fuel assemblies to be isolated from the plant and within the passive confines of an ISFSI. Current technology may allow placing spent fuel in dry storage within 5 to 7 years of removal from a reactor, depending on the fuel's heat generation. Dry storage is therefore generally considered compatible with the overall objectives of prompt decommissioning. Some plants will continue to operate their pools in conjunction with an ISFSI.

Continued operation of the spent fuel pools may be advantageous in situations where the dismantling is deferred, or where the shutdown reactor shares the site with operating units. While wet storage costs more, it does avoid the significant capital expense associated with procuring storage casks. Nevertheless, given a long-term storage period, the economics will eventually favor dry storage. Another key factor is the effect on the fuel pins from extended wet storage, particularly older stainless-steel clad fuel and assemblies with damaged and/or leaking pins.

The first dry-cask storage systems were designed solely for storage and are not qualified for transport. These casks must be brought back into the pool for assembly repackaging in a transport-qualified container. Or, the owner could build a dry transfer facility, e.g., a simplified hot cell, so that the pool(s) could be decommissioned. The decision depends on economics and practicality.

14.4 DRY FUEL STORAGE

Selection and implementing dry storage technology permits the timely decommissioning of interim facilities and the corresponding termination of the operating license. The release of the site from the restrictions of the operating license and the containment of spent fuel within a passive storage facility further reduces the owner's nuclear-based liabilities and costs. Current dry fuel storage technology reflects a maturation of the industry from the technical and licensing problems of the 1990s. Table 14.1 includes several approved dry-cask storage system designs. At present, the number of operating reactor sites using or actively planning to use dry spent fuel storage has expanded from two in the mid-1980s to more than a dozen sites in 2001. Correspondingly, the single-unit commercial sites actively pursuing prompt decommissioning are exclusively using or planning to use some form of dry storage to isolate spent fuel during decommissioning. Shutdown reactors at multi-unit sites, or older dormant nuclear facilities, tend to continue operating their wet storage pools.

Table 14.1 Dry Cask Storage Systems

Vendor	Storage Design Model*	Capacity (assemblies)
BNFL Fuel Solutions	W-21, W-74	21 PWR/64 BWR
General Nuclear Systems	CASTOR V/21	21 PWR
Holtec	HI-STAR 100	24 PWR/68 BWR
	HI-STORM 100	
Nuclear Assurance Corp.	NAC S/T	26 PWR
	NAC-C28 S/T	28 Canisters (fuel rods from 56 PWR assemblies)
	NAC-UMS	24 PWR/56 BWR
	NAC-MPC	37 PWR/89 BWR
Pacific Sierra Nuclear	VSC-24	24 PWR
Transnuclear	TN-24, TN-32, 32A, 32B	24/32 PWR
	TN-68	68 BWR
Transnuclear West	NUHOMS-24P	24 PWR
	NUHOMS-52B	52 BWR
Westinghouse Electric	MC-10	24 PWR

* 10 CFR §72.214. List of Approved Spent Fuel Storage Casks

The owner's desire to minimize financial risks through prompt remediation of the site, and the potential for additional delay in the transfer of spent fuel, usually justifies moving the fuel to an ISFSI. The construction/augmentation away from the reactor but still within the confines of the site allows decommissioning to proceed as expediently as practical, without placing the safe storage of spent fuel at risk. Moving the fuel, a large source term, also lets the owner more easily assess the extent of required and achieved radiological decontamination and site remediation. And it simplifies the termination of the operating license.

In developing a spent fuel management scenario, the precise dry storage technology need not be identified. Since all currently licensed or planned technologies rely on the same basic components and processes, the cost estimate need only accommodate the technical specifics of dry storage in a general sense. Current designs involve placing individual fuel assemblies within metal canisters that are then stored inside large metal or concrete modules, the modules then placed on concrete pads within the ISFSI to await DOE acceptance. Variations occur within this scenario among dry storage vendors, but the process and end result are uniformly similar. In the absence of site-specific data, the capacity of a selected cask can be used to determine the footprint of the ISFSI and the number of storage units required.

Table 14.2 identifies several cost considerations to be evaluated when preparing an estimate for either the construction of an ISFSI or the expansion of an existing facility to support decommissioning. Typically, when a facility or a portion of the facility is required to support continued plant operations, only those costs associated with the post-operation augmentation of the facility are assigned to decommissioning.

Table 14.2 Cost Considerations For On-Site Spent Fuel Storage (Dry)

	Existing Facility	New Facility
Selection of Dry Storage System (including capacity of canisters)		X
Siting, Design, Construction and Licensing		X
Expansion and Relicensing of Existing ISFSI	X	
Reactor/Fuel Building Modifications		X
Other Required Site Modifications and Facility Upgrades	X	X
Fabrication of Storage Liners and Overpacks	X	X
Fuel Inspection/Characterization	X	X
Failed Fuel Containerization	X	X
Cask / Module Loading for On-Site Storage	X	X
On-Site Transfer Equipment and Transporter		X
ISFSI Monitoring (including Security)	X	X
Transfer to USDOE	X	X
ISFSI License Termination/Remediation	X	X

14.5 WET FUEL STORAGE

The storage of spent fuel in a passive dry system is typically regulated by the decay heat generation of the assemblies. Assemblies discharged from the last core load will require a minimum of five to seven years of cooling before they can be removed from the pool and stored without active cooling. Consequently, the plant's storage pool will be active for a minimum period of time after shutdown.

Most utilities in the process of decommissioning have effectively isolated the spent fuel pool so that decommissioning operations can proceed in adjacent areas. Separating and isolating the "spent fuel island" is relatively straightforward at a pressurized water reactor (PWR) or a Mark III boiling water reactor (BWR) where the pool is located in an independent structure distanced from the reactor. Isolation is significantly more difficult when the pool is adjacent (and usually connected) to the reactor cavity where dismantling operations are being conducted. Examples of activities involved in isolating a storage pool are shown in Table 14.3.

Table 14.3 Elements of On-Site Spent Fuel Storage (Wet)

Activity	Cost Considerations
Licensing	Part 50/72 license amendments/conversion
Engineering	Spent fuel pool and control room modifications
Storage Pool Modifications	Cask loading pit reinforcement, cask drop mitigation, pool cleanout
Systems	Controls relocation and upgrade, cable re-routing, ventilation modifications, power isolation and independent supply, installation of local filtration and cleanup systems/ components, upgraded fire protection, pool monitoring equipment, and primary release point monitoring equipment, security system relocation
Building Modifications	Crane re-qualification, upgrade, and/or replacement, facility construction and furnishings for new equipment and operating personnel, cask transfer area modifications
Pool Operations	Systems operation and monitoring, power, lighting, ventilation, security, chemistry, fuel handling
Transfer to DOE	Cask handling, loading and sealing liners, loading and decontamination of transport cask
Decommissioning	Disposition of storage racks, processing of water, decontamination of pool liner and concrete (if necessary), removal of system components and support services, decontamination of building, confirmatory surveys, and conventional demolition (if necessary)

Since support services are borrowed from the operating organization, operating costs for a pool are generally indistinguishable from the costs to operate a main complex during decommissioning. However, as an isolated activity during a prolonged dormancy, the operating costs of the ISFSI can be significant. An operating organization will include certified fuel handler(s), maintenance, chemistry, security, radiological, and administrative personnel. Additional resources will be required during the periodic loading of the transport casks for relocation of the fuel to the DOE's geologic or alternative interim facility.

14.6 COST CONSIDERATIONS

The cost of an on-site ISFSI depends on such considerations as the dry storage technology, the supplier, the economics of bulk purchase, and site-specific contractual arrangements. To date, there is no consistent standard for pricing options in the dry storage industry. Indeed, the proprietary nature of such costs prohibits use of detailed cost values. These cost factors necessarily lead to the use of simplifying cost assumptions in developing the decommissioning cost estimate.

14.6.1 Direct Costs:

1. *Capital Construction and Procurement Costs*: These include the construction, loading, and operation of a dry storage facility included within a decommissioning cost estimate. Estimates are influenced by assumptions on the costs of essential components, such as the cost of canisters, overpacks, and supporting facilities, as well as collateral costs for operating personnel, essential services, property taxes, etc.

   For the purpose of calculating a decommissioning cost estimate, dry storage facilities are assumed to be located within the plant's currently protected area. Costs for capital expenditures (facility construction/augmentation, liners, shield overpacks, etc.), are generally accrued in the first 5 to 6 years following the cessation of plant operations, consistent with the cooling prerequisite for fuel transfer.

   Wet storage also involves capital expense, particularly if the pool is to be isolated and operated independently from adjoining power block structures. Costs can include system upgrades, physical modification to the facility/support systems, independent cooling and power supplies, and relocation of control, monitoring, and security systems. ISFSI isolation allows early decommissioning of surrounding structures; therefore, modifications must be made shortly after reactor defueling is completed.

2. *Waste classified as "Greater-than-Class C"*: This waste is radioactive material with higher specific activity than is permitted in shallow land disposal. The dismantling of reactor internals at most commercial nuclear units will generate radioactive waste designated as GTCC, as defined by 10 CFR 61 (NRC 1983). Although the material is not classified as HLW, the DOE has indicated it will accept title to this waste for disposal at its future repository. For estimating purposes, GTCC is typically packaged in spent fuel canisters and stored along with the spent fuel. Consequently, fuel storage development costs include the additional modules for GTCC. It is reasonable to assume that the DOE would not accept GTCC waste prior to the completion of the transfer of spent fuel. Therefore, until such time as the DOE is ready to accept this waste, it is also reasonable to assume that this material would remain in storage at the various reactor sites.

3. *Cask Loading and Handling Costs*: The engineering, design, licensing, and operations effort involved in loading spent fuel and GTCC material into storage/transport canisters includes the development of technical descriptions, design calculations, and equipment specifications for the equipment designed and/or supplied. Site-specific procedures are also required for loading, decontaminating, welding, testing, and transferring the canisters to the storage casks. Typically, the dry storage system vendor supplies design documentation and licensing submittals as part of the contract, subject to owner review and approval.

The vendor will typically supply the metal storage canister with the internal subassemblies for fuel assembly positioning, the transport cask with the associated lifting beam and/or frame, the canister vacuum drying and inerting systems, welding equipment, transporter or trailer, and a prime mover. In most instances, the vendor will also supply the labor force to load, seal, and transport the canisters. Typically, the process of off-loading the pool includes the mobilization and training of the vendor personnel, a dry run, the loading campaign, and a demobilization phase. The loading campaign includes the documentation of the loading process, canister welding, draining, drying, inerting, testing, and inspections. It should be noted that the owner is generally responsible for system and component upgrades to support loading operations. Such support typically includes single, failure-proof cranes, building modifications for cask accommodation and increased floor loading, support services, and maintaining water clarity for activities in the fuel pool and reactor cavity during the loading operations.

For a large campaign, an average of three to four casks per month can be used to estimate the duration for loading the canisters. However, at some sites, weather conditions may prohibit the movement of the fuel from the pool to the pad during certain months of the year and this constraint should be factored into the schedule.

4. *Decommissioning Costs*: These costs include the decommissioning of on-site fuel storage, i.e., those associated with license termination. For a spent fuel island, this process is consistent with that employed for the adjacent power block. Equipment is removed, the facility decontaminated, and the structure dismantled. Since the pool may have been operational for 60 years or more (in total), the potential is greater for the migration of contamination into the surrounding structure and local environment from liner leakage and/or fuel transfer activities than if the pool is shut down shortly after shutdown. This should be a consideration in projecting decommissioning costs for any facility or structure with long-term exposure to water-born contaminants.

   A dry storage installation will also need to be decommissioned, with the concrete overpacks used for on-site storage becoming surplus material. Prolonged exposure to the spent fuel may result in minor activation of the overpacks' interior surfaces. If so, the overpacks will require decontamination prior to demolition and disposal.

14.6.2 Indirect Costs: The NRC has proposed revisions to the current regulations governing insurance coverage for permanently shutdown plants (NRC 1997b). The proposed rule would allow licensees to reduce their on-site insurance coverage and off-site financial protection requirements based upon the reduction in radioactive decay of the spent fuel assemblies with the passage of time. This can result in significantly lower premiums, particularly once the fuel has been removed from the storage pool and the risk of off-site contamination is reduced.

14.6.3 Other Considerations: Two critical issues concerning spent fuel activities are schedule and demolition. The removal of spent fuel and other waste from the spent fuel storage pool is a critical path activity in most prompt DECON scenarios. With the spent fuel pool empty, the critical path shifts to the decommissioning activities associated with building cleanup and site dismantling. Because of this constraint, the transfer of spent fuel from the pool must be optimized and isolated so that decommissioning activities can proceed without affecting the remaining areas of the plant. Based on current industry designs for dry storage, the minimum amount of time to satisfy the canister and cask decay heat load design parameters is 5 to 7 years following reactor sub-criticality. Allowing for the actual loading of casks with the last reactor core, an additional six months is typically added to the schedule. With a higher-burn fuel and longer fuel cycles, some assemblies may require additional cooling, or the casks may require short-loading to meet the heat generation criteria (fewer assemblies loaded than the cask capacity).

Typically, the buildings are dismantled once the equipment has been removed and the power block structures decontaminated. Conventional dismantling relies on explosive charges to weaken structures and remove bulk quantities of heavily reinforced concrete. With the fuel residing in an ISFSI away from the power block, conventional demolition techniques can be employed. However, less destructive techniques, and potentially less efficient techniques must be considered if the fuel continues to reside in the plant's storage pool. Therefore, the estimator also must consider the location of the irradiated fuel and the timing of the building demolition when estimating the impact of a particular storage methodology.

14.7 ASSESSMENT OF OPTIONS

The inability of the DOE to accept spent fuel has resulted in the need to expand existing, or build new, interim fuel storage capability. For a number of plants, completing the transfer of fuel from the site is not expected until years after the plant stops operating. Consequently, most licensees have decided to accumulate funding for spent fuel maintenance and storage as part of the same trust fund used for decommissioning. Regardless of the actual funding mechanism, the owner needs to know the potential magnitude of this financial liability.

If dry storage is required for continued plant operations, most owners will chose to expand the facility to support decommissioning, allowing decommissioning to proceed on the adjacent facilities without constraint. For facilities with enough wet storage capacity, adding dry storage becomes an option during decommissioning. In any scenario, the pool will be required for active cooling of the assemblies until the fuel can be stored as dry -- five to seven years.

The selection of a wet or dry management system depends on several factors, including the preferred decommissioning scenario, the anticipated fuel storage period (based on the DOE acceptance rate), available capital, and future site operations. For wet storage, the estimator need only assess the modifications necessary to the existing facility for independent operation and any impact its presence may have on the intended scenario. For dry storage, the estimator will need to select a referenced system (in the absence of site-specific information). Storage cask capacity and the reactor discharge schedule can be used to determine the size of facility. It should be noted that during the wet cooling period, fuel could be transferred directly to the DOE rather than to the on-site facility, if possible. In most situations, the fuel will be transferred from dry storage directly to the DOE's geologic repository or other interim facility in its existing package. But fuel in first generation, storage-only containers may need to be repackaged before transfer.

All storage systems will incur operating costs. While the individual cost components may vary, maintenance, security, and radiological resources will be required to monitor operations and support periodic fuel transfer to the off-site destination. Eventually, both systems will require decommissioning. Cost considerations were identified in Tables 14.2 and 14.3; sample decision nodes are given in Table 14.4.

TABLE 14.4 Sample Decision Nodes

Plant Design	PWR Mark III BWR Mark I or II BWR
Decommissioning Options	Prompt Incremental Deferred
Site Configuration	Single unit Multi-unit (operating) Multi-unit (non-operating)
Site Disposition	Unrestricted release Restricted/controlled Continued operation
USDOE Pickup	During 1st six years < 10 years following shutdown > 10 years following shutdown
Financing	Available Limited Unavailable
License Termination	With fuel pool operational With independent facility
Owner	Single site Multiple generating sites Risk adverse Risk tolerant
Existing Fuel Storage Mode(s)	Wet Wet and dry

Chapter 15

TRANSITION PLANNING AND SITE CLOSURE

15.1 INTRODUCTION

Many activities must be performed just before or soon after a nuclear facility concludes its mission and enters the transition from operations to decommissioning. These activities include assurance of immediate and long-term safety for the facility systems, human resources challenges, systems configuration changes, regulatory changes, and stakeholder relations. Long-term stewardship and eventual site closure activities must also be considered and addressed.

15.2 TRANSITION PLANNING

Many tasks must be performed at the conclusion of the primary operational mission of a nuclear facility. They will vary somewhat based on site-specific circumstances and conditions, but all facilities will need to plan for the transition in the following areas:

- Placing the facility and systems into a safe shutdown condition
- Organizational realignment
- Near-term licensing and regulatory interaction with cognizant regulators
- Procedure, process, and program reductions
- Contractual realignment
- Financial, property tax, and industry fee reviews
- System reclassification
- Fuel or HLW storage
- Site characterization
- Facility modifications to support decommissioning
- Preparations for decommissioning activities
- Personnel and contractor training
- Asset recovery
- Stakeholder relations

Some specific areas (e.g., site characterization and stakeholder relations) are discussed in detail in other chapters of this handbook. These so-called transition activities may begin months, even years before a facility's future permanent shutdown is announced, funded, and planned. Typically, the more time available for transition planning before facility shutdown, the shorter the actual transition takes. This has value particularly for facility staffing issues. The transition period and its inherent uncertainty typically create substantial stress in the workforce. And if workforce reductions are anticipated after transition, promptly completing these activities will bring opportunities for overall project cost savings.

15.2.1 Placing a Facility into Safe Shutdown: Safe shutdown is a must, but it's also often a challenge to shut down systems and the facility so as both to comply with operating requirements and also support future decommissioning. In some cases the decision to shut down is unplanned, so there's little opportunity to optimize shutdown procedures. In these cases, the only criterion is to follow established operating protocols to safely shut down equipment and processes. When possible, ongoing processes (e.g., chemical processes conducted in a batch approach) should be concluded in order to allow for potential flushing and removal of radioactive materials.

The process of shutting down a nuclear facility should follow established protocols. Even though the facility will not be restarted, it is essential to maintain all systems in an operable condition until the appropriate safety and licensing changes have been approved by the applicable regulatory agency.

15.2.2 Organizational Realignment: The staffing mix required for facility operations varies as widely as the types of nuclear facilities. Small, single mission facilities may have only a handful of professionals as the entire staff. Nuclear power reactors typically have staffing levels of approximately 250 - 350 per reactor, including professionals, managers, and technicians engaged in activities, such as maintenance, operations, and ES&H. Large, multi-mission national laboratories have staffing in the thousands -- also with a wide range of technical, professional, and managerial posts. Although the number and type of personnel required for decommissioning also vary widely based on facility size and work scope, the matter is further confounded by requiring retraining and developing new skills for incumbent staff, or by changing facility staff.

15.2.2.1 Human Resource Transition Issues: The determination should be made early in the decommissioning process as to the probable approach for staffing, the two primary options being to retrain incumbent site staff to perform most tasks or to contract out. Both approaches have plusses and minuses . In many cases, there will be substantial local stakeholder input about this decision. (More on stakeholder interaction may be found in Chapter 8.) The owner retains the ultimate responsibility (through a general duty disclosure) for compliance with regulatory requirements and for the health and safety of the staff, the public, and the environment.

15.2.2.2 Incumbent Staff: The benefits of retraining and using incumbent staff include:

- No site-specific access training required
- Familiarity with site structures, systems, and components
- Familiarity with operating and safety envelope
- Familiarity with site procedures, processes, and programs
- Reduction in overall staff trauma

But here are the challenges:

- Project management rather than continuing routine operations
- A different mix of routine skills, such as pipefitting, welding, carpentry, demolition, and heavy equipment operation
- Typically higher use of skilled and unskilled craft labor
- Less reliance on professional support staff
- A sequence of infrequently performed tasks, some being of higher risk.

Several analyses should be performed regarding the use of a facility's incumbent staff as the primary work force for the decommissioning project. The first analysis is to compare the skill sets required with those of the incumbent staff. Developing an understanding of the skills required and staffing levels needed should be accomplished with the assistance of decommissioning professionals who have had previous experience with the tasks required for such projects. Once the skills and staffing levels are identified, these requirements are compared to the levels and skill of the existing staff. The result of this comparison (gap analysis) is the basis from which training should be developed for the incumbent staff. In virtually all cases, some training will be needed for the incumbents to acquire decommissioning skills. The costs associated with this training should also be determined, and an estimate of the time required. The projected training costs and time period are used to help determine the best cost-effective staffing approach. Even when the intent will be for the use of incumbent staff for the project, it is likely that some will be unable to effectively make the transition from operational to decommissioning functions. In these instances, human resources may be needed from external organizations to support potential staff recruiting and provide out-placement support for those unable to make the transition.

15.2.2.3 Contracted Staff: If the primary workforce will be contracted, the most critical early need is to beef up human resource support to be able to help the incumbent workforce find new jobs. In concert with this task is the development of skill sets and staffing needs for decommissioning, as discussed above. These staffing requirements will be a crucial element in the contract bid specifications. Most likely multiple contract firms will be required for the project, so the development of contract bid specifications is also an early challenge. Once the contracts are let, then the facility should plan for augmented site-specific training for the newly contracted facility personnel. This training should be timed to support the rapid influx of personnel.

15.2.2.4 Economic Comparison: Like many aspects of decommissioning, the decision to use incumbent staff or contracted staff may be supported by the use of a simple economic model. In both cases there will likely be costs for furloughing of staff, for contractor staffing, and for incumbent staff training. The various factors to consider can be noted in the following equation, which may be used for various staffing scenarios in order to optimize the costs.

$$\text{Staff Costs} = [\text{FTE-I}_i \times (SB_i + TC_i)] + (\text{FTE-O} \times \text{OUT}) + [\text{FTE-IN}_j \times (SB_j + IN_j)] + (\text{FTE-C}_k \times CC_k) + \text{DELAY}$$

where:

1. FTE-I_i is the number of incumbent staff remaining with a given salary
2. SB_i is the salary and benefits for the specific incumbent individual i.
3. TC_i is the training cost for the specific incumbent individual i.
4. FTE-O is the number of incumbent staff to be outplaced
5. OUT is the cost of outplacement and human resource services per outplaced incumbent individual
6. FTE-IN_j is the number of incoming individuals
7. SB_j is the salary and benefits for the specified incoming individual j.
8. IN_j is the in-processing and training costs per incoming individual
9. FTE-C_k is the number of contracted individuals j
10. CC_k is the full costs per contracted individual j
11. DELAY is the project costs (if any) due to needed incumbent staff training/qualifications.

In the case of governmental projects, the provisions of the Davis-Bacon Act (DBA 1964) affect the project labor cost. The particular project scope and schedule will determine the extent of work conducted under the provisions of the Act, which requires that each contract over a certain limit to set the minimum wages to be paid to various classes of laborers and mechanics employed under the contract. Under the provisions of the Act, contractors or their subcontractors are to pay workers employed directly at the work site no less than the locally prevailing wages and fringe benefits paid on projects of a similar character.

15.2.2.5 Developing the Organization: As stated above, in most cases there will be some facility personnel who are either unwilling or unable to staff the transitional phase. To ease the transition for these staff members, some facilities have provided programs and opportunities, such as those listed below, as mentioned in the decommissioning report of the Yankee Rowe plant at Rowe, Massachusetts (EPRI 1997):

- Local job fairs
- Preferential placement at other nuclear facilities
- Outplacement services
- Support for resume development
- Early retirement opportunities
- Augmented separation/severance packages
- Communication programs focused on the family of affected staff

Another challenge may be retaining key personnel throughout the process. Some facilities have used bonuses to ensure their availability through the project or at a future date, as detailed in the EPRI report on the decommissioning of the Fort St. Vrain Reactor, located at Platteville, Colorado (EPRI 1998).

There is no best project organization. An organization develops based on the type of facility, its history, the decommissioning approach selected, the available funding available, and many other variables. Its development starts with understanding the regulatory requirements expected during facility deactivation, decontamination, and dismantling. A reduced level of regulatory requirements may apply during periods of surveillance and maintenance. The regulatory requirements will provide the basis to determine numbers of technicians (operators, maintenance, and ES&H). The number of technicians provides the basis for numbers of supervisors and managers. In a typical example, one supervisor can direct the efforts of five to ten technicians, and managers may direct three to five supervisors or additional professional staff. The required number of professional staff is typically developed on a site-specific basis. This base staffing level accounts only for the resources needed to maintain the facility in accordance with the safety and regulatory envelope. This staffing does not account for personnel needed for the actual decommissioning work. In some cases, the regulatory requirements will decrease over time as key milestones are reached. As an example, after a sufficient period of radioactive decay, cooling for spent reactor fuel in storage is not required. In this case, elimination of the S&M requirements for the cooling systems would be reasonable. As these regulatory reductions are approved and implemented, then this base staffing level can also be reduced.

Job descriptions and responsibilities should be developed for the project's organization. Ensure at a minimum that a sufficient number of facility-knowledgeable individuals are retained in the organization (i.e., archiving). These individuals will be valuable in providing essential institutional knowledge to the project. The number of personnel used should also be developed in concert with the development of a site-specific cost estimate and project schedule. Regardless of the number of individuals on the project, key responsibilities will exist for all facilities. Figure 15.1 denotes a typical functional organization chart for a decommissioning project.

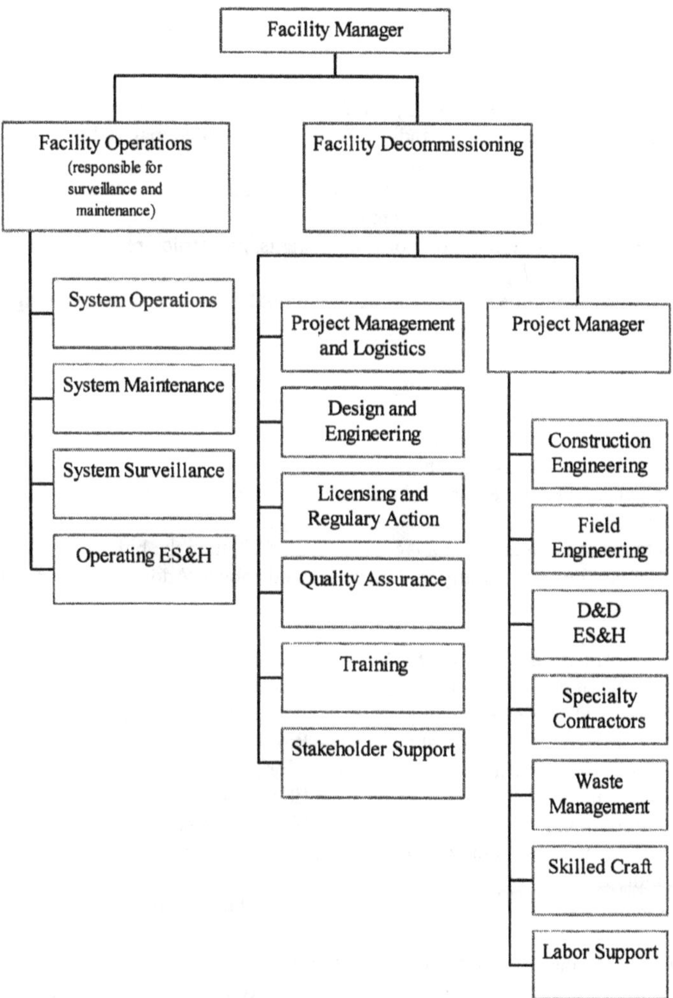

Figure 15.1 Sample Decommissioning Organization

The facility operations mode reflects the functions and staff required to comply with the regulatory requirements in effect, sometimes known as S&M requirements. As noted above, these staffing levels may diminish over time. Although not required, keeping operations staff functionally separate from the decommissioning staff helps ensure the meeting of regulatory requirements without potential impacts from decommissioning constraints.

In virtually all cases, decommissioning projects require some contracted support services. These may range from minimal augmented waste management services to the entire project being contracted to a DOC. Typical contracted services include:

- Processing, packaging, transportation, and disposal of radioactive material
- Decontamination and recycling of radioactively contaminated material
- Radiation protection staff augmentation
- Design and fabrication of special dismantling equipment
- Heavy loads management and transportation
- Dismantling and demolition of components, systems, and structures

The following attributes should be part of contractor bid qualifications when obtaining contracted services:

- Demonstrated experience in providing services on similar projects
- Cost and schedule compliance
- Technical and operational capability
- Ability to meet regulatory requirements
- Sufficient staffing available to meet both short- and long-term project demands

If the intention is to proceed under the management of a DOC, then the bidding and proposal selection process should be substantially more detailed and robust. Additional information needed from bidders includes:

- Scope of project per purchase order
- Implementation of quality programs
- Proposed craft staffing levels
- Project standard work days and hours
- Project controls and reporting
- Project documentation
- Verification programs
- Project administration
- Project organization for management team
- Project personnel qualifications
- Project budgets and schedules
- Work control process
- ES&H protocols
- Nonconformance control
- Proposed interface agreements between project and facility operations personnel
- Safety standards
- Records management
- Material procurement
- Oversight activities

For large contracted projects, the contractor organization should establish and maintain certain core management positions. Many of these functional positions are noted in Figure 15.2.

Project Director
Primary Facility Interface
Organize, Plan and Lead Team Efforts
Encourage Teaming with Others Onsite
Focal Point for Offsite Support

Site Project Manager
Day to Day Operations Responsibility
Primary Focal Point for Cooperation with Others Onsite
Create a Safe and Productive Workplace
Establish and Enforce Policy

Decommissioning ES&H Safety Manager
ES & H Planning and Implementation
Develop and administer ALARA Program
Specific Decommissioning Work Package Input
Coordination of Protective Measures
Member of Utility Radiation Protection (e.g., ALARA) Team
ALARA/Radiation Training
Safety Training
EPA and OSHA Reporting
Conduct of Subcontractor Audits and Surveillance
Emergency Services Coordination

Design & Engineering Manager
Decommissioning Work Packages
Technical Reports
Technical and Non-technical Analyses
Document Control
Construction Support

Human Resources Manager
Personnel Policies
Labor Relations
Training and Public Relations
Staffing Adequacies
Community Activities

Construction Manager
Project Execution
Promote Safety and Quality
Meet Budget and Schedule
Area Superintendents
Temporary Facilities

Project Controls Manager
Cost Integration and Control
Schedule Development
Management Information systems
Performance Analysis and Reporting
Schedule Interface with Others Onsite

Business Manager
Policy and Procedures
Payroll and Billing
Accounting
Procurement
Subcontractor Administration

QA/QC Manager
QA Policy, Plan and Procedures
Conduct of Subcontractor Surveillances
Documentation and Certifications
Interface with Other QA Onsite Organizations
Conduct of Quality Audits

Figure 15.2 Functional Positions for Contract Organization

15.2.3 Licensing: Many licensing activities will occur throughout the transition period. Some will be required and others should be undertaken to reduce costs. A brief listing of some of these required and suggested licensing and regulatory activities is provided herein. Additional information on regulatory actions necessary by the NRC is provided in Part 2 of this handbook.

15.2.3.1 Power Reactor Licensing Activities: Within 30 days of ceasing operations, the NRC requires a certification indicating the permanent cessation of operations. The same 30-day response period is also required following the removal of all nuclear fuel from the reactor. In addition, a

PSDAR is required within two years of a prematurely shutdown power reactor, and no later than five years prior to a planned shutdown.

Within two years after shutdown, licensees must submit a site-specific cost estimate for the decommissioning project and a description of how management will provide funding for the safe storage of irradiated fuel until title is transferred to the DOE. Licensees are prohibited from using the full amount of money that was accumulated during operations for the decommissioning process until the site-specific cost estimate is submitted to the NRC. Unless granted permission to the contrary, the site must be decommissioned within 60 years and the licensees remain accountable to the NRC until decommissioning is complete and the license terminated. To conclude its obligations, licensees must submit a license termination plan (NRC 2002).

15.2.3.2 Other Licensing Activities: Notification to the NRC of the transition of a non-power reactor facility from active to decommissioning status is required within 60 days of any of the following conditions:

- The license expires or is revoked.
- The licensee decides to permanently cease operations with licensed materials.
- Two years elapse since activities were conducted under the license.
- No principal activities have been conducted in a separate building, or outdoor area, for a period of 24 months and residual radioactivity is present that would preclude its release under NRC regulations.

For a non-power reactor, decommissioning must begin within 12 months of the above notification or a decommissioning plan must be submitted to the NRC. This plan can be configured as a supplement to the safety analysis report. In addition to this, 10 CFR 50.33(k)(1) requires licensees to submit a report indicating how reasonable assurance will be provided that funds will be available to decommission the facility.

15.2.3.3 Optional Licensing Submittals:

1. *Technical specifications*: Modify technical specifications to discuss limiting decommissioning conditions, required equipment, and surveillance and administrative requirements.
2. *Physical security plan*: If special nuclear material is removed from site, physical security can be reduced.
3. *Emergency plan*: Regulations may require an emergency plan. However, as radioactive materials are removed from the site, emergency planning may be reduced and ultimately eliminated when hazards no longer exist.
4. *Quality assurance plan*: With a reduced radiological risk, a reduced scope QA plan may be appropriate.
5. *EPA effluent permits*: Depending on planning results, revisions to effluent permits (e.g., NESHAPS, NPDES) may be required. New permits may also be required.

15.2.4 Procedure, Process, and Program Reductions: Once the cognizant regulators have made appropriate regulatory approvals, certain safety systems are no longer required. When appropriately configured for eventual decommissioning, the S&M requirements for these systems may be reduced and ultimately eliminated. The governing procedures for S&M tasks can also be eliminated. As approvals come for reduced scope security, emergency planning, quality assurance, or other programs, the governing implementing procedures should be correspondingly simplified.

Wholesale process/program re-engineering can actually help the incumbent staff change from an operating to a decommissioning mentality. This re-engineering continues support for operational requirements and yet allows the staff to prepare procedures that are appropriately reduced in scope and relaxed in terms of regulatory mandates. Typically, both the number and complexity of procedures are reduced. As the project progresses, more standard processes and/or procedures will be developed.

Decisions can now be made about which configuration control system is most useful. Typically, this will involve the use of some operating and maintenance procedures while others will be specifically developed to place systems and components into configuration for long-term storage or in preparation for decontamination. Throughout the review of programs, procedures, and processes, it's important to ensure that all license-required conditions are maintained until regulatory approval is received.

15.2.5 Contractual Realignment: Once the decision to permanently shut down the facility is made, the specific needs and staffing levels of these various contractor groups may substantially change. An example would be a contracted project in progress to increase facility production capability. Clearly, the result of the project is no longer needed. A typical approach for the physical work would be to ensure that all systems and components are placed in a long-term safe configuration, and then the physical work could cease. A separate review would be desirable to determine the most cost-effective contractual options. In another example, if a long-term contract was in place for general labor support, it is likely there would be some continued need. A critical review, however, would be appropriate to determine appropriate (and likely reduced) staff levels.

When permanent shutdown will mean reduction in permanent facility staff, additional consideration should be given to replacing contracted support with otherwise displaced facility staff. Although on the surface this may appear to be a more costly solution, it may bring several benefits not readily measured in dollars, including the continued support by individuals who may help provide facility history that can be valuable during early site characterization activities. The continued employment of facility personnel also supports stakeholder relation needs. In all cases, when substantive changes in facility contracts are considered, appropriate legal counsel should be consulted.

15.2.6 Financial, Property Tax, and Industry Fee Reviews: A variety of financial reviews should be undertaken upon facility shutdown, with an eye to substantial cost savings. If the facility is non-governmental, one essential review may provide for a reduction in property tax liability. The annual property taxes paid are based on a percentage of the assessed value of the property and facility. In the case of facilities that develop or generate a product (e.g., radio-pharmaceutical manufacturers or power reactors), once the operation ceases, and a clear case can be made that the value of the facility had been diminished. If the proper analysis is provided to the local taxing authorities, a reduction in facility taxes may be obtained.

Similarly, many other annual fees may be based on facility operations, which, once shut down, will merit regulatory fee reductions. If regulated by the NRC, differing annual fees may be assessed depending on the facility license type. For power reactors, participation in and payment of annual fees to industry groups should be re-evaluated. If continued membership in these organizations is considered, the facility manager should ensure that it provides value commensurate with a permanently shutdown condition.

Once the decision has been made to decommission, the available project funding sources should be evaluated. For NRC-licensed facilities, funding assurance is a legal requirement. Acceptable methods are discussed in Chapter 3 of this handbook. Two of the acceptable methods involve placing decommissioning funds into trusts. For these cases, the owner needs to determine the appropriate protocols and requirements to access the funds in trust. If the funding assurance is to be provided by a governmental agent (e.g., a facility owned by a state university), then the owner should begin the administrative processes to gain governmental project funding. This approach is also the one used for government facilities not licensed by the NRC, such as DOE sites. In some cases the administrative hurdles that must be overcome to secure governmental budget authorizations can be arduous. In the case of small facilities, the decommissioning may be accomplished within one budget cycle. In many other cases, however, the project will proceed over several years and several budget cycles. This can lead to limitations on project management and controls since funding only comes year by year.

15.2.7 System Reclassification: As long as an operating license is in place, most nuclear facilities have different levels of safety systems to operate and maintain. Systems directly used to maintain positive controls over the radioactive materials (e.g., accelerator interlock systems, reactor emergency core cooling systems, glove box ventilations systems) are typically designated as the highest level of safety. Accordingly, the license requirements for these systems are the most rigorous and stringent. The next level of safety is for systems that support the primary safety systems (e.g., secondary electrical systems for interlocks). The lowest level is for systems that support facility operations but do not have a direct effect upon nuclear safety (e.g., drinking water delivery, sanitary sewage disposal). When the facility is permanently shut down and placed in a safe configuration, the most safety-significant systems may become unnecessary. For example, from a safety basis, the permanently shutdown accelerator will no longer require beam-on safety interlock systems.

The re-evaluation of each system in light of the decommissioning plan is a beneficial task. In some cases there will be continued need for certain systems but only for finite periods of time. For example, a power reactor's spent fuel cooling systems may only be required for a few years, after which the heat load may be sufficiently low that specific support cooling systems are not required. Such a critical assessment should be conducted for each facility system, and its ongoing need and safety significance should be established. The beneficial outcome of this systems review and reclassification typically allows for:

- A reduction in the number of safety significant systems
- A reduction in safety significance of other systems
- The complete elimination of some systems

Once the systems analysis has been conducted and appropriately documented, concurrence will typically be required from the appropriate regulatory agencies before implementation. Upon approval, the following actions should be taken to benefit from the reclassifications:

- S&M requirements, implementing procedures, and costs may be reduced to a level commensurate with new classifications
- Systems designated as no longer required may be formally labeled as "abandoned in place" or removed, as appropriate
- Controlled facility drawings and other controlled documents may be revised and reduced to uncontrolled status

15.2.8 Spent Nuclear Fuel or HLW Storage: Further challenges will arise for facilities with spent nuclear fuel or HLW. Common practice is to manage the disposition of these significant radiological hazards as a discrete individual project with linkage to the decommissioning project. Early decisions need to be made for both the near-term (transition) and long-term (stewardship or site closure) management of these waste products to ensure success to the overall decommissioning effort. Additional discussion on differing near-term approaches and decision models is provided in Chapter 10 of this handbook. An overall discussion of the management of spent nuclear fuel and HLW is given in Chapter 14.

15.2.9 Site Characterization: The site's routine operational radiological and hazardous materials surveys can be valuable to the site characterization process. Facility operational records prepared to meet statutory and regulatory requirements are another important source of early characterization data. One of the first tasks in transition will be starting the HSA. The process summarizes the facility's history with respect to the management of radioactive and hazardous contaminants. Typical inputs include historic and present documentation and information from facility workers. Chapter 11 gives detailed information on the overall site characterization process.

15.2.10 Facility Modifications to Support Decommissioning: Throughout its operating life, changes or modifications are made to facility systems, structures, or components to improve safety, efficiency, or productivity, or to meet changing regulatory requirements. Modifications may even be in progress when a facility shuts down, or systems and components may be in various stages of maintenance. An evaluation must be performed for either case to determine the appropriate future action to take for each task. For in-progress modifications designed to improve safety, it may be appropriate to continue the task to its conclusion, based on its related safety significance. For modifications designed to improve efficiency or productivity, the most likely course of action would be to ensure that the system is placed in a configuration appropriate for long-term safety, then to cease any additional work. A similar approach is taken with in-progress S&M tasks. In any case, the systems must be in accordance with license conditions. This typically means that in-progress S&M tasks should be concluded and continued. If reductions in system S&M are deemed appropriate by regulatory authorities, the changes should be reflected in implementing procedures, and personnel should be appropriately trained before the reduced-scope tasks start.

Once the overall approach to decommissioning is selected (e.g., SAFSTOR or DECON), and a high-level project plan established, work could begin to determine which facility modifications could more effectively support the project. One aspect will be to determine if any structures can be permanently closed (typically auxiliary or service buildings). Even though such action may involve the consolidation and relocation of personnel, the closure of unnecessary buildings can provide significant cost savings. Another typical facility modification is to provide supplemental support for the movement of heavy or bulky equipment. This may require the installation of overhead rails or lifting lugs in the ceiling. In larger facilities, it may also require the change or installation of new cranes. This can be a significant modification that may even require structural or building foundation changes in order to install larger crane capability. As decommissioning work begins, additional modifications may be required. These may include the use of temporary, skid-mounted equipment, such as the following:

- Portable ventilation units, such as high efficiency particulate air (HEPA) vacuum units, that may be piped into plant ventilation systems
- Portable waste processing, filtration, or demineralization systems
- Portable heat exchangers or chiller systems
- New portable electrical power sources or transformers

Modifications may be particularly important for very old facilities with many "field run" electrical systems. It may be more cost-effective to completely shut down existing facility electrical power systems and establish ones systems. As decommissioning continues, the numbers and types of facility modifications will change continually. Proper configuration management, control, and documentation of these tasks help ensure the overall success of the project.

15.2.11 Preparations for Project Activities: In addition to the previously described tasks, several project evaluations and business cases should be conducted to determine the overall approach and project scope. Some of these evaluations were addressed in previous chapters. Critical evaluations that must be performed early in the planning process include:

- The evaluation of the overall approach (i.e., prompt DECON versus multiple versions of SAFSTOR)
- The decision to perform full or partial system decontamination
- The determination of facility end-state, preferably with the consideration of appropriate stakeholder inputs, as expeditiously as possible
- Evaluations for spent nuclear fuel or HLW storage approach (wet versus dry)

Once these decisions or business cases are concluded, the final decision, the decision basis, and associated information should be documented. This documentation is often useful for interactions with regulators and stakeholders. Other preparation activities occur in the transition process that should not begin until appropriate plans and protocols are established for their safe conduct. Most of these actions relate to the removal of hazards and preparing for later full-scale decommissioning tasks. Following is a nine-step checklist:

- Identification of engineering/regulatory actions needed to reduce ongoing S&M costs
- Removal of fuel to interim safe storage configuration
- Performance of chemical decontamination of key systems if business case deems prudent
- Commencement of work for long-term fuel storage
- Abatement of asbestos and hazardous materials (e.g., mercury, lead, PCBs)
- Remediation of known high-risk radiological hazards
- Deactivation of unneeded systems and structures
- Remove nonessential clean equipment
- Commencement of initial site characterization activities

15.2.12 Personnel and Contractor Training: Some level of additional training will be required for all personnel on a decommissioning project, whether incumbent facility staff or contractors. At a minimum, this training should include the following topics:

- Decommissioning approach
- Project management metrics -- schedule, scope, costs
- Site characterization activities
- Disposition of structures, systems, and components
- Integrated safety management topics -- ES&H plans and the RCRA

Further training may be required for contractors to safely work on the site. The level of training provided should be commensurate with the intended work scope for trainees. More information on required training for decommissioning personnel is in Chapter 6.

15.2.13 Asset Recovery: Most often, the system components at a facility undergoing decommissioning are considered liabilities. But in some cases they may actually be assets that can be reused. Throughout the planning process, concepts of life-cycle analysis and management should help determine possible reuse opportunities. Components may be transferred to other facilities with related nuclear missions, shielding may be used elsewhere, structures may be put to another use, etc. Transferring these materials doesn't typically save costs because of the expense of the extra care taken in removal and transportation; however, it can demonstrate prudent environmental stewardship.

15.2.14 Public Participation Planning: Interaction with the affected and interested public is critical. A public outreach and communication plan should be developed early in the planning process to begin communication and interaction with various stakeholders as soon as possible after shutdown is announced. Such a public outreach and communication plan is equally important in the case of a planned future shutdown. Common practice is to establish a community advisory panel for the decommissioning team. Participants are both a valuable voice for planning and a vital communication path from the project team to the affected communities. Tasks to be performed include:

- Prompt communication with stakeholders, prior to permanent shutdown if possible (discussed in detail in Chapter 8)
- Early development of a community advisory panel
- Planned interactions with media
- Assessment of community impact and development of mitigation strategies
- An increase in community communication and public relations programs

15.3 LONG-TERM STEWARDSHIP

In many cases, facilities will not be remediated to unrestricted release criteria for a long time, either because of the approach selected or because the nature and extent of contamination preclude prompt remediation. In commercial facilities, this extended period of storage is known as safe storage or SAFSTOR; in DOE facilities, it is stewardship. If the agreed-upon cleanup levels do not meet unrestricted release criteria at a DOE facility, it may continue to perform certain protective actions that are also considered part of a long-term stewardship program.

Long-term stewardship activities reflect the fact that remediation has been insufficient to allow for unrestricted release. These activities focus on containment of residual contamination, ongoing assurance of the structural integrity of the containment, ongoing monitoring of potentially affected dose pathways, maintaining documentation to support further remediation at a future date, and modifying the facility as needed to ensure interim safety and preparation for future remediation actions. Table 15.1, extracted from a DOE report on long-term stewardship (DOE 2001d), provides typical examples of such long-term stewardship actions.

Table 15.1 Examples of Long-Term Stewardship Actions

Media	Stewardship Actions
Groundwater (i.e., subsurface water): All contaminated subsurface water that has not yet been remediated to unrestricted release of agreed upon final release levels	- Verification of contaminant levels and ongoing monitoring - Use or deed restrictions and access controls as needed - Periodic program reviews for adequacy of protection
Surface Water/Sediments: All contaminated surface waters or sediments that have not yet been remediated to unrestricted release of agreed-upon final release levels	- Verification of contaminant levels and ongoing monitoring - Use or deed restrictions and access controls as needed - Periodic program reviews for adequacy of protection
Soil: All contaminated surface and subsurface soils that have not yet been remediated to unrestricted release of agreed upon final release levels. includes areas with residual wastes under engineered or other soil caps.	- Verification of contaminant levels and ongoing monitoring - Use or deed restrictions and access controls as needed. Access controls may be institutional, Also engineered, or both. - Ongoing maintenance of any engineered controls - Periodic program reviews for adequacy of protection
Structures: Buildings or other structures no longer used that are contaminated at levels above unrestricted release criteria or agreed-upon release levels.	- Verification of contaminant levels and ongoing monitoring - Use or deed restrictions and access controls as needed. May require supplemental security requirements - both engineered controls and institutional controls - Periodic program reviews for adequacy of protection

15.4 SITE CLOSURE ACTIONS

The key tasks for site closure involve the assurance that contaminant remediation has satisfactorily met free release criteria, that post-release requirements have been met or satisfactory mechanisms are in place to continue, and that project documentation has been concluded. In many cases, stakeholders have much that they want to say about what constitutes "satisfactory" cleanup. This is typically in addition to any regulatory cleanup criteria from such agencies as the NRC, EPA or DOE. Stakeholders often erroneously assume that cleanup means that all contaminants resulting from facility operations will be removed and that the environment will be fully returned to a greenfield condition. The document, *Accelerating Cleanup: Paths to Closure* (DOE 1998b) defines site cleanup as complete when the following five criteria have been met:

- Required activities at all site facilities have been completed, excluding any required long-term S&M
- All releases to the environment have been cleaned up in accordance with agreed-upon standards
- Groundwater contamination has been contained and long-term treatment or monitoring is in place
- Nuclear materials have been stabilized and/or placed in safe long-term storage
- Any residual wastes (e.g., legacy wastes) have been disposed of in an approved manner

Before a site reaches the closure phase, the facility end-state and its remediation criteria will be firmly established. A large task for site closure will be the completion of MARSSIM surveys and documentation showing both regulators and stakeholders that the remediation criteria were met. For NRC licensees this is known as the final site survey or license termination survey. These surveys may or may not be independently verified. The more complex the site or controversial the final release, the more likely that independent surveys will be needed. More information on this process comes in earlier chapters of this handbook.

PART 4

TECHNOLOGIES

Previous parts of this handbook have addressed regulatory requirements and operational aspects of decommissioning. These projects demand a high level of integration and life-cycle planning to safely navigate schedule and cost considerations. Just as a single phase of decommissioning cannot be conducted in isolation, so should no given technology be applied without considering its impact on the overall project. Regulatory compliance and the successful completion of projects depend crucially on deploying appropriate technologies. The choice of remediation technologies depends on the type, quality, and accuracy of characterization data; the planning and order of deactivation; the end-state selected for the facility; and the waste disposal options. Therefore, the technologies discussed here are described as they relate to the various elements of the decommissioning process, as follows:

1. Once the initial characterization has identified the hazardous chemical and radioactive material present, the project managers must identify the additional characterization surveys (both ongoing and final) needed to determine the degree of decontamination achieved and to identify the quantities and locations of any remaining contaminants.

2. A planner must be familiar with decontamination techniques to be able to integrate decontamination with other decommissioning activities. Some decontamination work will be required in any project, regardless of the planned end-state.

3. The bulk of facility decommissioning is in the segmentation of metal components and the cutting and demolishing of concrete structures that must precede building demolition. (Equally important to this activity is the decontamination to suitable levels of any components and structures that will remain beyond demolition.) Traditionally, dismantling refers to the physical disassembly (reverse assembly) of piping, pumps, tanks, etc. in buildings or other areas (e.g., hot cell, pump room, laboratory), while segmentation (size reduction) refers to cutting a piece of equipment into smaller parts to facilitate disposal. Although dismantling and segmenting are usually used for the equipment inside a building, they may also be applied to its walls. Demolition, then, refers to activities performed to completely raze a building, reducing it to rubble.

4. The safety of the workers should, as a priority, surpass all other objectives. While occasionally some may regard worker protection as an impediment to production operations, most successful project leaders recognize that worker protection goes hand in hand with the successful achievement of project objectives. Decommissioning a nuclear facility presents unique health and safety challenges. The traditional industrial safety hazards, often complicated by encountering toxic and hazardous materials, are further compounded by radiological contamination.

5. The potential of robotics and intelligent machines to meet the challenges of decommissioning has been recognized for some time. The primary reason for using such tools at nuclear

facility projects is to limit human presence in hazardous environments, thus dramatically improving safety and reducing worker exposure. Use of such equipment may also reduce costs, increase productivity, and improve access to otherwise inaccessible areas. While industrial robot systems have reached a level of maturity to provide high availability at low cost, they still must be customized to accommodate the demands of a complex radiological environment. This environment has special needs, primarily: 1) decommissioning activities are unstructured, nonrepetitive, and require continuous sensory feedback and supervisory oversight, and 2) loads and motion ranges on equipment may be exceedingly large.

6. Finally, numerous other technologies directly or indirectly support decommissioning, including such tools as software for tracking remediation, temporary utility units, and systems that perform multiple functions.

The goal here is to help achieve the safe, timely, and cost-effective cleanup of a site. Experience has shown, however, that projects exhibit unique challenges and that regulations continually change. Further, the advancement of technology requires that this text be supplemented with other information sources to remain current. There are a number of web sites covering various aspects of technology developments related to decommissioning. Technology descriptions and a list of websites related to decommissioning technology development are provided on the CD in this handbook.

Most decommissioning tasks can be performed adequately with currently available technologies. But technology is not static and new ideas and developments constantly add new options. While innovation can offer significant benefits to improving the safe and cost-effective execution of a project, its acceptance and implementation often depend on overcoming the potential cost and regulatory cost considerations.

This Part of the handbook establishes a foundation of technological information to help in the conduct of decommissioning projects and in staying abreast of future technology developments. Because decommissioning covers a broad range of objectives, the description of all available technologies would increase the size of this book to an impractical level. This Part therefore presents an overview of technologies that have been demonstrated or deployed at one or more of the Department of Energy (DOE) weapons facility cleanup projects, or that have been used in commercial decommissioning.

Field demonstrated technologies have shown variable degrees of success. While documented in this handbook, they should not be applied without seeking vendor input.

Chapter 16

CHARACTERIZATION

16.1 INTRODUCTION

Characterization is an essential and substantial part of decommissioning. An initial scoping survey must be done to identify personnel hazards and to develop a decommissioning plan. Detailed characterization data must be collected to determine the type and extent of contaminants before any actual decontamination or dismantling. Surveys must be done during decontamination to test the effectiveness of the efforts; they may be either the in-process type or post-decontamination, or both. A final status survey is required, and a confirmatory survey may also be required to document the facility end-state. In addition, should the facility be placed in some form of long-term or interim storage, it will be necessary to characterize the remaining radioactive material and hazardous chemical waste to support long-term stewardship requirements. Such extensive characterization efforts contribute substantially to the overall cost of a decommissioning project.

No single technology can address the full spectrum of requirements for facility and material characterization. Decommissioning managers must identify the appropriate suite of characterization methods needed to thoroughly characterize the facility throughout the life of the project. Often, characterization methods are used in combination with one another to take advantage of the strengths and compensate for the limitations of each method. State-of-the-art characterization technologies show one or more of the following competitive discriminators:

- Reduced labor to conduct the characterization activities
- Improved data management, including automated generation of data reports
- Practical application of standard laboratory practices in the field
- Improved data accuracy, precision, and detection limits
- Deployment of survey devices into inaccessible locations
- Reduced radiation/chemical dose by using remotely deployed instruments
- Improved operation
- Reduced training and education requirements
- Real-time or near real-time generation of data
- The easily read data, and interpretation thereof
- Acceptance of results by regulators

16.2 TECHNOLOGIES

The technologies in Table 16.1 were radiological survey technologies demonstrated as part of Large Scale Demonstration and Deployment Projects. Table 16.2 contains information on pipe inspection tools and Table 16.3 summarizes non-radiological survey technologies demonstrated at these projects.

Table 16.1 Demonstrated Technologies -- Radiological Survey Tools

Technology	Description	Capabilities	Limitations
Airborne and Ground-Based Laser-Induced Fluorescence Imaging	A fast and accurate uranium-characterization tool that can visually "see" uranium contamination.	Good screening tool for uranium contamination over large area. Can be used to quickly survey a large area or to discretely survey a 2' x 2' area at a time. Best suited as a rapid, screening tool to assist planning and operations.	Detection limited to uranium contamination. Not suitable for detection of uranium in sunlit areas. ***Currently cannot be used for regulatory purposes.***
Gamma-Ray Imaging Systems	Provides visual imaging of area radiation levels by superimposing color- coded dose rate data over a visible image of the surveyed area.	Automated survey procedure minimizes personnel exposure. System can be located outside of contaminated areas. Best suited for areas with higher radiation levels. 2-D and 3-D imaging systems available.	Cannot detect contamination down to free-release levels. System capita costs are relatively high.
Surface Contamination Monitor/Survey Information Management System	A platform to deploy a large variety of radiological detectors and efficiently manage the data stream from the detectors.	Applicable to detection of alpha and beta-gamma contamination on large, open floors. Generates color map showing quantity and location of contamination. Can quickly take 400 measurements per square meter.	Limited to surveys of smooth regular surfaces.
Laser-Assisted Ranging and Data System	Uses off-the-shelf detectors to identify the precise location and contamination levels found during contamination surveys.	Provides clear, concise, and comprehensive radiological surveys. Accurate correlation of contamination level to contamination location (within 2 cm). Data is electronically handled.	Works best on flat, regular surfaces. Detector must still be positioned and moved by hand.
2-D Linear Motion System	Robotic platform used to deploy a variety of tools on vertical surfaces.	Suitable for deployment of radiation sensors for wall characterization. Can be used to scabble or otherwise treat wall surfaces.	Works best on flat or slightly curved surfaces. Setup and takedown of the unit is time-consuming.
In-Situ Object-Counting System	Radiological monitoring system used to measure small levels of contamination on large objects or surfaces.	Real-time detection of gamma radiation. Identifies specific isotopes.	Not useful for use in areas with high levels of background radiation.
Mobile Automated Characterization System	Battery-powered, autonomous robot base with a laser positioning system used to detect radiological contamination.	Good for contamination surveys on large, open floors. Generates color map showing quantity and location of contamination. Best suited for repetitive, pre-programmed surveys -- does not require operator after pre-programming.	Not good for surveys in confined spaces. Limited to surveys of smooth floors (no walls, ceilings, or equipment).
Portable Liquid Scintillation Counter	A portable, single-tube liquid scintillation counter that can be set to respond to the low-energy beta radiation from tritium.	Eliminates the "down" time required to transport the swipes. Minimizes processing times by reducing handling. Increases productivity. Reduces labor costs. Provides real-time surveillance of dismantling work	Requires manual sample changing. During demo, it was observed that while operating on the 12-Volt battery, the unit became unstable when the battery became weak. Once recharged, the unit performed normally.

Table 16.1 (Cont'd)
Demonstrated Technologies -- Radiological Survey Tools

Technology	Description	Capabilities	Limitations
Long Range Alpha Detector	Air molecules are ionized by alpha particles and flow to an external detector for collection on an electrode.	Alpha contamination is measured on irregular surfaces and confined spaces.	Limited to measurement of alpha contamination.
Electret Ion Chamber	Air molecules are ionized by alpha particles and are deposited on electret which changes its voltage	Measurement of alpha contamination	Limited to measurement of alpha contamination
Global Positioning Radiometric Scanner System	Global positioning system and radiation sensors installed on a four-wheel-drive vehicle	Suitable for radiological characterization of land surrounding facilities	Limited to radiological characterization of areas accessible by vehicle
Remote Underwater Characterization System	Radiation sensor and camera attached to an underwater vehicle	Suitable for characterization of spent fuel pools and underwater research reactors.	Cannot directly measure inside underwater pipes
Field-Transportable Beta-Counting System	Instrument provides for fast detection and spectral analysis of beta emitters in the field.	Full analysis for beta radiation in 30 min. Provides quantitative and qualitative analyses.	Limited to detection of beta radiation.

Table 16.2 Demonstrated Technologies -- Pipe Survey Tools

Technology	Description	Capabilities	Limitations
Pipe Explorer TM	In situ piping characterization system uses inverted membrane to deploy detection/measurement devices.	Navigates around elbows and some obstructions up to a distance of 250 ft. Can survey two 40-in diameter pipes and ducts. Alpha, beta, gamma detectors; position locators; visual inspection with video record.	Cannot be used in pipes with standing water or debris. Radiological measurements can be affected by debris in pipes. Does not differentiate specific isotopes.
Pipe Crawler	In situ robotic piping and ductwork characterization system.	Can survey two 18-in diameter pipes and ducts. Maximum deployment distance of 250 ft. Detectors have ample sensitivity for release limits. Best suited for beta-gamma detection.	Cannot be used in pipes with standing water or obstructions. Cannot differentiate isotopes. Single crawler is suitable for a narrow range of pipe diameters (+/-0.5 in).
Pipe Inspection System	Identifies residues in process piping.	Quick visual inspection of short pipes.	Limited to straight, short runs of pipe (< 14 ft).

Table 16.3 Demonstrated Technologies -- Versatile Survey Tools (Nonradioactive)

Technology	Description	Capabilities	Limitations
Portable X-Ray Fluorescence Detector	Portable field characterization system provides nondestructive, real-time, elemental analysis.	Applicable to a wide range of material characterization (i.e., solids, liquids).	Not useful for detecting elements with atomic numbers less than 32.
		Good for simultaneous detection of elements ranging from sulfur to uranium.	Detection limits are over TCLP limit for most RCRA metals.
		Useful as a rapid, field-screening instrument to guide planning.	***Not a substitute for regulator-approved methods of sampling and analysis.***
Lead Paint Analyzer	Uses x-ray fluorescence spectra to quantify lead and 25 other elements	Real-time identification of presence of lead paint	***Not a substitute for regulator-approved methods for lead analysis***
Detection of Free Standing Liquids	Infrared imaging camera which detects thermal gradients	Detects liquids and other objects inside pipes, vessels, and tanks	Does not identify contamination
Alloy Analyzer	Uses x-ray fluorescence spectra; compares elemental composition to built-in library of alloy compositions	Real-time identification of metal alloy	Small possibility that XRF spectra will not identify an alloy
Vehicle and Cargo Inspection System	Container passes between gamma-ray imaging source and detector	Useful for imaging the contents of containers	Does not identify radiological or chemical contamination
PCB Analyzer	Polarized XRF detects chloride ion	Field screening for the possible presence of PCBs	Requires confirmation analysis for PCBs after detector determines that chloride is present.
X-Ray, K-Edge Heavy Metal Detector	Absorption of x-rays in heavy metals; x-ray transmission is diminished at binding energy of K-shell electrons	Non-destructive evaluation of containers with uranium and other heavy metal contamination	Not useful for characterization of confined spaces
Field Raman Spectroscopy	Provides rapid compound analysis in the field.	Suitable for characterization of homogenous mixtures such as liquids. Can characterize organic mixtures.	Not viable for analysis of radiological samples.

16.3 DESCRIPTION OF SELECT TECHNOLOGIES

Several innovative and improved characterization technologies are discussed in more detail in this section.

16.3.1 Radiological Survey Tools:

16.3.1.1 Surface Contamination Monitor: The surface contamination monitor from Shonka Research Associates can survey horizontal and vertical surfaces for alpha- and beta-gamma contamination. Mounted on a motorized cart, it consists of a position-sensitive gas proportional counter that can take 400 radiation measurements per square meter. Detector arrays ranging from 0.5 to 5 m wide are mounted on the cart. The monitor gives the operator a real-time visual indication of the activity level on a liquid crystal display. Survey data are automatically processed with the system's management software.

The software combines data from individual strips of detectors to create a uniform grid for the survey area. The data can be analyzed using a wide range of image processing algorithms, and the processed survey data can then be overlaid on facility drawings. The software automatically generates data reports that meet regulatory requirements for unrestricted release.

The surface contamination monitor demonstrated that it is about 5 to 6 times faster than using hand-held instruments for routine surveys and associated documentation. For free release surveys, it's expected to be at least 16 times faster than hand-held, mainly due to its automated report generation.

16.3.1.2 Gamma Ray Imaging Systems: Gamma ray imaging systems are available from AIL Systems (i.e., GammaCam), BNFL Instruments (RadScan 600 and 700), Cogema, and other manufacturers. Although the systems differ slightly, they generally use a sensor head containing both gamma ray and visual imaging systems linked to a portable computer for control, monitoring, and data management. The sensor head is typically mounted on a tripod or overhead crane for stability and can be located from a few feet to several hundred feet from the radiation source. The systems provide a color-coded image of the gamma ray radiation field superimposed on a black and white visual image, with the colors corresponding to relative strength of the gamma ray field. The major radiation source-term need not be in the operator's line of sight to use a gamma ray imaging system. These systems can image through walls, shields, doors, and other physical obstacles provided that the source is strong enough.

16.3.1.3 In-Situ Object Counting System: The field-ready in-situ object counting system (ISOCS™) from Canberra consists of a germanium detector with portable cryostat, a cart for transporting the detector, lead shielding and collimators, a portable spectroscopy analyzer, a portable computer, and calibration software. The steel-jacketed lead shielding surrounds the germanium detector to shield it from background radiation and to adjust the field of view between 30, 90, and 180 degrees. The detector rotates on the cart for alignment with the area to be characterized. The response of the germanium detector to a series of point sources surrounding it has been characterized using Monte Carlo code. The ISOCS™ determines the radioactive sources associated with spectroscopy peaks and the software automatically ascertains the relationships between the geometry of the radioactive source, the measured count rate, and the quantity of radioactive material. The ISOCS™ has nine standard geometry templates for correlating the count rate to the strength of the radiation source for a variety of geometries.

16.3.1.4 Field Transportable Beta Spectrometer: The field transportable beta spectrometer delivers both qualitative and quantitative information on beta emitters, such as ^{99}Tc, ^{90}Sr, ^{60}Co, and

^{137}Cs. It uses solid scintillation, coincident-guarded counting, and photo multiplier tubes to count specific radioactive elements. Software in the instrument analyzes spectra real-time; detection limits are in the 20 pCi range. Typically, results are available in 30 minutes or less. The baseline approach is to collect samples and analyze them in a laboratory, where results normally take days or weeks. The technology was developed by Argonne National Laboratory but is not yet available as a commercial product.

16.3.1.5 2-D Linear Motion System: The two-dimensional liner motion system (i.e., Wall Walker) from Pentek Incorporated is a remotely operated system used to control the movement of tools and instruments on flat, vertical walls. It may be used for radiation surveys, decontamination, and painting. It uses motorized pulleys with cables hooked to a holder for tools and instruments. The motor driven pulleys are attached to a wall using anchors, magnets, or vacuum force. A portable computer controls the movement of the holder at speeds up to 60 ft/min for tools and instruments weighing up to 350 lbs.

16.3.1.6 Laser-Assisted Ranging and Data System: The laser-assisted ranging and data system is based on the integration of a modified auto-tracking laser system used to conduct civil surveys with a radiological detection system. The system may be used on interior and exterior surfaces. Similar to manual surveying, the innovative system uses hand-held radiation detectors. Three dimensional position data and radiological survey information are recorded, sent from detector to receiving station, and then combined into electronic files to provide clear, detailed, and accurate surveys. These files are interfaced with geographical information system (GIS) software to produce reports and records. Supplementing text reports, survey data can be depicted graphically with color-coded radiological levels overlaid on computer-aided design (CAD) drawings or on digital photographs of rooms or facility sectors. The system is capable of interfacing with many different kinds of radiological detectors, allowing for a wide variety of applications.

Use of the laser-assisted ranging and data system yields well-documented clearance surveys of walls, floors, and ceilings of structures before demolition. The precision of the survey and quality of the documentation expedites regulatory review and may negate the need for confirmatory surveys. The system can be configured with large detectors on a mobile platform, which enhances productivity without compromising data quality when large surface areas require surveying. In addition, the system can be configured with high-level radiation detectors for investigating and documenting more extreme radiological conditions. In summary, surveying with this system produces more useable data with greater accuracy and reproducibility with respect to locating measured activity levels. Report clarity is high and use of the equipment is relatively easy to learn.

16.3.2 Pipe Survey Tools:

16.3.2.1 Pipe Explorer: The pipe explorer system, developed by Science and Engineering Associates, deploys a variety of survey tools to characterize the interior of pipe and ductwork. These tools include alpha- and beta-gamma radiation detectors, video cameras, and pipe locator beacons. The primary components of the pipe explorer are an airtight membrane, a pressurized canister, the radiation detectors, and a video camera. Pneumatic pressure in the canister causes the membrane to invert and propagate into the pipe with sufficient force to tow radiation detectors and video cameras. Contamination of the radiation sensors and video camera is eliminated because they are towed inside the tubular membrane. The explorer can be deployed in pipes with multiple bends from 2 to 40 inches in diameter up to 250 ft long. The main value of the explorer is its ability to free-release buried or embedded pipe without excavating the pipe.

16.3.2.2 Pipe Crawler: Pipe crawler from Radiological Services Incorporated is an internal pipe characterization system for conducting radiological characterization and free release surveys. It uses manually deployed, wheeled platforms (i.e., crawlers) to convey one or more arrays of Geiger Muller detectors into a pipe. Visual surveys with a video camera and tape recording system may be used in conjunction with or separate from the radiological surveys. The pipe crawler can be conveyed inside pipes ranging from 2 to 18 inches in diameter at distances up to 200 feet, including multiple 90 degree bends. Flexible fiberglass rods attached to either end of the wheeled platform are used to push and pull the crawler within the pipe. (Crawlers for pipe larger than 8-inch diameter use compressed air to assist in extending, retracting, and rotating the detector array.) Radiological measurements are taken in increments using the fiberglass rods to move the crawler in a stepwise fashion.

16.3.3 Versatile Survey Tools:

16.3.3.1 Portable X-Ray Fluorescence Spectrometer: X-ray fluorescence analyzers deliver real-time information on the elemental composition of a variety of materials, including surfaces, soils, liquids, and thin films. X-ray fluorescence results from excitation by a radiation greater than the binding energy of an element's electrons. The energy spectrum of the x-rays emitted by each element is unique since each element has a different electron shell configuration. The spectrometer measures the peak energies of x-rays emitted from samples to identify the elements in the sample. Also, because the intensities of the x-ray emissions are proportional to the number of atoms in the sample, it is possible to determine the concentration of each element. The baseline approach to analyze samples for elemental composition is to collect and analyze the samples in a laboratory, using acid digestion and atomic spectroscopy. Typically, results take days or even weeks depending on the workload of the laboratory. The portable XRF analyzers give real-time data in the field.

16.3.3.2 Vehicle and Cargo Inspection System: Scientific Applications International Corporation's vehicle and cargo inspection system was demonstrated to non-intrusively inspect fiberglass-reinforced plywood crates containing plutonium glove boxes and other transuranic materials. These plywood crates can range in size up to 12 x 7 x 30 ft. The unit uses a shielded 1.6 curie source (^{137}Cs) aimed at a linear detector to create an image as the crate passes between the source and the detector. The source and detector are mounted on a boom truck. An on-board computer creates a composite image of the crate's contents. This image enables prioritization of when to open the crates based on their contents, to identify crates containing mixed waste because of lead shielding, and to plan the safest approach to open each plywood crate based on the location of items inside the crate. This inspection system has been used by the U.S. Customs Service to inspect vehicles for drugs and other contraband at border crossings.

Chapter 17

DECONTAMINATION

17.1 INTRODUCTION

Decontamination is the removal of material from areas where it is not wanted. This chapter will deal with the removal of radionuclides, but the same or similar technology could be used for the removal of hazardous chemical material. The technology is similar to the cleaning of dirt, oil, or corrosion products except that radionuclides are associated with the material. Both cleaning and decontamination require similar technologies, methods, equipment, and procedures but the degree of cleaning may be different. The degree of cleaning is defined by certain cleanliness standards. Decontamination is based on cleanliness standards that typically revolve around a personal dose rate or contamination level associated with the component or surface being cleaned. For disposal of the cleaning material, the chemical composition of both the waste and the radionuclide content must be understood. The disposal method and packaging will depend on all of these items.

Decontamination is used to reduce the dose that workers may receive from a component or surface, to reduce the potential for airborne radionuclides, or to reduce the associated disposal cost. Some decontamination can allow for the reuse or recycle of the material, although this approach is currently less common. An example of reuse of material is the melting of slightly contaminated steels to be manufactured into shield blocks for a particle cyclotron. The shield blocks keep natural radiation sources from interfering with the acceleration of particles. Many tons of materials have been recycled for use in nonradioactive environments. In order to accomplish a use for "free-release," the material must not only be cleaned but also surveyed and found to be below the releasable level of the facility. The facilities that perform "free-release" are licensed by the federal government or the state in which they operate. The license and regulations specify the release limits of the facility.

This document is meant as an introduction to decontamination, both chemical and mechanical, with references to direct the reader to greater detail. The listing of any commercial companies should not be construed as a recommendation for the offerings of products or services.

17.2 PREPLANNING

Prior to the decontamination of any removable component or part, management should consider available options and the economics of the process. An engineering evaluation should include the following questions:

- Why is the component being decontaminated?
- Can it be free-released?
- Can it go straight to disposal?
- Which disposal site is required?
- What are the acceptable radiation and contamination levels at the disposal site?

- Is decontamination cost-effective?
- What decontamination method should be applied?
- Has the site decontaminated this type of part or component previously?
- What is the secondary waste generated?
- Is the secondary waste hazardous, radioactive, both?

17.3 OPERATIONAL CONSIDERATIONS

All decontamination efforts have a common objective: the removal of radioactive contamination from its present undesirable location. A variety of decontamination methods are available, regardless of the nature or extent of the contamination. An evaluation of several considerations will determine which specific methods should be applied. The weight given each will vary greatly with the situation, with some requiring no consideration. These considerations may be generalized as follows:

- The radionuclides present and their physical state
- The type of surface and substrate material bearing contamination
- The size, configuration, location of the contaminated surface and its relationship to other surfaces
- Accessibility of the contaminated area
- Desired condition of the contaminated surface upon completion
- Desired degree of decontamination to be achieved
- Functional nature that must be retained following decontamination of the surface
- Equipment and materials available for use in decontamination, including those present at the site and available from other sources
- Worker safety and radiation dose minimization
- Cost-effectiveness of various alternatives, including the relative costs of disposal and replacement versus decontamination and reuse
- Quantity and nature of waste that will be generated (including secondary waste from decontamination solvents, etc.) and its disposal cost
- Human resource and training requirements

The weight given these individual considerations will vary greatly with the situation, with some requiring no consideration in many circumstances.

Experience has shown the desirability and benefits of providing hands-on training for decontamination workers and their supervisors and foremen. This training may include the use of mock-up facilities, simulating as closely as possible the thing to be decontaminated and the methods and equipment to be used during decontamination. Such training should be performed with the personnel dressed in the actual anti-contamination clothing they will wear during performance of the work.

This training with mock-ups is generally at least as useful to the people charged with planning and scheduling the work, and with staging the equipment and material to the work area, as it is for the workers and their supervisors and foremen who will perform the decontamination activities. Such training activities are the best way to ensure that the safest, and most effective and cost-beneficial results will be obtained from the work. They also provide assurances for complying with the regulatory requirement for accomplishing the required work with resultant worker radiation exposures maintained as low as is reasonably achievable (ALARA).

The decontamination operation must be based on a detailed action plan. Key elements of an action program include:

1. Preplanning all work, including specific activities to minimize the release of contamination, such as the use of catch basins, tenting and venting through HEPA filters, and glove box construction, the strict observation of step-off pads and other controls, and adherence to tag out, drain down, flushing, and other maintenance prerequisites.

2. Minimizing the generation of contaminated items that must be transported through the site for cleaning or disposal, with the inherent potential for contaminating otherwise clean areas or items. Examples of this are the removal of all packaging materials prior to entry to a contaminated area, use of a "runner" to reduce the times dressing and undressing are required of maintenance and operation personnel, and careful selection of tools and equipment prior to the start of work so that items are not exposed to contamination unnecessarily.

3. Emphasis at all times on the necessity for each individual to observe and deal with any discrepancy in system function or procedural operation. This should include the immediate reporting to maintenance of any leakage or other equipment problems, the willingness of any individual to point out to another an error in procedure, open post-job debriefing, or any other activity that will correct or preclude errors in function of equipment or personnel.

The cost-effectiveness of such a program is easily realized in the reduction of labor and supplies used for decontamination, reduction in waste disposal costs, and reduction in total person-rem expended per year.

17.4 RADIOACTIVE CONTAMINATION

There are two types of radioactive contamination, smearable (removable) and fixed. Smearable contamination is typically removed by wiping the surface with a cloth rag, similar to the dust that is found in all homes. Just as dust is removed from a surface by wiping with a cloth sprayed with a chemical to help "hold" the dust, so might smearable contamination. Treating or attempting to remove fixed contamination also may generate smearable contamination. Some methods will loosen the fixed contamination, rendering it amenable to removal as smearable. Smearable wipes used to wipe the surface (typically 100 cm^2) to determine the contamination levels are based on this type of contamination.

Fixed contamination is held tightly to the surface and typically is associated with corrosion products on metal or concrete surfaces. In these cases, the contaminants have diffused into the material, or the radionuclides form some type of electrostatic or chemical bond with the surface materials. The removal of fixed contamination typically requires harsh removal techniques. Chemical dissolution of corrosion films or concrete scabbling is required to remove fixed contamination; these techniques will routinely result in the creation of smearable contamination.

Decontamination is normally performed for a specific reason, and the worker should understand both the reason and the form of contamination. If free release is the criteria, the leaching of radionuclides from crevices after the part has been surveyed and free-released is of concern. Many investigators have found that some contamination resides in the cracks and pores of metals and concrete. Although a survey for free release may find no radioactivity, it is often found again later -- the result of radionuclides leaching from the interior of the surface. It is believed that the radionuclides are not found during the final survey because they are imbedded deeply enough to be shielded from detection. Decontamination techniques are also performed to reduce the exposure of workers or avoid the spread of contaminants. In these cases, the survey requirements are somewhat less restrictive.

17.5 DECONTAMINATION FACTORS

The effectiveness of decontamination is measured by the decontamination factor (DF). The DF is the ratio of radiation level of the material or component prior to the decontamination application and the level of the same measured immediately after the application. A decontamination process that removes radioactive material will result in a DF greater than one. There have been processes that have actually left more contamination behind in certain areas; these unsuccessful processes have resulted in a DF of less than one. Reproducibility of the measurement must be ensured when measuring the radiation levels, as follows:

1. Standardize the procedure for measurement. Different technicians may hold the meter differently, point it in a slightly different direction, or not measure exactly the same specific area.
2. Use a recently calibrated instrument for the before and after measurements. Ideally, an instrument should be used whose calibration is good throughout the decontamination application.
3. Use the same exact location for each measurement. Each location should be marked so that, if necessary, a different person may perform a measurement in precisely the same location.
4. Determine a distance and direction from the surface that can be duplicated. All radiation measurements should be contact measurements in order to reduce the source term impact of surrounding and adjacent areas.

The following equation is the formulation that may be used to determine the percent of radionuclides removed from the surface (Remark 1989). A DF of 10 will result in a 90 percent removal of radio-nuclides. Likewise, a DF of 100 will result in a removal efficiency of 99%.

$$\text{Percent Activity Removed} = (1-1/DF) \times 100$$

A decontamination engineer will determine the most cost-effective DF, taking the objective of the decontamination into account. If it is being performed to reduce collective dose, both the surrounding radiation levels and work location must be considered. It is possible that a DF of 500 (99.8% removal) will result in decontamination effectiveness of only 5 person-rem, due to higher radiation levels in the surrounding area. However, if the work is being performed close to the area that was decontaminated, the person-rem reduction also may be large. The decontamination engineer in conjunction with the health physicist determines the optimum DF, taking into account the use of relevant resources.

Several types of electronic dosimetry can be used to help measure the person-rem effective DF. These have been used so that area radiation level measurements can be made instantaneously and person-rem effective DFs can be calculated. The electronic dosimetry, monitored at a central location, has been very effective in monitoring on-line radiation levels and it has assisted health physicists in limiting access to high radiation areas of the plant during the application. It is worth noting that past applications have shown that a plant's low-dose areas used as waiting areas have become the higher dose areas after an application. It is not that the radiation level has increased in the "low dose" area, but rather the levels in the remainder of the plant have decreased.

17.6 OXIDE DISSOLUTION

Corrosion products depend upon the type of environment in which it was generated. In BWR environments, the corrosion film closest to the coolant is typically hematite (Fe_2O_3) while the inner surface of the corrosion film is composed of a nickel-substituted magnetite ($Ni_xFe_{3-x}O_4$) where x is greater than 0 but less than 1.

In PWRs, where the water chemistry contains soluble hydrogen and therefore much less dissolved oxygen, the corrosion film is composed of a chromium and nickel substituted magnetite ($Ni_xFe_{(3-x-y)}Cr_yO_4$) where x and y are variable (Segal 1983). The sum of x and y are typically less than 1 in the outermost portion of the film but will increase as the depth into the film increases.

Due to the two types of oxides on the different reactor surfaces, different chemical solutions are required for optimum removal. A reducing condition is required for removal of a magnetite or nickel-substituted magnetite. However, if the film is composed of a high percentage of chromium, an oxidation environment is required to remove the protective chromium matrix.

17.6.1 Reductive Dissolution: The basis for the dissolution of corrosion films and therefore the mechanism for decontamination is the destabilization of the oxide lattice by attack with protons, as in the following reaction.

$$8H^+ + Fe_3O_4 \rightarrow Fe^{+2} + Fe^{+3} + 4H_2O$$

This is the basis for chemical decontamination methods based on mineral or organic acids. Typically, higher concentrations of these chemicals are needed to increase the kinetic rate of the dissolution reaction. To accomplish this with low concentration reagents, a reducing agent is used to help destabilize the film matrix. The chemical reaction listed below is the typical reaction that occurs with the use of a low concentration chemical reagent that has a reducing agent.

$$8H^+ + 2e^- + Fe_3O_4 \rightarrow 3Fe^{+2} + 4H_2O$$

It should be noted that there is no ferric ion (Fe^{+3}) on the product side of the reaction. The reducing agent reacts with the ferric ion in the oxide matrix and reduces the ferric ion to the corresponding ferrous ion (Fe^{+2}). This reduction destabilizes the oxide film and renders it more susceptible to attack by the chelating agent that is contained in the low concentration chemical decontamination solutions. There are three conditions that a reductive-dissolving reagent must fulfill:

- Rapid electron transfer mechanism
- Chelation to provide adequate metal-ion solubility
- Sufficient proton concentration to give rapid dissolution of oxide anions

17.6.2 Chromium Dissolution by Oxidation: The chromium found in corrosion products is in the plus three-oxidation state, very insoluble, and therefore must be either oxidized or reduced to render it water-soluble. It is significantly easier in power plant conditions to oxidize the chromium to the water-soluble chromate (Cr^{+6}) than it is to reduce it. The oxidation of the chromium can be accomplished in either an acid or a basic condition. The reactions are presented here.

$$2Cr^{+3} + 7H_2O \rightarrow Cr_2O_7^{-2} + 14H^+ + 6e^-$$

Acid Conditions

$$Cr(OH)_3 + 5OH^- \rightarrow CrO_4^{-2} + 4H_2O + 3e^-$$

Basic Conditions

These reactions only oxidize the chromium on the surface of the oxide film. They do not appear to leach into the oxide matrix and react with the chromium in the "layers" below the surface. Thus, during the course of a chemical decontamination where the goal is to achieve a maximum DF, several alternating steps of reductive dissolution followed by chromium dissolution must be performed.

Both acidic oxidation (NP) and basic oxidation (AP) of chromium have been used in nuclear plants, with at least one site using both. The acidic dissolution works best on stainless steel surfaces but has not been as effective on Inconel 600 surfaces. During the chemical decontamination of the channel heads of the steam generator at Indian Point Unit 3, the chromium oxidation step was initiated in the basic condition to remove the chromium from the Inconel 600 tube surfaces, and then was changed to acidic conditions to effectively oxidize the chromium on the stainless steel surfaces (Remark, 1986).

The composition of the basic solution for the chromium dissolution is approximately 1,000 ppm potassium permanganate, with the pH adjusted to 10.5 to 11.5 with sodium hydroxide. This is a relatively stable solution and typically does not have to be replenished during the 6 to 12 hour step. For the acid solution, the most common composition is 1000 ppm potassium permanganate, adjusted to a pH of 2.5 with nitric acid.

Framatome ANP, GmbH (formerly Siemens KWU) in Europe has used permanganic acid for the dissolution of chromium (HP). Permanganic acid is potassium permanganate with the potassium replaced by a hydrogen ion ($HMnO_4$). The acid potassium permanganate is not very stable in these conditions and must be monitored carefully during the course of the reaction. Replenishing both the nitric acid and the permanganate ion is not uncommon. The oxidation of the chromium during the AP, HP, or NP process results in the reduction of permanganate anions. This reduction product is manganese dioxide (MnO_2), a dark brown precipitate that is very fine in character. It appears to collect on the walls and surfaces of the components and tends to collect in sludge piles in low flow areas. This material should not be left behind in an operating plant because it will react with any kind of reducing agent typically present in the iron dissolution step to follow. An oxalic acid rinse is used to remove the manganese dioxide prior to the iron dissolution step.

17.6.3 Oxalic Acid Rinse: The amount of chemical required must be calculated prior to the oxalic acid rinse step. This calculation is performed by summing the amount of oxalic acid required to reduce the remaining potassium permanganate, as well as the manganese dioxide to manganous ion (Mn^{+2}). This calculation is performed by knowing both the initial concentration of potassium permanganate added to the system volume and the "ending concentration" of potassium permanganate. The amount (moles) of MnO_2 generated can be calculated using the following equation:

$$4H^+ + 3e^- + KMnO_4 \rightarrow MnO_2 + 2H_2O + K^+$$

The exact amount of oxalic acid can be calculated from the stoichiometric relationships presented in the following two equations, in which oxalic acid is reacting with the potassium permanganate and the manganese dioxide.

$$6H^+ + 2KMnO_4 + 5C_2H_2O_4 \rightarrow 2K^+ + 2Mn^{+2} + 8H_2O + 10CO_2$$

and

$$2H^+ + MnO_2 + C_2H_2O_4 \rightarrow Mn^{+2} + 2H_2O + 2CO_2$$

Note that the permanganate and the manganese dioxide destruction reactions require an acidic condition. For the oxalic acid to dissolve the manganese dioxide, the pH of the solution must be on the order of 3.5. In order to achieve this pH during the destruction phase after the oxalic acid is added, the solution must pass through a cation resin column in the hydrogen form. The cation resin column will remove the potassium and any manganous (Mn^{+2}) that was generated with the addition of the oxalic acid. These ions will be replaced by hydronium ions (H_3O^+) that will reduce the pH. The pH must be monitored often during this phase of the process. If the cation resin columns are continuously used, it is possible to reduce the pH below 3.0. While these chemical conditions will assist in the removing the corrosion film with the associated radionuclides, it also will lead to an aggressive condition that has been found to lead to IGSCC (intergranular stress corrosion cracking) on heavily sensitized SS 304. Research in the early 1990s determined that formic acid would be a good alternative for oxalic acid in the destruction of permanganate ion and the manganese dioxide (Remark 1993). The use of formic acid would eliminate any potential for the oxalic acid formation of intergranular attack (IGA) or IGSCC on heavily sensitized SS 304. The reaction between manganese dioxide and formic acid is listed below:

$$MnO_2 + 2CH_2O_2 \rightarrow 2CO_2 + Mn^{+2} + 2H_2O$$

Notice that no hydronium ions are required for this reaction so, in theory, it could be used at a neutral pH.

Carbon dioxide is a major reaction product of all of the destruction reactions. During a chemical decontamination process, the carbon dioxide must be vented off so that the system or component is not over pressurized. There also is concern that the carbon dioxide could react with the water in a pressurized system to generate bicarbonate.

$$CO_2 + H_2O \rightarrow HCO_3^- + H^+$$

The resultant bicarbonate anion would then result in additional anion resin requirements for the cleanup of the solution. Both research and an application during a 1995 full system chemical decontamination at Indian Point Unit 2 have shown that the carbon dioxide will not result in the formation of bicarbonate anion.

17.7 THE NATURE OF CONTAMINATION

The decontamination process is significantly simplified if the nature of the contamination is understood. Clearly, there is no one technology that will be applicable in all situations or for all types of contamination. While additional measurements made during the characterization process can provide further volume characteristics, it is also beneficial to learn as many content characteristics as possible beforehand, such as oily mixture, corrosion product film, diffused through surface imperfections, fixed in the surface substrate or smearable on its surface. In many cases, the process history may be helpful in identifying many relevant parameters.

Mechanical or chemical decontamination methods may be used at commercial nuclear power plant decommissioning projects when they can significantly reduce overall external radiation fields in high-level or confined areas, thereby minimizing potential worker exposure. The intact shipment of large reactor components may require extensive chemical decontamination to reduce their radiation levels to those acceptable to transportation regulations. If the eventual end state of commercial power reactor project is a return to greenfield, all sources of contamination must be removed and shipped off site.

Chemical techniques are used less frequently at DOE sites, where it is more common to use mechanical decontamination on structural surfaces that will remain in place for some future purpose. The typical goal at a DOE project has been to mechanically reduce surface levels in dpm per 100 square centimeters sufficient to eliminate requirements for contamination postings, protective equipment, or TLDs. When this technique is not cost-effective, it is still less costly for the DOE to send lightly contaminated components and materials to its own radioactive disposal sites than to attempt free-release levels with no permanent controls.

The commercial scenario of unrestricted disposal of slightly contaminated radioactive material is often a much more challenging task than the DOE's scenario of unrestricted release for reuse by personnel under DOE control. As a result, commercial decommissioning projects have provided extensive research into the formation of surface contamination layers in piping and components, and the examination and application of processes to reduce the levels of such contamination. Understanding the nature of the surface contamination is valuable in determining an appropriate decontamination process. In many cases, technologies used during operations can be applicable or modified for the decommissioning process. Newly developed, task-specific technologies for chemical decontamination can be expensive, not only in the purchase of special processing equipment but in the generation of additional liquid and contaminated waste requiring disposal. The cost-benefit aspects of chemical decontamination should be evaluated in correlation with a project's end-state goals and its health and safety requirements.

17.7.1 Radioactive Corrosion Products in Power Reactors: Prior to choosing the chemical decontamination process and steps required for successful application, the composition and characterization of the corrosion film must be known. The types of corrosion films found in boiling water reactors (BWRs) differ from those found in the pressurized (PWRs). BWRs operating under hydrogen water chemistry will also have an effect on the formation of the corrosion film. The main difference in the corrosion films is the concentration and oxidation state of chromium.

The normal BWR operating chemistry involves an oxygen concentration of several hundred parts per billion (ppb). In comparison, the normal PWR operating chemistry contains several parts per million (ppm) of hydrogen, which keeps the oxygen concentration in PWRs in the lower ppb range. This difference in oxygen concentration results in significantly different concentrations of chromium in the corrosion film.

During the operation of a nuclear power plant, high temperature water is recirculated through the reactor coolant system. In BWRs, the fluid typically is composed of high purity water with a small amount of oxygen (200 ppb). The PWR reactor coolant contains lithium, hydrogen, and boron but contains less than 10 ppb of oxygen due to the high concentration of hydrogen.

The corrosion film on the reactor coolant piping is composed of two discrete layers (see Figure 17.1). The inner corrosion layer, next to the base metal, is formed by the in-situ corrosion of the base metal. The characteristics of the inner layer are based on the chemical composition of the base metal, with some dependence on the recirculating solution chemistry. The inner corrosion layer is developed over the first two to six months of operation at temperature. Any corrosion layer developed at a lower temperature or in different chemical environments will transform as the temperature and solution environment is changed. As the outer layer of the corrosion film develops, the growth of the inner layer is substantially reduced.

The outer corrosion layer is formed from the deposition of corrosion products that are in the fluid flow as the coolant is recirculated throughout the coolant system, and from the precipitation of dissolved metals from the coolant flow. The composition of the film is governed by the source of the impurities. In PWRs, most of the corrosion product is from the slow dissolution of stainless steel and inconel that make up the majority of the surfaces exposed to the coolant surfaces. Cobalt also is a contribution to the corrosion film. The cobalt results from the erosion of stellite hard facing surfaces and is typically found as particulates as opposed to soluble.

The main isotopes causing the high radiation levels in reactor systems are ^{58}Co and ^{60}Co. The source of radioactive ^{60}Co is the neutron activation of ^{59}Co. The ^{59}Co is incorporated as a hard facing material in various areas of the reactor coolant system. It is also an impurity (100 to 1000 ppm) in iron-base and nickel-base alloys. Radioactive ^{58}Co is produced from the transmutation of ^{58}Ni by the reaction $^{58}Ni(n,p)^{58}Co$. The ^{58}Co activity is more significant in PWRs than in BWRs because of the high nickel alloys employed as materials of construction in their primary coolant systems (Remark 1981).

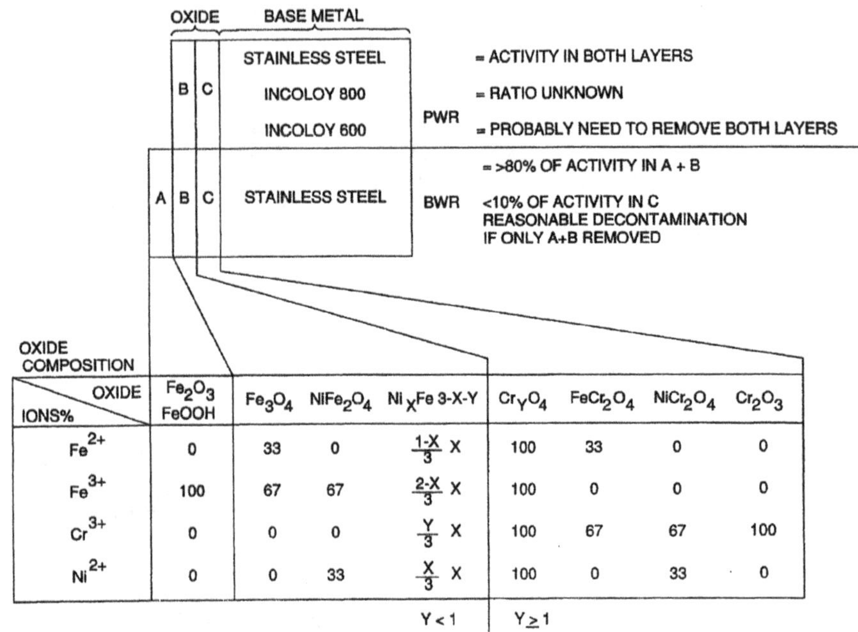

IONS% \ OXIDE	Fe_2O_3 FeOOH	Fe_3O_4	$NiFe_2O_4$	Ni_XFe_{3-X-Y}	Cr_YO_4	$FeCr_2O_4$	$NiCr_2O_4$	Cr_2O_3
Fe^{2+}	0	33	0	$\frac{1-X}{3}$ X	100	33	0	0
Fe^{3+}	100	67	67	$\frac{2-X}{3}$ X	100	0	0	0
Cr^{3+}	0	0	0	$\frac{Y}{3}$ X	100	67	67	100
Ni^{2+}	0	0	33	$\frac{X}{3}$ X	100	0	33	0
				$Y<1$	$Y \geq 1$			

Figure 17.1 Oxide Types in Light Water Reactor Systems

The concentration of chromium in the corrosion film is a function of the reactor coolant water chemistry. In a PWR chemistry environment, as the surface material corrodes, the chromium is oxidized to the plus-three oxidation (Cr^{+3}) state. Cr^{+3} is insoluble in water and is incorporated into the corrosion layer and helps to protect the stainless steel and inconel from further attack. In a normal BWR chemical environment, the chromium ion is oxidized to the plus-six oxidation state. Cr^{+6} is highly soluble in water and thus is solubilized from the corrosion film. In BWRs with hydrogen water chemistry, the chromium will remain as the $Cr+^{3}$ in the corrosion film due to the low concentration of oxygen in this environment.

The chromium concentration in the corrosion film will affect the type of chemical decontamination process used to remove the corrosion film and associated radionuclides. In a normal BWR environment, simple chelation chemistry can be used to obtain DFs in the range of 6 to 10. This is because there are few if any chromium ions in the corrosion film to help protect it from attack by the chelating agents. In a PWR or hydrogen water chemistry BWR environment, the Cr^{+3} in the corrosion film protects the film from attack by chemical chelants. If this chemical environment is encountered, a pretreatment process must be used to obtain acceptable DFs. The pretreatment chemicals typically used are alkaline potassium permanganate, nitric acid potassium permanganate, or permanganic acid. All of these treatments will further oxidize the Cr^{+3} in the corrosion film to a water soluble Cr^{+6}. This will then open the film for further attack and dissolution by the chelates used in the following step.

Several steps may be required, depending on the thickness of the corrosion film and the goal of the decontamination. Thus, an alternating chemistry of pretreatment followed by a chelation step may be required. If the goal is to have very high DFs in a normal BWR environment, then a multi-step process, including a pretreatment step involving some type of permanganate ion, is recommended.

17.7.2 Corrosion Product Transportation, Activation, and Deposition: While a nuclear power plant is in operation or in hot standby, the reactor coolant system is in a continuous recirculation state. This means the reactor coolant pumps are operating, the chemistry is in the proper concentrations, the temperature is at operational parameters, and water is flowing through the pipes at the required rate. Base metals are corroding and corrosion film is eroding due to operational parameters. The system develops equilibrium, so deposition of some of the soluble and insoluble material is occurring; however, all this is occurring at a very low rate. None of the corrosion or erosion processes are such that they will impact the safe operation of the plant. This equilibrium of corrosion products has a significant impact on the radiation levels that occur on out-of-core piping surfaces.

Both the corrosion and erosion products may deposit on the piping, vessel, and core surfaces due to the equilibrium that is continuously taking place. These materials are activated once they deposit in the vessel area and are exposed to neutrons. Once activated, the materials still participate in the equilibrium occurring in the reactor coolant system, and are either dissolved or eroded from the surface and then transported by the coolant to a different area of the system. If the surface is an out-of-core surface, the result is a radioactive particle outside of the core. This results in increased radiation levels outside the pipe. The radiation level in the piping increases as the radioactive material (crud) continues to deposit in these areas. This phenomenon was first noticed at the Chalk River Laboratory (in Canada). The technicians called it the Chalk River Undefined Deposit, hence its name, CRUD.

Some of the radioactive material is dissolved or spalled from the surface and deposited in another area, but some of it is incorporated in the deposit and held in place. Once the particle is held in place, it appears to be able to diffuse in the metal surface deposit. Thus, after several years of operation, radioactive material may be found throughout the deposit layer. The radioactive material is not only found in the loose deposits from solid deposition on the surface but also in the tightly held deposits underneath the loose deposits.

Radioactive materials also can diffuse into the pores between the metal grains, thus limiting the amount of decontamination that can be achieved with dilute chemical processes (Smee 1999). This fact must be considered when performing a chemical decontamination and when projecting the potential DF.

Several investigators have performed substantial work on corrosion product generation, activation, and transportation in both BWRs and PWRs. These references may be referred to for further information on this subject (Lister 1979, 1989, 1996) (Berry 1979) (Cunnane 1978).

17.8 CHEMICAL DECONTAMINATION PROCESSES

There are a number of reviews on the subject of decontamination. These include an older but very good treatise on chemical and mechanical decontamination edited by J. A. Ayres (Ayres 1970a).

The proceedings of a conference in 1979 on decontamination and decommissioning of nuclear facilities has valuable information (Osterhout 1980). The Electric Power Research Institute (EPRI) has published various review articles and proceedings from numerous decontamination conferences (Remark 1981, 1989), (Smee 1999), (EPRI 1998, 1991, 1993, 1995).

Chemical decontamination involves the removal of contaminated material by chemical agents that dissolve it or loosen its surroundings. The contaminant may then be flushed or rinsed away simultaneously or in a subsequent step. Chemical decontamination has the potential advantage of being able to remove accumulations from crevices, joints, piping, and other restrictions that may be accessible by chemicals but are generally encumbered for mechanical methods. On the other hand, the use of chemicals necessitates consideration of process equipment and conditions, as well as hazards and environmental issues that relate to both the process itself and the handling and disposition of resulting wastes. Additional chemical-resistant personnel protective equipment worn by operators may reduce worker productivity.

Prior to 1979, all of the chemical decontaminations performed were at government facilities and all used concentrated chemicals. Some of the earliest were performed at the Plutonium Recycle Test Reactor at the Hanford Reservation. After an experimental fuel element failure in August 1962, the reactor required decontamination. This reactor was of particular interest since it was the first chemical treatment of a complete reactor to reduce radiation levels. The process used was APOX, to treat the surfaces with alkaline potassium permanganate followed by oxalic acid (Perrigo 1966). It was found that a ferrous oxalate precipitate formed on the piping and component surfaces after application of the oxalic acid. This precipitate was difficult to remove and another application of a modified solvent was required. In the modified process, called APACE, the application of alkaline potassium permanganate was followed by citric acid with ethylenediaminetetraacetic acid (EDTA). This solvent removed the ferrous oxalate precipitate problem and achieved an overall DF of 4 to 16.

The N-Reactor at Hanford was another example of concentrated chemicals being used to remove the corrosion products from reactor coolant surfaces. In this case, the solvent most often employed was phosphoric acid. This reactor was decontaminated annually from 1967 to 1983, and resulted in DFs of 3 to 6 on the carbon steel surfaces.

The first nuclear reactor to decontaminate the primary piping and then put the used chemical on resin was the Shippingport PWR, decontaminated in early 1964 (Abrams 1966). The process used was AP-AC, an alkaline potassium permanganate process, followed by an ammonium citrate process. This decontamination was of great historical importance as it was the first application of dilute reagents and treatment of all spent solutions by ion exchange resin.

The amount of ion exchange resin was underestimated during this project and the estimated 700 ft^3 of ion exchange resin was expanded to 900 ft^3. The radiation levels at several areas of the reactor coolant system were reduced by a factor of 50 while other areas resulted in higher levels after the decontamination process. The higher levels were a result of undissolved crud being transported to low or no-flow areas. It appears that the diluted ammonium citrate solution was not aggressive enough to dissolve the crud in a timely manner. Thus, some of it was released to the liquid and deposited in those areas with little or no flow.

The Shippingport chemical decontamination was the first full-scale test of a procedure that essentially was a cyclic or continuous process requiring no large additional facilities for auxiliary operations, such as preparation and storage of solution, and storage and disposal of wastes. The effort merits special commendation in that it showed that the dilute chemical decontamination process is feasible, although much work still needed to be performed.

The first U.S. commercial application of a low concentration, chemical decontamination process occurred at Vermont Yankee in 1979. London Nuclear used the CAN-DECON process to decontaminate the reactor water cleanup system. The average DF obtained during this first commercial application of dilute chemicals was five, which resulted in the removal of 80% of the activity associated with the radiation levels on the piping surfaces. The only waste that resulted from this work was approximately 25 ft^3 of ion exchange resin (Beaman 1982) (LeSurf 1982).

The development of more efficient chemical decontamination processes was furthered by work funded by the Electric Power Research Institute (EPRI) -- a low oxidation state, metal ion (LOMI) process that was first applied at the Monticello Nuclear Power Station in 1983 (Remark 1984). This process was novel in that it used the vanadous (V^{+2}) ion to reduce the ferric ion to ferrous (the vanadous ion is oxidized to a vanadic (V^{+3}) ion during this reduction) in the corrosion films on interior surfaces in the reactor. This process was substantially faster than earlier dilute chemical processes. The DFs achieved during the decontamination of the entire reactor recirculation system were 25, with a range from 1.6 to 64.3. The process used was a LOMI-NP-LOMI.

Bechtel National performed the first major chemical decontamination of a steam generator channel head at Indian Point Unit 3 in 1985 (Nielsen 1986) (Greenman 1986). The entire steam generator channel head was decontaminated, as well as approximately 6ft into the steam generator tubes. A combination of AP and NP was required since this was a PWR, and therefore required the removal of chromium from both stainless steel and Inconel surfaces. The process was initiated with an AP step; after approximately 12 hours of contact time, the pH was changed to acidic conditions by adding nitric acid.

Figure 17.2 presents the number of BWRs and PWRs that applied chemical decontaminations from 1990 through 1998. The relationship between the two types (many more BWRs) is also typical of those performed prior to 1990. Figure 17.3 illustrates the type of chemical decontaminations that were performed at the various plants. The LOMI process has been used the most at operating commercial nuclear power plants (more than 70% of plant decontaminations during the past two years). It generates the most radioactive waste but is favored by most utilities due to its backing by the EPRI.

17.8.1 CAN-DECON and CAN-DEREM Processes: CAN-DECON is composed of EDTA, citric acid, and oxalic acid in a molar ratio of approximately 2:1:1. The CAN-DECON process was developed by AECL in the 1970s for use in CANDU heavy water reactors. This is a low concentration chemical decontamination process with the only radioactive waste generated being ion exchange resin. This is important when used for the decontamination of CANDUs since they are heavy water moderated reactors and cannot afford the loss of the expensive heavy water during the process.

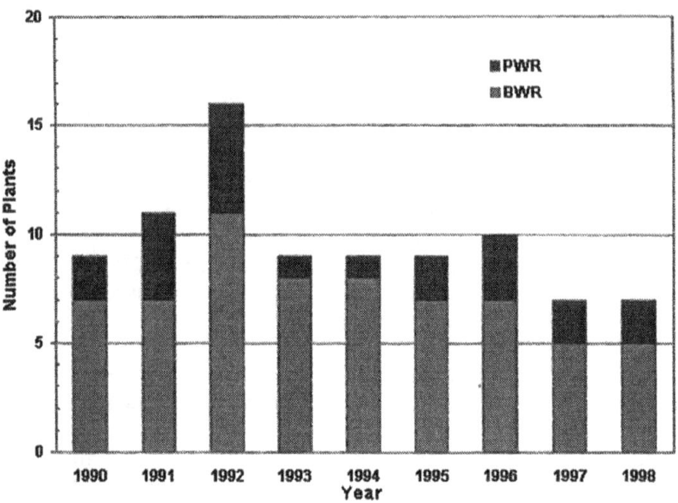

Figure 17.2 Ratio of BWRs and PWRs Performing Chemical Decontamination

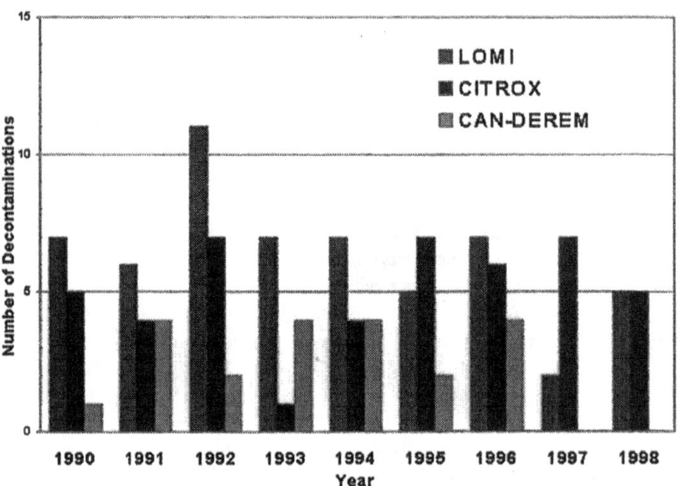

Figure 17.3 Types of Chemical Decontamination Performed

The CAN-DEREM process does not contain oxalic acid since aggressive testing indicated this acid in the CAN-DECON solvent might lead to IGA/IGSCC in heavily sensitized SS 304. Since then, only the CAN-DEREM process has been used on stainless steel systems. These processes start with a solid powder that is typically dissolved in a mix tank. Prior to adding any chemicals, the decontamination volume is heated to an operating temperature of approximately 200°F. After the chemicals are dissolved in minimal water they are added to the chemical decontamination volume. Enough chemicals are added such that the final concentration is on the order of 1,000 ppm, or 0.1% by weight. There also have been instances where the chemicals are added to the decontamination volume in a flow-through basket and are dissolved as the decontamination volume is circulated through the basket.

As the EDTA, citric acid, and oxalic acid dissolve in the decontamination volume, the oxide deposits begin to be solubilized and some crud burst may be observed due to the loosening of deposits by the decontamination chemicals. The low concentration of chemicals does not allow a large volume of deposits to be removed by the available chelant. Therefore, on-line regeneration is required to remove the iron from the chelant (EDTA).

Removal is accomplished by flowing the decontamination volume through cation exchange resin in its hydrogen form. Thus, as the iron is exchanged to the cation resin, it gives up to the solution two or three hydronium ions (H_3O^+), depending on the oxidation state of the removed iron. The DFs would be limited without this on-line ion exchange and regeneration of the active decontamination ingredients.

The recirculation and regeneration can continue indefinitely, however in practicality, the chemical decontamination is terminated with 12 to 24 hours after the addition of the chemicals. The termination of either process is a subtle judgment call by the process engineer. Typically, the iron concentration can be monitored during the chemical decontamination process and the removal rate by the cation exchange resin can be determined. This step can be successfully terminated as the soluble iron is reduced to the approximate value of the cation exchange resin outlet, assuming that the cation exchange resin has not been depleted.

On-line radiation monitors were set up throughout the piping system during Indian Point's full system decontamination. The dose rates were monitored continuously. If some were being reduced, the process was not terminated. Typically, all of the dose rates were stabilized significantly. Once the step was terminated, anion exchange resin was brought on-line with the cation exchange resin. Therefore, the flow path was through a cation exchange column followed by an anion exchange column. A mixed bed column could be used in lieu of these two columns.

The cation/anion resin columns, or the mixed bed resin column, remove all remaining cations and anions, including all of iron ions, EDTA, citric acid, and oxalic acid. The solution is passed through the columns until the water quality meets or exceeds the quality demanded by the site. Typically these specifications are on the order of a pH between 5 to 10, a total organic carbon (TOC) of < 10 ppm, and a conductivity of < 10 μmhos. The TOC is the hardest to achieve.

All of the organic acids used for chemical decontaminations are weak acids. Therefore, only a portion is removed as they pass through the ion exchange resin since their equilibria are determined by their concentration and the equilibrium constant. If the pH can be increased to between 7 and 8, it will push the equilibrium to the dissociated state where the organic acids can easily be removed by the ion exchange resin. The equation below is a generic equation for the dissociation of weak organic acids. The observer can see that the organic acid tends to be dissociated as sodium hydroxide is added to the solution. The $\bar{R}$ is the part of the organic acid that leads to the TOC concentration. As can be seen in the equation below, as the acid becomes dissociated, the $\bar{R}$ can be removed on the anion exchange resin.

$$RH + NaOH \rightarrow Na^{+} + R^{-} + H_2O$$

DFs of 5 to 20 have been observed with the use of the CAN-DECON and CAN-DEREM processes. Lower DFs will be found in BWRs if no oxidation step is used. Alkaline potassium permanganate results in lower DFs than nitric. Since the CAN-DEREM process is less aggressive, it can be used on both stainless steel and carbon steel surfaces without corrosion concerns.

17.8.2 CITROX Process: The citrox reagent was developed by R.D. Weed and was first used on the Plutonium Recycle Test Reactor (PRTR) located in the 300 Area of the Hanford Reservation (Weed 1968). The reagent was composed of citric and oxalic acids; thus, its name CITROX. The original formulation for the PRTR chemical decontamination was approximately 0.3 M (25,000 ppm) citric acid and 0.2 M (58,000 ppm) oxalic acid. Inhibitors were also added to this solution due to its corrosive nature. Early in the history of the dilute chemical decontamination, PN Services, Inc. modified this solution and applied it in the field at a concentration of approximately 0.2 weight percent. This product became known as PNS CITROX A that did not contain an inhibitor. PNS CITROX B did include an inhibitor and was used in carbon steel systems. This process is somewhat more aggressive than CAN-DEREM or CAN-DECON solvents due to its higher concentration and use of oxalic acid. The process remains popular for BWR and PWR component decontamination, but has not been qualified for application on full reactor systems in either.

The CITROX system is similar to the CAN-DEREM and CAN-DECON system in that it starts with a powder that is dissolved in a mix tank. After the chemicals are dissolved in minimal water the soluble chemical is slowly added to the decontamination volume. The rate of addition is such that the chemicals are added throughout one recirculation of the decontamination volume. These chemicals also could be added by flowing a bypass system through a basket or other container that holds the solid powder. Enough chemicals are added to result in a final concentration of CITROX in the total decontamination volume of 0.2 weight percent. Typically, the molar concentrations of oxalic and citric acid approximately equal. Depending upon the results of any laboratory tests, the ratio between the oxalic and citric acid can be slightly varied to optimize the DF.

The CITROX is also regenerated on-line. The on-line demineralization should be maintained at a high enough rate to keep the concentration of ferric ion to less than 10 ppm. Due to the elimination of the EDTA in the reagent, the ferric ion is not as strongly complexed in the CITROX solution and therefore is available for ferric ion corrosion. This corrosion is substantially reduced if the ferric ion in solution is maintained at concentrations of less than 10 ppm.

The bulk solution can be analyzed for oxalic acid, various metals, and pH during the decontamination operation. The effectiveness of the regeneration demineralization cation resin vessels should also be monitored. The decontamination process should be terminated once the iron concentration stops increasing and the on-line radiation levels stop decreasing. Other metals, such as chromium, should be monitored. If an increase in chromium concentration is observed in the decontamination solution after the initial injection, it is an indication that base metal corrosion may be occurring. At that time, the process engineer should evaluate the reactions occurring on the system surfaces and take appropriate action. At the termination of the CITROX process, the solution can be demineralized with cation and anion resin to return the decontamination solution to demineralized water.

All of the reagents added to the solution, as well as all of the dissolved material, are either cations or anions that will be removed on ion exchange resin. The weak organic acids are not totally dissociated at all pH units and therefore it will take some recirculation of the solution through the ion exchange vessels to totally clean the water. The time required to return the water to plant-specified conditions of 10 μmhos and 10 ppm TOC is typically 8 half-lives. The recirculation of the decontamination solution can continue indefinitely, but the chemical decontamination is typically terminated within 12 to 24 hours after adding the chemicals to the system volume.

17.8.3 CORD Process: CORD is an acronym for Chemical Oxidation Reduction Decontamination. It evolved from the OZOX process (Ozone/Oxalic Acid), which originally used ozone as the oxidizing agent but, because of technical difficulties, used alkaline permanganate for all field applications (Wille, 1986). The CORD process is actually a family of processes that uses permanganic acid ($HMnO_4$) as the oxidizing agent and oxalic acid as the dissolution and chelating agent. Ultraviolet (UV) light and hydrogen peroxide are used to reduce the concentration of the oxalic acid, thereby reducing total waste volume. All of the waste generated is ion exchange resin. Using the permanganic acid as the oxidizing agent also reduces the total ion exchange requirements.

The permanganic acid oxidizes the protective chromium in the corrosion film to the water soluble bichromate anion by the following chemical reaction:

$$Cr_2O_3 + H_2O \rightarrow 2HCrO_4^- + 2MnO_2.$$

Permanganate ion also tends to react with any oxidizable species contained in the water. This oxidation takes place in an acid medium. The reaction products of the permanganate are manganese dioxide, oxygen, and water. The process engineer must monitor the concentration of the permanganate ion during the permanganic acid step in the CORD process, keeping it in the range of 50-300 ppm for proper application. The application temperature is 190 to 200°F. Early in the reaction process, the permanganate concentration must be maintained at the higher end of the concentration range; as the step reaches the termination phase, the concentration can be maintained at the lower end. A high concentration of permanganic acid should not be added in the later stages since it does not increase the amount of soluble chromium but does add to the amount of waste generated.

The chromium concentration also should be monitored during the oxidation phase. The concentration will determine the termination of the process. It will increase rapidly and then the rate will fall off sharply. When the rate of increase has reached diminishing returns, the oxidation phase can be terminated. At this point, the excess permanganic acid and the resultant manganese dioxide must be destroyed prior to proceeding to the reduction step. The material is destroyed by reducing both species to the corresponding manganous ion (Mn^{+2}). The reduction is performed by adding stoichiometric amounts of oxalic acid to the oxidation solution. This step is not a separate step from the decontamination step. The addition of oxalic acid for the reduction of the remaining permanganic acid and the manganese dioxide is added by one addition to the solution.

The decontamination step is achieved by adding an excess of oxalic acid for an ultimate concentration of approximately 1,500 ppm. Without the addition of the excess oxalic acid, the pH will not be sufficiently acidic for the destruction reaction to occur. The pH must be in the range of 3.5 to 4.0. For other processes described, the pH is achieved by passing the solution through a cation resin that will remove the cations in solution, replacing them with hydronium ions (H_3O^+). Once the permanganate and manganese dioxide has been reduced to manganous ion, the oxalic acid will dissolve the magnetite in the corrosion film by the following reaction:

$$Fe_3O_4 + 4H_2C_2O_4 \rightarrow 3FeC_2O_4 + 4H_2O + 2CO_2$$

When the decontamination process is initiated, the cation columns must be placed on-line. The cation columns will remove all of the metal ions that have been dissolved by the oxalic acid and will regenerate the oxalic acid for further removal of the corrosion film. The regeneration of the oxalic acid will produce hydronium ions that will assist in sustaining the reaction. The temperature should be maintained at approximately 195°F for optimum conditions.

During this stage of the decontamination process, the metal concentration and the oxalic acid concentration should be monitored to maintain the chemistry in the optimum concentration. The reaction may not be completed if the amount of oxalic acid is substantially reduced at the same time that the iron concentration is reduced. In this case, the oxalic acid concentration may need to be spiked to optimize the reaction kinetics.

The termination of the reaction will occur with optimum oxalic acid concentrations, along with reducing metal concentration, especially iron. It must be noted that the iron concentration should be monitored both before and after the ion exchange resin. The iron removal rate can be calculated from these analyses. Since the resulting buildup in concentration of the metals' ions in the bulk solution from the dissolution of the corrosion film can be determined, the calculations will indicate the optimal time for termination.

At the termination of the decontamination step, the solution must be cleaned up to demineralized water quality. This can be accomplished in two ways with the CORD process. The original method of removing the oxalic acid was to use ion exchange resin. The oxalic acid was passed through an anion resin that removed the oxalate anion and the resultant hydroxide ions (two) released to the solution by the anion resin reacted with the hydronium ions from the oxalic acid and formed water. In the United States, the small amount of ion exchange resin resulting from this process was not a problem. In Europe, however, the off-site disposal of any ion exchange resin is difficult. For this reason, the inventors of the CORD process developed a decomposition method for the oxalic acid that uses hydrogen peroxide and UV light. The following reaction lists the products and reactants of the oxalic acid decomposition.

$$H_2O_2 + H_2C_2O_4 \xrightarrow{uv} 2CO_2 + 2H_2O$$

During this reaction the solution is passed through a light chamber using intense UV light to decompose the oxalic acid to carbon dioxide and water (Bertholdt, 1994). If a CORD process is applied in a closed system, calculations should be performed in order to determine the increase in pressure that may be seen due to the generation of the carbon dioxide gas. Depending on the venting system used, the carbon dioxide gas generated during the permanganate and the oxalic acid destruction process may exceed the system design pressure.

The anion resin required for a 10,000 gallon decontamination volume is typically in the range of 25 to 60 ft^3 per CORD step. Depending on the system being decontaminated and the loading on the decontamination volume surfaces, the generated cation and anion resin may fit into a single high integrity container (8-120) for disposal in the United States.

The process described above is for a single CORD application. This application can be repeated until the DF goal is achieved. A typical CORD process takes from 4 to 16 hours if ion exchange cleanup is employed but can be extended to 10 to 36 hours if the hydrogen peroxide/UV oxalic acid decomposition is used.

The decontamination process engineer must work closely with the outage scheduler to determine the optimum treatment method. The UV destruction process requires up to 20 additional hours. If the decontamination process is on critical path, it may be substantially cheaper to generate the ion exchange resin or use a combination of ion exchange and UV destruction to reduce the duration. Additional information regarding the CORD process is available (Bertholdt 1998a, 1998b).

17.8.4 LOMI Process: The Central Electricity Generating Board in the United Kingdom developed the LOMI chemical decontamination process in the early 1980s (Bradbury, 1983). LOMI, an acronym for Low Oxidation state Metal Ion, is considered to be fundamentally different from the other processes described previously. This is somewhat correct in that it employs a strong reducing agent (vanadous ion in solution chelated with picolinic acid) to attack the ferric ion in the oxide matrix. The vanadous ion is oxidized to vanadic ion while reducing the ferric ion to ferrous ion, as illustrated in the following reaction.

$$V^{+2} + Fe^{+3} \rightarrow V^{+3} + Fe^{+2}$$

Once the various metal species are solubilized, they are complexed with the picolinic acid that is added to the solution at the beginning of the step. The process is used at a pH of between 4 and 5; therefore, sodium hydroxide is also added to the concentrated chemicals prior to injection. This process is identified as reduction dissolution and is much faster than acid dissolution. Typical LOMI applications involve a completed reaction in approximately 6 hours. The vanadous ion appears to be unstable in a carbon steel system and therefore it is not recommended for carbon steel piping systems.

A typical LOMI process uses about a 6 millimolar solution of vanadous ion. A solution of picolinic acid is mixed in the decontamination volume prior to injecting the concentrated vanadous ion. The picolinic acid is mixed in a concentrated solution that is prepared in a manner such that six moles of picolinic acid are used for one mole of vanadous ion. Sodium hydroxide is also mixed with the acid prior to injection into the decontamination volume. Besides the obvious acid, the vanadous formate contains formic acid as an impurity from the chemical reaction that produces the vanadous ion. When dissolving the picolinic acid, add the sodium hydroxide to the mix tank prior to adding the picolinic acid. This mixing method will significantly reduce the dissolution time for the acid.

The molar ratio of picolinic acid to vanadium is rationalized by understanding that three atoms of picolinic acid are required to complex one atom of vanadium as well as one atom of iron. It is assumed that for every atom (or mole) of vanadous there will be one atom (or mole) of ferrous iron produced (see above equation). Some researchers have theorized that a lower molar ratio of picolinic acid could be used to maintain ion solubility, but the majority of utilities have maintained the higher molar ratios during the decontamination process.

Both the vanadous ion and iron concentrations are monitored to identify the termination point of the LOMI process. The vanadous ion can be monitored spectrophotometrically and will decrease with time while the iron concentration will increase as the corrosion film is dissolved. Even though the vanadous ion concentration may be zero, there is still a driving force for the dissolution of the corrosion film by the oxidation potential of vanadic/vanadate oxidation couple. Since this couple is also found to be somewhat corrosive, the reaction should not run for more than 60 to 90 minutes in this mode. By this time, the iron concentration should start to approach an asymptotic value that will result in the termination of the process. At termination, the solution will be diverted to a cation and anion resin column that will remove the cations and anions respectively.

There are three disadvantages of the LOMI process. One is that the vanadous ion is not stable in carbon steel systems. If there is a large amount of carbon steel in the decontamination volume, a different process should be used. The second disadvantage is the large amount of resin required for removal. Resin quantities can be somewhat reduced by reducing the molar ratio of 6:1 for picolinic acid to vanadous that is added to the system. However, the majority of the resin burden is from the three equivalents per mole for the vanadic ions that are in solution. The other decontamination processes do not have this resin burden. The last disadvantage is the cost of chemicals. The vanadous formate is made specifically for each client and has a short (6 month) shelf life. This is an expense that other processes do not have.

Even due to the resin burden and extra cost of chemicals, the LOMI process is the most frequently used by utilities within the U.S. The reason for its popularity is the reduced time for completion of the process. However, as the systems get larger, such as full system decontamination with a 100,000 gallon (378,000 liter) volume, the higher quantity of resin and cost of chemicals have pushed the economical evaluation away from the LOMI process. The only full system chemical decontamination process used to date on a commercial operating reactor was the AP/CAN-DEREM process.

17.8.5 DFD Process: The DFD, an acronym for decontamination for decommissioning, process uses fluoroboric acid as the active ingredient (Bradbury 1997, 1998). The DFD process was developed to remove the material's radioactivity from metallic surfaces by corroding the surface. In theory, this general corrosion on the base metal would release any radionuclides deposited in pits or crevices within the base metal. After several cycles of DFD (8 to 10), the component's radiation levels are reduced and the component may be able to be free-released.

The DFD process, licensed by EPRI, is multi-step and uses potassium permanganate to raise the oxidation potential of the solution to dissolve the protective chromium layer on some types of stainless steel. The process is initiated by heating the decontamination volume to 200°F. At that time, enough concentrated fluoroboric acid is added to the system to achieve a final concentration of 10mM (approximately 0.9 grams per liter) that results in a pH of approximately 2. The addition of the fluoroboric acid raises the pH of the solution due to the dissolution of the corrosion products. The solution is circulated through a cation resin column to maintain a pH level for optimum performance of the fluoroboric acid (2.0 to 2.4). The cation resin removes the dissolved metals from the solution, replacing them with the resin's protons (H_3O^+).

In order to remove the protective chromium layer (after the dissolution of the corrosion products has been substantially reduced), the flow through the cation columns is terminated and enough potassium permanganate is added to the decontamination volume to maintain an active concentration of the potassium permanganate of 100 ppm. Typically, a total of 300 ppm is required to oxidize all the exposed protective chromium to the soluble chromate anion.

The next step involves the dissolution of any MnO_2 that is generated by the reduction or decomposition of the potassium permanganate. This dissolution is accomplished by adding a stoichiometric amount of oxalic acid to the solution. The pH of this solution must be in the area of 3.5 to efficiently drive this reduction reaction. The pH can be reduced during this step by passing the solution through a cation resin column. This ion exchange resin will remove the metal ions in solution (such as manganous ion, Mn^{+2}) and replace them with hydronium ions (H_3O^+). Note that the Cr^{+3} can form a $Cr(C_2O_4)_3^{-3}$ compound with excess oxalic acid. The chromium ion will not be removed from this compound by the cation exchange resin and, thus, the chromium will tend to build up in solution during the course of a DFD application. The formation of this chromium compound is not a problem in other chemical decontamination applications because a mixed bed resin is inserted into the flow at the end of the oxalic acid destruction step. The anion portion of the mixed bed resin does an excellent job of removing this chromium oxalate complex.

If excess oxalic acid is present in the solution after all of the manganese dioxide is dissolved, a stoichiometric amount of potassium permanganate is added to the solution to remove it. With the cation resin column on line, all of the cations will be removed, achieving a chemical balance for the next fluoroboric acid step.

In summary, the DFD process consists of four discrete steps for each cycle.

1. Initiation: The first step initiates the chemical decontamination by adding the fluoroboric acid to the solution once the solution has achieved the DFD operating temperature of 200°F. The typical concentration of the fluoroboric acid is 10 millimolar. During this time, the solution is recirculated through both a filter and a cation resin column until the particulate material has been removed and the dissolved metal concentration is below 10 ppm. At this point the cation resin columns are removed from the flow path.

2. Oxidation: After the cation columns are removed from the flow path, approximately 200 ppm of potassium permanganate is added to the fluoroboric acid solution to solubilize the chromium in the corrosion film. This reaction is also performed at 200°F and continues until the potassium permanganate is reduced to manganese dioxide. This reaction requires two to four hours.

3. Destruction: The destruction phase of the process involves the addition of stoichiometric amounts of oxalic acid to dissolve the manganese dioxide. The cation resin columns are placed back on-line during this phase in order to maintain the optimum pH for the destruction and to remove the soluble ions (preparation for the next cycle). The purification half-life should be less than one hour, with 30 minutes being the preferred half-life. The purification half-life is defined as the time required for reducing the concentration of any species by a factor of 2 and assumes a 100% removal efficiency. It is calculated by multiplying the system volume by ln (2) and dividing by the flow rate through the ion exchange resin. The completion of this step can be determined by the measurement of the soluble manganese ions in solution. These ions should be dissolved and removed by the ion exchange resin. Therefore, the step can be terminated when the concentration of the soluble manganese ions approach zero.

4. Transition: The final step in each cycle returns the solution to the fluoroboric acid condition. In this phase, the oxalic acid concentration is determined by analytical methods and a stoichiometric amount of potassium permanganate is added to oxidize the oxalic acid to carbon dioxide and water. During this step, the cation resin vessels are again placed in the flow path to remove the dissolved cations. Once the cation concentration is below approximately 10 ppm, the cycle is complete and additional cycles may be initiated. Typically 6 to 12 cycles are required to achieve the satisfactory DF. The number of cycles required will depend upon the ultimate decontamination goal at each individual facility.

At the termination of the DFD process, the temperature is reduced to approximately 140°F and either mixed bed resin or a combination of cation followed by anion resin is required to remove the remaining dissolved metals along with the fluoroboric acid. The anion resin also will remove any of accumulated chromium oxalate complex.

17.8.6 Termination Criteria: All chemical steps are terminated by specific analytical termination criteria. Typically one or two parameters are used.

1. NP or AP Steps: The potassium permanganate steps are typically terminated by determining both the concentration of potassium permanganate and the concentration of chromium ions in the solution. As long as the concentration of the chromium ions is increasing, the step should continue. When the concentration levels off with sufficient concentration of potassium permanganate, then the process should be terminated. Obviously this assumes that all process parameters, such as temperature, remain in the optimum operational range.

2. Oxalic Acid Destruction Step: The concentration of both the oxalic acid and the soluble manganese ion should be monitored during the oxalic acid destruction step. The solution will go through several color changes. At the start, there still may be a significant concentration of potassium permanganate that produces a purple color in solution. As the oxalic acid is added, the permanganate will be reduced to manganese dioxide, resulting in a brown precipitate in the solution. When the solution turns water clear, the easily reduced manganese dioxide has been reacted, and the remaining oxalic acid will attack the manganese dioxide on metal piping surfaces and in the sludge piles in low-flow areas.

Once the solution turns water clear, the concentration of manganese should be determined. If the concentration of soluble manganese levels off at the same time as the oxalic acid approaches zero, there is a possibility that there is more manganese dioxide in solution but the oxalic acid concentration reached zero prior to its reacting with the manganese dioxide. Therefore, additional oxalic acid should be added to remove all of the solid manganese dioxide. At termination of this destruction step there should be 50 to 100 ppm of soluble oxalic acid in solution to make certain that all of the manganese dioxide has been dissolved. If it is not totally dissolved, it could interfere with the next addition of chemical and some chemical could be depleted if enough manganese dioxide is still in the decontamination system. Some plants have found a strange brown substance in the pipes during startup after a chemical decontamination; others have found a brown solid upon cutting out the pipe during replacement. It is difficult to remove all of the manganese dioxide during a chemical decontamination, but, depending on the conditions required after the application, a special effort should be made to remove as much as possible.

3. CAN-DEREM, CAN-DECON, CITROX and CORD

These decontamination steps are typically terminated by analyzing iron concentration or a specific radionuclide, or by monitoring the radiation levels on a specific section of pipe. Since these processes are regenerative, the soluble cations are removed by recirculating the solution through cation resin on-line. The iron concentration does not build up but rather is continuously removed. Therefore, as the concentration of the soluble iron in the influent of the cation resin column approaches the concentration of that in the effluent, little if any new iron is being dissolved. If the process conditions are optimal at this point in the process, the step can be terminated.

4. LOMI

The LOMI process uses an active ingredient, vanadous ion, to help drive the reaction to completion. To determine the termination point for the LOMI process, at least two parameters must be monitored. The first is the vanadous ion concentration, which can be monitored by the use of an UV/visible spectrophotomer. The iron concentration is a good choice for the second parameter. Another recommendation is to monitor total gamma activity for every sample and develop a graph of vanadous ion, iron, and total activity.

The goal of each LOMI decontamination is to have the concentration of iron and total activity level off prior to the oxidation of the entire vanadous ion. If the vanadous ion approaches zero while the iron and total activity concentration is still increasing, not enough LOMI was added to remove all of the activity in the piping system. Additional vanadous ion could be added at this time, up to a maximum concentration of approximately 10 millimolar.

When the concentration of vanadous ion approaches zero it is imperative that either more vanadous formate be added to the decontamination volume or that the demineralization to remove the LOMI chemical be initiated within an hour. After the vanadous concentration reaches a concentration of zero, the vanadium redox couple becomes corrosive to the material of construction. In order to reduce the chances for excess corrosion either the vanadium couple must be removed or additional vanadous ion should be added to change the redox couple to one that is not corrosive.

17.8.7 Recontamination: Prior to discussing the recontamination of core radiation levels, one must understand that the water circulating through a reactor system is in a state of equilibrium and the corrosion products on the piping is a part of that equilibrium. Any decontamination or chemistry change in the coolant will upset the equilibria and therefore result in changes in the system. Such changes include noble metal additions, a change from normal to hydrogen BWR water chemistry or other "improved water chemistries," or General Electric's zinc injection process (GEZIP).

When a chemical decontamination is performed, the corrosion products will tend to redistribute throughout the system to reestablish the equilibrium. The nuclear fuel has never been decontaminated with the reactor coolant surfaces and therefore there is a source of corrosion products available to reestablish the equilibrium. A report by Smee and LeSurf (LeSurf 1991) details the recontamination data up to 1990. A report by Smee has updated the recontamination data from 1990 to the present (Smee, 1999).

The best information for full system chemical decontamination recontamination comes from a study of Indian Point's Unit 2 chemical decontamination program (Parry, 1998). This plant was chemically decontaminated with the fuel removed from the vessel in 1995. Figure 17.4 is a chart of the radiation levels at the secondary side hand holes on the steam generators. The first bar in each section is the dose rate just prior to the decontamination. The second bar is the dose rate immediately after the chemical decontamination with the third bar representing the dose rate one cycle after the chemical decontamination. In three out of four of the steam generators there was no appreciable change in the dose rates one cycle after the chemical decontamination. But in Steam Generator 24 the dose rate increased by approximately 50 percent during the first cycle. SG24 was the equipment used for the crud burst at the start of the 1997 outage. This could have resulted in the higher radiation dose rates on that piece of equipment. Figure 17.5 illustrates the general area dose rates for the pre- and post-chemical decontamination, as well as the 1997 refueling. The general area radiation dose rates were substantially reduced during the chemical cleaning and were still considerably lower than the pre-chemical decontamination dose rates during the refueling outage.

The average DF obtained for the chemical decontamination was 7.8, determined by measuring the dose rate at 55 pre-selected points. The DF of 7.8 indicates a removal of 87 percent of the radionuclides. After one operating cycle the DF over the same 55 points was 6.4, indicating an overall reduction of 84 percent of the radionuclides.

17.9 MECHANICAL DECONTAMINATION PROCESSES

Many operators report that mechanical decontamination works best on large, regular surfaces that are readily accessible to the equipment and unencumbered by other structures. Contamination in crevices and joints and in other restricted access locations may be difficult or impossible to remove by mechanical methods. Other processes for radioactive decontamination exist that apply various methods, such as the use of lasers, sonic vibration, or thermal effects, but these methods often involve a special technique for affecting or enabling mechanical removal of the contaminant or its substrate.

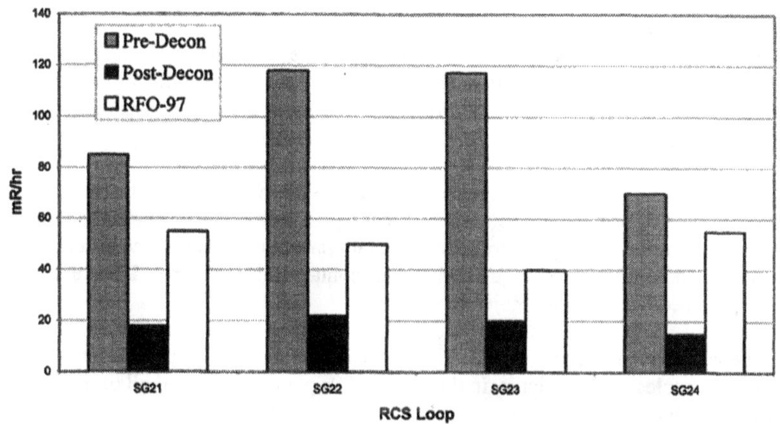

Figure 17.4 **Steam Generator Secondary Side Hand Hole Dose Rates at IP-2**

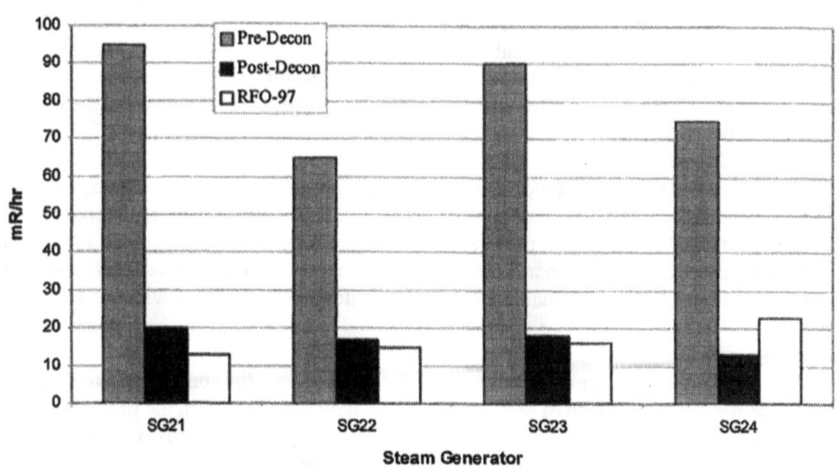

Figure 17.5 **RCS Loops General Area Average Dose Rates at IP-2**

There are a number of mechanical processes available to remove contamination from various types of surfaces. There also are several review articles describing techniques and specific tasks (Ayres 1970b) (Remark 1981, 1989) (Irving 1986). The techniques required in the removal of contaminated material are similar to the techniques required to remove dust, dirt, or corrosion products that are nonradioactive. The difference is that more waste may be generated in the cleaning of radioactively contaminated surfaces if cross-contamination should occur (the contamination of surfaces previously free of contamination). Additional supplies are required, such as rags or chemicals, to avoid such potential.

As dust becomes airborne, radioactive contamination also can become airborne. The surface of components can be wetted with water to reduce the possibility of airborne contamination. There are some aerosols that can be applied to heavily contaminated surfaces to prevent release in-air. Aerosols should be used in extreme measures for contamination control. To date, aerosols have just been used in decommissioning projects.

In the 1980s, there was a change in the direction of the use of mechanical decontamination techniques at nuclear power plants. In the early 1980s plants performed all of their own decontamination of parts and components and recycle centers were set up in Oak Ridge, TN. These centers accepted contaminated material for decontamination, volume reduction, and recycling. Since the establishment of these centers, the amount of mechanical decontamination performed at nuclear power sites has decreased.

Before performing decontamination, one must understand the reason for the decontamination. Is the component being decontaminated in order to reduce dose or the spread of contamination or is the part or component being decontaminated for free release? If free release is the criteria, one must consider the potential leaching of radionuclides from crevices long after the part has been surveyed and free-released. Investigations have shown that contamination residing in the cracks and pores of metals and concrete can resurface after an area is surveyed and cleared of any radioactivity. Subsequent evidence of contamination can only be the result of the radionuclides leaching from the interior of the surface, undetected during the final survey because they were imbedded deeply enough to be shielded.

External surfaces of piping and process equipment, building floors and walls, equipment, and tools may become radioactively contaminated from coolant leaks and routine maintenance activities. Much of this contamination is spread and deposited in a manner similar to that for ordinary nonradioactive dust and dirt in any building where industrial processes are performed. That is, all exposed surfaces become contaminated to varying degrees, in keeping with certain common sense principles.

17.9.1 Janitorial Decontamination: Since the mechanisms for contamination of this type of cleaning are similar to those of ordinary, nonradioactive dust and dirt, the processes, equipment, and procedures for janitorial decontamination have many similarities to those of good cleaning practices in standard industrial plants. The only significant differences are:

- Greater care is taken in monitoring and surveillance of the workers to prevent inadvertent overexposure to radioactive materials or ingestion and accumulation of radionuclides on exposed skin areas

- Decontamination equipment is modified to retain (for disposal) a considerably larger fraction of the contaminant materials removed
- Greater care is taken to prevent surface recontamination while performing remedial actions on remaining contaminated surfaces

The decontamination methods most commonly used to decontaminate surfaces and structural members of building/facilities within the nuclear industry are as follows:

- Hand wiping/scrubbing
- Water flushing
- High pressure water lancing
- Strippable coating
- Steam cleaning
- Cleaning with non-electrical conducting solvent

Additional methods include abrasive belts, abrasive grit blasting, grinders, shavers, and scabblers. Virtually all plants within the nuclear industry use at least the first two or three of these techniques for building and facility decontamination.

17.9.2 Advanced Recyclable Media System: The advanced recyclable media system, provided by Surface Technology Systems, is an open blast technology that uses a soft recyclable media. This media consists of a urethane foam matrix that may be manufactured in various grades of abrasiveness. The fiber media may be remade and/or reused up to 20 times and can clean almost any surface (wood, metal, or lead) and geometry, including corners and the inside of air ducts. The system is divided into three units: the media feed unit, the sifter unit, and the media remake unit. The media is propelled from the feed unit toward the surface to be cleaned by a portable blast unit. The used media is then manually collected and placed into the sifter unit. Large debris (> 0.25 in or 64 mm) and small, fine particles (< 1/16 in or 16 mm) are discarded as waste, and the remaining media may be used for media remake or may be fed back into the feed unit for recycling.

17.9.3 Concrete Shaver: The model DTF-25 concrete shaver, available through Marcrist Industries Limited, is a self-propelled, walk-behind electric powered shaving machine that uses diamonds impregnated on drum-mounted blades to shave concrete surfaces with extremely accurate tolerances. The shaver is suitable for flat (or slightly curved) walls and floors. The machine is fitted with a 10 in (23 cm) wide by 5 in (12.7 cm) diameter shaving drum onto which are fitted numerous Marcrist-patented, diamond-impregnated blades. The number of blades chosen depends on the surface finish required. The design for mounting the blades on the drum results in low vibration levels. Infinitely variable shaving depths from 0.004 in (0.01 cm) to 0.5 in (1.3 cm) can be achieved. The system is designed for use with a vacuum extraction unit for dust-free operations. The travel rate is infinitely variable, giving a high productive rate. The unit weighs 330 lbs (150 kg) and consumes 16 amps of 380 volt to 480 volt, 3-phase power, and has forward and reverse action. The depth of shaving is set by the use of a manual rotary wheel that is linked to a digital display. One set of shaver blades is rated for 156 hours of operation.

No special implementation concerns apply to concrete shaver technology. Industrial electrical power supply voltage (480 VAC) and circuitry are required. A high efficiency particulate air (HEPA) vacuum filtration unit is required. The concrete shaver is not appropriate for small, congested concrete floors and slabs where smaller hand-held tools (e.g., diamond wheel grinders) may be used. This technology is well suited for large open areas -- over 100 ft^2 (10 m^2) -- with few obstructions.

17.9.4 Concrete Spaller: Pacific Northwest National Laboratory's concrete spaller is a hand-held tool that may be used for decontaminating flat or slightly curved concrete surfaces, obtaining concrete samples, and in-depth removal from cracks in concrete. The concrete spaller includes a 9-ton hydraulic cylinder and spalling bit that runs from a hydraulic pump that expands in predrilled holes in the concrete. The result is removal of concrete chunks that fall into the attached metal shroud. The concrete spaller is more efficient than traditional tools, such as hand-held pneumatic scabblers and scalers. For example, the spaller is capable of spalling 0.23 ft^2/min (1.3 m^2/hr), compared to 0.20 ft^2/min (1.1 m^2/hr), for the baseline scabbler and scaler demonstrated at 1/8 in (0.3 cm) depth. The spaller is also capable of removing concrete at a greater depth than traditional tools.

Under optimum conditions, operating costs for the spaller would be 22% lower than the baseline scabbler and scaler. The concrete spaller is also easy to operate and has less vibration than the scabbler and scaler baseline tools. However, it is heavier and slower than a diamond grinder for depths to approximately 3 mm but removes concrete to greater depths faster. With regard to secondary waste generation, the spaller generates very little concrete dust but produces chunks of waste (contaminated concrete). Dust from pre-drilling holes must be controlled. Its performance measures at 13.6 ft^2/hr and removes surfaces to depths over 1/8 in.

17.9.5 Electro-Hydraulic Scabbling: The concrete electro-hydraulic scabbling process is a rapid, controllable way to remove surface layers of contaminated concrete while generating minimal secondary waste. The device delivers powerful shock waves to the concrete surface, originated by a pulsed, high voltage electric discharge between two electrodes. The hydraulic shock wave propagates through water between the discharge channel and the concrete, causing the concrete to crack and peel. Changing the pulse energy, shape, and electrode position controls the depth of scabbling. Water not only provides efficient transfer of energy but also acts as a debris retainer and transport medium. Water consumption is much lower than in conventional high pressure water jet techniques.

The electro-hydraulic scabbling process may be used to decontaminate deeply contaminated floors, walls, or ceilings. Benefits include reduction of waste volume subject to regulated disposal, reduction of health and environmental hazards associated with such processes, and reduction of costs due to lower energy consumption, higher processing rates, lower labor requirements, and lower mixed waste disposal costs.

17.9.6 Heavy Duty Roto Peen: This innovative concrete removal process uses heavy duty roto peen flaps to remove concrete and coatings from concrete floors. The flaps can be mounted on an EDCO Model CPM-4 floor unit. The system can further be outfitted with a Pb sentry. This proprietary system will shut off electrical power to the concrete planer should the detected vacuum drop below a safe threshold. Captive tungsten carbide shot, supported on flexible flaps, is rotated against the contaminated surface, mechanically fracturing coatings. The particles removed should then be simultaneously collected in a drum by a vacuum fitted with a HEPA filter. The concrete planer has a cutting width of 5.5 in (14 cm), with larger units available from EDCO.

The roto peen technology is able to remove a uniform layer of concrete and coatings from about 500 ft^2 (46.5 m^2) of floor area at a rate of approximately 71 ft^2 (6.6 m^2) per labor hour. In one instance, after removing the layer, five of the six contaminated locations were at or below background levels and the contamination for the sixth location was reduced from 70,000 to 16,000 dpm/100 cm^2. The depth of removal was about one-sixteenth of an inch. This achieved a uniform appearance on the roto-peened, finished surface. Approximately 2.1 ft^3 (0.06 m^3) of waste in the form of powdery concrete and print chips was generated during the demonstration. Due to the captive shot, only what was removed from the floor is what entered the waste stream. The vacuum system was sufficient to contain the dust generated; survey smears did not show any removable contamination.

17.9.7 Milling Decontamination: Pentek milling technology is a method for removing contaminated coatings on concrete and steel located on floors, walls, ceilings, and structural components. The system uses a hand-held roto peen scaler with localized exhaust. The scaler is equipped with 3MTM heavy duty roto peen flaps, each studded with rows of tough tungsten carbide cutters and mounted on a rotating hub. The debris removed by the roto peen scaler is simultaneously collected in a VAC-PACR, high performance HEPA vacuum/drumming system.

17.9.8 Remote Dry Ice Blasting System: Concrete surface contamination inside buildings is one of the most serious problems in nuclear facilities. Contamination typically consists of various radionuclides, heavy metals, and organic deposits resulting from leaks and spills. New technologies are required to remove concrete surface contamination so that buildings may be reused, or if they are to be dismantled, to permit disposal of the majority of the concrete as nonhazardous, nonradioactive waste.

This combined system, ROVCO 2, is based on an ANDROS remotely operated, six-wheeled vehicle designed for non-nuclear applications. It integrates two demonstrated technologies: a remotely operated vehicle (ROV) and a dry ice (CO^2) blasting system. The vehicle provides transport and power for all vehicle-mounted subsystems, including the CO^2 XY orthogonal end effector (COYOTEE), cryogenesis dry ice blasting system, and the vacuum/ filtration/ containment subsystems. The COYOTEE manipulates a specially designed vacuum containment workhead with the cryogenesis blasting nozzle to cover every point within a rectangular workspace. Since ROVCO 2 uses CO^2 gas, it has the potential to eliminate process waste resulting from the blasting material. Additionally, since the system is remotely operated, both personal protection requirements and worker exposure risks are reduced.

17.9.9 Remotely Operated Scabbling: Pentek's remotely operated scabbling technology, MOOSE, is designed to scarify large concrete floors and slabs. This scabbler is comprised of three integral subsystems: the scabbling head assembly, the on-board HEPA vacuum system, and the six-wheeled chassis. The scabbling head houses seven independent reciprocating tungsten carbide-tipped bits, with the pistons being driven by compressed air. Dust and debris are captured by the two-stage positive filtration HEPA vacuum system that deposits the waste directly into an on-board 23-gallon (87.1 liter) waste drum. Independent skid steering allows the unit to pirouette 360 degrees about its geometric center and it is capable of scabbling within six inches of walls

In one demonstration, the MOOSE successfully removed a nominal 1/8 in (0.3 cm) of concrete from 618 ft^2 (57.4 m^2) of floor area with a production rate of 130 ft^2/hr (12 m^2/hr) for a two-person crew. Operated remotely via a 50-ft (15.2 meters) power and control tether, the operator and control panel were able to be located outside the demonstration area, where the operator could work without the need for personal protective equipment. Because of the confined geometry of the floor, a second operator was required to remain inside the demonstration area, where he could be close enough to manage the tether and air hose. The MOOSE proved to be very maneuverable and easy to operate. Radiological contamination in the area prior to the demonstration included hot spots ranging from 1,000 dpm/100 cm^2 to 105,000 dpm/100 cm^2 fixed beta/gamma. After the demonstration, all but one hot spot was reduced to 3,500 dpm/100 cm^2.

17.9.10 Soft Media Blast Cleaning: The soft media blast system is a variation of the baseline high pressure water cleaning system. The kinetic energy of the pneumatically propelled soft media impinging on a surface provides the removal mechanism for the surface contaminant. The soft blast media, because of the high transport velocity, impacts the surface with high energy, but due to its soft structure, has very little bounce-back. On impact, the soft media absorbs and traps the contaminants and carries them away from the substrate for easy disposal. The use of the soft media blast system as a cleaning technology appears to apply to enriched uranium process residue. A major obstacle to the implementation of this is the noise level associated with the system in operation -- measured at 110 dBA.

17.9.11 Steam Vacuum Cleaning: The Kelly Decon System is a variation of the baseline high pressure water cleaning system. The kinetic energy of a pressurized super-heated water stream provides the mechanism for removal of contamination from surfaces sprayed with the water. The Kelly system differs from the baseline system in that it provides a superheated (up to 300°F = 149 °C), pressurized (250 psi = 175,765 kg/m^2) stream of water with a flow rate that may be varied between 0.4 and 2.0 gpm (1.5 to 7.6 L/min). However, the key difference is that most of this system's cleaning heads (the tools for administering the water spray to the surface) integrate a vacuum hood and return line, to capture and control the steam, water droplets, and dislodged contaminants generated when the water spray impacts the surface.

The primary application for the Kelly system has been the decontamination of rooms, pools, walls, large components, or similar applications related to large and/or smooth surfaces. An advantage of this system is that most of its cleaning heads incorporate vacuum collection of the clean medium. For these heads to be effective, they should be used on suitable planar surfaces (e.g., large flat surfaces that are compatible with the cleaning head designs). It should be noted that even the relatively flat tank segments are not totally compatible with the Kelly system. Many of these segments may contain the remnants of weld attachments that interfere with the heads designed for use on flat surfaces.

17.9.12 Water Blasting: The basic equipment to perform water blasting consists of a high pressure pump, a fluid transport hose, and a nozzle used for forming a controlled stream of water for impingement onto a contaminated surface. The nozzle typically is attached to some type of wand useful for controlling and directing the flow of water. The water blasting equipment frequently may be taken into the contaminated facility for cleaning of surfaces and equipment. In addition, water blasting equipment is frequently set up in fixed base operations that are used to process and decontaminate small equipment and disassembled members that are brought to the decontamination facility. This area may be equipped with sumps, curtains, and other devices designed to reduce the spread of contamination and collect residues.

The pressurized water used for blasting will involve some hazards. High energy water streams can be lethal, and industrial accidents with pressure washing have resulted in fatalities. Operator training is essential and minimum protective clothing should consist of eye protection, hearing protection, and gloves, but their prescription must be commensurate with the potential hazards. Water blasting can be effective in removing contamination that loosely adheres to a readily available surface. Frequently, the nozzle must be held a few inches from the surface to be effective and contamination in cracks and crevices or behind obstructions is not readily removed by water blasting. Close proximity, "line of sight" applications, are best for water blasting. Its productivity rate is approximately 100 ft^2/h for regular, readily accessible surfaces.

17.10 OTHER TECHNIQUES

In addition to chemical and mechanical decontamination techniques, there are those that rely on other principles. The following technologies have been successfully demonstrated and may be considered for application.

17.10.1 Electrokinetic Decontamination: ISOTRON has developed an electrokinetic process for in-situ removal of contaminants from porous concrete. This method provides a viable alternative to scabbling or physical abrasion. The technology developed for electrokinetic decontamination of concrete surfaces applies an electric field to induce migration of ionic contaminants from within the porous concrete into the decontamination unit. The components of the system are a proprietary surface electrokinetic extraction concept pad, electrolyte solution, and electrode. The electrolyte solution contains various complexing agents and other materials to promote formation of a soluble ionic complex of each specific contaminant present. The electrolyte solution is in contact with the concrete surface through the pad, a material similar to fabric or carpet that partially removes contaminants from the electrolyte solution and limits the solution's flow. All contaminants are collected in either the aqueous electrolyte solution and/or in the pad. Both of these may be treated and disposed of by conventional technologies. Current alternative configurations of the system are being evaluated to enhance the removal of specific contaminants from concrete.

The electrokinetic decontamination process provides an in-situ alternative to concrete decontamination, thereby eliminating physical or mechanical damage of the concrete, allowing for reuse of the structure or facility. This process generates minimal secondary waste and no airborne particulate matter common to conventional scabbling or physical abrasion techniques. Furthermore, compared to conventional systems, this process is capable of removing contaminants diffused deeply into concrete.

17.10.2 Laser Ablation: Over the years, as buildings and equipment surfaces became contaminated with dust or spills from uranium and other radioactive materials, or became contaminated with hazardous materials like PCBs, surface contamination was stabilized in-situ by applying coats of paint and, in some cases, strippable coatings. Most of the earlier paint used was lead-based. More recently, hydrocarbon-based and latex paints have been employed for fixation. Many facilities have been repainted many times, resulting in several coats that may vary in thickness and in type from one location to the next.

Industrial lasers have the ability to ablate coatings from metal and concrete surfaces in a controlled manner. The laser uses the proper combination of wavelength, pulse duration, energy, power densities on target, pulse repetition rate, and scan rate. This not only yields efficient removal of coating material from the surface and surface pores but also ablates material faster than a thermal wave can propagate into the substrate, preventing entrainment of surface contamination in molten substrate and leaving behind a cool surface. This characteristic also avoids damage to the substrate and reduces potential volatilization of contaminants. A gas, vapor, and particulate suction device captures the ablated material. A vacuum system draws the mixture of entrained air and gases, vapors, and particulate matter from ablation through filtering stages.

17.11 SUMMARY

Those experienced in the use of decontamination processes and techniques should be the ones who choose the appropriate technique for each activity. Decontamination engineers are not granted degrees in colleges or universities, but in the field after the experience is obtained by working with others on a process or technique. Critical tolerances can be damaged by an inappropriate application or destroyed by corrosion or abrasive mechanisms. If not performed properly, some techniques may generate more contaminated material than was cleaned by the process.

The generation of secondary waste depends on the process or technique used and affects the total cost of the process. Care must be taken in using any types of solvents in the generation of mixed waste (both hazardous and radioactive). The elements removed during the decontamination must be identified so that if a mixed waste is generated, it can be properly packaged and treated.

Is the decontamination cost-effective? Once the goal of the decontamination is determined, the process or technique is chosen to reach that goal. Any further activity after the goal is reached may not be cost effective. The cost of the decontamination equipment is a major factor in choosing a decontamination method. The initial choices should be those for which the individuals performing the activity have the equipment and the most experience. Consideration should also be given to performing a technique that may not be an optimum choice, but has the advantage of on-site equipment and personnel experience.

Chapter 18

DISMANTLING

18.1 INTRODUCTION

Most U.S. power reactors were built with one concept in mind: to produce electricity as efficiently, economically, and safely as possible. Plants were not designed to facilitate simple and efficient deconstruction. Since decommissioning was not planned by the original construction team, dismantling is a completely new process that must be learned for each unique facility. Fortunately, the process is becoming more predictable and manageable because of experience gained to date. Dismantling is probably the phase of decommissioning with the strongest visual impact. A major portion of the project involves dismantling, removal, and size reduction of concrete structures and metal components, such as vessels, pipes, conduit, and structural steel. Generally, the metal and concrete components have radioactive, organic, chemical, or heavy metal contamination, thereby presenting potential exposure hazards. The site and facility characterization data will show which and how much of those materials must be dismantled and removed.

Dismantling involves decontamination, removal of components and structures, packaging of wastes, transport of packages, and disposal in a controlled burial facility. But the planning should be performed in reverse order. Once characterization has identified the waste streams, the next step is to contact the waste disposal facility to learn and understand its waste acceptance criteria, including permissible package/container radioactive inventory (curies and dose), package size and weight requirements, and documentation. If the facility will not accept the waste form or content, alternative measures must be taken. When the package sizes, weights, and radioactivity limits are selected, the mode of transport and number of shipments may be determined -- truck, rail, or barge. The planner may then select the most cost-effective technology to segment piping, equipment, and structures to fit into the selected packages. The segmentation/removal production rates (tons/day) for each potential technique evaluated will determine the number of shipments per day or week and thereby set the overall schedule. Adjustments might be made as experience demonstrates improved productivity with other technologies.

This chapter will help the decommissioning planner choose appropriate technology for specific applications. Current dismantling, removal, and size reduction technologies include mechanical saws, circular cutters, abrasive cutters, diamond wire, explosive cutting, plasma arc torch, oxyacetylene torch, arc saw, abrasive water jet, and hydraulic shears. Backhoe hydraulic rams, conventional wrecking techniques, and explosives are used to dismantle large, thick concrete structures. The use of explosives requires a certified blasting expert for safety.

Many of these techniques are available as hands-on or remotely operated equipment. The selection of any of these technologies requires an evaluation of the following:

- Cutting speed
- Maintenance frequency
- Potential exposure of workers to radiation
- Limited accessibility in congested areas

- Dust emissions
- Generation of secondary waste
- Stringent rigging requirements
- Industrial safety issues associated with working at heights
- Spread of contamination
- Fire hazards
- Noise

Years of experience working with these technologies have yielded methods to effectively deal with these issues, and lessons learned have been shared among projects to improve performance.

New technologies often entail changes to old processes and accepted equipment. This has been the case with remote operations, where increased use is closely linked to safety and the diminished risk of working with hazardous materials. Opportunity to use remote technologies is not limited to waste removal but is also advantageous in other areas of mechanical operations.

Nuclear facilities are generally substantial structures built to withstand external and internal hazards. And the structures are often unique, so that no single method of dismantling and demolition can be all-encompassing. But there are principles that may help. Maneuverability and ease of transport play large roles since the dismantling tool or equipment often must fit within a confined space. Reach is also important, as some buildings may only require equipment with a two-story reach, while others are so large that other techniques are necessary, such as controlled blasting, which has been shown to be effective in bringing down structures quickly to permit the use of conventional size reduction methods.

This chapter describes the proven technologies used in nuclear facility dismantling, removal, size-reduction, and demolition activities, along with equipment or technique performance factors based on best available information, either from actual project-specific data or as quoted by the manufacturer or vendor. In either case, the selection of a technique for a specific application should be made after careful evaluation of the proposed technique and potential competing equipment. The planner should realize that vendor information may reflect ideal conditions, in a non-radioactive environment, or without the limitations of access space or restrictions of working in protective clothing and respirators. If possible, the best approach is to contact previous users of the equipment before deciding for or against it.

This chapter also encompasses innovative technologies that have been developed and tested under a DOE program of large scale technology development. The testing was performed under controlled conditions similar to those expected to be encountered in actual decommissioning projects. The compact disk (CD) provided with this handbook contains the detailed descriptions of these technologies, their performance data, and reference photographs. Once again, the decommissioning planner should evaluate these against the proven technologies described here, or against the baseline technologies used for comparison with innovative methods.

18.2 CURRENT TECHNOLOGIES

Dismantling methods are of two types: disassembly and segmentation. Disassembly generally means removing fasteners and components in an orderly, nondestructive manner (the reverse of assembly). Segmentation includes flame cutting, abrasive cutting, and cold cutting. Flame cutting technologies include oxyacetylene and other gas torches, carbon-arc torches, air or oxy-arc torches, plasma arc torches, and cutting electrodes. Most torches can be operated either remotely or by hand.

Abrasive cutting technologies include grinders, abrasive saw blades, and drilling machines. Cold cutting technologies include nibblers, shears, and cutters for bolts, pipes, and tubing.

The best method depends on the application. The following discussion presents dismantling activities for the two primary materials, concrete and metal. A final section will discuss component removal as a separate category due to the unique activities involved in removing large components.

18.2.1 Concrete: Methods for dismantling concrete structures (walls, floors, ceilings, and foundations) are diverse in that each has particular advantages and disadvantages related to cost, personnel exposure, and overall effectiveness. Some methods minimize waste while others are designed strictly for structural demolition (with no segregation of material, clean or contaminated). Each method has a justifiable application; what's best in a particular case depends on its unique circumstances and ultimate goal. Tables 18.1 and 18.2 summarize dismantling and cutting techniques for concrete.

Table 18.1 Dismantling Techniques for Concrete

Technique	Application	Methodology	Objective
Selective Cutting			
Impact Destruction	Concrete Sections	Inserting steel bit and applying reciprocating motion	Segment / Reduce to rubble
Abrasive Water Jet	Concrete Sections	Propelling abrasives combined with a high-velocity water jet	Segmentation
Diamond Wire	Concrete Sections	Threading or encompassing material and circulating wire	Segmentation
Blasting			
Controlled Explosive	Concrete Sections	Inserting explosive into pre-drilled holes in the concrete and imploding	Segment / Reduce to rubble
Bristar	Concrete Sections	Filling of pre-drilled holes with a compound that expands when it hardens	Segment / Reduce to rubble
Surface Erosion/Removal			
	Concrete Sections	Using saw and impact hammer to break up sections	Segment / Reduce to rubble

18.2.1.1 Selective Cutting: Selective cutting removes sections of concrete in planned increments or quantities. The object of selective cutting is to remove the concrete in an orderly, precise fashion to enable the application of additional techniques to accomplish specific goals. Some of these techniques and goals are:

- *Removal in sections*: enables decontamination of each section to permit free-release of the section after contamination has been removed

- *Removal of designated areas:* large surfaces may not be entirely contaminated; thus, it is only necessary to remove those affected areas
- *Removal of sections restricting access to contaminated zones*: selective cutting permits an orderly removal of such structures without cross-contamination.

Table 18.2 Cutting Techniques for Concrete

Technique	Cutting Speed	Liquid Waste	Solid Waste	Approximate Cost	Containment	HEPA Ventilation	Maximum Cut
Diamond Wire	7-9 ft/hr	3-5 gal/min	Material debris	$300/hr rental	No	No	Unlimited
Water Jet	5 ft²/hr	1.4 gal/min	Material debris	$174 K / unit	No	No	Unlimited
Controlled Explosive	Immediate	None	None	Varies	Yes	Yes	~ 6 in
Bristar	3 - 20 hrs	None	None	$210 / 44 lbs	No	No	1 ft
Flame Cutting	10 ft/hr	None	None	$200 plus gas	Yes	Yes	60 in
Thermic Lance	1"dia x 12" deep/min	None	None	$75/unit	Yes	Yes	1 ft
Rock Splitter	10 min/hole	No	Yes	$5000/unit	Yes	Yes	Unlimited
Various Saws	150 in/min	Yes	Yes	$700-$1500 /unit	Yes	Yes	1/3 blade diameter

1. *Flame Cutting*: Flame cutting of concrete is a thermite reaction process whereby a mixture of iron and aluminum powders is oxidized in an oxygen jet. The jet flame temperature (~16,000°F / 8,900°C) causes rapid decomposition of the concrete. The generated molten concrete is blown away from the kerf by the jet mass flow rate, keeping the cut clean. Reinforcing rods (rebar) are also segmented by the flame since the iron actually augments the thermite reaction. Normally, track-mounted flame cutters are capable of cutting through 60 inches of concrete with or without rebar. The major disadvantage of flame cutting is the generation of large quantities of heat, smoke, and toxic gas.

2. *Thermic Lance*: The thermic lance is an iron pipe packed with a combination of steel, aluminum, and magnesium wires through which a flow of oxygen gas is maintained. The lance cuts are achieved by a thermite reaction at the tip of the pipe, where all components are completely consumed. Temperatures at the tip will range from 4,000 to 10,000°F (~2200 – 5600°C), depending on the environment (e.g.: in air or under water). Lances can only be used in the hand-held mode in cutting metals and demolishing concrete. Their use is discussed in further detail in paragraph 18.2.2.1.3 of this chapter.

A single lance can burn a 2-inch diameter hole through 2 to 3 ft thick reinforced concrete in ~ 6 minutes. While the lance is best used for producing holes, a rock splitter is required to complete segmentation of concrete. The lance generates significant smoke in air and bubbles in water, so ventilation must be maintained to protect the worker and the environment.

3. *Rock Splitter*: The rock splitter uses a hydraulically operated expanding wedge placed into a drilled hole to fracture the surrounding concrete. The hydraulic cylinder drives the wedge into the hole at a pressure of about 7,000 psi and can develop splitting forces of up to 350 tons. Rebar in the concrete must be cut by other means. Limited access areas can be segmented easily by rock splitting since the unit is held manually. About ten minutes are required to drill and split each hole. Contamination controls required are minimal since only a small amount of dust is generated.

4. *Sawing*: A motor-driven diamond or carbide saw blade may be used to cut a kerf through concrete floors or walls. The blades are capable of cutting rebar if it is at right angles to the cut. Most concrete saws are track-mounted and operated manually, although newer technologies are evaluating remotely-operated systems. A normal thickness of cut is about 1/3 the diameter of the blade, with about 3 ft of concrete being the maximum thickness. Saw cut speed can reach 150 in^2 per minute, depending on the concrete composition. Since most concrete saw blades are water-cooled to prevent warping, the water is a secondary waste concern. Another concern is airborne contamination (dust and metal particulate).

5. *Bristar Demolition Compound*: Bristar™ concrete demolition compound is a mixture of limestone, siliceous material, gypsum, and slag. When mixed with water and poured in pre-drilled holes in concrete, the compound hardens and expands within 20 hours, developing a force of over 4,300 psi. This force causes cracks along a predetermined fracture line delineated by the drilled holes. Exposed rebar is cut by other methods. This process produces no secondary waste (except during the hole drilling) and there are no noise, gas, or airborne considerations. A hole two inches in diameter and one ft deep takes 2.5 lbs of Bristar.

6. *Explosive Cutting*: Used to segment metals as well as concrete, this method uses an explosive, normally RDX™, formed in a geometric shape specially designed and sized to produce the desired segmentation. Consisting of an explosive core surrounded by a casing of lead, aluminum, copper, or silver, the cutter produces a high explosive jet of detonation products and deformed casing metal. The shock wave cuts the target material.

   Explosive cutting, which must be performed by a licensed contractor, can be performed in air and under water on materials greater than 6 inches thick. But its use is limited since the blast may affect the structural integrity of surrounding structures or produce an uncontrolled spread of radioactive materials. It's therefore used in situations where other cutting methods are not practical or access is limited. Containment control is required because of the high levels of noise, smoke, and debris.

7. *Diamond Wire Cutting*: Diamond wire cutting is most often used to cut reinforced concrete monoliths. A cart-mounted unit drives a wire that carries diamond impregnated beads. The wire is either threaded through holes drilled in very large components or, if the component is

small enough, the wire completely surrounds the component and is field-spliced. The wire size and length depends on the application (normally 3/8 or 1/2 inch in diameter) with a cutting speed of 7 to 9 ft/hr. This technique has a ratio of 4 ft^2 of concrete cut per foot of diamond wire consumed.

The diamond wire must be water-cooled during cutting; the water is also used for flushing the debris from the cut. The water consumption, now a secondary waste, is normally 3 to 5 gallons per minute. Airborne activity is minimal as long as the cooling water and contained dust is periodically removed.

8. *Abrasive Water Jet Cutting*: The water jet cutting system uses abrasives (such as garnet) in a high velocity water jet to cut through large pieces of concrete. Achieving cutting speeds up to 5 ft/hr, the system consists of an intensifier pump, a nozzle head assembly, and an abrasive supply system. The nozzle head assembly can be located away from the supply system in order to permit remote manipulation. Nozzles on this system erode after about 50 hours of use and must be replaced. The main waste stream is water-generated at the rate of 1.5 gal/min. Airborne contamination is not a concern, but cross-contamination due to water spray and mist must be considered.

9. *Other Techniques*: Inefficiencies and narrow or specialized task application are the key reasons for not using various other cutting techniques. While they may be suitable on a small scale, they are not considered viable for the scope of a decommissioning project. Examples of these slower, more narrowly focused, techniques are:

   - Paving breaker: normally used for breaking up pavement
   - Chipping hammer and chisel: limited in application
   - Core stitch drilling: requires additional technology application, i.e., rock splitter and rebar cutter
   - Power saws: normally under-powered and inadequate blade technology

18.2.1.2 Concrete Blasting: Concrete blasting is a general term to describe indiscriminate removal of concrete, without regard for keeping the structure in workable form. Concrete blasting normally results in rubble that is not conducive to further decontamination. The decision to blast should follow the determination to dispose of an entire structure as either clean or as contaminated waste.

An acceptable method for removing a clean structure is to survey and release it in-place before demolition. This process is less time-consuming and radiologically more concise than attempting to survey rubble after demolition. The following criteria must be satisfied to dispose of an entire structure as clean material:

- The entire structure must be radiologically free of contaminants -- all smearable surfaces and all internal surfaces and penetrations.
- Destruction must not subject the structure to cross-contamination from other sources (e.g., allowing a clean wall to fall into an area with a contaminated floor).
- Material from the structure must be controlled as clean until removed from the radioactive waste area.

The determination to dispose of a structure as contaminated should be reached only after it satisfies the following criteria:

- The structure is completely (not superficially) contaminated.
- All efforts have been made to decontaminate the contaminated portions of the structure.
- Other options have been evaluated and discarded (e.g.: cutting out contaminated sections).
- No other viable alternative is available.

Controlled blasting is used to demolish concrete greater than 2-ft thick, provided it is not limited by the effects of noise and shock to adjacent occupied areas. Controlled blasting produces the high degree of fragmentation needed for heavily reinforced concrete. The exposed reinforcing bars must then be cut by other methods (torch or bolt cutter). The high level of dust in the work area must be carefully controlled. Controlled blasting is used to achieve maximum demolition at a minimum cost, but it is also used to reduce occupational exposure to personnel. Personnel can ensure a minimum spread of contamination by using the following preventive techniques:

- Using explosive charge sizes that minimize rock throw and dust generation while accomplishing the intended task
- Applying a three-element blanket:
    - Three layers of sealant (e.g., TURCO 5580-G) to the concrete seals in the contaminants
    - Layered tar paper and rubber-backed carpet to absorb the blast and limit rock-throw
    - Cover of all exposed surfaces and those beyond the perimeter with a protective material (e.g., Hypolon) to protect against contamination

Table 18.3 Removal Rates and Relative Costs for Controlled Blasting

Concrete Type	Removal Rate (yd^3/day)	Removal Cost (\$/yd^3) (1)
Massive Reinforced Standard Concrete (non-radioactive)	100 - 400	Low
Massive Non-Reinforced Standard Concrete (non-radioactive)	250	Med
Massive Reinforced Standard Concrete (radioactive)	4 - 6 (2) 100 (3)	High
Lightly Reinforced Standard Concrete (non-radioactive)	200	Low
Non-Reinforced High Density Concrete (radioactive)	6 - 8	Low
Lightly Reinforced Standard Concrete (radioactive)	6 - 8	High

Notes:
(1) High (>$500); Med (<$500); Low (<$200)
(2) Removal rates for radioactive areas include inefficiency due to personnel and restrictive contamination controls
(3) Removal rates are dependent on using equipment to load and transport

When compared to other concrete dismantling technologies, blasting generates a high level of radioactive waste, with no opportunity for separation of clean and contaminated. Table 18.3 lists typical removal rates and relative costs for controlled blasting. The operational characteristics are shown in Table 18.4.

Table 18.4 Operational Characteristics for Controlled Blasting

Item	Comment
Material Applicability	Massive concrete; reinforced concrete walls and floors
Maximum Depth/Thickness	Limited by drilling depth only
Bulk Removal Rate	Rapid
Vibration/Shock/Noise	Moderate with controls
Dust	Moderate using fog sprays
Missile Generation	Must use blasting mat
Contamination Control	Controlled with blasting mat and fog spray
Consumables	Drill bits, explosives, detonators
Safety (Industrialist)	Requires certified blasting technician
Relative Cost	May be high for small volumes of concrete

Following are details about the types of controlled blasting currently available.

1. *PETN (Pentaerytritol Tetranitrate)*: This explosive is primarily used in the form of a detonating cord. It is effective for surface spalling that requires the removal of very small sections of about 8 inches from an exposed surface.

2. *High-Velocity (85%) Gelatin Dynamite*: This explosive is used in shallow holes ranging in depth from 18 into 5 ft. It may also be used as a partial loading in selected holes when a concentrated energy is needed in a particular area. While the breakage created is excellent, this explosive produces high quantities of dust and crushed stone around the bore hole.

3. *Cast TNT (High Detonation Pressure Primers)*: This explosive is used to achieve a high degree of fragmentation with less of a heaving effect than with conventional dynamite.

4. *Binary Energy System*: This liquid explosive is normally mixed immediately before use and is not considered an explosive until mixed. It is used as a replacement for TNT during the spalling of surface contamination and to make shaped charges that can be used to punch holes in high-density concrete.

5. *Water Gel Explosives*: Gelatin dynamite is sometimes replaced by a water gel explosive containing a large amount of aluminum. It produces good shattering characteristics and larger rubble than other explosives. It's mostly used in areas of little or no reinforcing rods, where larger sized rubble is preferred.

18.2.1.3 Surface Erosion/Removal: Surface erosion/removal is the removal of concrete structures by gradual erosive techniques, such as scarification (also known as scabbling), or massive demolition. Selecting a particular technique depends on several considerations:

- Radioactivity associated with the target area
- Worker exposure
- Potential for release of radioactive particulate during demolition, via fluid leaks or airborne activity
- Porosity of the concrete (potential resistance to nondestructive cleaning methods)
- Actual volume of material to be removed
- Reinforcement (imbedded and superficial) of the concrete
- Accessibility of the structure to be removed

Typically, a concrete shield surrounding a radioactive source will consist of massive sections (2 to 20-ft thick) of standard (140 to 150 lbs/ft^3) of high-density concrete (magnetite or metal aggregate, 250 to 325 lbs/ft^3). Some concrete shields are also heavily reinforced to meet seismic design criteria. The technique selected will either remove the whole object or only the contaminated portions. Table 18.5 gives information on concrete removal and recommended applications and limitations.

The more practical methods for complete removal reduce the concrete to rubble. These methods are described below:

1. *Controlled Blasting*: Controlled blasting reduces concrete sections to rubble by controlling the actual explosion and its effect on material movement and adjacent structures. When applied to thick, massive, or heavily-reinforced concrete sections, the wave created by detonation separates the fractured boreholes and moves the material toward the structure's free face. The boreholes are pre-drilled in the concrete and are loaded with explosives. The various types of explosives and the method's operational characteristics were discussed previously.

2. *Wrecking Ball/Slab*: The wrecking ball is typically used for unreinforced or lightly reinforced concrete structures less than three feet thick. This 2 to 5 ton ball or slab, suspended from a crane boom, repeatedly impacts the structure and reduces it to rubble. While it is an effective demolition technique, the wrecking ball is not recommended for use on radioactive structures for its inaccessibility to internal structures and its inability to control dust generation or rebound impact.

   The wrecking ball has been used on decommissioning projects, but only after all radioactive material has been removed from a facility. The equipment and operating crew required make this technique costly and useful only in limited applications.

3. *Backhoe-Mounted Ram*: The backhoe-mounted ram is a backhoe arm equipped with an air or hydraulically-operated impact ram fitted with a chisel point. The chisel point impacts structural surfaces at a rate of about 600 blows per minute, at up to 2,000 foot-pounds of energy per blow, depending on the size of the ram head. This configuration normally keeps the operator away from the structure being demolished due to the reach (20-25 ft) of the backhoe arm. This technique is used extensively on many decommissioning applications because of its versatility and relatively low rental costs. Generated dust and noise pollution are adverse side effects that must be considered, but the overall effectiveness on concrete

Table 18.5 Concrete Removal Methods

Removal Method	Advantages	Disadvantages
Blasting	Fast material removal rate where adjacent material damage is not of concern	May cause vibrations and shocks (vibration control possible by limiting maximum charge per delay)
	Adaptable for cutting irregular surfaces	Requires oxyacetylene torch for cutting exposed reinforcing steel
	Adaptable to remote operation	Generates moderate quantities of dust that must be controlled
	Adaptable to control of size and weight of material removed	High noise level
	Selective removal of radioactive contamination possible	
Core Boring and Rock Splitting	Proven techniques used extensively in conventional mining operations	Reinforcing steel slows cutting speed and damages core drill
	Low noise level	Requires oxyacetylene torch for cutting exposed reinforcing steel
	Low vibration/shock level	Cooling water required for drilling may have to be filtered and recycled
	No toxic gases generated	Not ideally suited for irregular surfaces or cramped quarters
	Cooling water spray minimizes airborne particulate material	
	Moderate to slow cutting speeds	
	Controlled rate of material removal	
Flame Cutting	Concrete and reinforcing steel may be cut in one operation	Generates very large quantities of toxic gase and smoke
	Fast cutting speed	Consumes large quantities of bottled oxygen
	Adaptable for remote operation	Smoke can potentially spread gross contamination throughout control envelope unless adequately controlled
	Adaptable for pivoted circumferential cutting	Requires through starter-hole made by core drill to provide outlet gas-flow path
	Adaptable for cutting on irregular surfaces	
	No vibration/shock	

Table 18.5 Cont'd
Concrete Removal Methods

Removal Method	Advantages	Disadvantages
	Low noise level	
	Controlled rate of material removed	
Thermic Lance	Cuts both concrete and steel without difficulty	Generates moderate quantities of toxic gases and smoke
	Fast "hole-punching" speed	Requires rock jack to break away cut concrete sections
	Ideally suited for cutting irregular surfaces	Consumes large quantities of bottled oxygen
	Remote operation is possible for up to approximately three meters	Smoke can potentially spread gross contamination throughout control envelope unless adequately controlled
	No vibration/shock	
	Low noise level	
	Controlled rate of material removed	

structures (less than 2-ft thick) makes this process valuable for the removal of interior walls and floors. The removal rate of a backhoe-mounted ram will vary greatly depending on accessibility, radiological concerns, and the method used for removing rubble. More recent technology uses the same concept but allows for remote operation of the backhoe, thus reducing worker exposures. Such remotely operated units (e.g., Brokk) are available in a variety of sizes and have a number of decommissioning applications.

4. *Wall and Floor Sawing*: Cutting speeds for concrete saws vary, depending on the thickness of the concrete and the amount of reinforcing metal present. A thickness of up to three feet (with a maximum of one third of the blade's diameter) can be cut at about 150 in^2/min.

Saws can be operated manually or remotely, with a corresponding increase in cost for the remote systems. The structure and thickness of the concrete determine how often the blades must be changed, thereby affecting the overall cost. The reasons for using floor and wall sawing are as follows:

- Keeps disturbance of surrounding areas at a minimum
- Removes only affected areas of a structure, enabling waste minimization
- Facilitates handling of smaller sections that can be decontaminated
- Removes structural sections to gain access to affected areas without unnecessary disruption of the area
- Improves accessibility to limited-access areas.

Saw blades for floors and walls are usually water-cooled and that water must be treated as radioactive waste, along with the minimal dust generated by the process. The process of cutting can be slow, depending on the material being cut, but it is a useful technique. Concrete sawing is a cost-effective but limited application.

5. *Paving Breakers / Chipping Hammers*: Sometimes better known as jackhammers or pneumatic drills, paving breakers remove concrete by mechanically fracturing localized sections of the surface. A hardened tool steel bit (either a chisel or moil point) is driven into the concrete surface by a reciprocating motion at the rate of about 1600 blows per minute. The "hammer" itself is either a compressed air or hydraulic fluid pressure source that sends energy to the bit at the rate of 34 to 95 foot-pounds per blow. Paving breakers are primarily used on floors to remove small areas that are inaccessible to heavy equipment. Paving breakers are also used to scarify surface areas of concrete where contamination has penetrated several inches. Using the paving breaker generates large amounts of dust and noises, requiring environmental controls. Although a relatively slow process, paving breakers are cost-effective and have proven to be valuable equipment.

   Chipping hammers are similar in concept but are lighter (15-35 lbs), making them suitable for hand-held for use on walls and ceilings. Their rate of delivery is normally about 2,000 blows per minute, but the nail size is about half that of the paving breaker and their weight (up to 35 lbs) makes them cumbersome. They are used to scarify small areas of walls where contamination may have penetrated several inches. Continued use renders them impractical except for instances where there are no other options. Reasonably priced, they are effective for removing surfaces in a partially contaminated structure.

   Neither the paving breaker nor the chipping hammer is designed specifically for the complete removal of a structure, but rather as an aid to removing small pockets of contamination or to reach areas not otherwise accessible. Several of each type are found at decommissioning sites and they are considered essential tools for decontamination.

18.2.2 Metals: Table 18.6 presents comparisons of the methods available for segmenting metals in a decommissioning environment.

18.2.2.1 Thermal: Thermal segmentation of metals refers to the general technique of cutting without making direct contact. In direct contrast to mechanical cutting, thermal cutting uses a medium other than a cutting edge to sever the metal. Such medium may be a focused beam of high-powered light (laser) or a high-temperature flame (plasma arc). Some of the more important benefits of most thermal cutting technologies are that they can be:

- Performed under water or in air
- Operated remotely; their use on highly radioactive materials reduces worker exposure
- Used for extended durations of time and reduce overall cost (although setup and take-down times are generally longer)

The following subsections discuss various thermal cutting techniques used in the nuclear decommissioning industry.

Table 18.6 Application Characteristics of Metal Segmenting Techniques

Technique	Maximum Cut (1)	Secondary Waste	Speed	Cut (kerf) Width	Specific Applications	Operating Mode (2)
Plasma Arc	36" (M) 72" (P)	Dross, Particles, Fumes	Fast	12 mm	Thermal Shield	P, R, S
Oxygen Burner	36" (M) 72" (P)	Dross, Particles, Fumes	Fast	2 mm	Pipe, Plates	P, R, S
Thermic Lance	36" (M) 72" (P)	Smoke, Particles, Fumes	Fast	6 mm	Pipe, Plates	H
Explosive Cutting	36" (M) 72" (P)	Dust, Smoke, Debris	Fast	N/A	Large Pipe	R
Hacksaws and Guillotine Saws	24" (P)	Swarf, Dust	Medium	6 mm	Pipe	H, P, S
Circular Cutter	3" (M) 72" (P)	Swarf, Dust	Medium	6 mm	Thermal Shield	H, R
Abrasive Cutter	All (M) 2.5" (P)	Fine Particles	Very Fast	2 - 4 mm	Pipe, Plates	H, S
Arc Saw	36" (M) 72" (P)	Dross, Particles, Fumes	Fast	12 mm	Reactor Vessel Internals	R, S
Mechanical Nibbler	36" (M)	Dust	Medium	4 mm	Tanks	H, P
Hydraulic Shears	2.5" (M) 2.5" (P)	None	Fast	Varies	Tubing, Pipe	H, P
Band Saws	72" (P)	Swarf	Medium	6 mm	Plates, Grids	H, R
Reciprocating Saws	36" (M) 72" (P)	Swarf	Medium	2 mm	Pipe, Strainers	H, P
Laser Cutting	36" (M) 72" (P)	Dross, Particles, Fumes	Fast	Varies	Pipe, Plates	R
Bolt Cutter	2.5" (P)	None	Fast	Varies	Tubing	H, P

(1) (M) = Metal Thickness (P) = Pipe Diameter
(2) (P) = Portable (R) = Remote
(H) = Hand-held (S) = Stationary

1. *Plasma Arc Torch*: The plasma arc cutting process establishes a direct current arc between a tungsten electrode and any conducting metal. The arc is established in a gas, such as argon, that flows through a constricting orifice in the torch nozzle to the work piece. The constricting effect of the orifice results in high current densities and high temperatures in the

stream (17,500 to 42,700° F). The plasma is ejected from the nozzle at high velocity and combines with the arc to melt the work piece. The resultant molten metal is blown away by the generated velocity. Higher current densities and a narrower kerf have been achieved by using a water injection option instead of air, resulting in a high-quality cut surface and reduced smoke generation.

Most plasma arc cutting systems are automated for remote use. The equipment required include a torch travel system, air and gas supply, power supply (direct current of up to 1,000 amps), and remotely operated travel controls for the gas, arc, and mechanical mechanisms. The process will cut all metals, although torch use in air will penetrate thicker sections than if under water. Maximum cutting thickness and speeds for typical steels, using an automated system, are approximated in Table 18.7.

Table 18.7 Plasma Torch Performance Data

Material	Thickness (in)	Cutting Environment	Speed * (in/min)
Stainless Steel	1	Under water	100
	1 1/2		35
	3		6
	1/2	In-air	75
	1 1/2		25
	3		8
Carbon Steel	2 1/2	Under water	6
	5 1/2		3
	1/2	In-air	40
	1 1/2		20
	3		6

* Cutting speed is the maximum linear travel rate of the torch in the direction of the cut with complete single-pass penetration of the work piece.

The torches can be hand-held as long as the operator wears adequate protection (filter masks and fireproof clothing) and the fumes emitted are properly ventilated through HEPA filters. Torch operation cannot be hand-held for materials greater than 1.5-inch thick, due to the reaction force of the gas flow and the plasma jet.

Plasma arc systems are expensive to purchase -- greater than $100 K for a complete remote control system -- and are normally leased, with a certified operator, from a contractor. The consumption of gas (up to 600 ft^3/hr), the electric power supply, and replacing torch components add costs to the overall process.

2. *Oxygen Burner*: Also referred to as oxyacetylene cutting and oxytorch, the oxygen burning process consists of the flowing mixture of a fuel gas and oxygen ignited at the orifice of a torch. The fuel gas may be acetylene, methylacetylene polypropolene gas, propane, or hydrogen. Since this process refers to the rapid exothermic oxidation of the metal to be cut, only those metals that will undergo oxidation can be cut with an oxygen burning torch. These are generally the ferrous metals, including steel products such as sheet, plate, bar, piping, forgings, castings, and wrought iron products. Oxygen burning, although easily adaptable to an automated process, is normally hand-held and can be performed in air or under water. The underwater torch uses a special hollow, shielded, and insulated electrode along with an annular bell; the force of the compressed air clears the water from both the torch and the work piece. The oxytorch cuts metals in air up to 60-in thick, but the maximum under water capability is only 3.5 inches because of greater heat loss.

Due to the formation of refractory oxides (e.g., chromium oxide and aluminum oxide) with high melting point temperatures, an oxytorch is ordinarily unable to cut stainless steel, aluminum, or other nonferrous or ferrous/high percentage alloy metals. These metals can be cut only if the temperature at the nozzle point can be increased to overcome the high melting point. Representative cutting speeds for in-air carbon steel cutting by an oxygen burner are shown in Table 18.8.

Table 18.8 Oxygen Burner Performance Data for In-Air Carbon Steel Cutting (in/min)

Material Thickness	0.25	0.5	1	2	4	6	8	10	12	48
Hand-held	16-18	12-14.5	8-12	5-7	4-5	3-4	2.5-3.5	2-3	2	1.5-2
Automatic-Remote	20-26	17-22	14-18	7-9	7-9	4-6	4-6	3-4	3-4	2-3

Oxygen burners are relatively inexpensive (approximately $500), with gas being the major consumable cost. The gas consumption varies with the thickness of the metal being cut.

3. *Thermic Lance*: A thermite reaction lance is an iron pipe packed with a combination of steel, aluminum, and magnesium wires through which a flow of oxygen gas is maintained. All constituents of the pipe are consumed by a thermite reaction at the tip of the pipe, creating temperatures of 4,000 to 10,000° F, depending on the environment (in air or under water). A lance is typically 10.5 ft long with a diameter of 1/8 or 1/4 in, and is used only in a hand-held mode. Operators must be protected by fireproof clothing and a face shield or respirator.

The thermic lance can cut any metal, limited only by the ability to keep the kerf clean of the molten metal generated. Operable in air or under water, the lance is considered a gross manual cutting instrument and has limited application on highly activated and contaminated components. Although well suited for achieving cuts on irregular surfaces with minimum access, the lance produces significant smoke and therefore requires ventilation control.

The lance can achieve a metal cutting rate of a 1-inch diameter hole at the rate of 12 inches per minute, provided the molten metal is kept clear of the kerf. A lance is normally consumed when it reaches 2 feet in length. The oxygen flow is then stopped and a new lance is inserted into the holder. Several lances can be used with the same holder if the application warrants.

Lances are basically inexpensive; the components (lance holder, lance, and oxygen) can be purchased for less than $100. However, each lance burns for only 6 minutes and consumes about 60 ft^3 of oxygen. These factors must be evaluated for the intended use.

4. *Explosive Cutting*: Explosive cutting is a method of segmenting metal or other materials with an explosive formed in a geometric shape especially designed and sized to produce the desired separation of the work piece. The explosive cutter consists of an explosive core (normally RDX) surrounded by a casing of lead, aluminum, copper, or silver. The cutting is accomplished by the high explosive jet, detonation products, and deformed casing metal, which form a directed shock wave to cut the target material.

   Usable on virtually any material, explosive cutting has segmented materials up to 6 inches thick in air and under water. The technique is limited only by the effect of the blast on the mechanical integrity of the surrounding structures and the ability to control the spread of contaminants. Explosive cutting of contaminated components and systems may be used for the following unique applications:

   - Where simultaneous cuts must be made
   - Where other cutting techniques do not have sufficient access
   - In high radiation zones where long-handled tools are used to position the explosive cutters

   Explosive cutting is expensive if used extensively, mainly due to the need to hire qualified explosive contractors. The explosives themselves vary greatly in cost, normally $20-$50/ft of explosive, and the contractor services can be expected to charge $100-$200/hr. The total cost will vary widely, depending on the actual application.

5. *Laser Cutting*: A high-power gas, dynamic, continuous wave CO_2 laser generates 10.6 micro-millimeters (far infra red) of radiation that can be focused by using reflecting optics to produce power densities in excess of 106 watts/cm^2. This focused power melts and vaporizes the metal that is then removed from the fluid zone via a high-velocity, assist-gas jet. This assist gas (H_2, CO_2, N_2, or O_2), in the case of ferrous metals, produces an exothermic reaction that greatly increases the speed and depth of the cut. The laser cutting system consists of a generator, pumps, high voltage supply, gas supply and cooling system, beam handling optics, focusing optics, and cutting nozzle assembly. For reference, a 15 to 20 kW laser system would require a 270 kVA, 3-phase power supply.

A higher power CO_2 laser is capable of cutting metals up to 2-in thick, and produces a very narrow kerf and small heat-affected zones, minimizing the quantity of secondary waste products. Remote control operation of lasers normally operates at about 2 kW of power. Underwater operation is not feasible since the water would excessively diffuse the laser, making it ineffective. Table 18.9 lists approximate cutting speeds for high-power lasers based on material types and thickness.

Table 18.9 High-Power Laser Cutting Performance Data

Material	Thickness (in)	Cutting Speed (in/min)
Aluminum	0.50	100
Carbon Steel	0.50	60
	0.75	200
Stainless Steel	0.25	200
	0.50	55
	1.00	20
	2.00	2

A high-power laser cutting system able to cut steel and stainless steel 2 to 3 inches thick would cost more than $1M. Currently cost-prohibitive, it is the object of research to develop a less expensive laser cutting system.

6. *Arc Saw*: The arc saw is a circular, toothless saw blade that cuts any conducting metal work piece without touching it. The cutting action is obtained by maintaining a high current electric arc between the blade and the material being cut. The blade, made of any electrical conducting metal such as tool steel, mild steel, or copper, rotates at 300 - 1800 rpm, causing removal of the molten metal created by the arc in the kerf of the cut. The molten metal then condenses in the form of highly oxidized pellets as it is expelled from the kerf. The depth of the cut, up to 3 feet, is determined by the blade diameter. The arc saw can be operated in air or under water with equal success.

   This saw can cut any electrical conducting material. High conductivity materials (e.g., stainless steel, high alloy steels, aluminum, copper, and Inconel) produce the best results. Most materials are cut rapidly and cleanly, although carbon steel cuts produce slag buildup in the kerf, impeding the cutting rate. Other materials, such as magnesium, titanium and zirconium, will produce hydrogen gas when cut, resulting in the possibility of small localized ignitions.

   This system can be used under water or in air, but the former is preferred since in-air cutting produces significant amounts of smoke, greater noise, and a rougher cut. Cutting in air requires adequate ventilation controls to filter the resultant vapors. Underwater cutting produces bubbles that may carry contaminated particles and vapors to the water surface, increasing the radiological concerns on the operating level.

The arc saw has been used to cut solid metals up to 8-inch thick and complex assemblies up to 12-inch thick. Cuts can be made in vertical or horizontal planes with small reaction forces. The most pressing problem associated with the arc saw with regard to decommissioning is the size of the saw head assembly required to support a blade (up to 40-inch diameter) that can cut the wall thickness of a reactor vessel (up to 11 inch). Table 18.10 specifies system characteristics of the arc saw, Table 18.11 presents applications, and Table 18.12 lists attainable cutting speeds.

Table 18.10 Arc Saw System Characteristics

Saw head size (in)	7	12	16
Max Blade Diameter (in)	30	50	72
Max Depth of Cut (in)	9	15	30
Current Capacity (amperes)	6,000	15,000	25,000
Operating Voltage Differential (volts)	25	25	25
Weight of Head and Motor (lbs)	400	1,170	2,729

Table 18.11 Arc Saw System Applications

Material Applicability	All conducting metals
Material Thickness	Up to 36 inches
Relative Cutting Speed	Rapid
Material Geometry Limitation	None
Cutting Environment	Under water or in-air
Preferred Cutting Environment	Under water
Major Drawback	Special access to accommodate blade diameter
Cost	High

Table 18.12 Arc Saw Cutting Speed

Material	Cutting Speed *
Stainless Steel	1750 cm^2 / min
Tool Steel	1750 cm^2 / min
Mild Steel	1750 cm^2 / min
Aluminum	5000 cm^2 / min

* Area of cut surface parallel to the plane of the blade as a function of time

The arc saw, while potentially viable for segmenting components during decommissioning, has not been widely used. After being evaluated at the Shoreham Nuclear Plant, it was eliminated for the following reasons:

- Too large to fit between the bioshield and annulus for cutting vessel nozzles
- Difficult to deploy
- Cost too high
- Significant generation of smoke, fumes, and dust
- Industrial safety concern due to large amperes (20,000-40,000) required for cutting large bore piping.

Other decommissioning projects, including the Trojan Nuclear Plant, have evaluated the arc saw but opted for other, usually less expensive, tools. But the technology is still available and the concept is still being explored. Argonne National Lab is using a small arc saw for demonstration purposes and hopes to develop it into a viable option soon.

18.2.2.2 Mechanical: Mechanical segmentation of metals is a generalization for any technique that cuts by making direct contact with the surface and uses the cutting edge of the tool to abrade the material. These mechanisms may employ different power sources (hydraulic, electrical, air, water, or manual), but all must contact the surface to achieve desired results. In direct contrast to most thermal techniques, mechanical devices are basically inexpensive but labor intensive. Most tools are severely limited to thinner metals and small diameters, and only a few may be operated remotely. Since most of these techniques require operation (or set up) in a hand-held mode, the accumulation of personal radiation exposure is increased due to the proximity of the radiation source.

Despite some obvious drawbacks, mechanical devices for cutting metals are a necessary addition to the tool inventory of a decommissioning project. Many items simply cannot be segmented any other way. In addition, more elaborate tooling means higher expenditures and the cost factor weighs heavily in the selection of tool use. The following mechanical metal cutting techniques are currently in use in the decommissioning industry:

1. *Hacksaws and Guillotine Saws*: Hacksaws and guillotine saws use mechanical methods for segmenting metals by means of the reciprocating action of a hardened steel saw blade. Two major advantages of using these devices (as opposed to thermal) are that fire hazards are reduced and radioactive contamination control is simpler (absence of fumes and gases). These types of saws are usually selected for soft metals like carbon steel, aluminum, or copper, and are an obvious choice during decommissioning due to their low operating cost, high cutting speeds, and ease of contamination control.

   Both types of saws can be operated by clamping onto the pipe and allowing the unit's weight to advance the saw into the work piece. In addition, both types may also be air or electrically driven. Setup time for these saws is minimal and they may be configured to operate without operator assistance, reducing radiation exposure to workers.

   Portable powered hacksaws can cut piping up to 14 inches in diameter. An 8-inch diameter, Schedule 40 pipe can be cut in 6 to 8 minutes. Guillotine saws can cut pipes from 2 to 18 inches in diameter at a speed of about 1 minute for each inch of pipe diameter, based on

Schedule 40 pipe. Large stationary hacksaws may be set up to segment pipe for packaging. Weighing up to 5 tons, these units are electrically operated, can cut material as thick as 25 inches at cutting speeds of up to 17 in^2/min.

Portable hacksaws and guillotine saws cost between $4,000 and $5,000 for pneumatic or electric models, with blades costing $30 to $40 each. The larger stationary hacksaw units can cost anywhere from $25 to $100K, depending on the cutting diameter.

2. *Circular Cutter*: Powered pneumatically, hydraulically, or electrically, a self-propelled circular milling cutter moves on a track around the outside circumference of a pipe. The unit is attached to the pipe by a guide chain sized to fit the outside diameter. The cutter's blades are hardened steel and will cut metal 3 inches thick, with diameters ranging from 6 inches to 20 feet. After positioning, the cutter may be operated remotely to reduce personnel exposure. Circular cutters cost between $10,000 and $11,000, plus $170 to $300 each for the blades.

3. *Abrasive Cutter*: With the same power options as the circular cutter, the abrasive cutter is a disk formed of resin-bonded particles of aluminum oxide or silicon carbide reinforced with fiberglass for strength. It severs the work piece by grinding the metal away, leaving a clean kerf. Cutting rates for the abrasive cutter are about 1 inch every 7 seconds for a two-inch diameter pipe. Piping is limited to a 2-in diameter because of the reaction force between the operator and the work piece.

   Contamination control can be a big problem with the abrasive cutter because of the generation of sparks and the removal of the swarf particles in very small pieces. Normal operation requires a contamination control envelope and protective clothing for the workers, including respiratory protection. Within a contamination control envelope, larger units capable of cutting up to 8-in diameter piping may be situated to prepare long sections of pipe for packaging. A portable abrasive cutter is relatively inexpensive, about $600 to $1K, depending on setup, while a large stationary unit will run $15 to $20K.

4. *Power Nibblers and Shears*: A nibbler is a punch and die cutting tool that normally operates with a rapid reciprocation rate of the punch against the die, "nibbling" away at the metallic work piece. A shear is a two-bladed tool that operates on the same principal as a pair of scissors. Both tools are normally used for cutting sheet metal, but adaptations can be made to increase their cutting abilities to allow for cutting heavier materials.

   Shears and nibblers can be used in air or under water, although precautions have to be taken to control the air bubbles generated under water when the unit is equipped pneumatically. Heavy duty nibblers and shears cost between $1,800 and $2,500, with remote operating extensions at an additional cost.

18.2.3 Component Removal: A major activity in any decommissioning project is removing components. Every building in the facility has main and support systems often consisting of thousands of feet of piping, large numbers of pumps, valves, tanks, heat exchangers, and operating controls. Large components such as a reactor vessel, steam generators, and pressurizers present unique problems of removal and disposal. The planning and engineering for removal and disposal of every component must be performed to ensure a successful decommissioning program. Techniques for the D&D of structures and components were discussed in the previous chapters and sections of this handbook. This section presents specific techniques for the component removal.

18.2.3.1 Administrative Controls: Before any work starts, administrative controls must be established. The foundation for establishing these controls is as follows:

- Safety of the work force
- Preservation of plant equipment needed for continued use
- Occupational radiation exposure control (ALARA)
- Minimization of the spread of radioactive and hazardous contaminants
- Protection of the public from adverse health effects
- Performance of work activities in a cost-effective manner
- Waste minimization

To these ends, an administrative control program should accomplish the following:

- Isolation of electrical and pneumatic services from components before their removal
- Sealing of internally contaminated components to confine internal contamination
- Decontamination/dismantling of contaminated systems, structures, and components by:
    - In-situ decontamination
    - Removal and decontamination
    - Removal and disposal
- Removal of contaminated system supports in conjunction with equipment removal or decontamination
- Removal of contaminated systems and components from areas and buildings before structural decontamination (block shield walls, or portions of other wall, ceilings, or floors may be removed to permit removal of systems and components)
- Removal or decontamination of embedded contaminated piping, conduit, ducts, plates, channels, anchors, pumps, and sleeves during structural decontamination
- Consideration of localized or centralized processing and cutting stations to facilitate packaging of the components removed as large pieces
- Intact removal of small components, where feasible. This includes most valves, small pumps, small tanks, filters and heat exchangers. These components may then be decontaminated in whole or part and reduced in size to facilitate disposal or release
- Using existing systems and services to support decommissioning operations

These controls should enable adequate preparation for component removal.

18.2.3.2 Waste Considerations: Removal generates large amounts of radioactive waste, both irradiated and contaminated material, consisting of the components themselves as well as secondary waste from the removal actions.. Therefore, it is important to think about waste generation *before* the actual removal.

Radioactive waste resulting from decommissioning can be grouped in three major categories:

1. *Neutron-activated*: includes the reactor vessel, its internal components (e.g., core shroud, core support assemblies, and control-rod guide tubes), and the inner portion of the concrete biological shield

2. *Contaminated*: includes the piping and equipment in the facility structures such as the reactor containment, fuel handling, and auxiliary buildings. In addition, some concrete surfaces of these buildings are expected to be radioactive and will require removal and disposal

3. *Secondary and dry active*: includes the swarf material removed from the kerf of a cut, cutter materials used in segmenting such as abrasives in water jet cutting, and the contaminated tooling used to cut or segment the material. Dry active waste includes discarded contaminated items such as rags and wipes, tools, and protective clothing

Decontamination will reduce the volume of material requiring controlled disposal. Some building materials and components can be removed, decontaminated, and disposed of as clean waste by applying techniques discussed in Chapter 17. The secondary waste streams contribute little to the overall resulting radioactive volume. The equipment, once cleaned, could regain a certain intrinsic value as either a recyclable or reusable commodity. To this end, it can be cost-effective to plan and implement a vigorous decontamination effort to free-release as much material as possible.

18.2.3.3 ALARA Considerations: Before starting removal activities, it is necessary to determine the radioactive inventory of the plant components to choose the most effective method of removal. Specifically, the radioactive inventory must determine the following:

- Radioactive waste streams and ultimate disposition
- Decommissioning techniques to be used
- Exposures to the decommissioning work force
- Desirability of a delay period before removing components
- Need and advantages of performing in-situ or remote decontamination
- Feasibility of decontamination versus intact removal and direct disposal
- Non-occupational radiation exposures
- The key factors for an environmental impact assessment

Once the radioactive inventory has been established for all components to be removed, the data are used to compute the estimated person-rem exposure for each task. This occupational exposure data is used in planning activities subject to ALARA, i.e., work orders and radiation work permits, shielding calculations, and manpower scheduling.

18.2.3.4 Removal of Radioactive Inventories: Excluding spent fuel, radioactive inventories at the time of shutdown are of two types: neutron-activated components in and surrounding the reactor core, and radioactive surface contamination from fission products and activated corrosion products (either deposited on the inside of piping and equipment or attached to structural surfaces). The complete dismantling of the facility requires the complete removal of these inventories. Each of the following components can be removed, segmented, and disposed of in a controlled manner using the techniques described in this chapter.

1. *Neutron-Activated Components*: All neutron-activated materials are located in the primary containment building, which houses the reactor vessel and primary system. Most of the vessel internals and portions of the reactor vessel and the sacrificial (or biological) shield contain the neutron-activated materials. Due to the high activity associated with the neutron-activated components and structures, they must be disassembled and sectioned to fit within suitably shielded shipping casks (possible exceptions to this requirement will be noted).

2. *Reactor Vessel and Internals*: Reactor vessels for commercial facilities are normally either BWRs or PWRs. Each has unique characteristics that are dependent on the fuel design, generating capacity, manufacturer, etc. It is not practical to describe dismantling methods for every known reactor type. Rather, this chapter describes generic techniques applicable to the removal and disposal of a reactor and its internals, with the understanding that this data is intended to be generic and should be viewed as such.

   The dismantling and disposition of the reactor vessel and its internals can be accomplished by segmentation or as an intact removal (with or without the internals remaining in place). Before starting to remove vessel and internals, due consideration must be given to the activated/contaminated radwaste that will be generated. The high curie content of the vessel and its internals will greatly affect disposal requirements due to the need to classify the waste forms generated. It will require:

   - A more detailed and accurate analysis of the activation history and component composition and weight
   - The use of personnel shielding and remotely operated segmentation tools
   - Remote handling of segments
   - Shielded transportation canisters
   - Use of supplementary shielding
   - Additional waste stabilization

3. *Reactor Vessel Segmentation*: With the above information as a guide, the activities required to dispose of the reactor vessel and internals include:

   - Identify activated regions of reactor vessel and internals
   - Determine radionuclide inventories of vessel/internals by individual structures or regions. Use this data to provide composition, weight, and curie content for each component. Then apply the results directly to the determination of segment size, packaging requirements, etc.
   - Establish segmenting schemes to satisfy the individual cask or box loading requirements
   - Determine, based on the total number of components or segments, the total number of shipments required for disposal
   - Establish a contractual agreement with the selected disposal site to accept the waste; comply with disposal site requirements

   Loading requirements, subject to the constraints of curie content, package size and weight, and total weight of shipment, must be considered before dismantling.

18.3 DOE TECHNOLOGY PROJECT RESULTS

The DOE sponsored a Large Scale Demonstration and Deployment Project to develop and test new and innovative technologies potentially applicable to decommissioning but that had not been previously applied to its sites or to the nuclear industry. In some cases, the technologies had been applied in other industries, but performance and test data were not available for situations and applications expected to be encountered in nuclear decommissioning.

Results of this DOE program were reported in a series of documents called Innovative Technology Summary Reports, or in other summary documents. Each report presented the full range of problems to be addressed and the technology's advantages to the DOE cleanup program in terms of system performance, cost, and effectiveness. Most reports included comparisons to both baseline and competitor technologies.

The large volume of information provided in the dismantling category of these reports necessitated collecting the offerings on a CD. Accordingly, the CD provided with this handbook presents summaries of the DOE's dismantling technologies, as extracted from its demonstration program and subsequent reports. Each technology also includes links to its fact sheet, related technical documents, photographs, and points of contact

As in all information available in literature, from test programs, or from any secondary source, planners are cautioned to research technology and incorporate long lead times for required product demonstrations at the specific facility undergoing decommissioning. Never lose sight of the fact that the schedule of critical path activities (especially at a commercial decommissioning project) cannot be altered to suit a research and development program at the expense of the owner.

Chapter 19

WASTE MANAGEMENT AND TREATMENT

19.1 INTRODUCTION

Decommissioning and remediation projects demand an integrated, life-cycle approach from planning through waste management. Concepts that must be clearly understood in developing a comprehensive waste management program include the various categories of waste, waste minimization, regulatory issues, and differing management approaches. As with most aspects of decommissioning, waste management provides opportunities for optimization. No one solution works in all situations and facilities, but understanding the differing approaches will let the manager make prudent and efficient decisions to best support stakeholder needs. Management of tritium waste is sufficiently unique that it is discussed separately in Section 19.7

19.2 WASTE STREAMS

The DOE is, by far, the largest source of waste from decommissioning. Research, development, and weapons production are the source of large quantities of waste located at an extensive network of national laboratories and other facilities. Detailed characterizations, defining composition of both the waste matrix and the hazardous and radioactive contaminants, is not generally available. The legacy waste streams of the DOE Complex generally include many of the same hazardous and radioactive contaminants in the commercial nuclear industry, plus some unique waste streams that result from specialized processes.

Waste streams at nuclear power reactors may be broken down into the following categories:

1. *Surface contaminated items* become contaminated either internally or externally due to exposure to radioactive materials. Often the contamination is a result of leakage of fluids and subsequent plate out of radionuclides on the material. The distribution of radionuclides for this waste stream remains relatively constant unless an event, such as significant fuel failure, changes the distribution. Taking and analyzing representative samples from the typical waste stream generally can be used to determine the radionuclide distribution.

2. *Activated materials* generally consist of steel, concrete, and other items resulting from neutron activation. The radioactivity associated with this material is incorporated into its structure. Activated steel may have high specific activities with dose rates that may exceed 1000 Rem/hr. Each type of steel (e.g., carbon steel, 304 SS, 321 SS, rebar, satellite, or zircalloy) has a different chemical makeup and impurities. Due to the high specific activity, direct sampling and analysis of activated steel may not be appropriate. The use of a properly documented activation analysis may be the only option for identification and quantification of such activated metal.

The quantity and depth of activated concrete depends on the reactor's neutron leakage rate. The activated band of concrete in a bioshield seldom exceeds a depth of 4 ft from the vessel wall. Signature radionuclides associated with activated concrete include various isotopes of Eu. Lead shielding is another potential source. However, activated lead typically has a relatively low specific activity.

3. *Water processing streams*, such as filters and sludge, result from the processing of coolants and moderators.

Material accountability or classification by source methodologies is used to establish waste classification at commercial operations, universities, and hospitals. Generally, the license issued to these facilities is specific to the type and quantity of radioactive material authorized to be in the facility and either decay or shipment usually accounts for the radioactive material.

19.3 LEGAL REQUIREMENTS AND CLASSIFICATION

Waste management and treatment is significantly affected by regulatory requirements. The Atomic Energy Act (AEA 1954) and its numerous subsequent amendments govern the management of radioactive waste in the U.S. Nuclear power plants and all other operations outside the government are covered by regulations of the Nuclear Regulatory Commission (NRC). Historically, the DOE has been a self-regulating agency; its operations follow DOE requirements.

Facilities undergoing decommissioning often have radioactive and other regulated wastes. The two major federal laws governing management of these other regulated wastes are the Resource Conservation and Recovery Act (RCRA 1976) and the Toxic Substance Control Act (TSCA 1976). Whereas waste regulated under the RCRA is hazardous waste, a few waste constituents are regulated under the TSCA, such as PCBs and asbestos. In addition, various states are authorized to have their own respective RCRA regulations, making management of these wastes complex.

A class of waste requiring significant attention is known as mixed waste. Mixed waste (radioactive and hazardous) has constituents covered under the AEA and RCRA, or the AEA and TSCA.

19.3.1 Radioactive Waste Classification:

19.3.1.1 Low-Level Waste: Commercial low-level wastes (LLW) are classified as Class A, B, and C, based on the waste's concentration of radioactivity. Regardless of the waste classification, disposal facilities prohibit or limit the presence of pyrophoric and explosive agents. Similarly, the presence of pathogens and excessive free liquids require treatment before disposal. Typically, disposal facilities provide waste acceptance criteria establishing individual limits for each of these requirements and others. Typical waste acceptance criteria are listed in Table 19.1.

19.3.1.2 GTTC Wastes: Radioactive wastes with concentrations of radionuclides in excess of Class C waste are designated Greater Than Class C (GTCC) wastes. The quantity of GTCC waste within the country's commercial sector is small. Type 304 stainless steel has been used in commercial reactors in the active core region, and its constituent of ^{63}Ni may constitute a GTCC limiting radionuclide. Water processing filter media, including filter cartridges and resin, is used to remove impurities from liquids to maintain quality and clarity. The potential exists to concentrate radionuclides at levels that are GTCC.

Table 19.1 Classification of LLW

	Concentration Limit		
Radionuclide	**Class A**	**Class B**	**Class C**
Total of T ½ >5 y (Ci/m^3)	700	none	none
^{3}H (Ci/m^3)	40	none	none
^{14}C (Ci/m^3)	0.8	> 0.8	> 0.8
^{14}C in activated metal (Ci/m^3)	8	> 8	> 8
^{59}Ni in activated metal (Ci/m^3)	22	> 22	> 22
^{60}Co (Ci/m^3)	700	(2)	(2)
^{63}Ni (Ci/m^3)	3.5	70	700
^{63}Ni in activated metal (Ci/m^3)	35	700	7000
^{90}Sr (Ci/m^3)	0.04	150	7000
^{94}Nb in activated metal (Ci/m^3)	0.02	> 0.02	> 0.02
^{99}Tc (Ci/m^3)	0.3	> 0.3	> 0.3
^{129}I (Ci/m^3)	0.008	> 0.008	> 0.008
^{137}Cs (Ci/m^3)	1.0	44	4600
TRU with T ½ > 5 y (nCi/g)	10	10 -100	10 - 100
^{241}Pu (nCi/g)	350	> 350	> 350
^{242}Cm (nCi/g)	2,000	> 2,000	> 2,000

19.3.1.3 HLW and Spent Nuclear Fuel: High-level wastes (HLW) are generated during the chemical processing of spent nuclear fuel. Efforts are in progress in the U.S. for the finalization of a geological repository in Yucca Mountain, Nevada. In the interim, commercial facilities store their SNF on site, either in wet or dry storage. The DOE is making extensive efforts to progress to the vitrification of its HLW and process its SNF for eventual disposal at Yucca Mountain.

19.3.1.4 Transuranic Wastes: Transuranic (TRU) waste is normally the result of reprocessing spent nuclear fuel for recovery of ^{239}Pu. It may also be generated as a consequence of accidents and mishaps. In the U. S., TRU waste is defined as that having concentrations > 100 nCi/g of radionuclides of elements above 92 (uranium) with a half-life longer than 20 years.

19.3.2 Hazardous Waste: A complex statutory definition of hazardous waste was established under the RCRA (§ 1004(5)), paraphrased here as: a solid waste, or combination of solid waste that, because of its quantity, concentration, or physical characteristics, may cause an increase in mortality or illness, or pose a hazard to human health or the environment. The Environmental Protection Agency (EPA) developed a regulatory definition from this broad statutory definition that is described in 40 CFR 261 and incorporates both listings and characteristics (EPA 2003). The listings presented herein describe specific types of wastes considered dangerous enough to require regulation; the characteristics identify properties that, when present, pose such significant threats as ignitability, corrosivity, reactivity and/or toxicity.

19.3.2.1 Ignitability: Solid waste exhibits the characteristic of ignitability (per 40 CFR 261.21) if a representative sample has any of the following properties:

- It is a liquid, other than an aqueous solution containing less than 24% alcohol by volume and has a flash point less than 60° C.
- It is not a liquid, and is capable (under standard temperature and pressure) of causing fire through friction, absorption of moisture, or spontaneous chemical changes, and (once ignited) burns so vigorously and persistently that it creates a hazard.
- It is an ignitable compressed gas as defined in Department of Transportation (DOT) regulations (DOT 2003a).

Potential sources of mixed waste exhibiting the characteristic of ignitability include:

- Contaminated gasoline
- Contaminated cylinders of acetylene
- Uranium/Thorium turnings
- Uranyl nitrate

19.3.2.2 Corrosivity: Solid waste exhibits the characteristic of corrosivity (per 40 CFR 261.22) if a representative sample has either of the following properties:

- It is aqueous and has a pH of < 2 or > 12.5, as determined by a pH meter.
- It is a liquid and corrodes steel at a rate greater than 6.35 mm/y at a test temperature of 55° C, as determined by a test method specified in NACE Standard TM-O 169-2000 (NACE 2000).

Potential sources of wastes with the characteristic of corrosivity include:

- Wet cell batteries from emergency lighting
- Acids and bases from radiochemistry laboratories
- Acids and bases from water treatment facilities
- Undiluted cleaning compounds

19.3.2.3 Reactivity: Solid waste exhibits the characteristic of reactivity (per 40 CFR 261.23) if a representative sample has any of the following properties:

- It is normally unstable and readily undergoes violent change without detonating.
- It reacts violently with water.
- It forms potentially explosive mixtures with water.
- When mixed with water, it generates toxic gases, vapors, or fumes in a quantity sufficient to present a danger to human health or the environment.
- It is a cyanide- or sulfide-bearing waste that, when exposed to pH conditions between 2 and 12.5, generates toxic gases, vapors, or fumes in quantity sufficient to present a danger to human health or the environment.
- It is capable of detonation or explosive reaction if it is subjected to a strong initiating source or if heated under confinement.
- It is capable of detonation or explosive decomposition or reaction at standard temperature and pressure.
- It is defined as an explosive by regulations of the DOT (DOT 2003b).

Potential sources for wastes with the characteristic of reactivity include:

- Sodium at certain reactors
- Compressed gases
- Laboratory chemicals

19.3.2.4 Toxicity: Solid waste exhibits the characteristic of toxicity (40 CFR 261.24) if, using the toxicity characteristic leaching procedure, the leachate concentration exceeds the limits shown in Table 19.2.

Table 19.2 Maximum Concentration of Contaminants for the Toxicity Characteristic

Contaminant	Limit (mg/L)	Contaminant	Limit (mg/L)
Arsenic	5	Hexachlorobenzene	0.13
Barium	100	Hexachlorobutadiene	0.5
Benzene	0.5	Hexachloroethane	3
Cadmium	1	Lead	5
Carbon Tetrachloride	0.5	Lindane	0.4
Chlordane	0.03	Mercury	0.2
Chlorobenzene	100	Methoxychlor	10
Chloroform	6	Methyl Ethyl Ketone	200
Chromium	5	Nitrobenzene	2
ortho-Cresol	200	Pentrachlorophenol	100
meta-Cresol	200	Pyridine	5
pera-Cresol	200	Selenium	1
Cresol	200	Silver	5
2,4-D	10	Tetrachloroethylene	0.7
1,4-Dichlorobenzene	7.5	Toxaphene	0.5
1,2-Dichloroethane	0.5	Trichloroethylene	0.5
1,1-Dichloroethylene	0.7	2,4,5-Trichlorophenol	400
2,4-Dinitrotoluene	0.13	2,4,6-Trichlorophenol	2
Endrin	0.02	2,4,5-TP (Silvex)	1
Heptachlor and its epoxide	0.008	Vinyl Chloride	0.2

Potential sources of waste exhibiting the toxicity characteristic include:

- Mercury contained in thermostats, mercury pressure switches on plant systems and phone switch panels, thermometers, light bulbs, and manometers
- Lead sheet, shot, blankets, or bricks used for shielding, and as a component of paint
- Cadmium in neutron shielding, and as a component of paint and lubricants
- Maintenance shops and chemistry labs: solvents, canned spray paint, oil based paints, dye penetrant test kits, asbestos test kits
- Chromium in lubricants for motors and in high chromium alloys

1.

19.3.3 Other Regulated Wastes: This group of wastes includes those listed under the RCRA and under the TSCA.

1. *F-Listed Hazardous Waste*: This group consists of wastes from nonspecific sources, and may come from various industrial processes. The list includes solvents used in degreasing metal treatment baths and sludge, and wastewaters from metal plating operations.

2. *K-Listed Hazardous Waste*: Wastes from specific sources are included in this group. Examples of industries that may generate K-listed wastes are wood preservation, explosives manufacturing, production of pigments, and iron and steel production.

3. *P-Listed Hazardous Waste*: Discarded commercial chemical products, off specification products, and their residues in containers and elsewhere. The hazards associated with P listed waste are determined to be acute.

4. *U-Listed Hazardous Waste*: The same sources as for the P-listed waste constituents, except that those included herein are determined to be toxic.

5. *TSCA Waste*: In addition to wastes regulated as hazardous under the RCRA, two other waste streams regulated under the TSCA are of concern for waste management during decommissioning -- the PCBs normally found in transformer oils and asbestos, used as insulation in a number of facilities.

19.4. WASTE MANAGEMENT CONSIDERATIONS

19.4.1 Cost-Benefit Analysis: Economics is a primary driver in waste management. In its most rigorous form, the cost-benefit analysis encompasses several steps. Exercising a full cost-benefit analysis regimen can be time consuming and expensive, but accurately defining the problem (Step 1 here) often results in direct resolution. Comparing alternatives and near-term costs for decommissioning projects becomes a manageable problem if the following steps are considered:

1. *Define the problem*: The problem includes the set of conditions, materials to be treated, treatment alternatives, and potential goals. Proper definition may render the decision directly (with little further analysis) once the full set of options and drivers are known. For example, should the problem be defined as TRU waste, the packaging, waste acceptance criteria, and disposal site alternatives would be determined and driven exclusively by access to the federal government's TRU disposal site at the Waste Isolation Pilot Project (WIPP).

2. *Set goals and objectives in advance*: Before attempting any analysis and resolution, identify the project objectives, including criteria for success and the agenda for including as many stakeholders (explicit or hidden) as possible. Otherwise, an apparent resolution to the problem or approach to the project may trigger a new set of stakeholder debates regarding the meaning of the solution that were not initially included. Different stakeholders have different objectives, and all require initial consideration to derive a meaningful solution.

3. *Identify all reasonable means to attain the objectives*: The key is to offer all "reasonable" approaches without being too restrictive to be meaningful. For radioactive waste management projects, the number of acceptable alternatives will normally be limited to well under five (often under three).

4. *Analyze the benefits of the alternatives*: This analysis should be in terms of the objectives of the overall project and requirements for waste disposal. D&D or remediation project approaches (such as a leaching approach that is heavily based on application of chelants) that result in wastes with no direct route for disposal are of no use to anyone.

5. *Analyze the costs*: This is often the toughest step. However, in the case of radioactive waste management, quantification normally is reduced to a selection from among predetermined options (e.g., waste forms and disposal sites). Secondly, a firm analysis of cost, supported by pro forma analyses, is standard procedure for most corporate bid decisions (not only the go/no-go decision, but also the choice of projects on the basis of highest return among several alternatives).

6. *Specify the perspective*: This is likely to vary among the various stakeholders. This is discussed in further detail in Chapter 8.

7. *Perform all cost evaluations in terms of net present value*: Discount all future costs and benefits to current dollars.

8. *Analyze uncertainties*: Emphasize steps 4, 5, and 7, where assumptions will have been made regarding costs, appropriate discount rates for the risks involved, and the like.

9. *Address the ethical situation of the solution*: This might involve social or environmental justice, particularly for large, problem-oriented procurements.

10. *Interpret the results.*

19.4.2 Strategies and Controls: Waste management strategies and controls vary by the type of facility being decommissioned. The strategies and controls that apply to most facilities include the following:

- A plan detailing the radiological and chemical characteristics of the site and the items for disposal should be developed. This detail is needed for cost estimation and demolition planning.
- A plan detailing the waste processing or disposal facilities to be used should be developed. This detail is needed for cost estimation and packaging planning.
- A plan detailing the waste streams to be consigned to each facility should be developed. This detail is needed for cost estimation, demolition planning, and packaging planning.
- A transportation plan should be developed to identify licensed and approved transporters. This should include planning for over sized components, such as reactor vessels and concrete segments.
- On-site material handling requirements should be identified.
- The spread of radioactive contamination should be minimized during demolition.
- Hazardous waste constituents should be segregated from radioactive materials, if possible, during demolition.
- Fluids should be segregated from solid materials during demolition.

19.5 TREATMENT, HANDLING, AND STORAGE

Waste treatment must consider the significance of regulatory requirements, which are based on two fundamental approaches:

1. *Technology-based*: Many regulations dealing with hazardous and toxic wastes are based on the application of a specific technology. Although virtually all regulations make provisions for application of alternative technologies, in practice, proving equivalency with a chosen technology can be too elaborate to be practical. Obviously, engineering judgment is necessary to apply these technologies.

2. *Result-based*: Treatment of radioactive waste is largely performance-based by identifying a given goal in terms of human health or ecological risks balanced against an optimum or acceptable cost.

Because of fundamental differences in treatment of radioactive and hazardous (including TSCA) waste, treating mixed of wastes, i.e., wastes containing both constituents, poses unique problems. Whenever possible, avoid generating mixed waste! Most facilities contain a variety of materials that may be radioactive, including the following:

- Miscellaneous chemicals, such as caustics, acids, grease and paints
- Fluids, such as water, sludge, oil, solvents and antifreeze
- Asbestos in insulation, floor tiles and wall panels
- PCBs in paint, transformers and lighting
- Mercury in switches, lighting, and thermostats
- Lead used for shielding
- Steel in piping and structures
- Aluminum in conduit and cabinets
- Concrete structures
- Copper or brass in piping, wiring, and motors
- Wood in structures and cabinets

19.5.1 On-Site Material Handling: Decisions on material handling must be made before demolition. Major required decisions include the following:

1. *Expected disposition of materials*: Materials may be processed on or off site, or sent directly to disposal.

2. *Physical dimensions of material removed*: Sizes of components must be acceptable for transportation off site. Additionally, the component size must be acceptable to the receiving facility, be capable of handling by individuals removing them, and fit the selected container.

3. *Weight of material removed*: The weight of the components should be acceptable for transportation off site. Additionally, the weight must not exceed the capacity of on-site handling equipment, such as cranes, rigging, or forklifts, and must be acceptable to the receiving facility.

4. *Trained people*: People sufficiently trained and aware of waste packaging requirements should perform package loading. This may be accomplished by pre-job briefings, general employee training, or direct observance by appropriate personnel.

5. *Package loading*: Loading packages should occur inside buildings, if possible. Packages loaded in outside areas should be closed when loading stops to prevent the intrusion of moisture.

During demolition, the efficient use of on-site labor requires that materials be handled the least number of times possible. Additionally, the waste must be packaged to efficiently use space and be acceptable for the receiving facility. Packaging for the waste stream being removed should be available as close to the work area as possible. A generally acceptable sequence for removing components and equipment from an area is as follows:

1. Hazardous materials or hazardous substances, such as mercury, lead, PCBs, and asbestos are removed before general demolition; this prevents subsequent contamination of the waste with the materials, and in the case of asbestos removal reduces protective clothing and monitoring requirements for subsequent work.

2. The higher dose rate components are removed; this reduces total worker doses and should reduce worker control requirements.

3. Components that interfere with access are removed; this improves worker efficiency.

4. The area is cleared of internal components or contamination.

5. The structure is demolished.

After a package is loaded, the contents and exterior should be inspected. The contents should not be able to shift or contain freestanding liquid. The exterior should not be damaged. Required gaskets and closures should be installed.

19.5.2 Waste Storage: Most facilities have limited waste storage capacity. The physical size of the facility and the license activity limits of the radioactive materials limit the capacity. Storage on site may be either long- or interim-term storage. Long-term storage for a decommissioning site is generally limited to such items as spent fuel or GTCC wastes. Interim storage of LLW is often required to allow efficient use of transport vehicle capacity.

19.5.3 Waste Prevention and Minimization: The goal of waste minimization is to reduce the cost of waste disposal and to prevent the generation of GTCC waste. The major methods are:

1. *Preventing Materials from Entering Contaminated Areas*: The most cost-effective method of waste minimization is the prevention of contamination of materials. This effort includes the removal of packing material before entry into controlled areas, use of reusable protective clothing in controlled areas, elimination of equipment modifications, and use of contaminated tools and tool rooms.

2. *Preventing Contamination of Hazardous Materials within Contaminated Areas*: Many items in facilities contain hazardous materials, including: mercury contained in vials for pressure switches, thermometers and manometers; acids; lithium; lead and cadmium in dry and wet celled batteries; PCBs in lighting ballasts and transformers; and lead sheeting or bricks. The best method for dealing with this type of material is to remove the items before they become contaminated with radioactive material (i.e., become a mixed waste). Mixed wastes are generally 10 to 100 times more expensive to process and dispose of than radioactive waste.

3. *Decontamination of Materials for Unconditional Release*: This process involves the removal of radioactive contamination from materials and/or the survey of materials for free release. This process is fairly labor intensive yet has the capability to reduce or eliminate a large volume of materials going for processing or disposal. An important part of the process is the selection of materials for the survey and release process. Acceptable materials have all areas accessible for survey and are easy to decontaminate. The risk associated with this method is the unintentional release of low levels of licensed materials. The risks and cost have rendered this alternative unattractive for most applications.

4. *Transfer of Equipment or Materials to Other Licensed Facilities for Reuse*: This process usually involves multiple-use items, such as electric motors, crane parts, scaffolding, tools, shielding, ventilation equipment, and protective clothing, respirators and fire protection equipment. In this case, the material is not waste to the receiving facility.

19.5.4 Technology Selection: A technical process and treatment plan should be prepared to select a process technology. This plan should include a chemical and physical evaluation of the waste streams, and evaluation and selection of preferred treatment options versus direct disposal. If on-site treatment is favored, a process flow diagram and a facility conceptual design should be prepared. The following paragraphs detail the approach usually performed to develop this information.

19.5.4.1 Evaluation of Waste for Treatment versus Disposal: The waste streams for processing must be reviewed with the objective being to aid an evaluation of treatment options. One possible approach may be the categorization of waste streams into organic, inorganic, metal, lead, and "special or other." Obviously, the quantity of each waste stream will affect the technology selected.

19.5.4.2 Evaluation of Treatment Options: The first thing to do is develop the evaluation criteria. Objective criteria typically include: equipment and operating costs, status of development (maturity), size, complexity, throughput capability, volume reduction factors, number of waste streams applicable for processing, effluents, secondary waste generation, schedule constraints, waste form performance, permits needed, and ownership (patents). Other criteria may include regulatory and license issues, public and stakeholder acceptance, and financing capability.

The optimal treatment processes must be determined through an options evaluation and selection process. The potential processing options for each waste stream are ranked in accordance with the evaluation criteria. All waste streams should have at least one potential treatment option. Based on the cost-effectiveness of the treatment options and the relative ranking, a preferred set of treatment options must be identified for the entire group of waste streams. A priority is given to proven, commercial processes. A backup, noncommercial technology may be carried as a further option if there are significant advantages.

A list of objective and subjective evaluation criteria should be based on the anticipated functions and attributes for typical systems and facilities for processing the expected types of wastes. Each individual evaluation criteria is given:

1. *Weight*, which may vary in percentage depending on the importance of the individual criteria to the project, and

2. *A range of raw scores* reflecting the potential merits of the alternative:

Best	5
Above Average	4
Average	3
Below Average	2
Worst	1

The total score for each treatment process is then calculated for each technology by multiplying the raw score and the weights. An example of this process, with hypothetical weighting, is shown in Table 19.3

Table 19.3 Example Evaluation Criteria Weighting

Criteria	Weight (%)	Raw Score	Total Score
1. Cost			
a. Equipment	5	(1-5)	5% times (1-5)
b. Operating Services	2	(1-5)	
c. Operating Labor	3	(1-5)	
d. Operating Consumables	2	(1-5)	
e. Development	3	(1-5)	
2. Schedule	10	(1-5)	
3. Equipment		(1-5)	
a. Complexity	5	(1-5)	
b. Maintainability	5	(1-5)	
c. Reliability	3	(1-5)	
d. Operability	2	(1-5)	
4. Hazardous Processes	4	(1-5)	
5. Volume Reduction	8	(1-5)	
6. Secondary Wastes	4	(1-5)	
7. Number of Systems	6	(1-5)	
8. Demonstrated Technology	8	(1-5)	
9. Regulatory Acceptance	10	(1-5)	
10. Public Acceptance	10	(1-5)	
11. Process Flexibility	5	(1-5)	
12. Throughput	5	(1-5)	
TOTAL	100		

19.5.4.3 Costs: Following are the evaluation criteria for the selecting treatment options:

1. *Equipment*: Relative cost for this evaluation is limited to capital costs for the equipment; there is no consideration for performance testing or installation costs.

2. *Operating Services*: Relative cost is limited to utilities (power, water, air, etc.) required to operate the equipment.

3. *Operating Labor*: Relative cost is the manpower required to operate the equipment.

4. *Operating Consumables*: Relative costs are for the materials necessary to support the equipment operation, such as containers and process chemicals or additives, but not utilities. If routine and frequent replacement of parts were necessary, such as furnace firebrick or equipment liners, these parts would be considered consumables for this evaluation.

5. *Development*: If the preferred process is laboratory or pilot scale, a relative cost will be defined to upgrade the technology to a full-scale system (including demonstration testing evaluation and implementation).

6. *Schedule*: The ability to purchase equipment, manufacture, deliver, install, commission, and inactively/actively test within the schedule needed for the overall plan will be considered. Any required development time will be included in the overall schedule for delivery. If this duration fails to meet the already established schedule objectives, the system will score low.

19.5.4.4 Equipment:

1. *Complexity*: Equipment that requires elaborate control and monitoring with careful control of material input will rank lower than processing equipment that has a higher tolerance of material input variance, is simpler to control, and is a more robust process.

2. *Maintainability*: Low maintenance equipment or equipment that can be maintained in-situ with minimal impact on processing throughput will be ranked higher than equipment requiring long periods of shutdown, or complex maintenance operations that require fabricating temporary confinement barriers to repair or replace equipment.

3. *Reliability*: Reliability is directly proportional to the quantity of moving parts required to operate the equipment. The less number of moving parts within the waste zone will result in a higher rank. A higher specification of seals, bearings, motors, and pumps will reduce the probability of equipment malfunction. The justification in terms of cost is insignificant to the processing impacts. Proven technology will, therefore, be ranked higher.

4. *Operability*: This is directly proportional to the level of qualified staff required to operate the processing equipment. Operability is closely tied to product quality and acceptance. Equipment that is relatively simple to operate will rank higher than more complex equipment.

5. *Hazardous Processes*: Processes that are potentially hazardous to the operator or personnel within the building will be ranked lower. Equipment must be capable of meeting relevant regulatory requirements or it will not be considered further.

6. *Volume Reduction*: Volume reduction may result in an overall cost savings, with significant reduction in waste containers and a better use of available storage grounds. A high volume reduction will receive a higher score.

7. *Secondary Wastes*: Secondary waste will be generated by most of the treatment processes under consideration. The lowest generator of secondary waste will receive the highest score. The level of effort and costs to treat secondary wastes will also factor in the scoring. Effluents, such as decontamination materials, will be considered as secondary wastes and will also require processing.

8. *Number of Process Systems*: Any process system capable of processing more than one waste stream will attract a higher score.

9. *Demonstrated Technology*: Evaluation of the current status of available process configuration technology will be evaluated in relation to the existing system size, either lab, pilot, or full scale. A higher score will be awarded to the process technology with full-scale experience at the projected or required throughput.

10. *Regulatory Acceptance*: Assessment of technology or equipment and the resulting waste form will consider the ability to meet federal and state regulatory requirements, including disposal. If the process will not meet requirements it will be excluded from further consideration.

11. *Public Acceptance*: Assessment of technology or equipment will consider the ability to meet with public expectations. Process equipment for each waste stream or group of waste streams will be evaluated.

12. *Process Flexibility*: Advantage is gained if the process technology can adapt to changing waste characteristics. The process with tightly tolerant constraints will score less.

13. *Throughput*: The ability of the equipment or technology to process at desired or higher rates will attain a higher score than equipment with a minimum rate for a given size of equipment. Equipment must meet minimum process throughput requirements or it will not be considered further.

19.5.4.5 Conceptual Systems Design: Once the overall processes are selected, a process block flow diagram should be prepared to detail each waste stream and its corresponding treatment process. Then a process description should be provided for each step of the process and accumulate the information from the selected options evaluation. The effluents and secondary wastes are then identified and quantified for each process. For example, for an incinerator facility, scrubber solution and filter media are expected secondary waste streams that will require processing and disposal. Proposed processing methods, such as liquid waste treatment (evaporation, ion exchange, etc.) for these secondary wastes must be selected and the resulting end products specified.

The primary equipment necessary for the selected processing system(s) must be identified. The size of the equipment for the selected throughput should be estimated. The approximate costs, both capital and operating, should be identified, based on information from similar systems or vendor information. Costs at this stage will be about plus or minus 50%. Alternatives for a larger central facility must be identified.

The first step of the conceptual design will be to establish the general design criteria (conceptual design stage) for the facility. The facility functions will be identified, along with design requirements. The results of the process system functional analysis that is performed during the process evaluation should be used. Applicable codes and standards for design of equipment should be estimated and included. General arrangement sketches should be prepared at the earliest stages. The overall approach for the preparation of the treatment facility involves generating a series of facility design sketches that use the process treatment system design requirements to form the functional basis of the facility.

The facility design sketches, reflecting input from all members of the design team, depict location or area layout, process arrangement, utilities, shipping and receiving, sampling and testing, confinement features, process support areas, elevations, and typical cross-sections. This activity begins by making simple sketches of material flow moving through the facility, incorporating the size information for basic equipment and areas necessary for operation, control, shielding, shipping and receiving, and material handling. Once a simplified sketch looks like it incorporates all the basic features, a more complete arrangement sketch should be prepared.

Next, process flow diagrams must be prepared for each major system to be included in the facility -- both process systems and service systems. A written description of the facility should describe the major construction features, systems, equipment, material handling, operations, and maintenance. Facility service utility systems, such as heating, ventilation, and air conditioning (HVAC), service water, power, communications, drains, and service and instrument air, must also be identified and described. Finally, staffing requirements and shift plans should be included.

19.5.4.6 Transportation Options: Based on size, weight, and cost, radioactive materials may be shipped by air, truck, rail, ship, or barge. In some instances, a combination of methods is used. There are about 100 M U.S shipments of hazardous materials annually. Of these, about two million involve such radioactive materials as radiopharmaceuticals or compounds for medical research. Only a small fraction of these shipments contain LLW. Since 1972, there have been four transportation accidents that led to the release of radioactive material. In all cases, the releases were small (i.e., like a package opening) and the released materials were quickly repackaged. No injuries or deaths have ever been caused by a release from LLW in a transportation accident. Most is shipped to disposal facilities by ground transportation, regulated by NRC or DOT. NRC requires that radioactive materials be packaged for shipment to protect the public in case of an accident. The kind of packaging required depends on the amounts and types of radioactive elements in the waste.

19.5.4.7 Technology Alternatives: Selection of treatment options should be preceded by a detailed review of the physical and chemical characteristics of potential input waste and determination of the waste form characteristics required to meet the disposal site's criteria. Such waste could arise from various sources during operation, decommissioning, and decontamination. A brief discussion of each waste source relative to a treatment plan selection is presented herein. Such consideration is essential for establishing a cost-effective treatment plan.

To determine processing needs, the waste must be described and evaluated. Typically the composition of the waste varies widely from one source to another, but the various streams may be grouped into several major treatment categories based on common characteristics of the groupings:

1. *Organic (combustible) waste*: Examples are organic sludge, resins, PVC and other plastics, wood, paper, and cloth. These are always candidates for volume reduction, but such reduction may concentrate RCRA metals, resulting in a mixed secondary waste.

2. *Inorganic (noncombustible) waste*: Examples are inorganic sludge, ceramics, firebrick, cement, glass and glass filter media, desiccants, dirt, and salts. Only sludge offers an opportunity for volume reduction, but all will require some level of stabilization to meet disposal site criteria.

3. *Metal waste*: Examples are carbon steel, stainless steel, aluminum, copper, and other miscellaneous metals. There are some opportunities for decontamination and recycle, but the greatest opportunity is often to manage the stream as debris waste.

4. *Lead waste*: Examples are lead brick, lead shielding, lead blankets, and lead rubber gloves. These are always candidates for either recycle or macro-encapsulation.

5. *Special waste*: Special waste is asbestos, bulk mercury, graphite, or other material that exists in small quantities and will require special treatment.

When the waste has been described, one may evaluate technology alternatives with respect to cost and compliance as the focus for final selection. A typical overview of waste treatment process options versus waste streaming is shown on Table 19.4.

19.6 WASTE TECHNOLOGIES

Waste forms must comply with regulations for disposal either through the container itself (as a macro-encapsulation approach) or through a process to either solidify or stabilize waste in a noncompliant container. In many, if not all cases, the waste is pretreated, by chemical or mechanical decontamination or separation, before generating the final waste form.

19.6.1 Solidification vs. Stabilization: Solidification refers to a range of processes in which additives are added (in predetermined rations and even monitored through a process control program) to a given batch of LLW or mixed radioactive and hazardous waste. The waste is then converted to a single, solid form. Before solidification, the waste may take a variety of forms: liquid, slurry (liquid plus suspended solids), sludge (wet solids), or dry solid particles. However, solidification does not mean the waste has been stabilized.

Stabilization implies that the waste is structurally stable under the effects of disposal conditions, such as the overburden pressure, the presence of moisture, and microbial action. Stabilization may be the result of the waste form, conversion of the waste form (grout solidification), or the high integrity container (HIC). Solidification is accomplished by mixing the waste with a solidification agent or binder. The binder forms a monolithic solid by (1) reacting chemically with the waste; (2)

forming microscopic cells that encapsulate the waste; (3) coating and binding the individual particles of waste together; and/or (4) encapsulation of the waste.

Table 19.4 Waste Treatment Options

	Thermal				Oxidation/ Reduction			Chemical Process			Solification/ Stabilization				Mechanical Process				Debris/ Containers		
WASTE STREAMS	Steam Reforming	Incineration	Plasma	Melting	WETOX	Silver II	Catalytic Extraction	Chemical Decon	CACITOX	Concentration	Glass Vitrification	Cement Grout	Polyethylene	Epoxy	Ultra Compaction	Size Reduction	Shredding	DECON	HIC's	Strong Tight	Casks
Organic (combustible) Waste																					
Organic Sludge																					
Resins																					
PVC and other Plastics																					
Wood, Paper, and Cloth																					
Inorganic (noncombustible) Waste																					
Inorganic Sludge																					
Ceramics, Firebrick, and Cement																					
Glass & Glass Filter Media																					
Desiccants and Dirt																					
Salts																					
Metal Waste																					
Carbon Steel and Stainless Steel																					
Aluminum, Copper and other Miscellaneous Metals																					
Lead Waste																					
Lead Brick, Shielding, and Lead Blankets																					
Lead Rubber Gloves																					
Special Waste																					
Asbestos																					
Bulk Mercury																					
Graphite																					
Other Miscellaneous Sources																					

U.S. regulatory requirements encourage solidification, with specific conditions placed on waste packages containing liquids. Media that cannot be effectively dewatered and other liquid waste streams are solidified or micro-encapsulated. Solidification binds liquids in radioactive waste and is mandatory in: wastes that cannot be dewatered to 0.5% free standing liquid for a steel container or 1.0% for an HIC, liquid waste streams, and pyrophoric wastes, such as uranium metal.

Examples of what must be stabilized include Class A waste containing radionuclides with a half-life > 5 years with a concentration > 1 µCi/mL, Class B and C wastes, and some waste containing chelating agents. Furthermore, all LLW containing source, special nuclear, or by-product material acceptable for disposal in a land disposal facility must be stabilized. Table 19.5, from an NRC Technical Position on Waste Form (NRC 1991) presents the waste form requirements for Class B&C waste in accord with 10 CFR Part 61. Table 19.6 is a subjective list of commonly used solidification waste form properties.

Table 19.5 Waste Form Requirements

Criteria	Old Requirements	Current Requirements
Compressive strength	60 psi	500 psi
After thermal cycling	60 psi	500 psi
After irradiation	60 psi	500 psi
After biodegradation test	60 psi	500 psi
After immersion test	60 psi	500 psi* **
Free Liquids	<0.5%, pH 4.0 to 11.0	<0.5%, pH>9
Leach testing	L>6, 90 days	L>6, 5 days
Full-scale correlation	Simulated waste	Simulated waste, then compressive test

* If post immersion is <75% of original strength, immersion test must be performed for longer immersion periods (120, 150, 180 days).

** For bead resin, chelates, filter sludge, and floor drain wastes, seven-day immersion is followed by seven days of drying, then examined and compressive strength test run.

19.6.2 Macro Encapsulation: Macro-Encapsulation is used to stabilize a waste into an acceptable waste form. Traditionally, waste that is defined as debris by the EPA may be macro-encapsulated by surrounding it with a neat solidification medium or by placing it in an HIC to ensure that the waste form disposal criteria are met.

19.6.3 Incineration: Absent a significant content of halogenated materials, such as PCBs, organic waste is generally suitable for incineration. The presence of such halogens as chlorides results in:

- Generation of excessive amounts of acid gas (at a minimum, increasing maintenance requirements and, at the maximum, increasing the constraints and materials cost of the capital equipment)
- Production of excessive salt product (secondary wastes from chloride neutralization, complicating stabilization requirements)

Table 19.6 Waste Form Properties

Property	Portland Cement	Asphalt *	Unsaturated Polyester	Polyethylene	Glas
Product density, lb/ft3	90-125	62-90	69-81	70-86	150-175
Water-binding strength	High	N/A	Moderate-High	High	N/A
Free-standing water	Occasionally	Never	Seldom	High	None
Compressive strength, psi	500	N/A	750	1000	5000
Mechanical stability	High	Moderate	Moderate-High	Moderate-High	High
Flammability	None	Moderate	Low-Moderate	Low	None
Leachability	Moderate	Low-Moderate	Moderate	Low	Low
Corrosivity to mild steel	Protective	Non-Corrosive	Non-Corrosive	Non-Corrosive	Non-Corros

* Due to material cost or undesirable characteristic, asphalt solidification is not currently used.

A number of potential incineration processes may be considered for organic wastes, but the two design types with the most extensive radioactive waste processing experience are rotary kilns and controlled air units. A typical controlled-air incineration system, such as that deployed at the Oak Ridge facility, includes waste sorting, feed material preparation, a gas-fired incinerator unit (meeting or exceeding the time-temperature relationships necessary for the destruction of the wastes), ash handling and stabilization, and off-gas treatment. Typically, volume reductions of 200 to 1 are obtained from incineration. A separate or integrated oil burning system may be used to process contaminated oils and oily wastes, depending on the quantity and quality of material. The oil burning system is a simple system very similar to a home oil-fired furnace with a burner and ignition system. Solid and water separation systems may be employed to improve the quality of the oil. Here are the basic descriptions for each of the most important incinerator technologies:

19.6.3.1 Vertical Hearth Incinerator: Vertical hearth incineration is a batch process accomplished in two stages: first the waste is thermally treated by pyrolysis, then it is combusted. Waste is introduced at the top of the unit and falls on the waste ash pile at the bottom of the pyrolysis chamber. The ash is held in the top pyrolysis chamber by swivel arms and released to the thermal treatment chamber as determined by process monitoring. In the combustion chamber, the pyrolysis products are then combusted under controlled air supply at about 900° C.

19.6.3.2 Horizontal Fixed Hearth Incinerator: In a horizontal hearth incinerator, solid waste is fed to the primary chamber and burned at roughly 50 to 80 percent of the stoichiometric air requirement (starved air condition). This pyrolyzes the waste, thus emitting a volatile fraction, with the required heat supplied by partial combustion and oxidation of the fixed carbon. The resultant smoke and prolific products, consisting primarily of volatile hydrocarbons and carbon monoxide along with some combustion products, pass to the secondary chamber. Ash is typically moved across the primary chamber hearth by augers, hydraulic rams, or moving grates toward an ash chute, where the ash is collected in a container. Excess air is provided in the secondary chamber to ensure complete combustion. Liquid waste may be incinerated in either the primary or secondary chambers.

19.6.3.3 Fluidized Bed Incinerator: A fluidized bed incinerator is a vertical refractory-lined vessel containing a bed of an inert granular material. The bed is "fluidized" by passing air for combustion through a perforated plate at the bottom of the vessel. Size-reduced solid waste, sludge, and liquids are fed to the hot bed, where the high thermal mass and turbulent mixing action of the bed material rapidly transfers heat to the waste. Auxiliary fuel is often used to maintain bed temperature. A secondary chamber may be required to ensure complete combustion for organic hazardous wastes. Limestone is usually added to the bed to provide capability for in-situ acid gas scrubbing capability. A variation of fluidized bed technology is a circulating bed system, where higher air velocities cause high carryover rates. The carryover material is recovered with cyclones and returned to the system.

19.6.3.4 Rotary Kiln Incinerator: The rotary kiln incinerator is a cylindrical refractory shell mounted on a slight incline. Kiln rotation moves the waste through the kiln and enhances waste mixing. Rotary kilns normally require a secondary combustion chamber to ensure complete destruction of hazardous organic constituents. The primary chamber combusts solid waste to gases (pyrolyze); the gases are then combusted completely in the secondary chamber. Both primary and secondary chambers are generally supplied with auxiliary fuel systems.

19.6.4 Steam Reforming: Steam reforming is ideally suited for processing medical wastes and organic wastes exhibiting high activity levels. Steam reforming vaporizes and destroys organic compounds in either liquid or solid form, leaving behind a dry, mineral-like solid residue. It is performed in a steam-laden, oxygen-deficient environment to convert organic and biochemical wastes to CO, H_2, CO_2, and H_2O. The two-step process first employs an evaporation phase that breaks down and vaporizes organic compounds and water from the waste. Several evaporator designs may be used, including in-drum processing or a heated screw auger unit. Waste solids in the evaporator are not exposed to the high temperatures, where most metals and radionuclides would volatilize from the residue. The volatilized gases exit the evaporator and pass through a high temperature filter that removes any entrained particles. The particulate-free gas exiting the filter is then mixed with superheated steam and passed through a high-temperature reformer, where the organic vapors are destroyed.

As the process does not form the secondary pollutants and dioxins/furans associated with combustion, steam reforming is not classified as an incinerator by the EPA and is licensed for on-site or fixed-base operations. Steam reforming has also been successfully used to process GTCC and spent fuel pool wastes to destroy hydrogen bearing materials, rendering the processed wastes suitable for long-term dry storage.

19.6.5 Plasma-Driven Reactor: The process uses plasma, a high temperature (> 5,000° C), ionized, conductive gas created within the plasma torch by the interaction of a gas with an electric arc. This interaction disassociates the gas into electrons and ions, making both thermal and electrical conduction possible. The conductive property of the ionized gas in the arc region provides a means to transfer energy from the arc to the incoming process gas and in turn, to the process or furnace. This state is called plasma and will exist in the immediate confines of the arc within the torch. By the time the gas exits the torch, it will have largely recombined into its neutral (non-ionized) state. However, the gas will still maintain its superheated properties. Plasma systems offer a means of achieving the high temperatures required for the safe destruction of many regulated wastes, including PCBs, dioxin, DDT, furans, and other halogenated compounds.

19.6.6 Metal Melting: In this process, metal is melted in a furnace; and the waste, usually an oily film or oxide layer, rises to the surface and is stripped away in slag. Various admixtures may be used to strip out radionuclide impurities as well. The end product may be formed into useful products or recycled for use in radioactive service. The process activities include size reduction, melting in a furnace, slag removal, pouring into molds, and off-gas treatment.

19.6.7 Other Nonthermal Treatments: A variety of treatment processes are being developed to handle organic compounds and other waste streams without combustion. Considerable work is being done, and commercial systems may soon be forthcoming. Some specific waste streams may be more suitable for alternative processes than using incineration.

1. *Wet Oxidation Process*: The wet oxidation process operates under aqueous conditions and involves the reaction of contaminated organics with hydrogen peroxide in the presence of a catalyst at 100° C and atmospheric pressure. It is capable of treating a wide range of organic wastes with organic concentrations ranging from 20% to less than 1%. Examples of wastes include ion exchange resins, organic decontamination effluents (ethylenediaminetetraacetic acid, citrate, oxalate), tributyl phosphate, and pyridine.

2. *Supercritical Water Oxidation*: Organic molecules are converted to CO_2 and H_2O during this process. The solvent properties of water are reversed above its critical point -- organic liquids and gases become miscible while inorganic salts become insoluble. In addition, all organic material becomes unstable at such temperatures and pressures. By introducing oxygen (as air), complete oxidation of the organic species is achieved. Oxidation takes place through a series of related oxidation and hydrolysis reactions at temperature in the range 175-300° C, and pressures of 300-3000 psig. The reaction rate may also be enhanced by the use of appropriate catalysts. The process is versatile with regard to the range of compounds it is able to treat, with virtually complete destruction of even the most intractable materials, such as plastics and rubbers. While this process has been used for municipal and sewage sludge, it has not been applied commercially to radioactive wastes.

3. *Chemical Hydrolysis*: Hydrolysis refers to reactions involving solvation with water that may take place either in an acid or alkali medium.

4. *Electrolysis*: A variety of solid and liquid organic compounds may be converted to CO_2, CO, water and inorganic acids by using nitric acid electrolyte in a divided cell. The process is suited for treatment of mixed and hazardous wastes as it is a low temperature/low pressure, wet process, and the chemistry may be stopped simply by switching off the current to the cells.

5. *Ultra-Violet/Peroxide Process*: This process is sometimes used for organic molecules with poor adsorption isotherms, using activated carbon to convert them to CO_2, water, and other components. This is accomplished with peroxide doses that are usually triple the chemical oxygen demand of the wastewater and with considerable expenditure of electrical energy. The UV/Peroxide process is subject to negative interference due to iron and suspended or colored material in the wastewater that may absorb or disburse ultraviolet light.

19.6.8 Microbial Degradation: Microbial degradation processes have been developed at the pilot scale for treatment of biodegradable waste and for removing surface contamination of what is not biodegradable. Biodegradable wastes may be treated using an anaerobic reactor involving three discrete groups of microbes in a multistage process consisting of hydrolysis of waste to small chain intermediates, followed by acetogenesis to organic acids, and finally methanogenesis to carbon dioxide and methane (a possible energy source). Such a reactor could be inoculated with sewage sludge and would require the addition of nutrients containing nitrogen, phosphate, and appropriate minerals.

19.6.9 Separation Technologies: These technologies optimize waste management. Instead of trying to find a technology that treats a large volume of a complex mixture, the waste is segmented, with each portion being treated optimally using a specific technology. For example, a large volume of an aqueous waste may be optimally treated by separating a large volume of water and treating the remaining segment appropriately. There are a variety of separation technologies available for treatment of waste resulting from decommissioning activities. Some are discussed in detail in the attached CD.

19.6.10 Water Processing Technologies

19.6.10.1 Reverse Osmosis: Reverse osmosis systems use pressure to separate dissolved salts from water by forcing water to pass through a semi-permeable membrane. In the simplest form, a semi-permeable membrane separates a solution of high salt concentration from an area of low salt concentration but allows the passage of water through the membrane. Pressure is applied on one side of the membrane. The applied pressure is sufficient to overcome the ionic bond of the salts with water molecules and cause the water to pass through the membrane, leaving the salts near the surface of the membrane. Hence reverse osmosis is a candidate treatment process to separate dissolved solids in the 1 to 12 angstrom size range from water. The system employs a multitude of membrane types, 60 to 1200 psi operational pressures, and operates at temperatures below 65° C to treat relatively chemically inert solutions. The system may be composed of hollow fiber membranes, spiral wound membranes, or plate and frame membranes, depending on the water quality, capital available for purchase of a wastewater treatment system, and the degree of volume reduction required. The operation of such a system is impaired by the presence of excessive scale-forming salts, any bacterial growths, oxidizers, extremes of pH, oil, silica, and many organic compounds.

19.6.10.2 Filtration: Suspended residue can be separated from water by passage through a filter. The filter medium may be layers of fine granular material, such as rock, garnet, sand and anthracite, or a bag constructed of fabric with a known pore size, or cartridges made of similar material. Some membrane cartridge filters may filter to as low as 0.05 μm. Bag filters exhibit pore size as low as 1 μm while granular filters exhibit an effective pore size as low as 5 μm. Common problems with all types of filters involve formation of an impermeable layer on the surface of the filter that prevents passage of water through the medium. Operational pressures for the various media range from a few inches of water to as much as 300 psi. While the uniformity coefficient and mean grain size of granular media controls the effective filter pore size, the pore size of bag and cartridge filters is usually a manufactured component. Membrane and standard cartridge filters that exhibit an extremely concentrated distribution of pore sizes about the rating of the filter are deemed "absolute" filter, while those that exhibit a wider range of pores in the media are deemed "nominal." Granular filters for cleaning the material are usually backwashed at 15 - 22.5 gpm/ft^2. After many backwashing events, the granular material will lose its angular configuration and must be replaced. Some cartridge filters may also be backwashed. Turbidity below one nephelometric turbidity unit (NTU) is possible with all of the above filters.

19.6.10.3 Nano-Filtration and Ultra-Filtration: Nano-filtration and ultra-filtration use membrane separation systems that operate at lower pressures than reverse osmosis systems and higher pressures than suspended solids filtration systems. Nano-filtration separates dissolved and particulate material of nm size, while ultra-filtration removes material in the μm size. Nano-filtration may be used to precondition the wastewater for subsequent reverse osmosis treatment by removal of multivalent cations at pressures of 60 to 180 psi. Both units can use an ultra-filtration system operated at 30 to 80 psi to precondition wastewater for subsequent treatment by removal of large chain molecules, such as oil, surfactants and powdered carbon slurries. Both nano- and ultra-filtration membranes are subject to the same scaling, fouling and organic bonding problems as reverse osmosis membranes, but to a lesser extent.

19.6.10.4 Demineralization/Ion Exchange: Ion exchange is the replacement of an ionized cation or anion in wastewater by a more active ion upon passage of liquid through a column of bead or powdered material. While many specialized systems are available, in the nuclear field, the most common cationic and anion exchangers are hydrogen ion and hydroxide ion, respectively. However, in some applications, sodium or chloride ions may be the exchanged ion. In order to provide detention times sufficient to allow the kinetics of the exchange process to function, ion exchange columns must be designed as volumetric flows within the proper gal/ft^3 range. In order to minimize the short circuit of flows through the beds, applied flows must be maintained within the proper flow rate. In most cases, ion exchange is applicable for treatment of liquids with dilute to medium ionic strength; however, some applications are valid in solutions of high ionic strength. Feed to ion exchange columns must be pretreated by filtration to remove suspended material in order to avoid a rapid rise of head losses through the columns.

These systems work by forcing equilibrium, through concentration and chemical activity, in which the extreme majority of ions in wastewater are exchanged for a more acceptable ion from the ion exchange medium. Hence, the kinetics of the process is driven by chemical and physical parameters in a manner similar to a chemical equilibrium.

19.6.10.5 Evaporation and Distillation: Evaporation is employed when the object is to reduce the residue remaining for disposal as much as possible. The process may be combined with a condensation of evaporated water and reuse of the water. In that case, the process is called distillation. Evaporation can remove 1) the interstitial water between solid particles, 2) water bound to the surface of the solids and, 3) given enough heat, even the waters of hydration from solids. In most applications, only the first two types of water are removed.

Multitudes of evaporation processes are available. The more common types are thin film evaporators, multiple effect vacuum units, and drum dryers. All evaporators lose heat transfer efficiency due to the accumulation of dried solids with low heat transmission potential on the heated surfaces of the evaporator. It is not uncommon to initiate operation with an overall heat transfer efficiency of 85 %, but after a few weeks of operation, achieve only 60 % or less. Evaporation rate is directly proportional to the area of the liquid exposed to the gas phase in the evaporator or the surface renewal rate in a mixed evaporator. Hence, an unmixed drum dryer is the slowest of the evaporation systems. All evaporators employ heat and/or vacuum to elevate the wastewater temperature to the boiling point or lower the vapor pressure at which boiling will occur. The most efficient method of evaporation involves use of a heat exchanger to condense the evaporator gas phase and exchange the heat to the wastewater feed. Thus, in this case, the evaporation process is actually distillation.

Using raw wastewater as a heat exchange medium requires pretreatment to avoid fouling of the heat exchanger. Spraying a film of liquid on a heated rotating cylinder creates thin film evaporators. Upon evaporation of the liquid, a scraper blade removes the dried material from the cylinder, producing a residue. A multiple effect, vacuum-assisted evaporator employs several stages, each under vacuum, to achieve evaporation of the water from wastewater. Only a drum dryer and thin film evaporator is capable of producing a totally dry residue. With other evaporators, it's rarely possible to achieve total dissolved solids above 25 % due to solubility restrictions.

19.6.10.6 Solids Separation: This process involves gravity segregation of suspended solids from wastewater to form a slurry of 1 % to 5 % dry solid content. These processes may involve chemical coagulation and flocculation of micro-solids into large particles that will settle from the water in hours rather than days or weeks. A typical settling rate for metal hydroxide solids is 8 ft/hr, while a solid produced with lime will settle at up to 12 ft/hr. The keys to designing a successful solids separation device are:

- Avoiding overloading the solids flux capacity of the component
- Maintaining surface overflow rates below 325 gpd/ft^2 of clarification area
- Avoiding surges in flow or solid content of the wastewater
- Removing the settled solids slurry before bacterial action generates sufficient gas to float them to the surface.

Therefore, oils, light solids, or fine solids are not good candidates for solids separation unless chemically coagulated, flocculated, and polymerized into large particles. Typically, a gravity solids separation device will produce a settled solid slurry that is about 1% dry solids and an overflow that is near 30 mg/L total suspended solids (TSS). However, one separation device attaches micro-air bubbles to the solids particles to float them to the surface in a manner similar to the way an unattended straw floats upwards in a soft drink. The solids-enriched layer at the top typically exhibits 5% dry solids content (and oils), while the clarified bottom discharge layer will exhibit 100 mg/l TSS or less. Due to losses of air solubility at higher temperatures, a dissolved air flotation treatment should never be used for treatment of wastes over 90° F.

19.6.10.7 Sludge Dewatering: Sludge dewatering usually takes two steps. The first step is to increase the solids content of the wastewater to 2% dry solids or higher and then dewater the resulting sludge slurry to a dry solids content of 12% to 60%. If a clarifier outflow of raw wastewater with less than 1% TSS is to be dewatered, it is common to employ a sludge decant tank. A decant tank is actually a clarifier with an extremely low surface overflow rate, hence, the solids will compact to as high as 4% dry solids (with certain sludge types condensing to 10% dry solids or more). Sludge decant facilities always feed a mechanical sludge dewatering device. These devices use sand beds, centrifuges, rolling paper filters, vacuum filters, or sludge presses to separate the sludge. All of these processes either multiply the gravity effect or employ mechanical straining of solid particles from the water, with or without pressure in addition to atmospheric pressure. All dewatering processes dictate that the device receive slurries that have been properly conditioned chemically and physically.

19.6.10.8 Activated Carbon: Activated carbon has the unique property of attracting nonpolar organic molecules and rejecting almost all ionized species. Activated carbon can be obtained in a powdered or granular form. As the adsorption of the carbon varies with respect to each organic species, the isotherm for each species and each type of carbon is different. Generally, the short chain aliphatic molecules are not candidates for carbon adsorption; the best candidates for activated carbon treatment are the higher molecular weight aliphatic and molecules with ring structures. While the same isotherm will apply to a given activated carbon for a specific organic, the final equilibrium concentration of organic remaining in the wastewater is always higher using the powdered form than when using granular. Although granular, like ion exchange resins or zeolite, may also be regenerated upon exhaustion, neither medium is normally regenerated in the nuclear field. Certain organic molecules have poor adsorption isotherms using activated carbon. These can be decomposed into carbon dioxide, water, and other components using ultraviolet light and hydrogen peroxide.

19.6.10.9 Coagulation, pH Adjustment, Flocculation, and Polymerization: Suspended solids and some dissolved solids can be removed at 99% or higher efficiencies using coagulation, flocculation, and/or polymer addition in conjunction with pH control to the isoelectric point of the component to be removed. Coagulation is the addition of a chemical to wastewater that results in the conversion of dissolved components to micro-solids. In order to coagulate a dissolved solids component, the pH of the wastewater must first be adjusted to the isoelectric point via addition of acid, base, and/or a coagulation agent. The isoelectric point is the pH at which a dissolved component exhibits minimum solubility. Flocculation is the building of those micro-solids into larger clumps of solids known as floc particles. Polymerization is the further building of floc particles into larger clumps. Typical problems with these processes involve inadequate detention times for complete chemical reactions, selection of an incorrect isoelectric point, faulty calibration of pH controllers and fouling of sensor surfaces due to coating of sludge and oil on the sensing surfaces that inhibit quick exchange of water across the sensor barrier.

19.6.10.10 Solvent Extraction: A widely used process, solvent extraction is based on differences in the solubility of various materials. It may be used to separate components of different types of mixtures, such as two or more solids, a solid and a liquid, or two or more liquids. At the end of the application of this process, one or more components of the mixture are removed (extracted). Depending on the solubility of the compound to be separated, the process may be repeated.

19.6.11 Soil Washing: Soil washing is based on commonly available mineral treatment processes. It is one of the few techniques that can treat soils contaminated with organic compounds, heavy metals, radionuclides, and combinations of contaminants. Soil washing is a means of partitioning the contaminants in contaminated soil so that a large portion of the inlet soil is cleaned and discharged with levels below a specified limit, while the extracted contaminants are concentrated in the remaining, smaller portion of the soil for disposal. It does this through a combination of particle separation by size and/or density to be followed by decontamination of separated particles by extraction using appropriate solvents and surfactant. Feed rates of 1/2 ton to 1000 tons per hour are dependent on extraction chemistry, required retention time, and equipment size. The almost infinite combinations of site soil characteristics and contaminating chemical waste forms necessitate a treatability study, but the flexibility of the process allows a wide range of application.

19.6.12 Mining/Leaching Technology: This technology may be applied either to mined material or in-situ to recovered minerals from the ground. Extraction is performed by percolation of an appropriate solvent through the solids or ground. This technique is used to remove contamination from soils and is applied in heap leaching, soil washing, and the in-situ pumping of a solvent into the ground. For example, uranium ore is processed by leaching it with either a strong acidic or strong alkaline solution (leachate). The uranium can then be recovered from the leachate using either solvent extraction or ion exchange.

19.6.13 Mechanical Processes for Size Reduction: Decommissioning uses a variety of techniques, such as cutting, burning, and component disassembly to reduce the size of large items for further processing. Possible systems include simple hand-held, remotely operated, and robotic equipment.

Shredding breaks up material for further processing and to creates material of the proper size for handling and processing. Typically, two stages of shredding are used. In the first, a hydraulic industrial shredder tears the material apart with low RPM counter-rotating sets of teeth. The second stage uses a shredder with appropriate grid to ensure a relative uniform output of material. Some materials may be encapsulated directly after shredding if this proves to be the most cost-effective handling method.

Compaction has been used in the nuclear industry for years. Early compactors used a hydraulic ram to compress paper, rags, cardboard, and plastics in a drum. These compactors are called low force systems, with a capacity of up to about 50 tons. Other relatively low force systems, such as box, shredder, or screw compactors were also used. A super-compactor is a large hydraulic press that crushes a drum or other receptacle containing essentially all types of solid waste. The container is held in a mold during the compaction stroke, which sizes the container's outer dimensions. The compressed drum is then stripped from the mold and the process is repeated. Two or more crushed waste containers (pellets) are then sealed inside an overpack container for storage and ultimate disposal. Compaction is used to pre-compact waste material into drums before super-compaction. It may also be used to compress used drums and boxes for loading into a larger container for grouting or direct disposal. Table 19.7 provides data on the super-compaction of a variety of materials.

A compactor system may be mobile or stationary and supplied as a basic, manually controlled system. Or it may be connected with a minimum of auxiliary equipment to an elaborate computer-controlled system. Such a microprocessor-based inventory control system can then select drums for processing, measure weight and radiation levels, compress drums and place them in overpack containers, seal the overpacks, and record their content, weight, and radiation level.

19.6.14 Solidification and Stabilization:

19.6.14.1 Vitrification: Glasses of various compositions resulting from vitrification, an operation that has been performed safely for more than 20 years, have received considerable attention for the solidification of HLW. The capital and operating costs of glass systems have largely precluded their

Table 19.7 Typical Volume Reduction Factors for Super-Compaction

Material Description	Density (kg/m³) Initial	Final	VR Factor
Rags, Paper	261	1242	4.8
Newspapers, Magazines	200	1006	5.0
Plastic Sheets (Wadded)	126	1200	9.5
Hardwood	376	1142	3.0
Scrap Metal	781	3312	4.2
Small Electric Motors	968	3680	3.8
Concrete Paving Blocks	800	1296	1.62
Glass	378	1558	4.1

application to most decommissioning activities, but it's a useful technique and is applicable to other waste forms. It may be used to stabilize incinerator fly ash, boiler ash, bottom ash, and other wastes, such as furnace slag. To limit off-gas volumes, the melter uses electric joule heat as the energy source, although heating initially or after shutdown period may be accomplished using a gas (propane or natural) burner. The melter should be capable of adequately mixing the waste with any necessary additives to generate a homogenous vitrified waste form capable of passing the EPA's lab toxicity test and 10 CFR 61 requirements for an LLW waste form (Class A).

Briefly, glasses resulting from vitrification of radioactive waste are inorganic oxides that solidify upon cooling, typically forming an amorphous structure with a high melting point. Waste solids are generally incorporated into the glass structure as oxides produced during the high temperature processing conditions (1200° C) of the process. The amount of waste oxides that can be incorporated in glass is limited, particularly if a single-phase glass is desired. However, because of the processing conditions, a large volume reduction is achieved, particularly for combustible wastes.

19.6.14.2 Cement-Based Grout: Portland cement, with mixtures of blast furnace slag or pulverized fly ash in specific qualified formulations, may be used to stabilize specific waste forms. The process consists of taking the waste feed stream, size-adjusting by shredding, mixing the grout and waste material, allowing the mixture to cure, and sealing the drum. Various techniques are used to ensure proper mixing of the waste and grout. Cement-based grouts are a suitable waste form for most waste streams.

19.6.14.3 Polyethylene Encapsulation: Wastes not suitable for grout encapsulation due to high soluble salts (e.g., ammonium sulfate) can be encapsulated in polyethylene. Polyethylene is used to physically capture the wastes in a stabilized matrix in order to prevent the release of radioactive or hazardous materials after disposal. Operationally, particulate matter and dried sludge is mixed with polyethylene in a waste extruder. The extruder melts the polyethylene, mixes it with the waste, and drives off residual waste and volatile compounds. The product is then loaded into waste drums and cured.

19.6.14.4 Epoxy-Based Solidification Media: An epoxy-based solidification medium immobilizes heavy metals in RCRA hazardous mixed waste (incinerator fly ash). It allows a 3:1 volume reduction of incinerator fly ash using waste loadings of 40% and densities of 80 lbs/ft^3. This waste form may be used to solidify ion exchange bead resin and aqueous solutions. It is also acceptable for disposal at the commercial LLW disposal sites.

19.7 TRITIUM WASTE

Among all radionuclides that must be managed during decommissioning activities, tritium requires special attention. Tritium is a radioactive isotope of hydrogen, with a specific activity of approximately 10^4 Ci/g. As an isotope of hydrogen, its chemical properties are identical to those of ordinary hydrogen (^{1}H or protium), and it can combine with oxygen to form water. However, because its mass is larger than ^{1}H by a factor of three, its separation relies on reaction kinetics.

Tritium is produced by nuclear explosions, by nuclear operations at nuclear reactors and spent nuclear fuel reprocessing plants, and as a result of natural processes (e.g., cosmic radiation). Tritium is an activation product of deuterium (^{2}H) in heavy-water moderated reactors. It is also produced in smaller quantities in light-water moderated reactors and experimental fusion devices, and may also be produced in particle accelerators. As a fission product, most of the tritium is released during the reprocessing of spent fuel. Consequently, most nuclear facility operations must deal with the presence of tritium.

19.7.1 Properties: The correct notations for identifying tritium and other isotopes of hydrogen are ^{1}H (elemental hydrogen), ^{2}H (deuterium) and ^{3}H (tritium). Based on international convention, D and T are acceptable symbols for ^{2}H and ^{3}H, respectively. As of this date there is no other symbol for ^{1}H. However, in this chapter, H may be used for ^{1}H, as a practical although less accurate method. The properties of tritium are as follows:

- Half-Life 12.3 y
- Mode of decay Beta
- Maximum energy 18 keV
- Occurrence in waste HTO, DTO
- Beta range in air 6 mm
- Beta range in water 0.006 mm

19.7.2 Legal Limits: For the purpose of decommissioning (worker exposure, residual contamination, and waste disposal), the drinking water standard is of primary concern. The EPA has promulgated regulations for drinking water, assigning a concentration limit of 20,000 pCi/L as the maximum contaminant level. A closer look at the EPA's decision indicates its intention to allow an exposure of 4 mrem/y (Moghissi 1986). This dose limit, if applied to tritium, would result in values ranging from 80 to 90 nCi/L. The EPA does not explain the reasons for its decision; consequently, 20 nCi/L is used as the legal limit.

19.7.3 Waste Management Technologies: Tritium normally occurs as HTO. Numerous technologies propose to handle tritiated waste, as described in *Tritium* (Anspaugh 1973) and more recently by Jeppeson (1999). But practical considerations (including cost) limit the application of all but a few. It should be noted that processes for separating hydrogen and deuterium may also be used to separate tritium. But since tritium is radioactive, only simple, reliable, and leak-tight processes may be used for tritium separation.

Given its short half-life, controlled storage is likely to be the preferred approach to safely and effectively manage tritium-contaminated materials. But the other choices presented here may be necessary and must be considered. Safety and regulatory requirements, economic conditions, and stakeholder acceptance will dictate the selection.

19.7.3.1 Controlled Storage: The relatively short half-life of tritium (12.3 y) lends itself to storage as a means of natural decay. Storage could be defined as containment or hydraulic control of tritium-contaminated groundwater. Prerequisites for acceptable containment or hydraulic control of such plumes are as follows:

- A scientifically acceptable method (modeling) must be used to predict migration.
- Appropriate natural or engineered barriers or mechanisms must be available to ensure that tritium does not move beyond established boundaries.
- A monitoring system is established to ensure that the model prediction and the barrier are performing as planned.
- The concerns of stakeholders, particularly the personally or administratively impacted, are addressed.

Virtually every other facility with significant tritium groundwater contamination has to rely on containment or hydraulic control, and must ensure that the four criteria identified earlier are met. A more detailed description of the natural attenuation process is given provided in Section 20.4.3. The careful control and monitoring required of the natural attenuation process, when coupled with tritium's short half life, make natural attenuation an attractive option for site remediation while minimizing worker exposure.

Containment is by far the most useful and often the only feasible method for treatment of groundwater contaminated with tritium. An example at the Nevada Test Site helps demonstrate this point. About 900 explosions occurred there during the U.S. nuclear testing program (DOE/NV 1994). Some occurred at the surface, others were above, below, and at groundwater level. As would be expected, one of the key groundwater contaminants was tritium. A recent peer review demonstrated that the first three criteria are likely to be met, and stakeholder participation was then appropriately managed. This review suggested planning and implementing appropriate monitoring wells.

In another facility, Brookhaven National Laboratory, a migrating contaminated tritium plume was controlled through years of catching the leading plume downstream and pumping it to recharge upstream of the groundwater flow. This in effect increased the residence time in a controlled area, such that when contaminated water reached the site boundary, it was below the EPA's maximum contaminant level for tritium. Even so, protection of public health and safety was further assured through the elimination of wells and the installation of a potable drinking water system for the potentially affected area.

19.7.3.2 Water Distillation: Fractional water distillation is one of the oldest and safest methods for separating hydrogen isotopes. Initially, the process was used to separate H_2O and DHO from D_2O, the latter being used as moderator in a number of nuclear reactors. The technology has been also used routinely for separation of DTO from D_2O in the same reactors.

The vapor pressure difference between HTO and H_2O is greater than the difference between HDO and H_2O or between HTO and HDO. Therefore, separation of HTO from H_2O by water distillation is more efficient than the other two separation processes. The vapor pressure ratios for H_2O/HTO, H_2O/HDO, and HDO/HTO are shown in Table 19.8 as a function of temperature. This table is based on the data of Kalyanam and Sood (Kalyanam 1988), as calculated by Van Hook (1968).

Table 19.8 Vapor Pressure Ratios for Isotopic Water Species

Temperature K (°C)	H_2O/HTO	H_2O/HDO	HTO/HTO
298 (25)	1.095	1.075	
323 (50)	1.065	1.053	
333 (60)	1.056		1.0125
373 (100)	1.030	1.026	

The water distillation process may be used to enrich HTO (and HDO) in H_2O because the vapor pressure of H_2O is higher than that of water containing heavier isotopes of hydrogen. It has been known since the 1950s that the lower the distillation temperature, the higher the difference among waters containing various isotopes of hydrogen. Recently, detailed experimental results for the optimization of the distillation process have been developed. Note that although the distillation process may also enrich or deplete various oxygen isotopes, the ratio of masses of rarer isotopes (^{17}O and ^{18}O) as compared to ^{16}O is too small to significantly impact the hydrogen isotope separation process. Based on cost-benefit analyses, water distillation is not currently used at industrial levels for separating low concentrations of tritiated water from wastewater, nor would it be employed for typical ranges of groundwater contamination.

The process follows conventional distillations, as used in many applications in the chemical industry. It consists of an evaporator at the bottom and a condenser at the top of a packed column, as shown in Figure 19.1. The vapor feed stream is introduced to the column according to concentration; the higher the tritium concentration, the lower the point of feed. Water vapor is produced in the reboiler at the bottom of the column and rises up through the column packing. Tritium is concentrated in the water as it moves down the column, and the water vapor is depleted of tritium as it moves up through the column packing. The depleted water vapor is condensed as it leaves the column through the condenser, partially refluxed to the column and partially discharged.

The columns and packing material are made of treated stainless steel, copper, or copper alloy to improve wetting and minimize corrosion. The columns are operated under vacuum to minimize tritium leakage and to maximize separation factors. Separation factors range from 1.0281 at 373 degrees K (100° C) and 101 kPa pressure to 1.07 at 333 degrees K (60 °C) and 0.120 kPa pressure. In order to prevent fouling of the packing material, the feed to the water distillation column must be free of organic and other contaminants.

19.7.3.3 Cryogenic Distillation of Elemental Hydrogen: The proprietary cryogenic distillation processes developed by the Sulzer Company can separate tritium from both heavy water and light

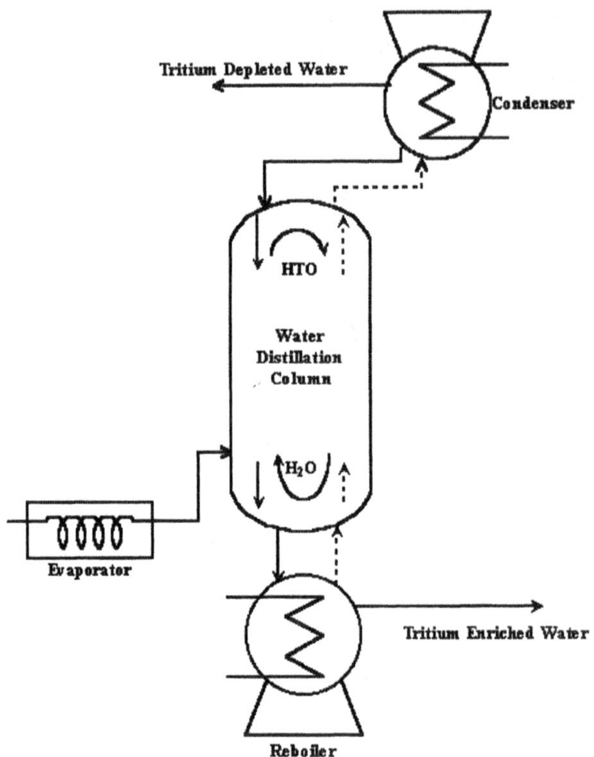

Figure 19.1 Water Distillation Process

water. These processes have been used for a number of projects. Figure 19.2 shows a simplified diagram of an operating tritium removal plant. The plant applies a vapor phase catalytic exchange to transfer tritium from heavy water (D_2O) into a deuterium carrier stream ($DTO + D_2 \rightarrow D_2O + DT$) and a subsequent cryogenic distillation. The process is supported by many subsystems:

- A feed purification system de-gases and purifies the entering heavy water
- Dryers remove the remaining heavy water from the deuterium before entering the cryogenic system.
- Adsorbers remove any high boiling gases, mainly nitrogen and oxygen.
- An electrolyzer produces the deuterium for initial filling and makeup.
- A recombiner provides the capability to burn hydrogen isotopes.
- A draining and purging system allows evacuation and purging of process systems.
- A tritium immobilization system is used to safely store and fill tritium containers.
- A cryogenic refrigeration system provides cooling of the distillation columns.

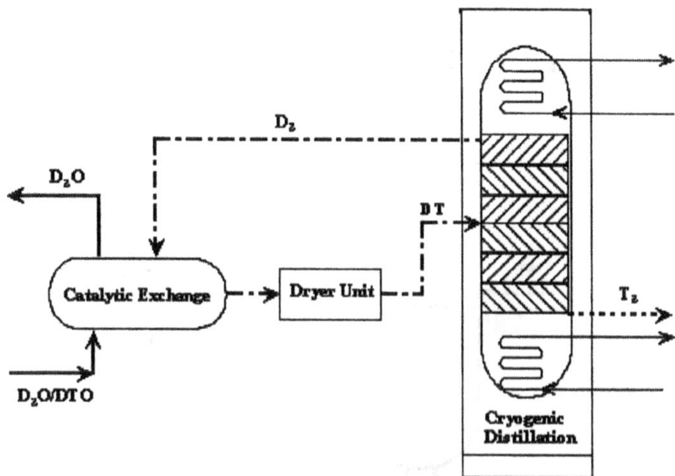

Figure 19.2 Sulzer - Tritium Removal Plants

19.7.3.4 Electrolysis Catalytic Exchange: The initial discovery of deuterium by Urey and the measurement of naturally occurring tritium by Libby were based on electrolytic enrichment of heavier hydrogen isotopes. The processes of cascading electrolytic process were dropped from consideration for preparation of D_2O for reactor applications because of safety reasons. However, it appears that recent developments have found favor for a process based on the combined electrolysis catalytic exchange process for separation of tritium at low concentrations. As mentioned earlier, during the electrolysis, tritium is enriched in the remaining liquid. The exchange process takes advantage of the catalyst-enhanced equilibrium reaction that favors the formation of HTO in the gas phase (Sienkiewicz 1988).

$$HT_{(g)} + H_2O_{(l)} \rightarrow HTO_{(1)} + H_{2\ (g)}$$

An early version of this process was used to remove HTO from liquid wastewater to reach a concentration level equal to drinking water standards. The process is based on conventional chemical engineering unit operations, as shown in Figure 19.3. It consists of countercurrent gas/liquid exchange columns with packed catalyst beds, an electrolytic cell, and a recombiner. The catalyst beds are ceramic packing material coated with the hydrophobic catalyst (platinum and Teflon). The process has been evaluated to calculate various aspects of its requirements, including flow rates and cost.

19.7.4 Management Selection: Due to the short half-life of tritium, controlled storage is likely to be the preferred approach to safely and effectively manage tritium-contaminated groundwater. If necessary, however, the other three choices must be considered: water distillation, electrolysis, and catalytic exchange. Safety requirements, economic considerations, regulatory compliance, and stakeholder acceptance will combine to dictate the selection.

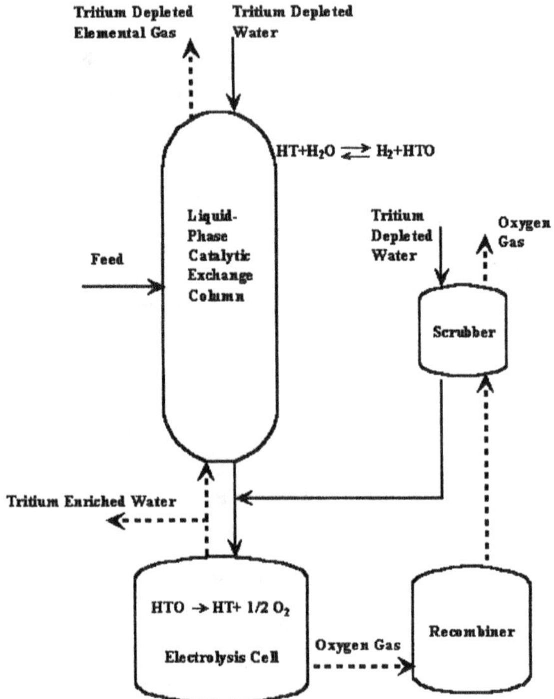

Figure 19.3 Combined Electrolysis Catalytic Exchange

Chapter 20

SOIL AND WATER CLEANUP

20.1 INTRODUCTION

Many U.S. nuclear power plants have permanently shut down and are in some phase of decommissioning. They include research reactors, other research facilities, and a small number of other commercial nuclear facilities, such as three commercial spent fuel reprocessing plants. There are hundreds of DOE nuclear facilities, i.e., reactors, fuel reprocessing plants, research facilities, and laboratories that supported a wide variety of operations using hazardous substances in the production of nuclear weapons. The DOE has restored many of its nuclear sites and cleaned up many hazardous waste sites. The decommissioning of commercial, private, and DOE-owned sites has resulted in the development of effective cleanup technologies and practices.

Radioactive and hazardous waste constituents found in the soil and groundwater during decommissioning got there because of spills, system upsets, maintenance activities, and outdated work practices. Only a few chemical processes support electricity production at commercial power reactors, but nevertheless, activities at these sites released contaminants to the environment. An example of such an activity is work performed in on-site analytical laboratories. These laboratories were often the source of radioactive and chemical agents that reached the environment via sink and floor drains. At some of the older facilities, both radioactive and chemical contaminants were found in leach fields where the contaminants tended to accumulate.

The primary radioactive contaminants in the soil or groundwater at a commercial site are ^{137}Cs, ^{90}Sr, ^{60}Co, and ^{3}H. The source of tritium in the groundwater is generally water leakage from the spent fuel storage pool that tends to occur over many years of operation. Insoluble radioactive contaminants tend to be trapped in the soil and are not easily transported by groundwater movement, therefore tend to remain in localized areas.

Commercial nuclear power plants also require many support activities associated with normal operations, such as equipment and vehicles maintenance, electrical transformers maintenance, paint disposal, degreasing and cleaning operations. These conventional industrial operations were often sources of soil and groundwater contamination. These chemical agents are found in the soil, such as fuel oil, lubricating oil, chromates and other corrosion inhibitors, lead, other heavy metals, PCBs, ethylene glycol, paints, and solvents. Depending on the site, these contaminants could originate from spills, releases, or outdated handling and maintenance practices.

Every site in the DOE complex is contaminated to some extent with radioactive and other hazardous materials; many also have similar subsurface contamination. In certain facilities subsurface contaminants also include isotopes of transuranic elements, including isotopes of plutonium and significant quantities of uranium and tritium. Contaminants may also be present as mixtures of metals and radionuclides, organic complexing agents, and solvents. In comparison to commercial power reactors, the production of weapons materials by its very nature often required working over many decades with large quantities and varieties of chemicals and radioactive materials.

20.2 CLEANUP STRATEGIES

The objective of remediation is to reduce residual radioactivity and chemical hazard in soils and groundwater below a level established by governing regulatory agencies. Systematic approaches to determining the type of cleanup technology that suits the local site conditions are available in several publications found at both DOE and EPA web sites. Technologies have been developed to remediate specific contaminants. The remediation technology, whether the contaminant is organic or inorganic, may be highly specialized, emphasizing the need for selectivity. Technology selection logic use the characteristics of the site or media (in this case soil and groundwater) to help reduce the number of potentially applicable technologies. These rules are generally structured so that nonapplicable treatment processes may be screened out quickly. The EPA has identified a suite of soil treatment technologies and practice guides for implementing those technologies. Used separately or in conjunction, the three primary remediation strategies are:

- Destruction or alteration of contaminants
- Extraction or separation of contaminants from environmental media
- Immobilization of contaminants

Technologies that destroy contaminants by altering their chemical structure are thermal, biological, and chemical treatment methods. These destruction technologies may be applied in-situ or ex-situ to contaminated media.

Treatment technologies commonly used to extract and separate contaminants from environmental media include soil treatment by thermal desorption, soil washing, solvent extraction, soil vapor extraction, and groundwater treatment by either phase separation, carbon adsorption, air stripping, ion exchange, or a combination. Selection and integration of technologies should use the most effective contaminant transport mechanisms to arrive at the most effective treatment scheme. For example, more air than water may be moved through soil. So, for a volatile contaminant in soil that is relatively insoluble in water, soil vapor extraction would be a more efficient separation technology than soil flushing or washing.

Immobilization technologies include 1) stabilization or solidification, including vitrification, and 2) containment technologies, such as placement in a secure landfill or construction of slurry walls. Because no immobilization technology is permanently effective, some type of maintenance must be planned for. Stabilization technologies are often proposed for remediating sites contaminated by metals or other inorganic species. These concepts about site remediation strategies and the representative technologies associated with them are summarized in Figure 20.1.

The figure makes clear that once a strategy is selected, the range of acceptable technologies is small. Further, no single technology can usually remediate an entire site, since multiple contaminants, both organic and inorganic, and soil and groundwater, are involved. Several treatment technologies are usually combined at a single site to form a treatment train. Soil vapor extraction may be integrated with groundwater pumping and air stripping to simultaneously remove

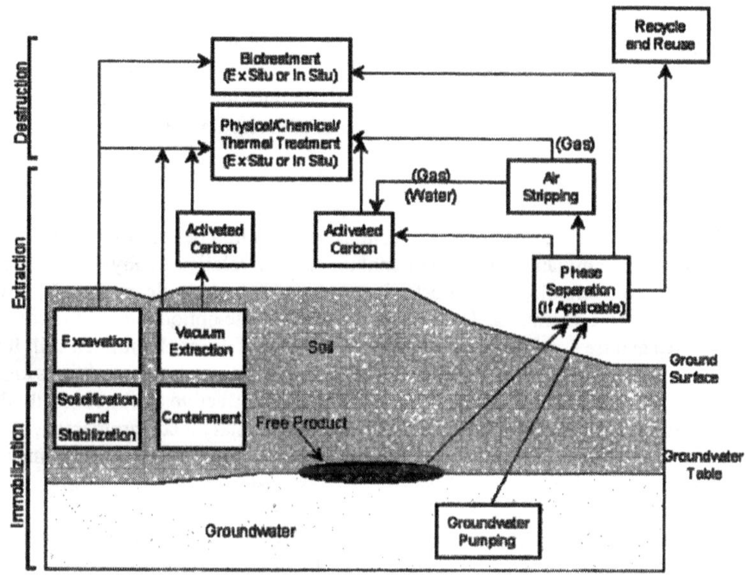

Figure 20.1 Classification of Remedial Technologies by Function

contaminants from both groundwater and soil. Emissions from the extraction system and the air stripper may be treated in a single air treatment unit. An added benefit is that the airflow through the soil is likely to stimulate or enhance natural biological activity, resulting in a potential biodegradation of contaminants. In some cases, air is injected into either the saturated or the unsaturated zones to facilitate contaminant transport and to promote biological activity.

The main advantage of in-situ treatment is that it allows soil to be treated without being excavated and transported – bringing potentially an enormous reduction in both cost savings and any environmental disturbances. However, in-situ treatment normally takes longer, and there is less certainty about the uniformity of treatment because of the variability in soil and aquifer characteristics and because the effectiveness of the process is more difficult to verify.

Bioremediation techniques are destructive, directed toward stimulating microorganisms to grow and use the contaminants as food and energy sources. Generally, this means providing some combination of oxygen, nutrients, and moisture, and controlling the temperature and pH. Sometimes microorganisms adapted to degrade specific contaminants are applied to enhance the process.

Biological processes are typically low cost. Contaminants can be destroyed, and often little to no residual treatment is required. However, as said before, the process takes more time and it is difficult to determine whether contaminants have been destroyed. The subject is considerably more complex than it may first appear. For example, the rate of biodegradation of some compounds may be higher than others in a mixture. As a general rule, an increase in the number of atoms of chlorine in a molecule leads to a decrease in the biodegradation rate. In certain cases, some compounds may be broken down into more toxic by-products during the bioremediation process. An example of such a case is tri-chloroethylene, which is converted to vinyl chloride during the biodegradation process. For in-situ applications, appropriate control techniques are necessary to ensure that these by-products are contained. Treatment of groundwater is addressed more fully in Section 20.4.

20.3 SOIL CLEANUP

The approach to soil cleanup is tailored to each site based on a number of criteria, including the nature and extent of contaminants, location of contaminants in relation to physical interferences, practicality, maturity of technology, and cost. The criteria also include cleanup goals, future use of the site, and regulatory requirements. The decision-maker has the option to treat the contaminated soil in-situ, remove and treat, or simply excavate and dispose of the contaminated soil. It is possible for a combination of these processes to be used at the same site since local conditions and the specific contaminants are typical factors in the selection of an effective treatment method. Appropriate characterization of the area to be cleaned up will ensure that all contaminants will be identified and will significantly enhance the probability of success. Several soil treatment technologies are described here.

20.3.1 Excavation and Disposal: Excavation consists of digging up contaminated soil and disposing of it in a licensed facility. Excavation and off-site disposal (with or without solidification/stabilization pretreatment) at a landfill have been performed extensively at many sites and are applicable to the complete range of contaminant groups. Even with the recent availability of new soil treatment technologies, excavation and removal is the primary soil cleanup and treatment method for decommissioning.

Excavation is also the first stage in all ex-situ treatments. This approach is applicable if the following criteria are met:

- The majority of soil contamination is found near the surface and is easily removed
- Excavation is fast as compared to other treatment technologies
- Excavation will ensure that contaminants are removed

Factors that may limit the applicability and effectiveness of the process include:

- Generation of fugitive emissions that may pose a problem during operations
- Disposal costs impacted by the distance from the site to the nearest applicable disposal facility
- Depth and composition of the media requiring excavation
- The affected community's acceptance of transportation through populated areas
- Limitations of disposal options for certain wastes, such as mixed or transuranic

Some pretreatment of the contaminated media is usually required to meet land disposal restrictions. Confined disposal facilities are engineered structures enclosed by dikes and designed to retain dredged materials. Such facilities may have large cells for material disposal and adjoining cells for retention and decantation of turbid, supernatant water. A variety of linings have been used to prevent seepage, the most effective being clay or bentonite-cement slurries; but sand, soil, and sediment linings have also been used. However, contaminants may migrate by several pathways, including effluent discharge to surface water, rainfall, surface runoff, leachate into groundwater, volatilization to the atmosphere, and dike uptake.

20.3.2 Soil Vapor Extraction: When properly designed and operated, soil vapor extraction is a safe, low maintenance process that is very effective at removing volatile organic compounds (VOC) from the unsaturated zone. Contaminants can be removed from the saturated zone with the addition of an air sparging system. Neither technique requires excavation of the contaminated soil. The extracted vapors usually require treatment, but costs for treating extracted vapors and liquids are low as compared to the costs of technologies requiring excavation. The technologies are relatively simple to install, may be used effectively in combination with other treatments, and are effective under a variety of site conditions. Soil vapor extraction and air sparging are good choices at sites contaminated with solvents, such as perchloroethylene, trichloroethylene, and fuels. Extraction is best used at sites with loose unsaturated soil, such as sand and gravel.

20.3.3 In-Situ Air Stripping: In-situ air stripping technology was developed to remediate soils and groundwater contaminated with VOCs both above and below the water table. This technique employs horizontal wells to inject (sparge) air into the groundwater and to vacuum-extract VOCs from vadose zone soils. The system combines two somewhat innovative technologies, air sparging and horizontal wells, with the baseline technology of soil vapor extraction to produce a more efficient in-situ remediation system. The horizontal wells provide a more effective access to the subsurface contamination. The air sparging process eliminates the need for surface groundwater treatment systems and treats the subsurface in-situ, directly attacking the problem of subsurface contaminant retention.

20.3.4 In-Situ Bioremediation: In-situ bioremediation, sometimes referred to as in-situ gaseous nutrient injection, remediates soils and groundwater contaminated with VOCs both above and below the water table. It involves injection of air and nutrients (sparging and biostimulation) into the groundwater and vacuum extraction to remove the VOCs from the vadose zone, concurrent with their biodegradation. This injection system combines three technologies: air stripping, horizontal wells, and bioremediation via gaseous nutrient injection with the baseline technology of soil vapor extraction to produce a more efficient in-situ remediation system.

The air sparging/gaseous nutrient injection process eliminates the need for surface groundwater treatment systems and treats the subsurface, both unsaturated and saturated zones, in-situ. It stimulates the growth of indigenous microorganisms in the contaminated zone to degrade VOCs. Extraction may be combined with the injection process to strip the higher concentration, more removable contaminants from the subsurface. The injection/extraction system can be designed to meet site specific needs. The types of sites most likely to apply this injection system will contain moderately permeable, relatively homogenous sediments contaminated with VOCs, especially if both an extraction and injection component is used. However, the presence of clay strata does not preclude its use. In fact, the bioremediation component may be well applied to enhance degradation and/or removal of VOCs from lower permeability zones. The application of in-situ bioremediation to a specific site must consider the following factors:

1. Bioremediation via gaseous nutrient injection must have been demonstrated to remediate soils, sediments, and groundwater contaminated with VOCs both above and below the water table.

2. The geometry of horizontal well treatment must conform to typical subsurface contaminated zones, which are often relatively thin but laterally extensive areas.

3. It may not be well suited for extremely low permeability sites if both injection and extraction are used. Some permeability is required to deliver the nutrients to the indigenous microorganisms.

4. It could be most effective when used in conjunction with in-situ air stripping; that is, if the stripping is applied first to quickly remove high concentrations of contaminants from source areas, and then the bioremediation is applied as a polishing step to remove contaminants present at lower concentrations. Under other conditions, only the bioremediation would be used, thus minimizing the contaminants removed from the subsurface that would then require treatment as off-gas at the surface.

20.3.5 In-Situ Soil Flushing: In-situ soil flushing is the extraction of contaminants from the soil with water or other suitable aqueous solutions. Soil flushing is accomplished by passing the extraction fluid through in-place soils using an injection or infiltration process. Extraction fluids must be recovered from the underlying aquifer and are recycled when possible. Injecting a solvent mixture (e.g., water plus a miscible organic solvent, such as alcohol) can enhance in-situ flushing into a vadose zone, a saturated zone, or both to extract organic contaminants. This process, known as co-solvent flushing, may be applied to soils to dissolve either the source or the plume emanating from it. The co-solvent mixture is normally injected up gradient of the contaminated area, and the solvent with dissolved contaminants is extracted down gradient and treated above ground.

Recovered groundwater and flushing fluids with the desorbed contaminants may need treatment to meet appropriate discharge standards before recycle or release to local, publicly owned wastewater treatment facilities. To the maximum extent practical, recovered fluids are reused in the flushing process. The separation of surfactants from recovered flushing fluid for reuse in the process is a major factor in the cost of soil flushing. Treatment of the recovered fluids results in process sludges and residual solids, such as spent carbon and spent ion exchange resin; these wastes must be appropriately treated before disposal. Air emissions of volatile contaminants from recovered flushing fluids are collected and treated, as appropriate, to meet applicable regulatory standards. Residual flushing additives in the soil may be a concern and should be evaluated on a site-specific basis. How long the soil flushing process takes depends on subsurface conditions and contaminant concentrations.

20.3.6 Soil Solidification: The solidification process produces monolithic blocks of waste with high structural integrity and low solubility. The contaminants do not necessarily interact chemically with the solidification agents (typically cement/ash) but are often mechanically locked within the solidified matrix. Stabilization methods usually involve adding chemical binders (cement, silicates, pozzolans, etc.) to limit the solubility or mobility of soil constituents, even though the physical handling characteristics of the waste may not be changed or improved.

Vitrification (molten glass) processes are solidification methods that employ heat up to 1,200° C to melt and convert waste materials into glass or other glass and crystalline products. The high temperatures destroy any organic constituents with only a few by-products. Such materials as heavy metals and radionuclides are actually incorporated into the glass structure, generally a relatively strong, durable material that is resistant to leaching.

20.4 GROUNDWATER CLEANUP

Radioactive contaminants and hazardous waste constituents exist in groundwater throughout the DOE Complex and at many commercial nuclear sites. The primary remediation goal is to prevent public access to groundwater contaminated above drinking water standards. The National Research Council analyzed performances of many cleanup technologies.

20.4.1 Pump and Treat Technology: This technology continues to be a commonly applied cleanup technology for the general remediation of groundwater. Wells are installed into the path of the contaminated groundwater and it is pumped to the surface where it is treated to remove contaminants. Then clean groundwater is either re-injected to the groundwater via a well in a suitable location or, under certain regulatory conditions, discharged as surface water.

Despite its popularity, limitations of pump and treat systems should be recognized before a decision is made to apply this technology. They may be expensive to operate due to the installation of regulatory compliant wells, the high volume of water being pumped, or the equipment and consumable materials associated with some treatments. They may also take years, if ever, to meet cleanup goals. These systems are not always effective since pumping depresses the groundwater level, leaving residuals in the soil. After the groundwater level returns to its normal level, these residual contaminants dissolve a phenomenon called "rebounding." Consequently, rebound tests must be performed frequently in the first few years after a system is turned off, and after major precipitation or flooding events. The National Research Council analyzed the performance of conventional pump and treat systems and determined that a wide range of results are achieved using this method, as depicted in Table 20.1.

The first step of any groundwater remediation project consists of gathering enough background site information and field data to make assessments of the remedial requirements and possible cleanup levels. The outcome of this effort is the determination of whether cleanup or containment will be the most appropriate remedial action. If cleanup is chosen, the level of cleanup must be determined. If containment is chosen, the groundwater pumping should use hydraulic barriers to prevent off-site migration of contaminant plumes.

The next component consists of the design and implementation of the groundwater pumping system, based on the data evaluated in setting goals and objectives. The criteria for well design, pumping system, and treatment depend on the physical site characteristics and contaminant type. It is not uncommon to find many wells extracting groundwater at the same time. Actual treatment may include the design of a train of processes, such as gravity segregation, air strippers, and carbon systems tailored to remove specific contaminants. The key component of groundwater extraction is a monitoring program to verify its effectiveness. Monitoring the remediation with wells and piezometers allows for iterative adjustments to the system in response to changes in subsurface conditions caused by the remediation.

Table 20.1 Performance of Pump and Treat Systems

Result	Explanation
Unequivocal Failure	Fails to contain subsurface sources of contamination and to clean up the plume of dissolved contaminants emanating from source areas
Prevention of additional exposure to contamination	Contains subsurface sources of contamination and prevents the plume of dissolved contaminants from increasing in size
Reduction of additional exposure and significant shrinkage of area affected by the contamination	Contains subsurface sources of contamination and possibly reduces the amount of contaminant mass in the source areas; cleans up part or all of the plume of dissolved contaminants to health based standards
Unequivocal success	Fully removes sources of contamination and cleans up the plume of dissolved contaminants to health bases standards

The final component is determining when to turn the system off. For contaminants regulated by the EPA, levels established under the Safe Drinking Water Act (SDWA 1974) are usually the prevailing levels for groundwater. Termination requirements are based on the cleanup objectives defined in the initial stage of the remedial process. The termination criteria also depend on the specific site aspects revealed during remedial operations. Although pumping for containment implies no treatment, the following treatments usually follow pumping in pump and treat systems.

20.4.2 Bioreactors: Contaminants in extracted groundwater are put into contact with microorganisms in attached or suspended growth biological reactors. In suspended systems, such as activated sludge, contaminated groundwater is circulated in an aeration basin. In attached systems, such as rotating biological contractors and trickling filters, microorganisms are established on an inert support matrix.

1. *Adsorption/Absorption*: In liquid adsorption, solutes concentrate at the surface of a sorbent, thereby reducing their concentration in the bulk liquid phase. The most common adsorbent is granulated activated charcoal. Other natural and synthetic adsorbents include forage sponge, lignin adsorption, sorption clays, and synthetic resins.

2. *Air Stripping*: VOCs are partitioned from groundwater by increasing the surface area of the contaminated water exposed to air. Aeration methods include packed towers, diffused aeration, tray aeration, and spray aeration.

3. *Granulated Activated Carbon and Liquid Phase Carbon Adsorption*: Ground water is pumped through a series of canisters or columns containing activated carbon where dissolved organic contaminants are adsorbed. Periodic replacement or regeneration of saturated carbon is required.

4. *Ion Exchange*: Ion exchange removes ions from the aqueous phase by the exchange of cations or anions between the contaminants and the exchange medium. Ion exchange materials may consist of resins made from synthetic organic materials or inorganic and natural polymeric materials. After the resin capacity has been exhausted, resins may be regenerated for reuse.

5. *Precipitation, Coagulation and Flocculation*: These related processes transform dissolved contaminants into an insoluble solid, facilitating its subsequent removal from the liquid phase by sedimentation or filtration. The process usually uses pH adjustment, addition of a chemical precipitant, and flocculation.

6. *Separation*: Separation processes seek to detach contaminants from their medium (i.e., groundwater and/or binding material that contain them). Many processes can perform ex-situ separation of waste streams: distillation; filtration, ultrafiltration, and microfiltration; freeze crystallization; membrane pervaporation; and reverse osmosis.

7. *Sprinkler Irrigation*: Contaminated groundwater trickles through a filter bed while the microorganisms attached to the filter medium degrade the organic contaminants.

20.4.3 Natural Attenuation Processes: Natural attenuation processes occur at all sites but vary depending on the types and concentrations of contaminants and the hydrologic and geologic characteristics of the site. Natural attenuation may reduce the potential risks posed by site contaminants in one of three ways:

- Contaminants may be transformed to a less toxic form through destructive processes (e.g., biodegradation, radioactive decay)
- Potential exposure levels may be reduced by lowering concentration levels (e.g., dilution, dispersion)
- Contaminant mobility and bioavailability may be reduced by sorption to the soil or rock matrix

Natural attenuation processes for reducing organic contaminant levels are currently best documented at petroleum fuel sites. Organisms in the soil and groundwater break down chemicals through biological degradation processes into by-products that are often nontoxic and harmless. For example, under appropriate field conditions, the compounds benzene, toluene, ethyl benzene, and xylene may degrade naturally through microbial activity and ultimately produce nontoxic end products, such as CO_2 and H_2O.

Chlorinated solvents, such as trichloroethylene, represent another class of common organic contaminants that may also biodegrade (generally via reductive dechlorination) under certain environmental conditions. However, hydrologic and geologic conditions favoring biodegradation of chlorinated solvents may not necessarily occur at a given site.

Radionuclides decay over time. Unlike organic contaminants, radionuclides have a predictable rate of decay. The specific half-lives of radionuclides allow for accurate prediction of the time required for reducing their radioactivity to an acceptable level. Tritium is a contaminant of concern in groundwater at many nuclear sites. Due to the relatively slow movement of groundwater, the low concentrations of tritium and its relatively short half-life, natural attenuation is a common option for remediating this type of contaminated groundwater. Section 19.7 contains a more extensive discussion of the management issues related to the treatment of tritium wastes.

Other natural processes may also effectively reduce the concentrations of mobile and toxic forms of nondegradable inorganic contaminants. The movement of metal ions and radionuclides is attenuated in the subsurface via sorption to mineral surfaces or soil organic matter and occasionally through volatilization. In addition, oxidation and reduction (redox) reactions may transform the valence states of some inorganic contaminants to less soluble (and therefore less mobile) forms, or to forms that are less toxic (e.g., hexavalent to trivalent chromium).

Contaminant immobilization through natural processes is contaminant and matrix dependent. Some metal ions, including radionuclides, often have only little interaction with the matrix and therefore can move with ease through the subsurface. Furthermore, sorption may be reversible, depending on the contaminant and method of attenuation, i.e., it either becomes a permanent fixture within that particular matrix or maintains the potential for re-release.

Even though some organic and many inorganic contaminants cannot be destroyed or transformed through natural attenuation processes, they are diluted and/or dispersed as they move through the subsurface. Unlike contaminant destruction or transformation, dilution and dispersion do not lead to a reduction in contaminant mass, but rather a reduction in contaminant concentration.

Natural attenuation means relying on nature's processes, within the context of a carefully controlled and monitored site cleanup, to achieve site-specific remedial objectives within a time frame that is reasonable compared to that offered by more active methods. The key to its acceptability as a remediation method is verification of its effectiveness. Monitoring is required to ensure performance objectives are being achieved as expected. Should monitoring detect an unacceptable migration of contamination, contingency measures must be implemented to prevent unacceptable risks to human health and the environment. Therefore, the critical component of any remediation by natural attenuation is monitoring.

PART 5

DECOMMISSIONING PROJECT EXPERIENCE

Organizations embarking on a nuclear decommissioning project can tap into a wealth of experience at both commercial and government facilities in the U.S. Some of this experience began in the early 1960s at demonstration nuclear power plants like Hallam (Nebraska), Piqua (Ohio), BONUS (Puerto Rico), and Elk River (Minnesota). The federal government began doing decommissioning at many of its sites at the end of the Cold War, addressing the cleanup and closure of weapons facilities and research and development sites. Since that time, decommissioning has advanced in complexity and size to reflect current regulatory rules and guidance, waste disposal issues, technological advances, and public input and concerns.

This part of the handbook summarizes the stories of several facilities undergoing decommissioning, with examples of developing technology, problems encountered, lessons learned, and models for management planning and implementation. The summaries include a cross-section of facility types to show the different approaches each facility applied to deal with site-specific situations. The specific decommissioning projects and facility types are:

Rancho Seco	Commercial Nuclear Power Plant
Big Rock Point	Commercial Nuclear Power Plant
Haddam Neck	Commercial Nuclear Power Plant
Yankee Rowe	Commercial Nuclear Power Plant
Maine Yankee	Commercial Nuclear Power Plant
San Onofre Unit 1	Commercial Nuclear Power Plant
Barnwell Fuel Processing Plant	Commercial Fuel Reprocessing plant
Iowa State University Research Reactor	University Research Reactor
Hanford C Reactor	Department of Energy (DOE) Facility

These summaries describe the reasons for decommissioning the facility, management approach, technologies applied, and lessons learned. In many cases, the technologies were those used in earlier decommissioning programs or were adapted from other industries. This proven technology approach results in assured performance and successful programs. The summaries also describe several unanticipated problems, including bankruptcy of one of the major contractors, and issues related to vessel transport and disposal scheduling, state-versus-federal site release and clearance criteria, and site restoration. They also describe contractual arrangements between the owner and contractors, problems encountered, and how they were resolved.

Sharing this type of information furnishes valuable, experience-tested guidance for future projects, and also enhances the credibility of the decommissioning process for stakeholders and the general public. As experience grows, the industry's data can be expanded to cover a wider range of facility types. The information here is current as of 2003. As additional work is performed, the project planner should visit the websites of specific projects to get the latest information.

Chapter 21

MAINE YANKEE ATOMIC POWER PLANT

21.1 INTRODUCTION

The Maine Yankee nuclear electrical generation plant is located on an 820-acre site in Lincoln County in the Town of Wiscasset, within the mid-coast region of the State of Maine, about 27 miles northeast of Portland. The plant is owned by a consortium of ten New England electric utilities and was operated by the Maine Yankee Atomic Power Company (MYAPCO). During its 24-year operating lifetime, Maine Yankee generated and distributed more than 125 million MW-hours of electrical power.

The construction permit for Maine Yankee was issued on October 21, 1968. The operating license, issued on September 15, 1972, authorized operation of the pressurized water reactor (PWR) facility until October 21, 2008, with power levels up to 75% rated thermal power. Commercial operation began on December 28, 1972. The license granted in June 1973 allowed for operation at 100% rated thermal and at power levels up to and including 2440 MWt.

Figure 21.1 The Maine Yankee Site

Various amendments to the operating license were subsequently issued, authorizing the power station to operate, for a period, at power levels up to and including 2,700 MWt, which corresponds to a nuclear steam supply system (NSSS) output of 2,715 MWt and a gross electrical output of approximately 931 MWe. On January 3, 1996, the Nuclear Regulatory Commission (NRC) restricted power operation to 2440 MWt (90% of the currently rated licensed power), pending reviews and assessments of the effect of a loss of coolant accident (LOCA) from a small piping break.

Maine Yankee ceased power production on December 6, 1996, to address cable separation and other issues. By July 1997, the reactor had been completely defueled, with all spent fuel in the pool. In August 1997, the Board of Directors voted (for economic reasons) to permanently cease power operations and to start the decommissioning process. Maine Yankee certified to the NRC that operations had permanently ceased and that all reactor fuel had been removed. The company submitted the Post Shutdown Decommissioning Activities Report (PSDAR) and started discussing environmental impact. In November 1997, a meeting was held in Wiscasset to hear public comments on the PSDAR. The site-specific Decommissioning Cost Estimate and a PSDAR update were submitted a year later.

Maine Yankee was a three-loop, pressurized water reactor with a power rating of 2,700 MWt. Its NSSS was supplied by Asea Brown Boveri/Combustion Engineering. The secondary plant consisted of three Asea Brown Boveri turbines, one high pressure and two low pressure, coupled with a 950 MVA Westinghouse electric generator and associated auxiliary systems. The site also included ancillary facilities used to support normal plant operations: warehouses, administrative office buildings, security structures, an environmental sampling complex, a substation, and a fire protection system. The major structures were the reactor containment, primary auxiliary building, fuel building, turbine building, service building, and circulating water pump house.

The reactor containment was a steel-lined, reinforced concrete cylinder with a hemispherical dome and an essentially flat reinforced-concrete foundation mat. It served as a confinement barrier during plant operations and now provides adequate radiation shielding for any defueled or decommissioning condition. The turbine building housed the turbine generator, the component cooling water heat exchangers and pumps, and the two diesel generators. The service building contained the main control room, switchgear rooms, shops, and employee facilities. The primary auxiliary building housed equipment used for purifying and processing water from the reactor coolant system. The fuel building provided space for the storage of spent fuel and waste disposal equipment. The circulating water pump house contained the circulating water and service water pumps.

21.2 DECOMMISSIONING PLANNING

After the shutdown decision, Maine Yankee immediately established efforts in the following areas:

- Decommissioning vision and general planning
- Employee relations and staffing
- Community relations
- Assurance of adequate decommissioning funds
- Regulatory changes

Maine Yankee chose the DECON method for decommissioning and decided to return the site to a greenfield condition (except for the necessary fuel storage period). This decision was based on owner, community, and other stakeholder desires. The vision was "Green in Seven," implying complete cleanup in seven years.

Maine Yankee realized that decommissioning activities are best performed as a project and not in an operating environment. Therefore, most activities would be performed by contract organizations, skilled in construction and project functions, who could work in a focused demolition environment. This necessitated significant Maine Yankee employee staffing reductions, which were communicated quickly and continually. An on-site career center and its significant effort to find outside jobs for displaced workers helped avoid employee uncertainty. Nearly all found new jobs before they were discharged. But since many job skills were needed for various lengths of time (a few months to several years), various methods were used to avoid losing them prematurely. As the project progressed, contractors were encouraged to hire certain former employees because of their skills and knowledge of the site.

A simultaneously developed decommissioning project plan, including skill and staffing requirements, called for preparing the site for decommissioning and performing a detailed site radiological and environmental characterization. Site preparation began with isolating the spent fuel from future decommissioning activities and creating a spent fuel island so most buildings and equipment could be declassified and abandoned, and the plant could be drained and de-energized. Detailed activities included:

- Simplifying and relocating the control room
- Initiating a detailed system layup and declassification process
- Removing asbestos from accessible pipes and equipment
- Preparing the analyses for future storage of spent fuel
- Streamlining site programs and procedures to match the shutdown decommissioning environment, including revisions in modification processes and the quality assurance and training programs.

The Maine Yankee decommissioning trust fund (DTF) was inadequate because the plant's income stopped before the budgeted end of its operating license. And the reality of the Department of Energy (DOE) still being unable to accept the site's spent fuel would compel additional funds. Maine Yankee requested, and ultimately received, adjusted decommissioning funding from the Federal Energy Regulatory Commission (FERC). This significant effort, requiring additional estimates, took several months. The rate case settlement involved some unique features, such as a future "true-up" mechanism to account for any proceeds to Maine Yankee from its litigation with the DOE over the spent fuel contract (MY 2000).

As part of the overall plan, Maine Yankee chose to hire a Decommissioning Operations Contractor (DOC) on a fixed-price basis to handle as much of the decommissioning, waste management, and associated activities as possible. One prime contractor in charge of coordinating all decommissioning activities would minimize communication and coordination problems among subcontractors and reduce the extent of Maine Yankee oversight. The plan called for the DOC to begin once site preparation activities were complete. As part of the bid and scoping process, potential DOCs were invited to witness site characterization and other preparation activities, thereby gaining the knowledge necessary to make a fixed-price bid.

21.3 CHARACTERIZATION

Additional site characterization (radiological and environmental), associated modeling, and final site survey planning took place as part of the decommissioning process. Final site survey activities began for areas that were to be released early (backland areas), for the ISFSI location, and for the industrial area and structures to be demolished first.

21.4 DECOMMISSIONING ACTIONS

Although decommissioning activities took place during the preparation and planning stage, full scale decommissioning did not start until the DOC was chosen and on site. After a detailed selection process based on technical capability (including industrial and nuclear safety criteria), price, and the degree of risk transferred from the owners, a DOC was contracted in September, 1998 but terminated in May 2000 because of its financial difficulties in another business area. The search for another DOC began, but Maine Yankee finally decided to self-perform.

The initial activities had involved setting up a DOC-style administrative, physical, and project management infrastructure. This process, including establishing subcontracts and further detailed planning, took several months. Physical decommissioning work did take place during this time, most notably the physical reduction of source-term in the plant. Preparation, construction, and loading of the independent spent fuel storage installation (ISFSI) was also part of the DOC scope. Transferring its fuel to an ISFSI was determined to be the most cost-effective option for safe storage; building and loading the ISFSI became the project's first critical path item..

Large component removal began in 1999 and included such clean components as transformers and heat exchangers; contaminated components included reactor coolant pumps, other coolant system components, pressurizers, and steam generators. Vessel internals were segmented from late 2000 through early 2001 (grit blasting technique). The reactor head was segmented and shipped at mid-year; the vessel was shipped to Barnwell, SC, in June 2003. The containment crane was removed in December 2002; containment and primary auxiliary building are in progress.

Throughout these decommissioning activities, the ISFSI and casks/canisters (provided by NAC-UMS) were being designed, licensed, and built. Service building demolition began in 2001; ISFSI loading, including greater than Class C (GTCC) waste, began later that year. Turbine building demolition was completed in 2002.

Radioactive waste from Maine Yankee was, and is, being shipped to Barnwell, SC, Envirocare, UT, or Memphis, TN (GTS Duratek for the steam generators). The projected quantity is about 130 M lbs, of which 86 M lbs are concrete and 19 M lbs are soil. Solid waste (clean and noncontaminated) is projected to be 100M lbs, of which 53M lbs is concrete. As of April 2004 more than 63% of all waste had been shipped, most of it by rail.

As of June 2004, the occupational radiation exposure was about 510 person-rem out of a projected dose of about 1100. For completed radiological work and corresponding dose estimates, only about 50% of the estimated exposure was expended, largely because of early source-term reduction. The cumulative recordable incident rate for occupational exposure was 2.19, compared to the construction industry average of greater than 7.

ISFSI construction was completed as of the summer of 2001. Loading of GTCC and spent fuel into the on-site fuel storage containers of the ISFSI began in September 2001 and was completed in March 2004. Reactor vessel removal and shipment allowed the start of major demolition and of the final radiation and chemical surveys needed to release the site for unrestricted use.

21.5 REGULATORY AND STAKEHOLDER ISSUES

The shutdown of a nuclear plant has significant impact on the surrounding communities and local stakeholders. Concerns include safety, employment, local tax base, future disposition of the site and spent fuel, and final cleanup standards. Maine Yankee began frequent discussions with a wide range of stakeholders and a Community Advisory Panel (CAP) to address these and other issues. The CAP, chaired from the outset by a local elected official (State Senator), included local officials and residents, state representatives, educators, environmental activists, and Maine Yankee. The first meeting took place shortly after the shutdown announcement and continued every four to six weeks. As soon as the decision to permanently shut down was announced, Maine Yankee started implementing the necessary and desirable regulatory (primarily NRC-related) changes and made required federal submittals (10 CFR 50). Various additional NRC submittals were made as required or desired to establish the new condition of the plant, facilitate decommissioning, and reduce costs. Initial submittals included a revised SAR and technical specifications, followed by submittals for decommissioning-appropriate revisions to emergency and security plans, training and fitness for duty requirements, QA plans, and reduced insurance requirements.

Considerable effort and discussion took place about the site's radiological and environmental end-state. Although the NRC criteria for radiological conditions would clearly be met, the State of Maine and various local entities wanted tighter standards. Ultimately, an agreement specified a limit of 10 mrem whole body dose, with no more than 4 mrem attributed to groundwater. Environmental criteria were worked out with the State of Maine, with the involvement of the Environmental Protection Agency (EPA) through the state. Final versions of the License Termination Plan (LTP) and the Quality Assurance Program Plan (QAPP) were submitted for approval at mid-year 2001; these incorporated NRC radiological cleanup standards and EPA closure requirements, respectively. Agreement was reached with intervenors in September 2001. After extensive review, the NRC approved the LTP in February 2003.

As an example of how the ultimate criteria selected can alter the scope of a decommissioning project, the Maine Yankee Atomic Power Plant (permanently shut down in December of 1996) originally developed decommissioning plans and cost estimates based on the NRC unrestricted release criteria of 0.25 mSv (25 mrem) per year to the maximum exposed individual. The intent was to allow rubblized building concrete meeting acceptable residual radioactive material limits to be reused as subgrade fill. Subsequent interaction with key stakeholders, including the State of Maine, resulted in a revised decommissioning plan and cost estimates in support of unrestricted release criteria of 0.1 mSv (10 mrem) per year from all pathways, and 0.04 mSv (4 mrem) per year from the groundwater pathway with the removal of all rubblized concrete. This change resulted in more than a significant increase in projected radioactive waste volume (MY 2001).

The CAP was involved throughout this decommissioning phase to facilitate the decision process and to ensure that various stakeholders could voice their opinions to Maine Yankee. MYAPCO worked with the local communities and stakeholders on site redevelopment. This process continues as a team effort with stakeholders and outside professional consultation, including a redevelopment web site. Maine Yankee will donate 200 acres of land (originally a farm site) to a local educational foundation for conservation and environmental teaching purposes.

21.6 EXPERIENCE AND LESSONS LEARNED

Since planning is critical to success, sites with an early, unplanned shutdown are at a disadvantage. Maine Yankee was able to do some decommissioning planning (primarily in the regulatory area) early because it anticipated the potential for shutdown. All operating sites should implement and maintain certain decommissioning plan elements well before they shut down. Record keeping, characterization of spills, remediation during operations, and other issues could be costly to recreate. Early planning of the basic decommissioning strategy and associated administrative and regulatory actions could reduce decommissioning cost and schedule significantly -- especially if shutdown comes early.

Community involvement and communication is critical and should occur early and often. Maine Yankee was not able to anticipate all the issues with the State of Maine, even though the state was represented on the CAP. Early planning and communication with regulatory authorities (including any agencies involved with permitting, applications, or legislative authority) must be specifically addressed and managed. The impact on the local community is also significant. Early and open discussion of impact resolution is critical to successful community relations.

NRC regulatory submittals should be prepared before shutdown and initiated as soon as possible. The majority of these submittals create economic benefits to the owner and the ratepayer -- the longer the wait the greater the cost. The NRC will not necessarily act expeditiously when approvals are needed, as it may find that issues at operating plants are more important. However, most of the experiences from Maine Yankee are now captured in the NRC by rule or precedent, which should speed the approval of appropriate submittals.

The decision to hire a DOC should be made on a case by case basis. There may be advantages to not using a DOC (self-performance, use of smaller scope decommissioning contractors, etc.). This decision depends on many factors, including the specified pace of decommissioning, the extent of pre-planning and site preparations, the availability of skilled contractors, existing decommissioning skills on-site , and proven technologies or techniques. Close monitoring and oversight are necessary for any contractor hired. Even though a contract may transfer monetary risk, a contractor's failure to perform could hurt the project and project owners.

Proper preparation of the site is important, regardless of the contract approach. Isolating spent fuel, determining the desired end-state (and when), managing employee issues, implementing appropriate regulatory requirements, communicating with stakeholders, and good project planning are all always necessary.

Since appropriate planning and source reduction significantly reduce radiation dose, the rule should be to "remove the hot stuff first." This premise and practice resulted in the vast majority of activities being accomplished at lower than the projected dose. (Early in the project, during initial asbestos removal, the overall activity exceeded dose projections primarily because of unanticipated insulation adherence to components.)

Waste disposal (both radioactive and hazardous) is complex and requires careful planning and implementation. Outside stakeholder concerns may also surface here, especially on transportation. Although nuclear sites are typically cleaner, from a hazardous waste perspective, than many other industrial sites, the planning, characterization, implementation, and regulatory approvals are significant and require the project organization's special attention.

Industrial safety takes continual attention. The Maine Yankee project has a good overall safety record, but several near misses and occasional elevated accident rates have caused some work stoppages. During decommissioning, which is a very different work environment than operations, all site personnel, including contractor management, must focus on safety. Its importance should be emphasized in employee and contract incentive structures.

Even though a nuclear plant is in decommissioning and no longer operating, many aspects of the nuclear regulatory environment continue. Because many workers may not have nuclear experience, radiation protection issues may arise. Since the working environment involves frequent changes in personnel and supervisory staff (typically contractors), an environment is created to surface worker concerns. It is recommended that the plant staff maintain firm control of the radiation protection and worker concern programs; any delegation must involve careful plant staff oversight.

New technologies used in decommissioning and fuel storage may involve unexpected costs. For instance, the grit blasting technique used in internal segmentation was relatively new and required considerable time to debug. However, because this debugging was done off site, the on-site job was completed with radiological dose much lower than planned. However, the job started late and some equipment breakdowns during the work caused further schedule delays. The spent fuel storage system purchased by Maine Yankee involved enough new equipment and fabrication techniques that the schedule was delayed several months. In general, expect delay in any activity or process involving equipment or techniques that the vendor has not used before. For both cases above, it is apparent, that the next use of the processes will likely go much smoother, since the vendors will have refined their technology.

In general, Maine Yankee has been able to achieve cost and schedule goals with two exceptions. The DOC termination caused schedule delays and increased costs. Commitments associated with end-state conditions have increased costs and could delay the schedule. But all stakeholders accepted these commitments as "the right thing to do." Due to its highly public profile, a decommissioning project is not complete until all, including stakeholders, agree that it is complete. Guiding all concerned parties toward a successful end-state requires that teamwork be established as a primary project goal.

Chapter 22

RANCHO SECO NUCLEAR PLANT

22.1 INTRODUCTION

Rancho Seco was a single unit, 913-MW electrical, Babcock & Wilcox design, nuclear power plant owned by the Sacramento Municipal Utility District (SMUD). The plant is 25 miles southeast of Sacramento, California on 2,480 acres of rolling terrain. Commercial operation of this PWR began in 1975 and ceased in 1989. The NSSS, housed within a hardened containment building, consisted of two independent primary coolant loops (each containing two reactor coolant pumps and a steam generator), an electrically heated pressurizer, and connecting piping. Major structures included two cooling towers, reactor building, administration building, auxiliary building, training and records building, nuclear service electrical building, interim onsite (waste) storage building, and support buildings (security, warehouses, and an unfinished technical center). An ISFSI was added in 1995.

Figure 22.1 The Rancho Seco Site

22.2 DECOMMISSIONING PLANNING

After a troubled operating history, Rancho Seco shut down in June 1989 as the result of a voter referendum. Due to a minimal decommissioning fund balance, the decision was made to enter an extended period of SAFSTOR to allow radioactive decay and the accumulation of enough funds for dismantling. The intent was to have the project be performed by a DOC in 2008. In 1991, the decision was made to place the spent fuel into dry storage, allowing the plant to enter a "hardened" SAFSTOR condition and reducing the required staff significantly. But delays caused the fuel transfer to fall behind schedule.

The baseline decommissioning cost estimate used a value of $405/ft^3 for disposal at the planned Ward Valley, CA disposal site. While this value is no longer valid, and the site opening is unlikely, it was used for comparison. The Envirocare disposal facility in Clive, UT, became available as an option for disposal of low-level wastes (LLW) in 1995. With a waste cost significantly below that estimated for Ward Valley, the Envirocare facility yielded significant savings for disposal of such LLW as steam and cooling systems in the turbine building that were contaminated from system-to-system leaks. This building was selected for initial dismantling activities because of the large volumes of potentially contaminated materials and the very low activity levels expected, allowing for minimal radiological controls on the work. Given the wait for fuel movement and the possibility of significant cost savings, a three-year incremental decommissioning project was proposed to dismantle the turbine building systems and a portion of the tank farm systems. The project was approved for 1997, with annual renewals based on performance.

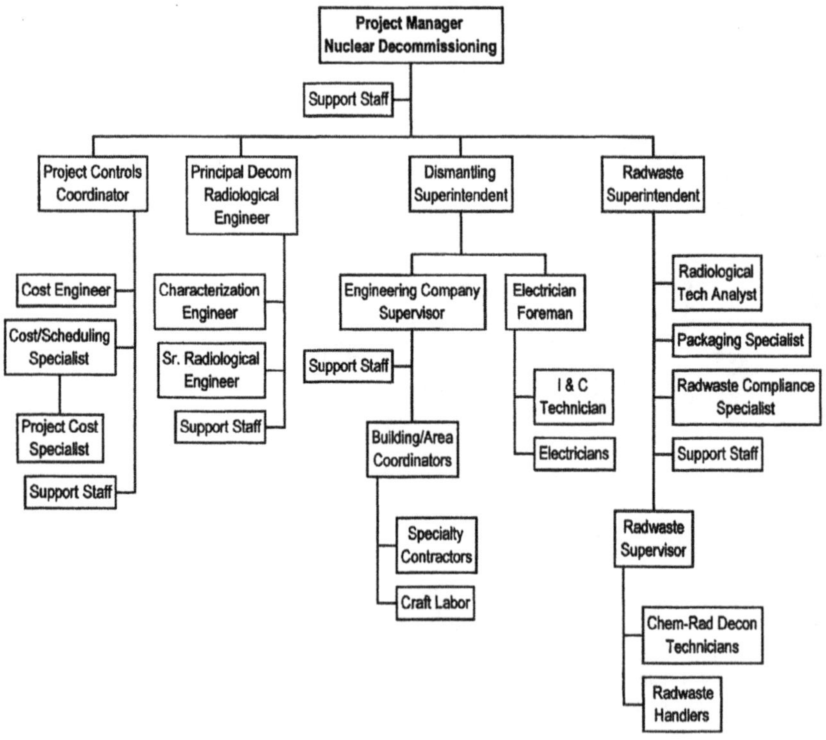

Figure 22.2 Decommissioning Project Team

Based on the success of the incremental project, on July 1, 1999, the SMUD Board approved continuing decommissioning to completion. The approval of the entire project was necessary to allow appropriate planning for long-lead items. Insufficient funding slowed initial dismantling, increasing only after the spent fuel was transferred to dry storage, thereby alleviating licensing requirements. Current planning shows a 2008 completion date for dismantling. No building demolition is planned beyond that necessary to allow license termination, but temporary buildings are being removed to simplify final radiological surveys.

Recognizing that decommissioning is to continue to completion, the staff was reorganized to meet the concomitant challenges. The transition was complete after fuel was moved and a licensed operations staff was no longer needed. The dismantling project team of about 65 is shown in Figure 22.2. A plant staff of about 100 supports the team with engineering, maintenance, radiation protection, quality/licensing activities, and document control.

Because the original cost estimate used a system-based approach that did not lend itself to project tracking, a new one was developed based on expected tasks sequenced by area. This estimate was loaded into a scheduling program to develop the entire project schedule and the required man-loading. Tracking tools were developed for schedule and cost to monitor progress. The approximate schedule is listed in Table 22.1.

Until 2000, job planning was barely ahead of ongoing work. But once the decommissioning project was approved, it became a priority to develop the planning process to support a complete detailed schedule. A high-level eight-year schedule was developed as a framework. Then, the detailed schedule was slowly extended to include the next two years. The transfer of spent fuel was the biggest unknown; it continued to slip, causing many challenges until completed in August 2002.

Table 22.1 Major Item Schedule

Activity Description	Start	Finish
Auxiliary Building System Removal	Sept. 1999	Dec. 2003
Reactor Building Preparation	Jan. 2000	Dec. 2000
Reactor Bldg Sys. Removal (non-RCS)	Jan. 2001	Dec. 2003
Remove RCP Motors	May 2001	Oct. 2001
Move Spent Fuel to ISFSI	Apr. 2001	Aug. 2002
Spent Fuel Pool Decontamination	Sept. 2002	Nov. 2004
Remove Reactor Coolant Piping	Jan. 2002	May 2002
Remove Reactor Coolant Pumps	May 2002	Dec. 2002
Remove Steam Generators	May 2005	July 2005
Remove Pressurizer	Jan. 2004	Mar. 2004
Remove Outside Tanks	Feb. 2003	Oct. 2003
Remove Underground Piping	Jan. 2004	Dec. 2004
License Termination Plan Preparation	Jan. 2004	Oct. 2004
Reactor Vessel Internals Removal	Jan. 2005	July 2006
Reactor Vessel Cut-up or Shipment	July 2006	May 2007
Building Decontamination	July 2005	Feb. 2008
Partial Site Release	Oct. 2006	Oct. 2008
Perform Final Survey	Jan. 2031	Dec. 2031

With the exception of fuel movement, most items have met or surpassed their planned completion dates. Dismantling work has been under the budgeted task amounts, but the savings will not cover fuel project overruns. Allowing planning to develop over time permitted evaluation of alternative methods for dismantling and the incorporation of lessons learned from more mature projects.

22.2.1 Spent Fuel Pool Island: A small self-contained cooling and water cleanup system was designed and installed to allow the removal of normal plant cooling systems and the isolation of the spent fuel pool. The system is unique in that it uses a refrigeration system for cooling. All of the equipment sat next to the pool, except for the condensing unit that was just outside the building. The system included a filter and a portable demineralizer unit.

The pool had a relatively constant leak rate of approximately 50 g/d that was directed to the liquid radwaste system. To avoid treating this water for release, a return system that passed through a filter and demineralizer was also installed. This system also stopped the reduction in boron concentration in the pool (resulting from subsequent replacement with demineralized water). Water with significant boron or tritium has been processed and released. Liner removal began in mid-2003.

22.2.2 Liquid Radioactive Waste System Reconfiguration: More than half the equipment in the auxiliary building is part of the liquid radwaste system. It was reconfigured to a smaller, simpler system to allow for removal of a significant portion of the auxiliary building equipment. Most of the tanks, pumps, and valves, both evaporators, and all of the filters and ion exchange vessels were abandoned. To accomplish this, accumulated water was processed with a (leased) reverse osmosis system and all remaining resin was transferred to high integrity containers (HICs) for storage and eventual shipment for disposal. Further accumulations of water were processed with a portable ion exchange system before release.

The remaining liquid radwaste system now consists of three sumps and pumps, a holding tank and its recirculation pump, a connection area for temporary equipment, including filters and media vessels, and a drum drying system for disposal of the concentrates and sludges. The system will be further downsized in time, and planning for sampling under the pool is in progress.
Key Rancho Seco procedures developed to control the dismantling process include:

- Plant Modifications (defines and controls Design Change Packages) (RS 2002a)
- Removal of Tools and Equipment From Controlled Areas (defines and controls the use of Inaccessible Surface Contamination Evaluation Forms) (RS 2002b)
- Control of Decommissioning Projects (defines and controls Integrated Decommissioning Packages and Decommissioning Work Permits) (RS 2003)

Systems and portions of systems would be turned over for removal under an Integrated Decommissioning Package. Isolating operating systems would use a Design Change Package. Removal would be done with Decommissioning Work Permits assigned on an area-by-area basis. The permits define exactly what is to be removed, the order of removal, and any hazards or hazardous materials in the area and how they are to be handled. Permits are used for worker briefing and clearance preparation, including system draining; the operations group then marks the pipe and equipment items that are ready for removal.

22.2.3 Planning for Large Components: The major components are expected to be removed between 2003 and 2006 and sent ultimately to Envirocare, with some sizing for transportation. Current plans call for the steam generators to be cut in half by wire saw, with each half to be capped and shipped by rail. Steam generator penetration closures are being installed and the necessary transportation exemptions are being pursued. The pressurizer, expected to ship intact in 2005, will travel by rail. Detailed characterization of the vessel and internals is complete. Internals will be segmented, with the GTCC waste placed within a module in the ISFSI. Class B and C internals will be segmented, packaged, and stored until Envirocare obtains its Class B&C license or until other disposal is available. The vessel and other Class A waste will go to Envirocare after segmentation and packaging. Remaining components include tanks and the spent fuel cooler. The underground piping removal began in the spring of 2004.

Current planning and budgeting calls for no building demolition beyond that required to achieve license termination. The utility plans to further develop the site with other energy projects. Two 1 MW photovoltaic units are operable, with another under construction. Construction is currently underway on the first phase (500 MWe) of a 1000 MWe combined-cycle, natural gas powered plant on the property that will make use of the switchyard and water supply. Final state approval was recently granted for the gas-fired plant.

22.3 CHARACTERIZATION

Initial measurements showed, with a few exceptions, that contamination levels in the turbine building components were low, allowing minimal radiation protection. No significant activity existed in the building itself. When measured with a thin window pancake probe, most components indicated only a few hundred counts per minute, if any, thus lowering the protective clothing requirements to work gloves and personal coveralls; personnel monitoring was simply a matter of frisking at work area exits.

The most contaminated components in the secondary plant were found to be the high pressure turbine and the moisture separator/reheaters. These components had loose contamination levels up to 20,000 dpm/100 cm^2 in isolated locations. Spraying a fixative on the surfaces before handling controlled the loose contamination. With only a few exceptions, dose rates on components were less than 0.1 mrad/hr.

22.3.1 Survey for Release: One of the most effective cost saving measures is surveying material for release as noncontaminated. Most material was expected to be free of contamination, but all system components in the turbine building were required to be surveyed as they were removed and size-reduced. Some system components required minimal surveys due to their history of no internal contamination, but steam systems required extensive survey of all surfaces.

Inaccessible surfaces require components to be cut open for survey or disposed of as contaminated. When inaccessible components were expected to be clean, a procedure was followed that allowed for an evaluation of its projected contamination levels. An Inaccessible Contamination Evaluation required system knowledge, a survey of accessible areas, and sometimes a sampling of inaccessible areas by destructive means. Items released under this program included the third and fourth point feedwater heaters, most of the auxiliary boilers, and the outer turbine covers. Major portions of the condenser and the fifth and sixth point feedwater heaters were also released in this manner.

Two different incidents caused the radiation monitors at the scrap yards to reject truckloads of scrap metal. In one case, a small area was apparently missed in the survey process. In the other, a second survey of the material by special means showed that a very small amount of activity (below levels that could be detected by normal survey procedures) was distributed over many pieces, causing the scrap yard monitors to alarm. In either case, the activity levels were not high enough to cause concern over unmonitored exposure to the public.

The corrective actions included survey procedure revisions including a grid map of materials, initialing the grids when released, and extensive training on the procedures. A belt monitor was purchased to allow automated monitoring of small items. Also, a greater portion of the material that might contain very low levels was sent for decontamination, and a truck monitor was purchased. The truck monitor performs a final check before release to ensure that aggregate quantities can be evaluated. Although these measures resulted in increased cost due to increased survey and decontamination time, the free release of material still remains the most cost-effective disposition. As of the end of 2002, the survey program has been responsible for the recycle of approximately 8.5M lbs of material that might have otherwise been sent for disposal.

22.3.2 Owner-Controlled Area Survey: A radiological survey of utility property beyond the security fence was performed to determine background levels, and to bound the impacted area. The purpose of the survey was to meet the intent of the planned rulemaking for partial site release as "not different from background" and may be used for that purpose. This survey also supported the planned addition of a gas-fired unit on utility property.

22.3.3 Historical Site Assessment: An Historical Site Assessment (HSA) was performed to support a final survey plan and license termination. The assessment included extensive document reviews for radiological incidents and interviews of selected long-term employees. Data was correlated to plant locations and eventual survey units.

22.3.4 Safety and ALARA: An active safety program resulted in no contractor lost-time accidents in 2000, 2001, and 2002, and only one OSHA-recordable injury in 2000, three in 2001, and one in 2002. On the utility side, one lost-time accident and two recordable injuries were counted in 2000 and no lost-time accidents and one recordable injury in 2001, significantly better than previous years. For 2002, there were one lost-time accident and four recordable injuries.

A high-level estimate of dose was completed (an ALARA assessment), along with the project cost estimate, but a detailed estimate is made annually, based on tasks to be performed. An active ALARA program (and a decade of decay) has kept the site dose significantly below that of other large reactor decommissioning projects. Through 2002, and including dose from fuel movement, decommissioning activities have resulted in site dose of 88 person-rem.

A program to remove hot spots in the auxiliary building (mostly in valves) was performed in 2000. These hot spots caused many rooms to be radiation or high radiation areas. The valves were removed before major work, simplifying work and at a significantly lower exposure. A similar program was completed for the reactor building in 2001.

22.4 DECOMMISSIONING ACTIONS

Significant preparations were needed before actual dismantling could start: contracts for disposal, shipping, waste processing, and labor; engineering for the abandonment of systems and components and the necessary isolations from active systems; and procedures governing dismantling. When all was ready, site personnel began dismantling activities.

An interdisciplinary team of loaned employees, including people from radiation protection, operations, maintenance, and engineering, was formed to manage the work and the waste. Specialized waste and decommissioning personnel supplemented the group. It then took more than a year to get the required contracts and additional specialized employees in place before waste could be shipped. Contracts were required for: waste processing, disposal, and shipping; contract labor and equipment; asbestos abatement; lead paint abatement; and specialized personnel. All these contracts were competitively bid, resulting in long lead times.

Piping and equipment items were marked with the system that they came from to help radwaste personnel determine disposition. Through-wall or underground piping stubs were marked with the system and line number to facilitate later removal or decontamination. Although previous characterization work had shown that most turbine building systems would be nonradioactive, except for the cooling water system, main steam, auxiliary steam, first and second point heaters, reheaters, and the turbine itself, all systems were removed to simplify final survey activities.

Actual dismantling began with demineralized water systems, the condensate polishers, and chemical addition systems. Next were feedwater heaters and reheaters, followed by the turbine, feed pumps, and finally the condenser. Most dismantling of lightly contaminated components used standard oxyacetylene torches. Completion of the turbine building occurred in mid-1999, with outside work continuing in the tank farm and pipe chase areas.

System removal began in the auxiliary building in September 1999 and will continue in the auxiliary and reactor buildings into 2004. Planning is in progress for decontaminating embedded piping Tank removal has been the most challenging task. Work in the reactor building began in late 2000 with a major cleaning to allow improved access to highly contaminated areas. A polar crane upgrade to radio control was completed. A temporary building was placed over the equipment hatch to allow an alternate access point and a laydown area for equipment processing and packaging. Major items completed to date include the removal of the reactor coolant pumps and motors, the normal and emergency ventilation units, and large bore reactor coolant piping. Small bore piping and electrical removal is in progress and is expected to be completed in 2004.

22.4.1 Decontamination: A booth was installed to allow grit blasting of lightly contaminated materials. It was found to be very economical for high density carbon-steel materials and was used on a large portion of the steam system components and piping that were found to be contaminated. Components were pre-sized to fit in the booth and have all necessary surfaces exposed. Once blasting was completed on contaminated surfaces, a complete survey was performed for material release. Since the more contaminated materials were sent directly to packaging for disposal, few items failed the survey after blasting. Approximately 8.5 M lbs of material have been successfully cleared and sent for recycling. About half this amount was decontaminated by grit blasting.

Some components that were mostly clean, but could not easily be decontaminated or surveyed, were sent off site for processing if deemed economical to do so. These included portions of the moisture separator/reheaters and the first and second point heater tubes.

22.4.2 Packaging and Disposal: Disposal was the remaining option for those items that could not be free-released, decontaminated on site, or economically sent for processing, Disposal cost is mostly a function of dollars per cubic foot. Therefore, packaging efficiency or density of the burial package is the most important factor driving disposal cost. To qualify at Envirocare for standard disposal, waste had to have one dimension that was no greater than 10 inches. This required that most waste material be cut to meet these criteria. Material that did not pass release criteria was transferred to a staging shop where it was torch-cut, sized, and shaped for packaging efficiency. Another innovative approach, placing smaller pipe in larger pipe, resulted in extremely heavy containers. One 100 ft^3 box exceeded 30,000 lbs; many were greater than 20,000 lbs.

Most waste from Rancho Seco goes to Envirocare. High density waste is packaged for disposal at Envirocare of Utah. Through 2002, approximately 104,000 ft^3 of boxed, drummed, and oversized waste were shipped. Most LLW is shipped for processing based on weight. Approximately 1,132,000 lbs were shipped for processing through 2002. Some high-level waste (HLW) (<100 ft^3 of Class B&C waste) was generated and is currently stored. Approximately 700 ft^3 of high activity resin is also in storage. Disposition of this waste will be to Envirocare when licensed, or to other sites that might become available. There is estimated to be 133,000 ft^3 of LLW remaining.

22.4.3 Hazardous Materials: Issues involving hazardous materials -- asbestos (in insulation, wiring, and roofing), lead in painted surfaces, and residual chemicals in process systems -- continue to have a major effect on the project. Also, additional health physics measures were required because of the thermal cutting of various metals. The original decommissioning cost estimate underestimated the expense for hazardous material handling and disposal.

The removal of asbestos from steam piping and components was a major effort. The standard procedure is to survey an item for activity, tent the area, and remove the asbestos. If the pipe or component could be surveyed and released, the whole section went to the asbestos disposal site, minimizing the effort; this led to the practice of glove-bagging the area of pipe to be cut and surveying the inside after removal, leaving the asbestos on the outside of the pipe. If large bore pipe were internally contaminated, it was moved to a central asbestos tent and remediated in batches.

Essentially all paint has been treated as lead-based. Where torch cutting was required, a lead contractor removed the paint in a band where the cut would occur. If possible, cutting was done with saws or machining devices, with the lead paint chips held in place by applying ordinary shaving cream to the cut area. Composite sampling of paint chips has yielded results of less than 1 ppm of polychlorinated biphenyls (PCBs), significantly less than the allowable limit.

Chemical addition systems were flushed and released by a contract service before removal. In spite of this, chemicals were found in released piping, requiring re-cleaning before disposition.

22.4.4 Tank Cleaning: Radwaste tank cleaning before removal is a high exposure but necessary activity. Some tanks have a thin layer of activated oxide material, some have significant sludge from sumps, and others have fine particle activated material. Each presents a set of problems for removal and waste handling. Early work on low activity tanks was done by washing the material to a sump, but this only allows the problem material to be revisited when the sump is cleaned.

The first high activity tank to be cleaned was the miscellaneous waste tank containing sump waste and a significant quantity of sludge. The sludge was washed to a bag filter system by high pressure water lance. Bag handling caused more dose than anticipated and left many filters to process. This was not an acceptable alternative for higher activity tanks. A method developed for the next tanks to flush the sludge to a high integrity container containing previously discharged resin allowed for low exposure handling and activity, averaged over the resin waste, with no increase in waste volume.

22.4.5 Tank Cutup: Tank removal was a significant obstacle. Small tanks and vessels were removed whole if possible. Larger ones had to be sized for removal from rooms and cubicles. Tanks were either stainless steel or carbon steel with paint and an interior coating. Paint and coating was removed manually along cut lines and the tank was cut by torch. For stainless tanks a variety of methods were used, including saws, machining, and plasma torch. Plasma torch cutting was the most efficient, but created smoke that plugged filters and required tenting the local area to control fumes. Concern for airborne heavy metals required health physics (HP) support. Each tank was evaluated to determine the best method.

22.5 UNIQUE ISSUES

22.5.1 Dry Fuel Project: The original decision to move the fuel to an ISFSI would allow the plant to go to a hardened SAFSTOR condition, thus minimizing the staff, and therefore the cost, during the dormancy period before deferred DECON. SMUD decided that a transportable dry cask system was needed to allow the fuel to be transported to the DOE without replacing it in a fuel pool for repackaging. Since no such system existed, SMUD decided to develop and purchase a unique, large-scale, canister-based system. The contract for the design, licensing, and fabrication of a transportable storage system was signed in 1992. In 1995 the ISFSI was constructed, and fabrication of the cask and associated equipment began. But quality issues forced a work stoppage in 1996, and a new supplier resumed the design and license work the following year. Work was completed to allow a May 2001 start to fuel movement, with completion in August of 2002.

The transportable storage system consists of a transportation cask, 21 dry shielded canisters, 22 horizontal storage modules, and a multi-axle trailer. The cask serves for on-site transfer and as a transportation overpack for the canisters, seal-welded at both ends and holding the spent fuel in a structural array. The horizontal storage modules are thick, reinforced-concrete, storage bunkers for storing the canisters. The twenty-second module will store GTCC waste from the vessel internals. The ISFSI was licensed to allow nonfuel-bearing components, such as control rods, burnable poison rods, startup sources and retainer clips, to be stored with the fuel, minimizing the need for separate handling and storage.

22.5.2 Miscellaneous Issues: Most of the early decommissioning activities occurred outside or in fairly open structures. This works well when the weather is good. Large components were segmented in the open-air turbine building and crane laydown areas. But it was too hot for workers and survey technicians on some summer afternoons, and radioactive components had to be moved indoors when the winter rains came.

Material handling was a major problem given the large volume of potential waste materials and the number of processing steps each item had to complete before release or packaging for disposal. As major components were removed, the work areas became filled with material being cut up and awaiting survey, decontamination, or size reduction. It was difficult to balance the personnel needed to perform all these functions so that material did not back up. This problem diminished with the arrival of more radiation protection technicians and decontamination staff.

22.6 OUTCOME AND LESSONS LEARNED

The slow increase of activities from an incremental project to a full decommissioning program has allowed time for innovation, and trial and error, in the process. While the approval for full decommissioning required catching up in the planning process, much of that planning was completed within 2001. The project goal was to have a three-year lead on planning. Another goal was to determine the ultimate disposition of the major components, thereby allowing detailed planning to begin.

Try not to be the first to do a major activity. If the dry fuel project were to start today, it could be completed in a couple of years, but it has caused major cost and schedule issues that could have been avoided in retrospect. The long, slow dismantling effort has allowed for the incorporation of lessons learned from other projects.

Industrial safety and health physics staff are paramount during decommissioning. It takes constant management attention to minimize the extra hazards of adding a construction environment on top of a nuclear dismantling environment.

As mentioned earlier, dismantling cannot use the same procedures and controls as those for operating and maintaining the plant. As many procedures as possible should be simplified or removed, since their original purpose or requirements may no longer exist. Failure to eliminate superfluous programs might cause cost overruns just as significant as those caused by delays in system or component removal.

Some costs might be avoided. If possible, transfer major components to other operating units on the site or to other licensees, thus avoiding disposal cost. For Rancho Seco, these transfers included the high-pressure turbine rotor and two MSR tube bundles sent to another nuclear power plant.

And finally, project personnel should be willing to work themselves out of a job. It is difficult for a staff who operated the plant to change its focus to dismantling. If the mentality to get the job done cannot be directed and maintained with remaining personnel, hire a DOC.

Chapter 23

TROJAN NUCLEAR PLANT

23.1 INTRODUCTION

The Trojan Nuclear Plant is located in Columbia County, Oregon, about 42 miles north of Portland. The 634-acre site includes a recreational area/park, various office buildings, and an industrial area enclosed by a security fence. Portland General Electric (PGE) is the majority owner and has responsibility for operating and maintaining Trojan. The Bonneville Power Administration, a power marketing agency under the DOE, is obligated through net billing agreements to pay costs associated with Eugene Water and Electric Board's share of Trojan, including decommissioning and spent fuel management costs.

Figure 23.1 The Trojan Site

Construction of Trojan began in February 1971. It achieved initial criticality in December 1975 and began commercial operation in May 1976. The PWR reactor output was rated at 3411 MWt with an approximate net electrical output rating of 1130 MWe. The plant operated for 14 fuel cycles and approximately 3300 effective full-power days. The NSSS was a four-loop, pressurized water reactor designed by Westinghouse Electric Corporation.

PGE performed a least-cost planning analysis in 1992, as required by the state's Public Utilities Commission (PUC). As a result of this analysis (favoring natural gas and regional hydroelectric surplus power) and other factors, Trojan permanently ceased operations. It was shut down on November 9, 1992. On January 27, 1993, after approximately 17 years of operation, PGE notified the NRC of its decision to permanently cease power generation. The NRC amended the operating license to a possession-only license (POL) on May 5, 1993.

The decision was predicated on both financial and reliability considerations. These factors included escalating inspection and repair costs associated with steam generator tube cracking, regulatory uncertainties concerning steam generator inspection and repair, availability of inexpensive replacement power from the southwest, and inexpensive and plentiful natural gas supplies. In March 1995, the Oregon Public Utilities Commission reviewed PGE's closure decision and granted 87% recovery of capital investment and 100% of the decommissioning costs.

23.2 DECOMMISSIONING PLANNING

PGE reviewed the decommissioning alternatives described in NUREG-0586 (NRC 2002), which defines the three alternatives of DECON, SAFSTOR, and ENTOMB. The ENTOMB option was considered less desirable since it would result in a decommissioning period greater than 60 years. Consequently, PGE selected several DECON and SAFSTOR implementation methods for detailed review and analysis. Evaluation of the alternatives considered cost, LLW volume, occupational radiation exposure, and public radiation exposure. The evaluation of SAFSTOR was based on deferring the final dismantling of the facility until 2018 -- a period of 25 y. The time frame was based on the schedule for transfer of spent nuclear fuel from PGE to the DOE. The results of the 2018 SAFSTOR were extrapolated to show estimated results for a 60-year SAFSTOR (maximum allowable by current regulations). The results of the evaluation were as shown in Table 23.1.

PGE chose the DECON alternative for decommissioning since, given prompt site remediation and release for unrestricted use, it would minimize financial uncertainties associated with waste disposal and other decommissioning costs. And yet it could still take advantage of the 60 y decommissioning time limit specified (10 CFR 50.82(a)(3)). The DECON alternative would also allow the use of experienced plant personnel.

Table 23.1 Estimated Costs For Decommissioning Alternatives

	DECON	25 Year SAFSTOR	<60 Year SAFSTOR
Cost (millions, 1993 dollars)	$197	$250	$293
Low Level Radioactive Waste Volume (ft^3)	313,000	313,000	149,000
Occupational Radiation Exposure (person-rem)	591	245	233
Public Radiation Exposure (person-rem)	4.8	4.8	2.2

Once operations had ceased, Trojan immediately began the transition from an operating plant to a permanently shutdown facility. This included simplifying processes, reducing the regulatory burden, and adjusting staffing. Procedures were reviewed and evaluated for continued use and applicability. Trojan began its de-licensing process by reviewing and evaluating the regulatory requirements for operations under for their applicability to a permanently shutdown plant. Table 23.2 lists the major de-licensing milestones.

Table 23.2 Major Milestones for De-Licensing

Action	Date
Possession-Only License	05/05/93
Certified Fuel Handler Tech Spec Change	05/06/93
Defueled Fire Protection Program	09/22/93
Defueled Emergency Plan	09/30/93
Defueled Security Plan	09/30/93
Quality Assurance Program	10/28/93
Radiological/Environmental Tech Specs	12/06/93
Decommissioning Plan, submitted	01/26/95
Permanently Defueled Tech Specs	03/31/95

PGE also began an aggressive de-staffing process. Staff reductions started with the immediate dismissal of almost all of the contractors (approximately 350 people). Regular employee staffing levels were reduced from over 900 to about 150 Trojan personnel, using a stair-step approach. Layoffs coincided with completion of specific milestones and work activities.

Activities that facilitated the transition to a permanently shutdown facility included:

1. *Assessment of the functional requirements for plant systems, structures, and components*: the identification of those plant systems, structures, and components required to support safe storage of the spent fuel, support spent fuel pool cooling, and facilitate ongoing plant activities.

2. *Deactivation and removal of plant systems, structures, and components*: Those that were not required were deactivated, drained, and removed as appropriate. A comprehensive plant layup program was developed and implemented. A total of 97 out of 120 systems were deactivated. Related procedures were updated and 1,200 drawings were revised to reflect the changes.

3. *Redefinition of regulatory basis for the defueled plant*: The conversion to a POL rendered many of the existing provisions of the technical specifications inappropriate. PGE submitted a request to the NRC in July 1993; it was supplemented in March 1994 to revise the specifications. The Final Safety Analysis Report (FSAR), revised and retitled as "Defueled Safety Analysis Report," was submitted in October 1993. The NRC issued its Amendment to Facility Operating License (Possession Only) in March of 1995. This amendment allowed the specifications to reflect the plant's permanently defueled condition, and the regulatory requirements and operating restrictions to ensure safe storage in the spent fuel pool. Additional in-house licensing documents were also revised to reflect the plant's defueled condition.

23.3 CHARACTERIZATION

The Trojan site characterization is an ongoing process similar to that described in the "Manual for Conducting Radiological Surveys in Support of License Termination" (NRC 1992). The site characterization will be completed in three phases, as follows:

1. *Phase I, Scoping Survey and Site Characterization*: This phase is completed. It was used to characterize the current radiological status of the facility, and to estimate the site source-term and isotopic mixture to support decommissioning cost estimates and decision-making. It was also used to determine the location and extent of contamination outside the radiologically controlled area, and to collect background information to help facilitate release of the site for unrestricted use.

2. *Phase II, Radiological Survey*: This phase involves routine surveys in support of PGE's current license. Phase II is used to help support facility decontamination and dismantling, and uses the existing radiation protection program and procedures.

3. *Phase III, Final Radiation Survey*: This phase will demonstrate that the radiological condition is within the site release criteria to support license termination. Areas that were not, or could not be, surveyed during Phase I will be surveyed in later phases.

23.4 DECOMMISSIONING ACTIONS

23.4.1 Large Component Removal: The first major decommissioning project, the Large Component Removal (LCR) Project, was completed in November 1995. This project entailed removing the four steam generators and pressurizer from the containment building, preparing them as their own shipping containers, and transporting them via barge and overland transporter to the U.S. Ecology LLW facility near Richland, WA.

Planning for the decommissioning project began after Trojan was permanently closed. While evaluating options, PGE personnel determined that performing the LCR scope of work as soon as possible could provide significant cost and strategic advantages within the overall process. It was estimated that approximately $4.5 million in cost savings could be realized by performing the LCR project before the end of 1995. This was based on favorable rates negotiated with the burial facility and efficient use of remaining plant personnel. The scope was well defined as the removal of only five components, yet it maximized burial volume (and future cost savings) due to their large mass. (The scope of work to be performed before decommissioning plan approval had precedent at Yankee Rowe and at other plants that had replaced steam generators under provisions of 10 CFR 50.59.) Removing these components would also improve in-plant logistics for future decommissioning activities. Finally, the LCR project would allow retention of knowledgeable plant personnel and provide them valuable experience for future activities. Based on these advantages, the PGE Board of Directors approved the project in February 1994.

Since the LCR project was performed under 10 CFR 50.59, the NRC did not formally approve the project even though it allowed the use of decommissioning funds. Yet, greater than normal regulatory oversight was involved because:

- The project was being performed before Decommissioning Plan (DP) approval.
- The existing regulations did not apply directly to the shipments.
- Water and land transport issues arose as a consequence of court actions in response to cases brought by both Trojan and Yankee Rowe stakeholder opponents.

After Trojan was shut down, Oregon's Energy Facility Siting Council began developing state rules for decommissioning. A provision was incorporated into the Oregon rules to allow the LCR project before state approval of the DP, with the stipulation that a project plan be submitted to the Council before starting the project. The state's decommissioning rules were adopted in June 1994, after which the LCR Project Plan was immediately submitted for approval. The plan was scrutinized by an independent technical reviewer hired by the Council and a Special Advisory Group comprised of representatives from the local community, PGE, and other interested parties. The Council approved the plan and adopted new Oregon rules specific to the LCR Project on November 17, 1995. The new rules had a stipulation that PGE submit a Transportation Plan to the Oregon Department of Energy for approval before shipping the components. The Transportation Safety Plan, submitted in May 1995, was approved in August. This ended a two-year public process to obtain final approval of the LCR shipments.

The LCR project was scheduled to begin with engineering preparations during the fall of 1994, and to end approximately a year later. Disposal of the steam generators and pressurizer was a critical path item since it had to be completed by the end of 1995 to fully realize the estimated cost reduction. The last large component of the project, the pressurizer, was placed in the burial trench on November 1, 1995. Certain HP goals were established for the project. A comparison of the radiation protection goals and actual results is in Table 23.3.

Table 23.3 Predicted and Actual Radiation Protection Requirements

Description	Goal	Actual
Personnel exposure	114 rem	53.4 rem
Personnel contamination	3	7
Radioactive waste buried	55,000 ft^3	57,800 ft^3
Internal uptakes	0	3

The LCR project was a success not only in technical, schedule, and safety aspects, but also in the budget aspect. The project came in under budget by nearly \$1.7 M, including loadings. This was achieved in spite of encountering costly court delays and some other unanticipated contractor costs. Costs associated with the project were estimated in the January 1994 budget (fully allocated costs) at about \$18.5 M. After completion, actual costs were about \$16.8 M (fully allocated).

The budgeted estimate of \$18.5 M represented a savings of approximately \$4.5 M compared to the cost of removing the components during the decontamination and dismantling phase (\$22.9 M). The cost reduction resulted from favorable radioactive waste disposal fees that applied under an existing agreement between PGE and U.S. Ecology, and from the use of existing Trojan personnel. Cost overruns occurred primarily due to court ordered delays and contractor scope changes. Using an actual total cost of about $16.8 M, the final savings totaled approximately $6.1 M.

23.4.2 Reactor Vessel and Internals Removal: The DP, as approved by the State of Oregon and the NRC, described the separate removal of the reactor vessel and the internals. The plan provided for the internals to be removed from the vessel and segmented. Segmentation would be performed under water in the reactor cavity. Segments classified as LLW and suitable for near-surface burial would be packaged and shipped to U.S. Ecology. Those classified as GTCC waste would be placed in containers fabricated to standard fuel assembly size and initially stored in the spent fuel racks in the spent fuel pool.

The Reactor Vessel and Internals Removal (RVAIR) project was a complex project that coupled innovative approaches with numerous precedent-setting regulatory approvals. It involved the intact removal of the vessel (with its internals) from the containment building, preparing it for transport and shipment. The successful experience with the LCR project led PGE to its innovative approach for this disposal. The original DP said the vessel would be removed intact or sectioned. If removed intact, the vessel could serve as its own shipping container, or possibly require certification as an exclusive-use container. The upper head would also be disposed of as intact or sectioned. Based on the radioactivity contained in the vessel and internals, it was proposed to be shipped as a one-time, exclusive use, Type B shipping package, in accordance with 10 CFR 71. PGE's evaluation indicated that the package could be classified as Class C waste for shallow land disposal, in accordance with NRC and State of Washington regulations.

Radiation Protection: Personnel exposure is tracked via a radiation work permit (RWP) that governs work in radiologically controlled areas. The goal for personnel exposure was 67 person-rem for the entire RVAIR project; the recorded exposure was 72.225 person-rem. Personnel, tracked via radiation work permits in high dose areas, would often exit the area and neglect to sign in on new lower dose area permits while they continued work in the containment. This additional time charged to higher dose areas was the key factor in the increase of recorded worker dose.

Project Budget and Cost: The project came in under budget by $4.2 M, including labor, material, and administrative overhead. This was achieved in spite of encountering regulatory delays that postponed shipment of the package by approximately one year. The project was budgeted for an estimated amount (fully allocated costs) of $26.1 M. The actual cost was about $21.9 M. Reduced in-house costs from using less outside contractors saved money; emerging issues and technologies requiring less contingency funds than anticipated retained unspent money. Improvements on the budget were attributed to:

- Experienced project manager and core group with management skills and knowledge
- Project team with specific expertise and experience in removal of large components
- Specialty (job-specific) contractors with experience from previous projects at Trojan
- Contract methodology (i.e., fixed-price, time and materials) specific to each contractor's scope of work
- Integrated, matrixed, project organization, with each participant having a vested interest in the successful completion of the project

23.4.3 Spent Fuel Debris Cleanup: The characterization of all radioactive materials stored in the pool was a prerequisite for transferring its contents to dry storage in the ISFSI. The characterization process included the separation, segregation, and inventory of the spent fuel pellets and fragments, nonfuel-bearing components, and LLW and GTCC waste. It was also necessary to process and separate the material from the organic and inorganic filter media used over its operating life. Once the processes were completed, the remaining work included:

- Disposing of the LLW
- Transferring the miscellaneous pool inventory into appropriate storage canisters
- Returning the material to the pool to await appropriate disposition.

A waste processing specialty contractor conducted underwater segregating, size reducing, and packaging of the waste materials in the pool and adjoining transfer canal from April to December 1997. The fuel-bearing and GTCC materials containing organic materials were packaged into 40 specially designed process cans and removed from the pool using specially designed, shielded handling equipment. The wastes were processed on site in the mobile steam reformer to remove the hydrogen gas-producing materials. The processed wastes were seal-welded into eight dry, inert capsules and stored in the pool to await the availability of dry storage casks. Other wastes were segregated for packaging. LLW was removed from the pool for disposal. Nonfuel-bearing components were sized and packaged separately in a process container for temporary storage in the pool. The contractor also removed a spent fuel pellet from a storage cask and removed sludge contaminated with spent fuel from a separate cask. Additional scope included segregating and processing wastes collected from cleaning the spent fuel pool bottom and segregating lead shot from all wastes. All steam reforming and underwater equipment was disassembled, decontaminated as required for transportation, and shipped by the contractor by February 1998.

23.4.4 Independent Spent Fuel Storage Installation: The largest remaining project on the critical path to terminating Trojan's license is the successful transfer of the contents of the spent fuel pool to dry storage and the completion of the ISFSI project. Other associated tasks include oversight of the design, licensing, procurement, and construction of spent fuel canisters and handling equipment, and oversight of the loading. The purpose of the ISFSI is to:

- Reduce the contribution of the spent fuel pool to air, water, and noise pollution, and to reduce solid or hazardous waste
- Facilitate future decommissioning activities
- Reduce fuel management costs
- Enable eventual fuel transfer to a DOE or other off-site facility

In keeping with the spent fuel management program, in September 1994 a bid specification was developed and sent to prospective bidders to request proposals to license and construct the ISFSI. The bid evaluation process included technical, cost, and commercial evaluations, as well as a risk-adjusted, life-cycle, cost-economic evaluation. The evaluation showed that a dual-purpose system was the most favorable option for satisfying Trojan's interim spent fuel management needs. In June 1995, PGE contracted with a dry storage vendor to design, license, fabricate, and supply a dual purpose, dry fuel storage system. The NRC issued the site-specific 10 CFR 72 storage license for the ISFSI in March 1999. All 34 concrete storage casks were fabricated on site and completed by August.

An incident occurred during the loading of spent fuel into the first basket. After loading eight fuel assemblies, the water in the cask became so cloudy that operations were stopped. Once the water clarity was restored (by use of temporary pumps, demineralizer, and filters), the fuel was removed from the basket and placed back into the racks in the pool. Subsequent investigation determined that the coating on the basket internals did not perform as originally anticipated -- it had been subjected to a qualification process prior to use but failed in actual application. The vendor's evaluation determined that the best path forward was to change from a coating to an electrolysis nickel-plating process for the basket carbon steel, which would require a license amendment. The vendor chose to withdraw its 10 CFR 71 transportation license application for upgrade. At this point it was anticipated that loading operations would be delayed until late 2002.

A technical evaluation to determine the best path forward for completing ISFSI activities resulted in two additional dry storage vendors being asked to assess their systems for use at the Trojan site. Their evaluations included proposed costs and schedules for each vendor to furnish the necessary design, licensing, and fabrication services needed to complete the remaining ISFSI activities. The interface evaluations indicated that either system would be compatible with Trojan spent fuel and the existing ISFSI components (i.e., concrete casks, transfer station). Both systems, however, would use new canisters and would require equipment design modification. In addition, site-specific license change applications would require submittal to the NRC for review and approval.

PGE performed detailed technical, commercial, cost, and risk evaluations for each of the dry storage vendor proposals. This thorough review ultimately ensured that two additional vendors could provide a feasible, technically sound, and licensable alternative for completing the ISFSI project. On March 6, 2001, the contract with the original vendor was terminated; a contract was awarded to the new vendor on March 7. Contracting for a "pool to pad" undertaking, the new vendor would:

- Manufacture and deliver 34 multi-purpose canisters (MPCs), a transfer cask, and all required ancillaries
- Prepare ISFSI implementation procedures
- Conduct personnel training
- Make regulatory demonstrations (NRC dry runs)
- Load the MPCs into the concrete casks currently stored on the ISFSI pad

Since Trojan would have to submit a license amendment request to the NRC reflecting the different design and components, the new vendor would also help to prepare the request while submitting its own license amendment to reflect the eventual transportation of the spent fuel components. The NRC licensed the new fuel storage and shipment systems as "generic systems."

The dry fuel storage project was successfully completed in nine months. Loading of the first MPC began in December 2002. Work was temporarily suspended after loading the 33rd MPC to allow for the spent fuel racks to be removed from the pool. The rack removal effort included vacuuming and removal, preparing them for shipment, and cleaning and inspecting the pool. Rack removal took approximately four weeks to complete. The 34th (final) MPC was subsequently loaded and placed on the ISFSI pad in September, 2003. Two MPCs were loaded in less than 5 days; about half of them were loaded in 5 to 6 days. The project was completed safely, within budget, and ahead of schedule (ANS 2003).

23.4.5 Auxiliary and Fuel Building Remediation: Areas in the auxiliary and fuel buildings were determined to be contaminated, which will require the surfaces to be remediated before final disposition of the facility. The auxiliary and fuel buildings include elevations of 5, 25, 45, 61, 77, and 93 feet, and the spent fuel pool itself. The total interior surface area is 520,000 ft^2; the total estimated to require remediation is 104,000 ft^2.

The surface remediation program began in April 1999 in the auxiliary building, starting at the 5-ft elevation. This area was selected because equipment had been removed and the rooms were accessible. Several factors were considered in scheduling surface remediation activities:

- Area availability, based on equipment removal
- Active systems
- Activities and equipment associated with the spent fuel pool and ISFSI loading operations.

A characterization program was developed and implemented. Characterizations in the auxiliary and fuel buildings involved evaluating the extent of contamination and radionuclide distribution to determine the derived concentration guideline limits (DCGL). The characterization results are used to determine the type and degree of remediation required. Characterization samples consist of core bore samples and surface samples. A number of areas have been found to contain contamination deep within the concrete, requiring floor and wall removal.

Trojan's basic approach to the surface remediation process is to remediate ceilings and walls first, and remediate floors last to reduce the chance of cross contamination. Remediation of the ceilings and walls involves wiping away all loose contamination, performing operational surveys, and then removing fixed contamination found during that survey. This is an iterative process that continues until the contamination levels are reduced to below the DCGL. Remediation of the floors involves removing the floor epoxy paint first, followed by operational survey of the concrete floor, and then removal of fixed contaminated concrete that is above the DCGL. This process is repeated until all fixed contamination is removed to below the DCGL.

The equipment used for surface remediation has changed and evolved as new technologies are developed or more efficient techniques are invented. Continual research, evaluations, and manufacturer demonstrations are performed and/or solicited to identify more productive and efficient equipment. Equipment evaluations have also considered site-specific conditions, including area accessibility and type of material to be removed. All remediation equipment is adapted to a vacuum system to ensure dustless operations, thereby preventing cross contamination of surfaces.

The auxiliary and fuel buildings remediation project should be completed in 2004, following decommissioning of the spent fuel pool. The pool is currently being drained in preparation for decontamination and ultimate removal of the pool liner, which is expected by the end of 2003.

23.4.6 Containment Remediation: The potential extent of activated and contaminated concrete in the containment building was evaluated to help determine the best remediation method. It was first thought that removing the activated concrete and scabbling the surfaces would be economical and sufficient to place the building in a condition for final survey. However, based on the potential for contamination migration deeper into the concrete through cracks, and a cost evaluation that compared selective removal and scabbling to total removal of concrete and structural steel, it was determined that total removal would be the most cost-effective and efficient method. Containment remediation required the removal of all materials necessary to support the final survey and ultimately, release of the facility. Remediation activities included:

- Removal and disposal of all structural steel
- Demolition and disposal of all concrete
- Decontamination of the containment liner and remaining components

The project was divided into two phases. The first phase removed the majority of activated concrete around the vessel cavity -- essentially all activated concrete between the 67- and 45-ft levels. The second phase consisted of removing the remaining activated concrete at the lower levels, and contaminated concrete and structural steel. The structure now houses only miscellaneous portions of the polar crane, polar crane attachments, and the containment liner.

The containment remediation project was managed and coordinated by a Trojan project manager and personnel. The actual demolition work was awarded to a concrete demolition specialty contractor in the form of a fixed-price contract. The activated and contaminated concrete that did not meet the established criteria was packaged and shipped to U.S. Ecology. Contaminated concrete that met the mandated criteria was boxed for shipment via rail car to a processing facility in Tennessee, where it was processed and the "clean" concrete released for landfill disposal.

Schedule: Remediation activities began after the vessel and the temporary lifting equipment were removed from containment in July 1999. From July through November, most activity consisted of removing structural steel and obtaining characterization samples. In November, the containment was prepared for placement of the excavator, a new waste processing building was constructed on the south side, concrete removal/burial boxes were fabricated, and the concrete demolition specialty contractor was mobilized. A railroad spur was constructed on PGE property (outside the industrial area) to run close to an older warehouse and yard that were modified to be a waste processing and shipping facility. The entire remediation project was completed by January of 2001.

Waste Estimates: It was estimated that 560 boxes of radioactive waste would be shipped to U.S. Ecology, and 2040 boxes of waste containing radioactivity levels exempt from regulatory requirements would be shipped to Tennessee, ultimately for landfill disposal. The actual number of boxes sent to U.S. Ecology was 1019; 2279 boxes were sent to Tennessee. Over 25 M lbs of rubble were removed from the containment building.

License Termination Plan: Trojan's LTP was prepared in accordance with the requirements of 10 CFR 50.82, and the Regulatory Guide 1.179 (NRC 1999). The LTP is maintained as a supplement to the Defueled Safety Analysis Report, in accordance with 10 CFR 50.82(a)(9)(i). This plan demonstrates that the remainder of the decommissioning activities at the site will be performed in accordance with the regulations in 10 CFR 50.82, will not compromise security or the health and safety of the public, and will not significantly effect the quality of the environment.

Decommissioning Plan: PGE submitted its DP and Supplement to the Environmental Report in January, 1995; they were approved by the NRC in April 1996. The DP was submitted and approved in accordance with the NRC's rule governing decommissioning and termination of license (10 CFR 50.82). This rule has since been revised. The revised rule, specifically 10 CFR 50.82(a)(9), requires all power reactor licensees to submit an LTP for NRC approval either prior to or with the application for license termination, at least two years before the date for termination. Trojan's LTP satisfied those requirements and was approved by the NRC in February 2001 -- the first one since the requirements were originally established in 1996.

Final Survey Project: The Final Survey Plan describes the process used to demonstrate that the Trojan facility and site meet the radiological criteria for license termination. This plan incorporates the release criteria of 10 CFR 20.1402 for unrestricted use of the site. The final survey encompasses structures, land areas, and plant systems that, as a result of licensed activities, are identified as contaminated or potentially contaminated. The majority of these are located within the industrial area. The final survey does not include the ISFSI area – it was previously surveyed and the results of that survey are documented in its own report. The final survey also does not include monitored gaseous and liquid plant effluent discharge pathways. As confirmed by characterization results, the monitoring program documents compliance with the ALARA criteria associated with monitored releases.

Final survey activities are planned, scheduled, and tracked as a part of the overall decommissioning planning process. The schedule depends on the progress and completion of several activities, including the transfer of spent fuel to the ISFSI. Survey activities are planned and discussed with the NRC and the Oregon Office of Energy sufficiently in advance to allow scheduling inspections. Survey of the containment building continues, with project completion scheduled for late 2004.

23.5 LESSONS LEARNED

Steam reforming of spent fuel pool debris proved to be an effective and successful method of processing HLW to meet long-term dry storage requirements. The steam reformer successfully removed organics, moisture, and hydrates from sludge, fine metallic dross, and mixtures containing bits of spent fuel pellets, deteriorated organic filters, and miscellaneous debris. The processing equipment provided a reliable, repeatable means to determine the end of run for each batch without having to perform sampling on individual batches. Eliminating individual batch sampling saved considerable expense and personnel exposure to radiation.

The underwater segregation of spent fuel pool debris was ultimately successful in sorting out and sizing:

- Spent fuel pellets
- LLW and nonfuel-bearing, highly activated metallic components
- Black filter flange material
- Lead shot

Unexpected materials and conditions required innovative engineering support to develop specialized sorting procedures and devices. The successful resolution and implementation of all segregation issues required the combined cooperative efforts of PGE and its specialty contractor.

Chapter 24

BIG ROCK POINT

24.1 INTRODUCTION

Big Rock Point, constructed on the shores of Lake Michigan near the resort town of Charlevoix, Michigan, began operation as a research and development reactor. In January 1960, Consumers Power (later Consumers Energy) submitted its formal proposal for a construction permit and operating license. The permit was granted four months later. Ground breaking for the plant took place July 20, 1960, and the plant was officially completed by September 1962.

Big Rock Point was a relatively small plant (67 MWe) that made large contributions to the nuclear industry. As the world's first high power-density boiling water reactor (BWR), one goal was to demonstrate that nuclear plants could generate electricity economically. In 1977, Big Rock Point set a world record for BWRs by operating for 343 consecutive days.

Figure 24.1 The Big Rock Point Site

Figure 24.2 shows a schematic of the plant as it existed during operation. Big Rock Point was a direct cycle, forced circulation BWR capable of producing 240 MWt and 75 MWe. The containment is a spherical steel vessel 130 ft in diameter. The sphere extends 27 ft below and 103 ft above grade. It was designed for 41.7 psia internal pressure, with the design-basis loss-of-coolant event pressure rating of 37.7 psia. The sphere was habitable during operations, allowing for routine inspections and equipment maintenance. There are three entry/exit points in the containment sphere: the personnel, equipment, and escape locks.

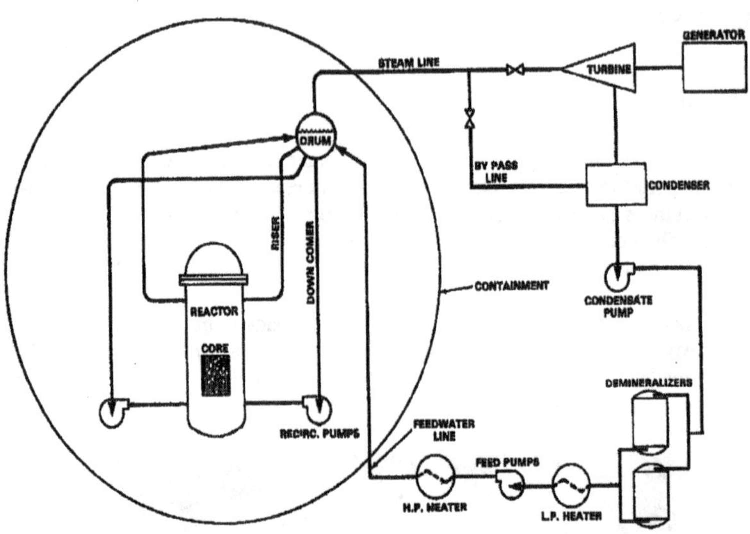

Figure 24.2 Schematic Diagram of Big Rock Point

The emergency core cooling system was designed to inject cooling through the ring or nozzle spray lines in the event of a loss of cooling accident (LOCA). It consisted of diesel and electric fire pumps, a fire/system water supply distribution system, a nozzle spray line, and a ring spray line. Since there is no suppression pool or wet well in the dry containment, a post-incident system was required for long-term cooling. This system allowed recirculation and cooling of the water accumulated in the lower portion of the containment after it rose above a certain level. This core spray recirculation mode automatically added emergency makeup water to the spent fuel pool.

The reactor depressurization system was installed in 1976 to allow the use of a low pressure core spray system during transients caused by small and intermediate LOCAs. This system, similar in concept to the automatic depressurization used in modern plants, varies significantly in design in that it discharges directly into the steam drum cavity since there is no suppression pool. The plant was also equipped with an emergency condenser similar in concept and design to the isolation condensers in other BWRs.

24.2 DECOMMISSIONING PLANNING

On August 29, 1997, after 35 years of operation, the plant was shut down. Consumers had decided that it was not cost-effective to continue operations and chose to immediately dismantle the plant after first considering a long storage period before dismantling. The goal was to restore the site to greenfield by 2005. The reactor was defueled in September 1997. Certification of permanent shutdown status authorized the start of major decommissioning activities. Consumers Energy chose to use plant employees as the main work group because of their extensive operating experience and the plant/equipment knowledge. Management also determined the need to secure additional outside help when in-house expertise was lacking.

In 1999 Consumers Energy awarded a consortium a $35 million fixed-priced contract for major component removal and building demolition, including the following activities:

- Removal and shipment of the reactor vessel to Barnwell for disposal
- The development of a safety analysis report for the design and fabrication of a Type B container to meet the 10 CFR 71 requirements for vessel shipment
- Removal of the steam drum, emergency condenser, and other tanks and heat exchangers
- Decontamination and dismantling of the containment sphere, turbine, and other buildings
- Removal of foundations and buried piping

24.2.1 Schedule: The seven major phases of restoration activities at Big Rock Point were identified as follows:

1. *Drafting the DP and preparing for shutdown*: Employees successfully operated Big Rock Point for two months after the shutdown announcement was made in June 1997
2. *Mobilization and hazard reduction*: This phase included the removal of all unnecessary hazards, such as oil, batteries, and asbestos from site
3. *Systems/Vessel internals removal and spent fuel storage*: Reactor internals and other systems were removed and shipped for disposal. Spent fuel bundles (441) and other GTCC waste were loaded into eight dry fuel storage canisters from November 2002 to May 2003.
4. *Reactor vessel and steam drum removal*: The vessel and steam drum were both shipped and disposed of in late 2003. The reactor vessel was shipped to Barnwell; the steam drum was shipped to Envirocare. The remaining buildings will be dismantled and debris shipped for disposal during this phase in 2004 and 2005.
5. *License termination*: The majority of plant property will be released for unrestricted use.
6. *ISFSI operation*: Dry fuel storage installation operation is expected to extend to 2011.
7. *Fuel shipment and ISFSI dismantling*: The decommissioning estimate calls for the shipping of fuel to the DOE's site to begin in 2010, be completed by 2012. The ISFSI dismantling should also be completed within 2012.

24.2.2 Occupational Safety: In August 1997, Big Rock Point reached 20 years without a lost-time accident. To continue this level of safety during decommissioning -- a project conducted in a significantly different industrial safety environment -- the following was established:

- A new site accident prevention manual was written to incorporate many of OSHA's standard construction requirements.
- A new industrial safety team was formed, consisting of a certified safety professional, an industrial hygienist, and employees with plant knowledge.
- All employees, both utility personnel and long-term contractors, had to wear the same hardhat and use the same Accident Prevention Manual.
- A safety committee was formed consisting of representatives of all major departments and groups.
- Industrial hygiene was given increased visibility and increased funding.

- Annual safety days were held where employees have a chance to focus on improved safety practices and new equipment.
- Periodic site safety stand-downs were held to re-emphasize safety. (National safety motivational speakers have been brought in annually in an effort to instill a strong safety culture at the site.)
- All work planning documents incorporated safety checklists, pre-job briefings strongly emphasized safety emphasis, and post-job critiques reviewed on-the-job safety practices.
- All exit interviews included questions about nuclear, radiological, and industrial safety.

Ultimately, site employees accomplished 23 years without a lost-time accident before a broken ankle ended the streak.

24.2.3 Spent Fuel: The 441 spent fuel bundles were stored in the spent fuel pool until 2002, and then loaded into transportable canisters for dry storage. The heat generated by the fuel had decayed sufficiently to meet the licensing and design requirements of the site's ISFSI. All assemblies were loaded into seven canisters, placed in concrete storage casks, and then transferred to the ISFSI in 2002 and 2003. The canisters will remain there until the DOE takes custody of the spent fuel.

24.2.4 Comparison between Planning and Actual Actions: The original critical path through removal of spent fuel remains credible, and actual progress may put the project slightly ahead of the baseline schedule if dry fuel storage equipment deliveries are timely. All dismantling work identified to be completed before fuel transfer has been completed. Following fuel off-load, major component removal started with the removal and shipment of the steam drum and reactor vessel. Barring any unforeseen difficulties or surprises, restoration activities should finish on schedule. Major planned/actual comparisons follow:

1. *Overall project*: Barring any unforeseen difficulties or surprises, restoration activities should finish on schedule.

2. *Work Packages*: Due to the ongoing development of the decommissioning work packages (DWP) over the past few years, there is very little difference between the planned work package and execution of work in the field. Before working a DWP, a walkdown is performed by the DWP lead, a construction supervisor/lead, and operations and radiation protection personnel to verify and validate that the plan is still applicable. This allows any necessary changes to be made to the package before work begins and is especially useful if the DWP was planned much earlier than the actual work date.

3. *Cost*: Table 24.1 shows estimated versus actual decommissioning costs in seven major categories. Three areas are responsible for most of the cost increase: dry fuel storage, project management, and waste removal contracts. Pursuant to Michigan's requirements to prepare and file decommissioning cost estimate updates with the state's Public Service Commission at three-year intervals, an estimate update was prepared for submittal in early 2004. Through December 31, 2002, the Big Rock Point decommissioning fund remained funded to the requirements set forth in 10 CFR 50.75(f)(1).

4. *Radiation Exposure*: When shutdown was announced in 1997, the already completed estimate of a five-year SAFSTOR option with subsequent dismantling was used as the basis

Table 24.1 Estimated vs. Actual Decommissioning Costs

	Dollars in Thousands			
Category	2001 Estimate in 2000 $	2001 Estimate YOE $*	Actual Cost Through 2000	Cost Remaining
Project Management	133,467	19,022	85,596	47,871
Major Component Removal	34,024	(262)	5,233	28,791
Systems Dismantling Projects	12,780	(1,515)	11,629	1,151
Dry Fuel Storage	58,604	40,057	14,453	44,151

* Escalated to year of expenditure (YOE) dollars

for the occupational exposure estimate. This estimate most closely aligned with the immediate dismantling of the plant and included estimates for conducting a chemical decontamination of the primary system. The occupational exposure estimate was reverse-calculated to remove the radioactive decay time incorporated into the five-year SAFSTOR estimate, establishing an exposure estimate of 700 person-rem for immediate dismantling.

While the actual decommissioning project dose goal remains at 700 person-rem, the radiation protection department provided a new occupational exposure ALARA target of 550 person-rem in 1999. The basis of this new target was founded on the collection of radiological survey data previously not available due to operational conditions of the plant, and most significantly on radiological data following the primary system decontamination. The ALARA goal was revised January 1, 2004 to 640 person-rem due to more than expected dose expended during the removal of the reactor vessel and other equipment. Dose expended at the end of 2003 was 569 person-rem.

5. *Waste*: Table 24.3 shows the estimated versus actual waste generated in 22 major categories from August 29, 1997 to December 31, 2003.

6. *Federal and State Regulatory Concerns*: Moving from an operating into a decommissioning status does not result in less regulatory oversight. It can actually increase scrutiny and involve more agencies. Brief descriptions of regulatory issues addressed by the Big Rock Point team are presented herein.

    a) PCBs: When Big Rock Point ceased operations in 1977 there was insufficient regulatory guidance from the EPA for disposal of waste containing PCB coatings. Due to employee efforts, Barnwell's waste acceptance criteria were revised in July 1998 to allow PCB bulk waste disposal in accordance with applicable regulations.

Table 24.3 Nature and Quantities of Estimated vs. Actual Waste

Type	Estimate (lbs)	Actual (lbs)
Dry Active Waste	450,000	407,384
Asbestos	324,000	182,231
Contaminated Metals	3,496,114	3,783,212
Activated Metals	2,086	66,3800
Contaminated Concrete	797,640	920,499
Activated Concrete	234,900	88,445
Contaminated Lead	56,800	226,415
Class B Material	10,000	0
Class C Material	15,000	4,250
Contaminated Soil	524,700	102,775
Contaminated Decommissioning Equip.	300,000	0
Potentially Clean Metals	4,779,450	125,560
Potentially Clean Concrete	84,668,220	4,928,100
Other Clean Material	As Requested	2,566,080
Clean is Green	As Requested	1,381,348
Resin/Sludge	100,000	90,700
Filters	20,000	12,500
Radioactive Sources	500	228
D007 - Chromium Mixed Waste	0	483
D008 - Lead Contaminate Debris	0	2000
D006 - Spent Degreasing Compound	0	380
Miscellaneous Mixed Waste	0	100

b) Archeology: The National Historic Preservation Act (NHPA 1996) requires that a phase I archeological study be performed for any federal undertaking. Holding a license from the NRC qualifies as a federal undertaking. Because an archeological study was not required when Big Rock Point was constructed, this study was necessary during decommissioning.

c) Wetlands: Decommissioning tasks that included construction of new facilities (such as the ISFSI and parking lot expansion) required working with the State of Michigan to ensure wetlands were not disturbed.

d) Underground Storage Tanks (USTs): A Michigan state law requires removal and closure within nine months of ceasing to use a UST.

e) Mixed Waste Storage: Careful planning and waste minimization will keep the site within its limited storage capacity.

f) Industrial Safety and Hygiene: A site-specific health and safety manual was developed to integrate state OSHA and NRC regulations, thereby complying with two differing sets of regulatory requirements.

g) Asbestos in Water: At shutdown, state regulations did not adequately address the disposition of plant water containing asbestos. A specific agreement was reached with Michigan's Department of Environmental Quality to remove asbestos fibers to meet the state's drinking water criteria before discharging the water.

h) Material Release Criteria: NRC regulations do not adequately address radiological requirements related to release criteria for large amounts of building debris. The approach chosen by Consumers Energy involved the application of 10 CFR 20.2002, which allows the majority of debris to go to a local landfill. The application was approved by the NRC and state agencies in early 2002. More than 7.6 M pounds of debris were disposed of in the local landfill by the end of 2003.

24.2.5 Proposed End-State: The proposed end-state of the site is the unrestricted release of the property as a greenfield site. Big Rock Point is located on 580 acres of mostly forested land, with approximately 10 acres used for plant operations. The vision is to leave the site in a condition similar to when the plant was constructed. All above-ground buildings, transmission lines, piping, tanks, and structures will be removed, as will all underground piping, cables, tanks, storm drain systems, and contaminated foundations. All soils identified as being contaminated radiologically or otherwise will be remediated. The final site survey will be conducted under the greenfield condition, meeting the requirements of the NRC's criteria for license termination.

24.3 FEDERAL AND STATE REGULATORY CONCERNS

Moving from an operating plant to a decommissioning status does not result in less regulatory oversight. It can actually increase scrutiny, and by various agencies. Brief descriptions of regulatory issues addressed by the Big Rock Point team follow:

1. *PCBs*: When Big Rock Point ceased operations there was insufficient regulatory guidance from the EPA for disposal of waste containing PCB coatings. Due to employee efforts, Barnwell's waste acceptance criteria were revised in July 1998 to allow PCB bulk waste disposal in accordance with applicable regulations.

2. *Archeology*: The National Historic Preservation Act (NHPA 1996) requires that a phase I archeological study be performed for any federal undertaking; holding a license from the NRC qualifies as same. This study was required during decommissioning since an archeological study had not been required during construction.

3. *Wetlands*: Decommissioning tasks that included construction of new facilities (such as the ISFSI and parking lot expansion) required working with the State of Michigan to ensure wetlands were not disturbed.

4. *Underground Storage Tanks (USTs)*: A Michigan state law requires removal and closure within nine months of ceasing to use a UST.

5. *Mixed Waste Storage*: Careful planning and waste minimization will keep the site within its limited storage capacity.

6. *Industrial Safety and Hygiene*: A site-specific health and safety manual was developed to integrate state OSHA and NRC regulations, thereby complying with two differing sets of regulatory requirements.

7. *Asbestos in Water*: At shutdown, state regulations did not adequately address the disposition of plant water containing asbestos. A specific agreement was reached with

Michigan's Department of Environmental Quality to remove asbestos fibers to meet the state's drinking water criteria before discharging the water.

8. *Material Release Criteria*: NRC regulations do not address the release of very low level radioactive material in detail. The regulatory approach chosen by Consumers Energy involves the application of 10 CFR 20.2002, which allows the majority of decommissioning debris to go to a local landfill.

24.4 STAKEHOLDER ISSUES

A number of issues were addressed after the decision was made to cease commercial operation and enter into immediate decommissioning. A communication process was used to inform employees, the affected shareholders, the regulators, and others with a stake in the operation and decommissioning of the plant.

24.4.1 Employees: The first step was to announce the shutdown in a way that would give employees the most information possible and leave them feeling proud of the work they had done. Since the decision was to announce the shutdown in June and not actually cease operations until August, management had to ensure that employees were able and willing to safely continue commercial operation.

On the day of the closing announcement, the company president, senior nuclear officer, and plant manager met with employees in group meetings. Other managers were simultaneously conveying the message to those required to remain at the plant. Employees were given as much information as possible about the reasons for closing the plant and plans for site decommissioning. Senior management also emphasized the proud history of Big Rock Point and its employees. The information was conveyed orally and in writing to let employees take it home to their families. All employees received an organization chart with departments and jobs (no names), a timetable for placement activities, instructions on how to bid for jobs that would be available, and numbers to call for information and employment policies, including the announcement of an enhanced severance and retention bonus plan, and enhanced educational assistance benefits.

Following those events and throughout the summer, meetings were held on outplacement activities, financial planning, employee assistance, and pension and savings plans. The company's severance plan was enhanced to ensure that those who did not find positions within the company would stay until the shutdown was completed and they would be released. All non-union employees were offered a retention plan that accumulated at two weeks per quarter for all quarters employed on the project, if they stayed until release.
Because union employees already had an employment guarantee, they were not offered the same option. However, concessions were given in the form of collapsing all job descriptions into six jobs within one occupational group. Salary increases were given for this increase in flexibility. Grievances have been essentially non-existent during decommissioning.

Another major effort was to integrate the 200 to 300 specialty contractors who arrived on site into the existing 200-person staff. This required ensuring that employees and long-term contractors understood the plant priorities and were performing as an integrated team. All site personnel were integrated into a unified management structure and performed post-job assessments as one group.

24.4.2 Greater Community: Big Rock Point has enjoyed community support throughout its 40 years of operation. Community support extending into the restoration phase is attributed to three major factors:

1. *Continued Commitment to the Community*: Both employees and contractors were encouraged to remain active in the community during the decommissioning of the plant. This continued commitment was demonstrated by the many volunteer activities of plant employees, philanthropic support from the plant, and the formation of local organizations.

2. *Increased Communication with the Community*: Considerable effort has been expended to ensure the community remained informed about restoration activities at Big Rock Point. Such efforts include the publication of a quarterly community newsletter, an expanded site newsletter provided to opinion leaders, speaker bureau activities, and the involvement and education of the plant's Citizen Advisory Board.

3. *Continued Economic Activity from Big Rock Point*: Because of an initial increase in plant employment following shutdown, the local community has not experienced a significant decline in economic activity. The full impact of the closing will become more apparent as employment levels decline following completion of major projects over the next three years.

Decommissioning must not come as a surprise to stakeholders, and regulators must also be kept informed every step of the way. Plant staff ensured that the NRC, the Public Service Commission, and state environmental and safety agencies would be kept informed of events. The NRC held its local meeting for public comment on the LTP in August 2003; no objections were raised concerning the plant's releasing the site for unrestricted use.

24.5 UNIQUE ISSUES AND EXPERIENCES

1. *Bulk Material Release*: Radiation protection management identified problems associated with release of materials having been decontaminated. The challenge was to comply with regulatory requirements and put into production a large scale building debris survey and disposal program. Without clear NRC guidance to licensees on detection levels associated with radiological survey of bulk materials (such as concrete), the licensee is subject to a moving target and takes on total risk management for the final deposition of waste material. As noted earlier, the 10 CFR 20.2002 exemption request was approved, allowing demolition debris surveyed to be clean or below detection limits to be sent to a local industrial landfill

2. *Decommissioning Power System*: Visits to other retired nuclear plants showed it was not always obvious which plant systems were energized and which were not. To ensure employee safety, a new substation and distribution system were installed to power plant equipment. The new system is colored yellow, identifying it as "live."

3. *Chemical Decontamination*: Big Rock Point, like several other plants, used chemical decontamination to clean a full, major plant system. The decontamination on the primary coolant system removed approximately 405 Ci of radioactive material from the interior of

equipment and piping. This activity reduction of approximately 90% significantly reduced radiation exposure to employees during the dismantling process.

4. *Spent Fuel Pool Cleaning*: The removal and disposal of nonfuel items from the spent fuel pool used several unique approaches. Because of a lack of room in the pool, the empty vessel was used as a work area to reduce waste volume. Use of the vessel brought the following benefits:

   - Dose reduction due to water shielding and working from a raised platform
   - Fewer work restrictions as compared to working over fuel
   - The ability to perform intrusive and destructive work without clouding the fuel pool water

5. *Resin Disposal*: Employees involved in this project used robotics to clean radioactive waste tanks, thus reducing radiation exposure. Contaminated water was used to clean the sumps and tanks, essentially eliminating the creation of new contaminated wastewater.

6. *Occupational Safety*: In November 2000, a shattered ankle from a slip on ice ended the lost-time record at 23 years. Still, the industrial safety performance to date had been exceptional. There have been no major injuries attributable to poor work practices. However, four serious injuries have been recorded: two broken ankles from slips on ice, one hernia, probably caused by unrelated factors, and one strained back caused by an employee swatting at a bat. Due to a declining trend observed in 2001, a thorough review of the Industrial Safety Program was underway by July of that year.

24.6 LESSONS LEARNED

Many lessons were learned during the planning and execution of Big Rock's decommissioning. These are categorized in five major groups, as follows:

1. *Community Involvement*: Frequent communications in all media are necessary to ensure that messages are heard and understood. The community should be informed of activities through local publications, plant tours, news media, and citizen advisory boards. Because plant closings are normally a time of high stress and change, it is imperative to inform the community of upcoming actions. Since informed employees are the best ambassadors, they should be encouraged to participate in community awareness, thereby demonstrating that the plant remains committed to its stakeholders.

2. *Employee Concerns*: A variety of employee programs must be offered and extended as long as possible, including training, job search and relocation assistance, and financial planning. Experience has also shown the success of including family members in as many sessions as practical. Employees must be given as much time as possible to understand and accept changes when they will affect their employment status.

3. *Decommissioning Operation*: A sense of teamwork and the integration of contractor and utility personnel will require a dedicated group of facilitators and change agents. Celebrations of milestones are a good way to build employee cohesion and pride in a job

well done. Similarly, clear priorities and expectations are required and must be frequently emphasized to ensure that the transition is made from an operation to a decommissioning mindset. The difference between the level of effort associated with decommissioning and that of operating personnel requirements, roles, and responsibilities must be understood. The core staffing levels should not be underestimated or the personnel reduced too early. Similarly, the site facilities should be evaluated before shutdown to ensure that temporary office space and site laydown requirements are not underestimated. Furthermore, many overhead costs during operations are covered by the corporate operating budgets. These overhead costs can be significant after shutdown and must be accounted for in the funds to be gathered in the trust fund.

4. *Waste and Associated Regulatory Requirements*: All potential waste should be surveyed and characterized as early as possible. This includes extensive inventories of existing waste, of the spent fuel pool, of potentially contaminated concrete (to determine the depth of contamination and the extent of activation) and of structural steel (to determine its coatings). Lead-based or other hazardous paints, PCBs, asbestos-laced coatings, and mastics should also be included in the survey since they could lead to significant disposal costs. One caution is related to underestimating the level of effort associated with design, licensing, and procurement of an ISFSI. The regulatory process may be long and cumbersome, and the number of dry fuel storage vendors and qualified fabrication shops is limited.

5. *Equipment and Safety*: An attempt should be made to understand the capability of existing plant equipment to support heavy load movements, such as dry fuel casks. Does the crane have sufficient capacity to handle heavy loads? Is it single-failure-proof if required? Are any other significant plant modifications required? On the other hand, many times it is no longer necessary to maintain equipment with surgical precision. Demolition techniques can create hazards not experienced during operations. Heavy load rigging and movements harbor potentially life-threatening circumstances. A decommissioning team must adapt to this new environment and perform work in a manner that maintains safety and schedule in harmony. A constant vigilance and emphasis on safety is also required. Plant conditions change daily during decommissioning, and the type of work being performed is entirely different from that performed during operations.

Chapter 25

HADDAM NECK ATOMIC POWER PLANT

25.1 INTRODUCTION

The Haddam Neck Plant is located in Middlesex County on the east bank of the Connecticut River, about 21 miles south of Hartford, in Haddam Neck, Connecticut. The Connecticut Yankee property consists of 563 acres, of which 4.5 acres are fenced-in power plant structures. Most of the property remains open forest.

Figure 25.1 The Haddam Neck (Connecticut Yankee) Site

In 1962, ten investor-owned companies established the Connecticut Yankee Atomic Power Company (CYAPCO), agreeing to share the costs and the power from a proposed nuclear plant to be built in Connecticut. The PWR facility was designed to produce 1,825 MWt and 590 MWe of power. The site was selected in 1963 and construction began in 1964. While under construction, a holding company, Northeast Utilities, was created by several of the investor-owned companies. Northeast Utilities managed the Haddam Neck plant throughout its history. The plant generated power for the first time in August 1967 and began commercial operation in January 1968, just six years from its original concept. During its 28 years of operation, the plant produced more than 110 billion kilowatt-hours of electricity, setting several industry operation records and two world records.

Before the decision to permanently shut it down, the plant was in a forced outage that had begun in July 1996 to address safety and maintenance issues. In August 1996, the company announced it was updating its economic analysis of the plant, and in October indicated that the plant was no longer economically viable. By December it was decided to permanently shut down, based on a potential savings of $100 million or more by the early closure of the plant before its licensed operating period expired in 2007. On December 5, 1996, the NRC was notified of the permanent cessation of operations and the permanent transfer of fuel assemblies from the reactor vessel to the spent fuel pool.

In August 1997, the PSDAR was submitted to the NRC and was subsequently accepted. In January 1998, CYAPCO transmitted an updated FSAR reflecting the plant's permanent shutdown status, and in June the NRC amended the facility operating license to reflect this condition. In October 1999, the operating license was amended again to reflect the decommissioning status and long-term storage of the spent fuel in the pool. Additional licensing basis documents were also revised and submitted to reflect long-term fuel storage, including a defueled emergency plan, a security plan, QAPP, and operator training programs.

In April 1999, CYAPCO selected a DOC to perform the decommissioning activities while it continued to perform spent fuel pool island operations and oversee all activities. By 2003, CYAPCO took over all DOC activities.

25.2 DECOMMISSIONING PLANNING

25.2.1 Waste Estimates: One of the key issues in decommissioning planning was estimating expected waste, including decommissioning-derived radioactive waste. Table 25.1 shows the results of estimates for both noncontaminated and radioactive waste.

25.2.2 Managing Certain Hazardous Materials: The original plant insulation contained 6% to 8% asbestos with a calcium silicate binder. Asbestos insulating material was identified on many plant systems and in most areas and buildings. Whenever maintenance activities required the removal of insulation containing asbestos, it was replaced with asbestos-free material and labeled accordingly. Most of the systems originally covered with asbestos insulation now have portions covered with asbestos-free insulation. Other asbestos was found in the roofing or in the gaskets and valve packing of numerous plant systems. Licensed contractors removed the asbestos and took it to permitted landfills for disposal. Materials contaminated with both asbestos and radioactive materials will be disposed of as LLW.

Paints with a lead base had been used to coat steel components, some concrete structures, and underground steel piping. During the operating life of the plant, some of the lead-based paints were re-painted with several coats of non-lead paint. In other cases, lead-free surfaces had been spot-covered or completely re-painted with lead-based paint. The lead-based paints on components that cannot be recycled would be removed, processed, and disposed of by authorized and licensed contractors. The potential for paints to contain PCBs was identified and would be managed according to federal and state regulations. The amount of PCB and lead waste to be produced by decommissioning had yet to be determined.

Table 25.1 Waste Estimates

Item	Quantity
Noncontaminated	
Lead batteries intended to be recycled (lbs)	115,200
Asphalt (lbs)	2,758,468
Mercury or devices containing mercury (lbs)	260-280
Materials containing asbestos (ft^3)	3,636,000
Contaminated	
Process water (gallons)	475,200
Activated concrete (lbs)	604,000
Contaminated concrete and asphalt (lbs)	4,739,253
Activated metals (lbs)	28,000
Structural steel (lbs)	2,187,700
Metal piping (lbs)	1,864,000
Conduit (lbs)	396,000
Cable tray (lbs)	1,202,000
HVAC (lbs)	380,000
Equipment (lbs)	5,683,048
RCS piping (lbs)	350,000
Decommissioning equipment (lbs)	300,000
Reactor coolant vessel (lbs)	836,248
Steam generators (lbs)	2,094,000
Pressurizer vessel (lbs)	256,000
Reactor coolant pumps (lbs)	656,000
Contaminated soils (lbs)	4,903,900
Contaminated ion-exchange resins (ft^3)	1,320
Dry active waste (paper, plastic, clothing) (lbs)	600,000
Sewage sludge (gallons)	33,334
Lead (lbs)	246,000
Contaminated oil (gallons)	600

25.2.3 License Termination Plan: The LTP describes the decommissioning activities, the process for performing the final status surveys, and the method for showing that the site meets the release criteria for unrestricted use. It contains specific information on:

- The HSAs and site characterization
- Remaining decommissioning activities
- Site remediation plans
- Final status survey design and implementation plan
- Dose modeling scenarios
- Update to the site-specific decommissioning cost estimate
- Supplement to the environmental report

Termination of the NRC license and environmental closure of the plant site are closely related; their completion will allow the site to be released for future use. The LTP describes the processes to be used in meeting the requirements for terminating the NRC license. A site environmental closure

plan will describe the processes to be used for nonradiological cleanup and release of the site. An integrated approach to site release processes will be used whenever practical.
The LTP was submitted to the NRC for review in July 2000. It is being maintained as a supplement to the plant's updated FSAR to support the application for a license amendment.

25.2.4 Final Status Survey: The primary objectives of the final status survey are to:

- Select/Verify survey unit classification.
- Demonstrate that the level of residual radioactivity for each survey unit is below the cleanup criteria.
- Demonstrate that the potential dose from small areas of elevated activity is below the release criterion for each survey unit.

The purpose of the Final Status Survey Plan is to describe the methods to be used in planning, designing, conducting, and evaluating final status surveys at Haddam Neck to show that it meets the radiological criteria for unrestricted use. The plan will be consistent with the MARSSIM (NRC 1997), yet will allow for the use of advanced technologies as long as the survey quality is equal to or better than traditional methods described in the manual. Since MARSSIM is not readily applicable to complex nonstructural components within buildings, the criteria in Regulatory Guide 1.86, Termination of Operating Licenses for Nuclear Reactors (NRC 1974), will be applied to nonstructural components and systems at the time of final status survey. Components meeting these criteria may be released for unrestricted use after completing the final status survey for the corresponding area. The plan also describes methods and techniques used to implement isolation controls to prevent recontamination of remediated areas. The survey plan also incorporates measures to ensure that final survey activities are planned and communicated to regulatory agencies to allow the scheduling of inspection by these agencies if desired.

25.3 CHARACTERIZATION

Characterization planning started with the HSA, conducted along MARSSIM guidelines and the ASTM standard for environmental site assessment (ASTM 1994), the latest version of which was published in 2000 (ASTM 2000). The HSA process began the effort to fully characterize, remediate, and survey the site in preparation for license termination. The process presented in MARSSIM includes a mechanism for reevaluating decisions reached during the initial phase of the process and changing them if the data warrant.

The site characterization report issued in January 2000 summarized historical information and data. A subsequent HSA issued in August 2002 gave more detailed information and survey data, facilitating access to additional information for further clarification of the site's HSA. The assessment consisted of a review of site historical records regarding plant incidents, radiological survey documents, operations and maintenance records, plant modification documents, and both routine and special reports submitted to various regulatory agencies. Interviews with site personnel, both past and present, were conducted. Reviews of historical site photos and extensive site walkdowns yielded a thorough understanding of events that affected site decommissioning.

Division of the site into initial survey areas was completed in parallel with the HSA process. These areas were created based on their anticipated MARSSIM classification, with updates reflecting new pertinent information. Therefore, the effort expended on the information assessment required to complete this HSA bounded both the areas impacted by operations and the development of the initial classification of areas within that scope.

25.4 DECOMMISSIONING ACTIONS

Planning for large component removal began in the spring of 1997, about six months after the announcement of shutdown.. The first major decommissioning activity began in late summer of 1999. All components were off site and disposed of by the end of August, 2001. The project entailed removing the four steam generators and the pressurizer from the containment building. Each component was prepared for transport as its own shipping container, with the steam generator lower assemblies placed in personnel barriers to prevent access.

Due to a low water level in the Savannah River, the components were stored on site for about 17 months, and a new transport plan was formulated. The steam generator lower assemblies were ultimately sent (two per barge) to Port Royal, SC, and transferred to a rail car for final transport to Barnwell for burial. All assemblies were disposed of by June 2001. The pressurizer was shipped via overland transporter to a local rail spur and transported by railcar to the Envirocare facility in Clive, Utah. Subsequent burial was in July 2001.

The steam generators were removed in two steps. First, each was cut, separating the steam dome from the lower assembly to facilitate removal through the containment's equipment hatch. As each dome was cut, it was prepared for shipment and treated with a liquid "lockdown" material as a precaution to prevent against spreading loose surface contamination. Once the domes were removed from the containment building, they were shipped off site by flatbed truck to the rail spur and then shipped by train to Memphis, TN for further cutting. The cut sections were transported to Bear Creek, TN for melting. The second step was to cap the lower assemblies in place, treat them to prevent potential loose surface contamination from spreading, remove them from the containment building, and store them in individual personnel shields at the large component storage area to await shipment. After the generators, the pressurizer was removed from the containment building and stored with the large components.

25.4.1 Reactor Vessel and Internals Removal: The internals segmentation plan was completed in August 1999. All equipment to support segmentation was delivered to the site by January 2000. The vessel was lifted and placed within its container in October of 2003 and shipped to Barnwell in January, 2004. The vessel head was shipped separately via overland transport to Envirocare.

Two options were available for disposal of the reactor internals: ship with the internals intact, or remove and segment them. However, due to the Barnwell burial site restrictions, the internals could not be shipped integral to the vessel. Management decided to cut them under water in the reactor cavity, using an abrasive water jet system that used a high-pressure water and garnet grit mix. Segmented portions were classified as GTCC or LLW. The segmented portions that were GTCC would be placed in 64 specially fabricated, fuel assembly sized containers, temporarily stored in the spent fuel racks in the pool, and eventually moved with the fuel to an on-site ISFSI. Ultimate transfer of both fuel and GTCC material will be to a federally approved HLW repository. The LLW segments were placed in the vessel to be grouted in place and shipped as one piece, thereby saving both burial volume and cost. The original completion date for the project was May 25, 2000, with

GTCC canisters stored in the spent fuel pool by June 8. Due to problems with the cutting process, segmentation was not completed until October 20. The canisters were moved to the spent fuel pool in April 2001.

A filtration skid, containing pumps, filter vessels, and associated piping and valves, was located under water in the reactor cavity. Again, due to the project's length, components began to fail. The system was not modular, so an easy removal and replacement of these items was not feasible. Separate pumps had to be placed into the cavity and attached to the system remotely. Both the ALARA goals and project schedule were significantly affected by failure to control cavity conditions.

25.4.2 Radiation Protection: The original exposure estimate for the vessel segmentation project, as established in the PSDAR, was 90 person-rem. Actual radiation exposure received during vessel segmentation was 110 person-rem. An additional 84 person-rem is projected for project demobilization. The current radiation exposure estimate for the vessel removal is 74 person-rem.

25.4.3 Project Results: Personnel exposure is in jeopardy of not meeting the PSDAR exposure goal. Approximately five times the estimated garnet was required to complete internal segmentation, significantly increasing the amount of radioactive waste being shipped off site. On the other hand, the Haddam Neck safety record has been excellent. As of September 2003, the site has not had a lost-time accident in 1,155 days.

25.5 LESSONS LEARNED

Originally, CYAPCO management had planned to use site personnel to coordinate and perform decommissioning, including large component removal. However, in 1998, it decided to take an oversight role and chose a DOC to do the work. All detailed planning for large component removal was turned over to the DOC. However, CYAPCO employees resumed the contracting and oversight of all activities in mid-2003. Work activities have gradually resumed during the transition period, but the project schedule is still more than two years behind.

The internals segmentation process turned out to be a research and development project with many lessons learned. Almost from the start, the abrasive water jet delivery system encountered difficulties. Cutting the internals took much longer than anticipated due to the time required to maintain the system, the failure of its capture system, and the material being harder to cut than expected. The clarity of the water degraded to such a point that cutting required frequent work stoppages until visibility improved. Originally, a herculite curtain was placed in the reactor cavity to contain the grit and debris within a defined cutting area. But the curtain failed, due to its lengthy duration in the water, its loose seal at the barrier interfaces, and its being cut by the water jet. Grit and debris flowed through the entire cavity, creating the need for an extensive cavity cleanup.

Taken in its entirety, large component removal was a success, from both safety and technical perspectives. (Budget data are proprietary information and hence not presented herein.) There were minimal delays in physically moving the components.

Chapter 26

SAN ONOFRE NUCLEAR GENERATING STATION, UNIT 1

26.1 INTRODUCTION

The San Onofre Nuclear Generating Station (SONGS) Unit 1 is located on the coast of Southern California, in San Diego County, next to operating Units 2 and 3. The SONGS site is located entirely within the Camp Pendleton Marine Corps Base under an easement granted by the U.S. Government. Southern California Edison (SCE) and San Diego Gas & Electric (SDG&E) are joint owners.

Figure 26.1 The San Onofre Site (Unit 1 Only)

The 410 MWe PWR began commercial operation in January of 1968. In April 1992, SCE submitted an application to the NRC to modify the Unit 1 operating license to possession-only. The NRC issued the license amendment in October. SONGS 1 discontinued power generation in November, at the end of its eleventh fuel cycle. The reactor was defueled and the fuel moved to the on-site pool. The license became effective in March 1993, after SCE certified that operations had been permanently terminated and that all fuel from the core was in the pool.

26.2 DECOMMISSIONING PLANNING

After an evaluation of whether to continue delaying or to proceed, it was decided to proceed with active decommissioning in the summer of 1998. By December, SCE advised the California PUC and the NRC of the intention to decommission SONGS 1 and submitted an updated PSDAR. The PSDAR public meeting was held in February 1999. SCE and SDG&E requested access to the DTF to proceed with the project. The PSDAR had estimated the cost to complete decommissioning at $459 million in 1998 dollars, as presented in Table 26.1. A detailed cost and schedule management system was developed to track activity progress.

Table 26.1 Estimated 1998 Costs to Decommission Unit 1

Activity	Cost
Preliminary Planning	$ 9,600,000
Dismantling Activities (including large component removal)	
- Decontamination	11,025,000
- Removal	133,044,000
- Packaging	4,256,000
- Shipping	4,625,000
Staffing	131,333,000
Other	46,281,000
LLRW Burial	74,089,000
Spent Fuel Storage	44,519,000
Total Cost to Decommission	$ 458,772,000

SCE elected to use a matrix project approach that would effectively use the talent and skill sets of its current workforce to manage and support the project instead of contracting a DOC. Most SONGS 1 personnel were transferred to positions in the two operating units. A small, dedicated staff was assembled to manage the project while other employees and contractors would be assigned on an as-needed basis. Specialty contractors with decommissioning knowledge and demolition experience would also be used as needed.

The California PUC approved the request to use the DTF in June of 1999. The decommissioning project began in July, with modifications to physically separate Unit 1 from Units 2 and 3, preparations for diesel generator building demolition, and the restoration of the Unit 1 gantry crane. The SONGS 1 Decommissioning Project will be conducted in three phases. During Phase I, the facility will be removed and an ISFSI will be placed in operation. Phase II includes stored fuel monitoring until the DOE takes possession of the fuel, currently expected to be complete by 2026 and Phase III will involve removal of the ISFSI and any remaining foundations. SONGS 1 NRC License Termination is scheduled concurrent with the termination of SONGS 2 and 3 NRC licenses.

SCE also submitted an application for the ISFSI at SONGS 1. Once the certificate of compliance was issued, the unit's spent fuel could be placed in dry fuel storage. When the decommissioning project began in the summer of 1999, there were 207, 70, and 118 Unit 1 fuel assemblies stored in the SONGS 1, 2, and 3 spent fuel pools, respectively. All assemblies stored in the Unit 1 spent fuel pool must be moved and placed in dry storage before its pool and associated building may be decommissioned. Due to the physical location of the spent fuel pool support systems and equipment, such as the spent fuel pool heat exchangers, component cooling water pumps, and their associated piping systems, SCE determined that it would not be economical to isolate or "island" the SONGS 1 spent fuel pool.

The ISFSI pad was completed in 2002. It will hold 31 concrete horizontal storage modules, 18 for SINGS 1 fuel and GTCC waste. Each module weighs about 200 tons and will contain a fuel canister with up to 24 fuel assemblies. SCE plans to remove all the Unit 1 spent fuel from Units 1, 2, and 3 pools by the spring of 2005 and store it in the ISFSI until transfer to the DOE. The first canister is scheduled to be loaded by the end of September 2003 (ANS 2003). The 118 SONGS 1 fuel assemblies that were stored in the SONGS 3 pool were removed and placed on the ISFSI in January 2004. (ANS 2004)

After commencing the SONGS 1 decommissioning, detailed engineering analysis were conducted on most of the major projects. As a result, the total cost of the decommissioning effort was revised to $622 million in 2001 dollars. Although the scope of the decommissioning did not change, the cost of large component removal/disposal and the reactor vessel internal segmentation were under-estimated significantly below the cost available in the market when the projects were ultimately bid. In addition, fuel storage cost increased substantially since the cost to design and build an ISFSI in a high seismic environment, such as exists in Southern California, and was also not sufficiently understood. The increase in the cost estimate from that included in the PSDAR resulted primarily from the above aspects of the SONGS 1 decommissioning.

26.3 DECOMMISSIONING ACTIVITIES

SCE created a new entry point for the Radiologically Controlled Area (RCA) in 2000 that allowed the demolition of a major building north of the containment. Crews also demolished the diesel generator building, dismantled the turbine generator, relocated the control room, and established a cold and dark status, whereby the existing electrical systems were de-energized, and a clearly identified temporary distribution system was installed to support decommissioning and SAFSTOR activities.

Segmentation of vessel internals began in April 2001 and ended in 2002. In May 2001, the security boundary to the SONGS 1 Industrial Area was modified so that separate entrances could be used for Unit 1 and the SONGS 2 and 3 protected area. Additional major milestones for the year 2001 included the completion of the ISFSI prototype canister, fuel canister ASME certification, and demolition of the control building and east feedwater heater deck.

The large components from containment were removed through an opening in the top of the containment sphere and its enclosure building. Before cutting any openings, an analysis was conducted to ensure that structural integrity would not be compromised by the openings or by any use of the building for support or attachments as part of the removal operation. The building roof was completely removed and access openings were cut in the roof of the containment sphere. The vessel and head, steam generators, and pressurizer were removed with a heavy lift crane. After the large component removal work was completed, the top of the sphere was enclosed with weather covers to allow further decommissioning activities to take place within.

In early summer 2003, the project began demolition of the sphere enclosure building. Due to its close proximity to the fuel storage building and the height of its top outer wall (above ground level), planning required specific precautions to ensure the safety of personnel, the spent fuel building, and its contents. The building was cut, using a diamond wire, into blocks weighing an average of 18,000 pounds. Each block is seismically restrained until lifted from its cut location to the ground by a specially configured crane. About 18% of the building was cut into 251 blocks from the areas near the fuel storage building. The remaining portions will be removed by more traditional demolition techniques. (ANS 2004).

Each of the large components (except the vessel) was shipped to Envirocare of Utah for LLW disposal. The reactor vessel is temporarily stored on site in its shipping container pending final disposition.

26.4 CHARACTERIZATION

Table 26.2 presents the total dose results for the SONGS project through 2003. Based on both the actual project dose to date and the estimates of future activity, it is anticipated that the project will not exceed the PSDAR estimate of approximately 1100 person-rem.

Table 26.2 Total Dose

Date	Person-rem (actual)
1999	16.260
2000	71.214
2001	57.785
2002	61.214
2003	36.146

26.5 SAFETY PROGRAM

The decommissioning team has implemented an industrial safety program with special emphasis on contractor safety. A division safety committee meets monthly. A behavior-based leadership observation program is in place with a focus on industrial safe work practices and leadership in the field. Teamwork through effective communication between work groups is a daily expectation, and the work groups meet each day for a review of key jobs in progress to ensure effective communication. The first point of discussion is always safety.

Table 26.3 summarizes the nature and quantity of waste shipments for the project as of April 2003. Non-contaminated concrete from clean building demolition, such as the diesel generator building, is being crushed and recycled.

Table 26.3 Waste Shipment Summary

Plant Location	Rad/Haz Matls (pounds)	Clean/Recycled
Admin/Control building	6,150,711	9,981,961
Backyard	32,393	1,324,000
Containment	300,000	1,854,389
Diesel generator building	477,493	15,767,050
HazMat	476,922	0
Intake	5,339	312,030
Large component removal	1,832,445	0
Reactor vessel internals	192,225	0
Sphere enclosure building roof	106,532	4,571,000
Turbine	2,942,410	6,448,410
Yard	54,955	184,000
Total	14,138,295	38,888,451

The focus the program is continual safety improvement. Contractor safety programs are reviewed during the bid process and contractors are held accountable to implement their site-specific safety plans and company injury, illness, and prevention programs. Contractors are also expected to submit site-specific safety plans. A monthly and quarterly self-assessment program ensures that safety performance in key areas is understood and that proper actions are taken to ensure a continued safe and healthy work environment. One OSHA-recordable event occurred on the project in 2002, and another in the fall of 2003, both involving contractors. The entire decommissioning team, including all contractors and SCE personnel, has subsequently worked 228 consecutive days without an OSHA-recordable injury as of June 1, 2004. The team had achieved 253 consecutive days in 2001.

26.6 TRANSITIONAL FACTORS

26.6.1 Spent Fuel Storage Approach: All Unit 1 spent fuel must be placed in dry cask storage in order to decommission its spent fuel pool, located in the center of the facility. Transnuclear, Inc. technology was selected from a competitive bid process to design an ISFSI for a high seismic environment. The vendor submitted a generic safety analysis report for its design and the NRC issued a certificate of compliance, effective February 5, 2003. SCE constructed its ISFSI and will store its spent nuclear fuel under an existing 10 CFR 50 license.

The Transnuclear ISFSI design consists of stainless steel dry storage canisters containing up to 24 spent fuel assemblies each. The canisters are positioned horizontally and installed into reinforced-concrete, horizontal storage modules. Shield walls will be incorporated into the ISFSI due to the close proximity of site boundaries and ongoing site support activities.

26.6.2 Proposed End-State: In addition to completing all requirements for NRC license termination, all Unit 1 structures and facilities will be removed to a depth of at least 10 feet below ground. The scope of the remaining below-ground structural removal will be at the discretion of the government.

26.6.3 Socio-Economic Impacts: The figures in Table 26.4 reflects the end of commercial operation, followed by placement of the unit in SAFSTOR. Spent fuel storage manpower figures for 2005 - 2007 are included in the decommissioning figures since current plans are to store the spent fuel in the ISFSI until accepted by the DOE. The number of employees during operations versus during decommissioning (including licensee employees and contractors) is as follows:

- Annual number of employees during operation of Unit 1 was 424
- Annual average of employees for decommissioning from 1999 to 2007 is expected to be 318

Table 26.4 Work Force Size Timed with Decommissioning Phase/Year

Year	Full Time Equivalents	Activity
1993	328	Preparation for SAFSTOR
1994	306	Preparation for SAFSTOR
1995	130	SAFSTOR and spent fuel storage
1996	119	SAFSTOR and spent fuel storage
1997	115	SAFSTOR and spent fuel storage
1998	115	SAFSTOR and spent fuel storage
1999	97	SAFSTOR and spent fuel storage
	40	DECON
2000	49	SAFSTOR and spent fuel storage
	348	DECON
2001	57	SAFSTOR and spent fuel storage
	310	DECON
2002	75	SAFSTOR and spent fuel storage
	324	DECON
2003	85	SAFSTOR and spent fuel storage
	414	DECON
2004	85	SAFSTOR and spent fuel storage
	286	DECON and spent fuel storage
2005	354	DECON and spent fuel storage
2006	257	DECON and spent fuel storage
2007	180	DECON and spent fuel storage

Table 26.5 shows the approximate expenditures (actual and projected) for labor and materials.

26.7 DECOMMISSIONING ACTIONS

26.7.1 Modeling of Remaining Subsurface Material: Computer models developed by the NRC (DandD) and the DOE (RESRAD) require that residual radioactivity in soil be modeled as if it all exists at the surface (the top 6 inshes). Additionally, the building occupation models assume that rooms are available and occupied by either residents or workers. In contrast, most commercial nuclear power plants, once dismantled, can have significant subsurface structures remaining, such as foundations and yard drains. And any contaminated soil will likely be well below the surface because leakage from components and buildings will probably occur in the lower elevations of the plant rather than at the surface. Moreover, any contaminated soil remaining after excavation would be below the surface. The end result is that the models preferred by the regulator have limited usefulness for power plants.

Table 26.5 Labor and Materials Costs – Actual and Projected
(Year of Expenditure Dollars)

Year	SAFSTOR Preparation	SAFSTOR & Spent Fuel	DECON	Total
1993	49.9M			49.9M
1994	37.6M			37.6M
1995		12.9M		12.9M
1996		13.4M		13.4M
1997		12.1M		12.1M
1998		12.9M	1.6M	14.5M
1999		11.1M	13.2M	24.3M
2000		8.1M	58.4M	66.5M
2001		8.7M	76.3M	85.0M
2002		8.1M	96.3M	104.4M
2003		7.7M	101.4M	109.1M
2004		7.9M	56.4M	64.3M
2005			77.1M	77.1M
2006			38.0M	38.0M
2007			29.2M	29.2M
2008 *			18.6M	18.6M

* Does not include future expenditures to decommission the ISFSI

The computer codes use conservative assumptions, including all possible means for a member of the public to be exposed to radiation. The alternative of using site-specific analyses results in a significantly higher level of regulatory review, thereby increasing the time required for LTP approval and the subsequent costs. SONGS has performed analyses to demonstrate that subsurface contamination is bounded by surface contamination values. Therefore, there has not been a significant impact to the project. However, SCE management continues to support ongoing EPRI research to develop generic subsurface contamination evaluations.

26.7.2 Clearance Standards for Solid Materials: There is no clear NRC standard for the clearance (unrestricted off-site disposal) of solid materials that may have minimal volumetric or surface contamination. The determination of what may be disposed of as nonradioactive and what will require disposal at a licensed facility should be evaluated in terms of cost, since disposal at the licensed site will be much greater than at an ordinary state-administered landfill site. Ultimately, off-site-disposal of such debris resulting from demolition and other activities must be resolved with all concerned regulators and other parties. The EPA, state, and local regulations, as well as the NRC's, may very well apply to certain debris and its proposed disposal sites.

The consequence of not addressing this issue is significant because this lack of a definitive standard will most likely significantly increase costs and possibly shift the financial responsibility for disposal from utility customers to shareholders. One approach to addressing this issue is to remove structures before license termination, and thus avoid the issue of residual radioactivity to the greatest extent possible. The SONGS 1 project will avoid the majority of the clearance and license termination issues by removing all above-ground and most below-ground structures (in accordance with the site easement agreements, as discussed later). The end result will be a near background level of contamination, thereby avoiding the above issues.

26.7.3 Waste Stream Determination Process: The Gamma Cam system was used early in the project to assist in identifying hot spots in elevated dose rate areas. This system provides a black and white image with a color overlay showing the relative dose rate intensity in the viewing area. Having completed the imaging process, a portable multi-channel analyzer system was used to obtain isotope information on each hot spot/area, thereby aiding in the waste stream determination process. The next task was to determine how many areas/piping systems matched (similar isotopic distribution), with the goal of reducing labor costs by narrowing the number of piping components needing separate sampling. It was found that the variance in gamma emitters was substantial, even within the very same piping system. It became apparent that heat was also a key factor in contamination deposition, as were flow rates, weld joints, and torturous flow paths. The result of using both systems (gamma can and multi-channel) was a relatively economical evaluation of which systems needed separate waste stream determinations. The work force could then gain access to those systems, obtain samples for analysis at an independent laboratory, and plan for removal in an efficient manner.

26.8 UNIQUE ISSUES

The SONGS site is located within Camp Pendleton under an easement granted by the U. S. Government. Under the terms of that easement, the plant owners are required to remove all materials from the plant site, including residual radioactivity, except as requested by the government. In addition, the off-shore circulating water intake and discharge conduits are located on land used under an easement granted by the California State Lands Commission. As it currently stands, the easement requires that the conduits be removed in their entirety before the land is returned to the owner. Both of the above two requirements were taken into account for the decommissioning cost estimates for all three SONGS units. If the property owners enforce the above requirements, the scope of this project would be greater than other decommissioning projects.

The SONGS site is relatively small compared to other commercial nuclear facilities. The three units are located on 78 acres of a 257-acre site that is crossed by a major interstate highway. Moreover, the plant site itself was carved from the surrounding coastal bluffs. As a result, decommissioning is very limited in the amount of land available for equipment maneuvering, laydown areas, and space to manage, handle, and package waste created by the dismantling of structures. Activities are carefully orchestrated and planned to not interfere with one another. LLRW and other waste is closely managed to ensure that waste does not "stockpile" and interfere with other decommissioning activities. And since the site has limited space available for the ISFSI, it is being built on a "footprint" that was created by removing certain Unit 1 buildings.

The SONGS 1 decommissioning is being conducted on the same site as Units 2 and 3, which continue to operate. SCE conducts the project so as to ensure that there is no impact or disruption at the operating units. In addition, site resources are diverted from the decommissioning project as needed to support the other units' refueling outages and operational requirements. It should be noted that the NRC licenses of all three SONGS Units cover the same real estate. As a result, terminating the SONGS 1 license would provide no benefit to the project since all regulations would still remain applicable.

The systems supporting the SONGS 1 spent fuel pool can continue to support the pool during the decommissioning process until the fuel is removed. SCE determined that the existing spent fuel pool cooling and cleanup systems were satisfactory to use during decommissioning. These systems will not be affected by planned activities until the fuel is placed in the ISFSI.

26.9 LESSONS LEARNED

It is necessary to continually apply aggressive oversight to numerous contractors and vendors to obtain quality and thoroughness of both services and product delivery. SCE discovered that they might not always appreciate the mindset necessary to provide an acceptable quality product when associated with the nuclear industry in general, and specifically with radioactive wastes. In addition, it was noted that several companies, with or without experience in the nuclear industry, still did not fully understand that dismantling a former nuclear generating station is significantly different than the demolition of other structures. Despite this caution, the project is demonstrating that the facility can be removed from its site with the same rigorous and safety conscious environment as when it was operating.

Due to the large quantity of material to be packaged for disposal, there can be substantial labor savings during dismantling by using very large packaging. Minimizing the number of cuts in the work area saves dose (ALARA), allowing for the near-direct loading of 18-foot piping sections instead of 7-ft sections. Additionally, disposal at Envirocare allows for the re-use of inter-modal containers for rail shipment. By not having to buy smaller, one-time use containers, labor is reduced in cutting the material down to size, and costs are avoided with greater disposal weights but less containers for disposal.

The obstacle, however, is that the very limited workspace makes it difficult to store various sizes of waste material. Material must be packaged as it is cut out, with the result that the optimal mix of debris sizes is not available to make an efficiently packaged waste container. SCE is researching second tier segmentation solutions since packaging efficiency is less than desired.

Since SONGS 1 was constructed in the 1960s, decommissioning planning assumed that asbestos would be found in piping/building insulation, and designed safety controls to handle the materials safely. Indeed, more asbestos than expected was found, including that on the outer surface of the underground diesel fuel storage tanks.

Chapter 27

IOWA STATE UNIVERSITY RESEARCH REACTOR

27.1 INTRODUCTION

The UTR-10 was an Argonaut-type reactor using uranium enriched to 19.75% in 235U in a graphite-reflected, water-moderated core. It was designed and built in 1959 by the Advanced Technology Laboratories division of the American Radiator and Standard Sanitary Corporation. Initial reactor criticality came on December 31, 1959; operations officially ceased on May 15, 1998. The reactor was housed in the Nuclear Engineering Laboratory located on the west edge of the main campus of Iowa State University (ISU) in Ames, Iowa. The facility is a two-story, three-level, brick building built in 1934 by the Department of Agriculture. It was deeded to the University in 1946. The reactor room housed the reactor (enclosed in a concrete biological shield), the process pit, the fuel storage pit, and a five-ton bridge crane.

Reactor in operation, circa 1970

Bioshield after removal of upper closures, thermal column, and graphite

Figure 27.1 10 MW Research Reactor at ISU

In 1991, the reactor fuel was changed to low-enrichment uranium. The reactor was controlled with four window shade-type Boral control rods. Heat from fission was removed from the primary coolant by a heat exchanger that used city water. The reactor was designed to be inherently safe -- it would automatically shut down if there were a loss of electrical power or if parameters important to safety were exceeded. The reactor was provided with multiple experiment features, including beam ports, thermal column, shield tank, internal reflector, rabbit tube, and radiation cavity. Figure 27.2 depicts the layout of the UTR-10 facility.

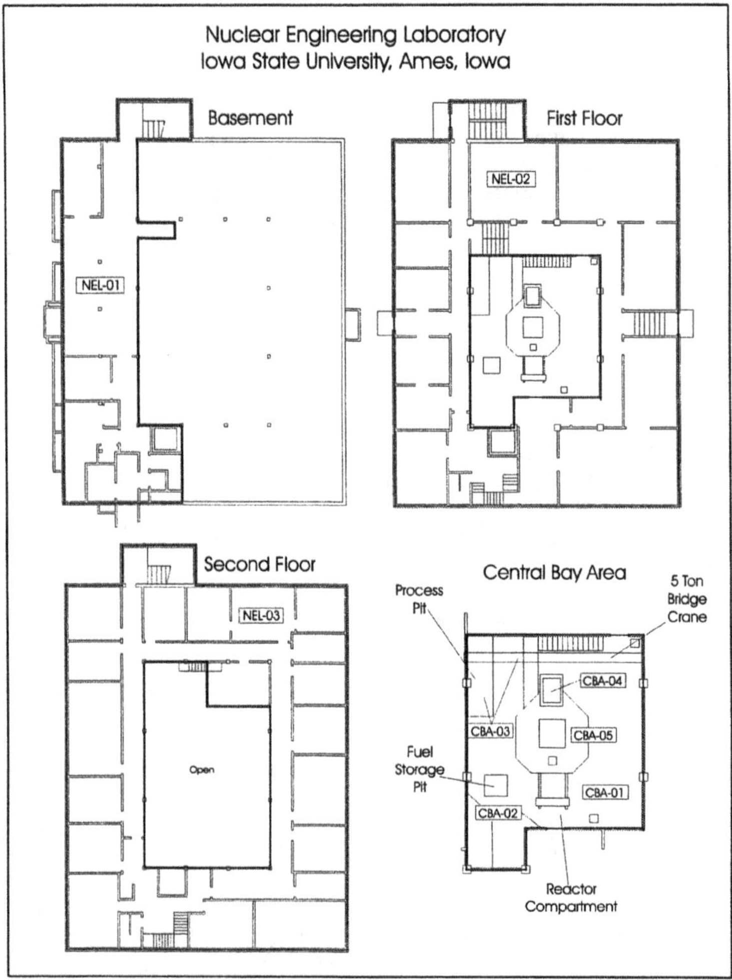

Figure 27.2 Layout of The UTR-10 Facility

27.2 DECOMMISSIONING PLANNING

Decommissioning was divided into three phases:

- Phase I consisted of two parts: 1) develop and implement a comprehensive characterization of the facility and prepare a characterization report, and 2) develop and submit a decommissioning plan (DP) describing the methods to be employed to safely dismantle, decontaminate, and release the facility to applicable criteria

- In phase II, decommissioning activities were implemented in accordance with the NRC-approved plan

- Phase III consisted of a final site survey to verify that residual radioactivity is below the release criteria, and preparing a final report.

27.2.1 Alternative Selection: ISU selected DECON for decommissioning the UTR-10 reactor after evaluating three decommissioning alternatives: DECON, SAFSTOR, and ENTOMB. DECON was preferred for the following reasons:

- The ISU UTR-10 was a small reactor, rated at 10 kW power level.
- Volume of radioactive waste generated from decommissioning activities was estimated to be small.
- Commercial radioactive waste disposal facilities were accessible to the project.
- Radioactivity was limited to the activated materials in the reactor room structure.
- ISU personnel familiar with the reactor were still available.
- The area was intended for other programs and facilities after unrestricted release.

The DECON alternative involved removing all spent fuel assemblies, source material, radioactive fission products, and radioactively contaminated material and components that exceeded the release criteria. The DP consisted of several tasks oriented towards dismantling the reactor vessel, bioshield, and the reactor systems and structures. Dismantling and removal activities adhered to all applicable regulations, and were in accordance with the ISU policy of ALARA applications.

27.2.2 Decommissioning Organization: As the owner of the facility, ISU chose to use a turnkey approach and retained a decommissioning contractor. The DOC project manager, who reported to the ISU project manager, was responsible for preparing the site for decommissioning and implementing the appropriate dismantling activities.

27.3 PROJECT SCHEDULE AND COST

The project began in July 1998, following the award of the decommissioning contract. Phase I was completed on schedule, culminating in the submittal of a DP to the NRC in January 1999. Delays in fuel shipment and regulatory review resulted in a seventeen-month interruption of the project. The NRC issued its Decommissioning Order in May 2000; decommissioning of the UTR-10 began in June and was concluded on August 4. The NRC performed confirmatory surveys in September, and the reactor room restoration was completed by early 2001.

The decommissioning cost estimate for the project was $1.4 million. The State of Iowa, through the ISU, provided the necessary funding. The project was successfully completed within the budgeted cost.

27.4 CHARACTERIZATION

In preparation for decommissioning, the DOC conducted a radiological survey of the reactor facility. A characterization/sampling plan was prepared and activities were initiated at the fuel storage pit in July 1998. The final surveys were concluded in September 1998. Characterization followed the MARSSIM guidance.

Elevated exposure rates ranging from 27 to 54 mr/hr were observed in areas that served as storage for radioactive sources and materials. Exposure rates up to 23 mr/hr were detected in the southwest stairwell and in an area above the east side of the reactor room. The remainder of the facility produced exposure rates consistent with the outdoor levels near the facility, which ranged from 12 to 28 mr/hr.

1. *Surface Contamination*: Surface contamination was limited to a small area on the process pit floor, and averaged 9000 dpm/cm^2. No removable surface contamination was detected on the surface of the reactor core. Elevated measurements that were observed in the core were due to activation.

2. *Reactor Core Activation*: Contact exposure rate readings collected on the east beam port plug and the central thermal column stringer indicated that the area of activation extended radially to approximately 40" x 34" from the core centerline. An analysis of four materials collected from the core (concrete, graphite, aluminum, and steel) indicated an estimated total volume of radioactive waste of 1200 ft^3.

27.5 DECOMMISSIONING TASKS

1. *Defueling*: The UTR-10 reactor was defueled in 1998; the fuel was placed in storage at the site for shipment to the Savannah River Site.

2. *Facility Modification*: The first stage of the decommissioning involved several facility modifications to create a larger workspace in the reactor area and facilitate removal of the reactor components. The north stairs, platform, and stairs to the reactor top were removed using flame torches. None of these items were contaminated. As pieces were sectioned, they were surveyed, unconditionally released, and removed for disposal as demolition waste or scrap metal for recycling. Access to the reactor top during the balance of the project was via temporary ladders.

3. *Removal, Remediation, and Disposition*: Each reactor closure was removed and staged for survey and remediation. The radioactive portions of each closure were removed using chipping hammers, drilling/splitting, and torch cutting in preparation for unconditional release criteria. Upon removing the activated portions, the closures were moved or oriented to a low background area for final survey. Following unconditional release, each closure was loaded for disposal as demolition waste. These items included the thermal column door, shutdown closure, operating closure, and experimental plugs (with activation through approximately the lower twelve inches).

4. *Tank and Sump Equipment Removal*: The tank and associated piping for the reactor water was removed to the point where it entered the concrete wall. Embedded piping was abandoned in place or removed as necessary. All reactor water equipment was surveyed, unconditionally released, and disposed of as recyclable metals or demolition waste.

5. *Graphite and Core Component Removal*: Graphite was removed beginning with that located at the south end of the bioshield. As each stringer was removed, it was numbered, segregated for survey, and eventually surveyed in a low background area. Sections of graphite containing detectable radioactive material were segmented and disposed of as radioactive waste. The remaining graphite was disposed of as demolition waste. Core mechanical components, activated support structures, and internals were removed from the core, segmented as necessary, and packaged as radioactive waste.

6. *Water Tank Demolition*: The water tank at the north end of the reactor was scored horizontally and vertically with a diamond blade concrete saw. It was turned into rubble using a hydraulic ram. Since the tank walls were outside the reactor activation zone, this material was surveyed, unconditionally released, and disposed of as demolition waste.

7. *Radiological Containment and Dust Control*: The planned sequence of dismantling was to remove all reactor components first, and have the concrete bioshield demolished last. Following the removal of the reactor components and graphite, a more definite activation profile was determined, using a combination of radiological surveys and concrete coring. The contamination levels inside the cavity were not expected to warrant the use of contamination control containment during removals; this turned out to be the case, and only nontransferable activated material was present, so no containment or dust control measures were required. Continuous air samples were taken in the work area, at the south end of the reactor room, and on the roof (where large fans vented fumes from the heavy equipment and tools).

   The primary means of dust control during concrete demolition was water applied in a low-volume mist at points where dust was produced. When necessary, containment was ventilated with additional high efficiency particulate air (HEPA) filters, either as an added measure or to prevent cross-contamination of office/occupied areas and noncontaminated areas of the reactor room. Excess water from misting activities was collected and re-used as necessary. Subsequently, remaining water was sampled and unconditionally released.

8. *Hydraulic Splitting*: A pattern of vertical holes was placed throughout the bioshield using rock drills. Drilling depths varied from four to eleven feet, with the intent of providing a clear demarcation of the interior activation zone. The holes were then filled with an expansive grout and allowed to cure for approximately twenty-four hours. The grout created a series of fracture lines, both on the exterior surface and within the non-activated portions of the bioshield.

9. *Demolition*: The bioshield was demolished using a BROKK. Care was taken to remove only the non-activated portions of the shield, as delineated by the grout. Misting and

ventilation were used to control dust during this portion of demolition. Air samples and daily surveys of the work area confirmed that no cross-contamination had occurred. Nonradioactive portions of the bioshield were disposed of as demolition waste.

Work was stopped upon approaching the activation zone, and a HEPA-ventilated containment was constructed around the remaining portion of the bioshield. Demolition was then completed in the manner described earlier. After removal of the bioshield, portions of the underlying floor area that had been activated were removed and, like the remaining bioshield material, disposed of as radioactive waste.

10. *Pit Excavation*: A small pit consisting of activated soil had been identified directly underneath the reactor centerline. This area was excavated until no detectable radioactivity remained. The removed material was disposed of as radioactive waste.

27.6 RADIOLOGICAL RELEASE CRITERIA

The rule on license termination (10 CFR 20.1402) gives radiological criteria for releasing a site for unrestricted use -- a maximum Total Effective Dose Equivalent of 25 mrem per year from residual radioactivity above background. Application of ALARA is also a requirement. Instead of performing statistical tests for the overall demonstration of compliance with the regulations, an adequate number of measurements were made to ensure compliance with release limits.

The DCGLs used to screen individual measurement values were based on the results of radioanalytical data. ISU site-specific DCGLs were established by adjusting the generic limits to account for hard-to-detect radionuclides that cannot be measured with typical survey instruments. Table 27.1 summarizes the bulk material sample (Part 61) analysis results and the calculated hard-to-detect nuclide (HTDN) ratios used to determine the DCGLs.

Table 27.1 Bulk Material Composite Sample 10 CFR 61 Analytical Results Summary

Sample	Identified Detectable Radionuclides	Radionuclide-Specific Dand D Limit (dpm/100cm^2)	Fn/Ln (Fraction/ Limit)	Identified HTDNs	Radionuclide-Specific DandD Limit (dpm/100 cm^2)	Fn/Ln (Fraction/ Limit)
Composite Smear Sample	None	N/A	N/A	None	N/A	N/A
Reactor Graphite	None	N/A	N/A	None	N/A	N/A
Reactor Concrete	^{60}Co ^{152}Eu ^{59}Fe	7,040 12,700 88,300	4.09 E-06^{-5} 3.76 E-05 1.30 E-06	^{3}H ^{55}Fe	1.23 E+08 1.10 E+04	9.76 E-10 2.06 E-07
Reactor Steel	^{60}Co-60	7,040	4.36 E-05	^{65}Zn	48,100	1.44 E-05
Reactor Aluminum	None	N/A	N/A	None	N/A	N/A

A DCGL was developed to address the contaminated soil that lay beneath the former location of the reactor housing. A value of one was used to develop screening values for each of the five nuclides identified in the analysis of soil samples taken during the remediation phase of the project. The samples were from a localized area of activated soil at the core centerline. The only input to the code was the nuclide soil concentration (1 pCi/g). All other input parameters were left at the default values. In addition, the resident farmer scenario was applied. Given the use of default parameter values and the application of the resident farmer scenario, the screening values appeared to be conservative.

Because the affected area was small, and continuation of activation products under the concrete slab was ruled out by survey, the remaining soil was removed and packaged as radioactive waste. A composite sample from the perimeter of the excavation and a sample at the reactor centerline were taken and analyzed to ensure that no radwaste would remain in the soil. Results indicated no licensed radioactive materials present.

The Data Quality Objectives (DQO) process was implemented in the final status survey plan using a graded approach, as recommended in MARSSIM. The final area classification of the laboratory was subdivided into the MARSSIM Classes 1, 2, and 3, as follows:

- Class 1: Areas with known contamination in excess of the DCGL before remediation, based on characterization survey
- Class 2: Areas with a potential for radioactive contamination or known contamination before remediation, but were not expected to exceed the DCGL
- Class 3: Any affected areas not expected to contain any residual radioactivity, or to contain levels of residual radioactivity at a small fraction of the DCGL.

Since the characterization report had shown that only Class 1 areas contained contamination that could be confirmed, these areas received the highest degree of survey efforts during the final survey.

27.7 QUALITY ASSURANCE

MARSSIM requires additional quality provisions for field surveys and collection of radiological data. The primary goal of the program was to limit the introduction of errors into the analytical data by identifying and specifying the surveying, sampling, and analytical methodologies. In addition, its purpose was to ensure that the samples were consistently collected, analyzed, and reported, and that the quality of the resulting data could be independently evaluated. All personnel directly involved with performing of the final status survey and data quality review were trained to a competency level commensurate with the activity being performed.

Chapter 28

BARNWELL NUCLEAR FUEL PLANT

28.1 INTRODUCTION

The Barnwell Nuclear Fuel Plant (BNFP), located in Barnwell, South Carolina, on a 1622 acre property about five miles southeast of the Savannah River site, was built in the early 1970s to reprocess spent nuclear fuel from commercial power reactors but was never used for this purpose. However, extensive testing used natural uranium as a surrogate material, and research and development work in the laboratories used plutonium and other transuranics (TRU).

Figure 28.1 Barnwell Nuclear Fuel Plant

Allied General Nuclear Services (AGNS), a partnership among today's Honeywell International, Shell, and Chevron, owned the facility. In 1983, AGNS halted plant testing and ended the research and development program. It then partially decontaminated facilities and deactivated the plant. Significant radioactive contamination remained within most processing equipment and some facility surfaces, as well as inside laboratory glove boxes and fume hoods.

A site characterization during decommissioning revealed that contamination totaled approximately 600 kg of uranium and 10 g of TRU, which far exceeded the quantities estimated to be present before the decommissioning began -- 5 kg and 1 g, respectively -- as reflected in the plant radioactive material license issued by the State of South Carolina -- an agreement state under the Atomic Energy Act (AEA 1954). In 1997, AGNS decided to move forward with decommissioning so the partnership could divest itself of the plant. The decommissioning concept involved releasing the property with the structures in place so the plant could be turned over to the community for use as an industrial park.

28.2 DECOMMISSIONING PLANNING

AGNS hired a team of two contractors, which developed a strategy based on its previous experience with planning the decommissioning of a heavy water components test reactor and a naval shipyard. Phase I included:

- HSAs and a characterization of the plant for residual radioactivity
- Development of site-specific cleanup guidelines
- Preparation of DPs for each of five parts of the plant

The plant was divided into five distinctly different areas. The process for HSAs, characterization, and developing cleanup guidelines followed MARSSIM. Site-specific cleanup guidelines were developed using the RESRAD and RESRAD-BUILD codes. The DCGLs facilitated achieving the cleanup criteria of less than 0.15 mSv (15 mrem) per year, with no more than 0.04 mSv (4 mrem) per year from groundwater, in commercial-industrial use of the property. Multiple calculations for different parts of the plant yielded guidelines that minimized decontamination of facility surfaces and plant equipment. The guidelines applied to both surface radioactivity and concentrations in soil, concrete foundations, and filter media.

The team completed the five DPs over four months. The plans for natural uranium contaminated pipes incorporated an innovative approach to determine if internal contamination levels fell below the cleanup guidelines. Using correlations between external contact dose rates and surface contamination levels inside the piping, this approach made it practical to determine with a simple external scan survey whether piping and ventilation ducts met the cleanup guidelines.

The South Carolina Department of Health and Environmental Control, rather than the NRC, issued the facility license. Therefore, as regulator, it approved the cleanup guidelines and the DPs. Phase I (planning) was completed in April 1998. Phase II, the actual decommissioning work and final radiological status surveys, began in June 1998 and was completed in May 2000. The site license was then terminated and the area was turned over to the state as an industrial park in June 2000.

28.3 CHARACTERIZATION

The HSAs began with review of the plant records, which were found to be complete. Inspections and walkdowns of the various facilities followed, with photographs were taken in each area. Team members interviewed key former employees. Technicians took radiological scoping surveys to help identify needed characterization. Plant characterization techniques employed include:

- Detailed sample and analysis plans
- Radiation scan measurements

- Surveys of facility surfaces and equipment
- Laboratory analysis of smears and material samples
- Opening equipment for internal surveys
- In-situ gamma spectroscopy of installed equipment.

The characterization program was completed in about four months. The team documented the assessments in five separate reports and briefed the regulator on the results.

The in-situ gamma spectroscopy used the in-situ object counting system (ISOCS ™). This system characterized sealed glove boxes in the laboratories and other sealed equipment in the separations building. In some cases, ISOCS data were also used for final status survey purposes, such as for a large sintered metal filter encased in lead shielding.

During Phase II, plutonium contamination was found where it was not expected (based on process history and the HSAs). This situation led to an investigation and a supplemental sample and analysis program to determine the extent of the contamination.

28.4 DECOMMISSIONING ACTIONS

To accomplish the decommissioning, AGNS hired the same people who had characterized the plant and planned the decommissioning. The prime contractor provided project management, engineering, health physics, waste management, and quality assurance support. The second contractor provided planning, work control, and radiological control support, along with decommissioning workers.

The eight-story steel-frame uranium hexafluoride building was the first structure decommissioned. The process took approximately 10 months. This building, with its 16 labs with 37 contaminated glove boxes and numerous contaminated fume hoods, proved to be the most challenging facility from a radiological controls standpoint. One large glove box contained PUREX pilot plant equipment heavily contaminated with plutonium. Radioactive contamination levels exceeded 1,010 dpm/100 cm^2 total alpha. Working inside a specially designed containment tent, workers carefully dismantled the equipment and the glove box itself. About 90% of the TRU waste associated with the project came from this one glove box. Another in the same laboratory stood 25-ft high. It had the same excessive contamination levels and was dismantled. Workers removed most standard glove boxes intact, separating assemblies where two or more were joined into single three-foot long units. These were sent to a nearby facility for super compaction, along with the fume hoods.

To screen waste associated with this work (to determine the presence of TRU waste), technicians used field instruments with specially designed radiation shields calibrated for this application. This technique proved to be effective. Materials identified as probable TRU waste filled fifty 55-gallon drums. Subsequent nondestructive assay showed that 33 of them actually contained TRU waste.

Decommissioning of the separations building presented a challenge because of its complex systems and equipment, some containing significantly contaminated with uranium. Piping exceeding cleanup guidelines was removed. One vessel was flushed with nitric acid to remove residual contamination, but most of the piping and equipment had internal contamination levels below cleanup guidelines, based on the dose rate survey technique previously described.

The most unexpected discovery during the project involved the 300,000 gallon underground waste tanks intended for eventual storage of HLW. Two of the three tanks were known to be contaminated with natural uranium and were expected to contain a small quantity of water. Instead, one tank was found to contain some 5,000 g of organic and aqueous mixed waste, with a high concentration of uranium. The other tank was found to be floating: rainwater had infiltrated the concrete tank vault and filled the space around it; approximately 70,000 g of water had entered the tank itself. Altogether, some 100,000 g of radioactive liquid waste in the plant was processed by evaporation.

28.4.1 Final Radiological Status Surveys: The final status survey plan followed MARSSIM protocols. The property was divided into approximately 160 survey units. Surveys focused on ensuring that the derived cleanup guidelines were achieved. Experience and professional judgment rather than MARSSIM statistical tests determined the number of measurements.

Surveying began as soon as practical. In the uranium hexafluoride building, for example, a floor would be surveyed right after being decontaminated. The results would then be compiled into a chapter of the final status survey report, and presented in draft form to regulators. This process enabled the regulators to be able to complete their independent final survey of each area of the plant right after those of the decommissioning team. The radioactive materials license was terminated in May 2000, after completion of the last confirmatory surveys.

28.4.2 Waste Quantities: The project disposed of about 8500 ft^3 of LLW. About 3500 ft^3 of this total was noncompacted waste shipped directly to Savannah River (including 900 ft^3 of stabilized residues from the radioactive liquid processing). The rest of the LLW was compacted with an overall volume reduction factor of about 5.8. The reduction factor for the laboratory glove boxes and fume hoods was higher.

About 244 ft^3 of TRU waste was shipped to Savannah River for later disposal at the Waste Isolation Pilot Plant (WIPP). About a third required packaging in Type B container. Mixed waste, totaling 460 ft^3 of mostly HEPA filters containing cadmium, was shipped for disposal to Envirocare of Utah. A small amount of the radioactive waste contained asbestos, while the amount of nonradioactive solid waste associated with the project was small.

28.4.3 New Technology: The project did not use new technologies, but made use of existing technologies in innovative ways. The ISOCS ™ was used extensively, effectively and non-intrusively characterizing sealed glove boxes and sealed equipment in the separations building. This application represented the most extensive characterization use of in-situ gamma spectroscopy up to this time, and may have been the first use of in-situ gamma spectroscopy for final status survey purposes. RESRAD-Build modeling was used more than in any earlier decommissioning project. This modeling took into account the possibility that at some future time piping in the facilities might be removed for salvage purposes, and showed that the 0.15 mSv per year criterion would be achieved even if this were to happen. The DCGLs produced by RESRAD-Build also facilitated the use of external dose rate surveys to ensure that piping remaining in place met the release criterion.

The extensive use of dose rate surveys for releasing in-place uranium-contaminated piping and ventilation ducts was an industry first. This process, developed for a wide range of piping sizes, made use of gamma survey instruments (μrem and μrad survey meters). Radiation levels on the piping exterior of no more than 5 μrad /hr above background indicated that internal contamination levels fell below the DCGL. In cases of open-ended piping, this survey technique was supplemented with one of two other types of measurements. One involved the use of a pipe probe (a Ludlum 44-6 side window GM detector) inserted in an open pipe end. The other was a beta dose rate measurement at the open end, with the beta dose rate being related to the average internal contamination level using Monte Carlo calculations to estimate beta efficiency. The combination of RESRAD-Build DCGLs for uranium contamination and the dose rate survey technique saved millions of dollars.

28.5 UNIQUE ISSUES

Dismantling the two large special glove boxes presented unique ventilation and containment tent design situations. A wall behind the PUREX glove box had to be removed to facilitate installation of a specially designed tent. The need for worker entry to the glove box to dismantle equipment required careful planning and a variety of radiological controls. Another unique issue involved the floating HLW tank. Once this condition was identified, a detailed structural evaluation had to be performed before further entries could be made into the tank vault.

28.6 OUTCOME AND LESSONS LEARNED

The BNFP project is the largest commercial nuclear facility decommissioning project undertaken in South Carolina. The approach, a success in terms of efficiency, included detailed planning, developing site-specific cleanup guidelines, close liaison with the regulator, and the use of small crews of radiological control technicians to perform the fieldwork. The amount of waste generated was small for a project of this scope. The final costs were far below the owner's early estimates.

The waste management plans developed early in the project served well, but in some ways the process for handling waste was made more complicated than necessary. Fortunately, a specialist improved the efficiency of the program in the early stages of fieldwork. LLW was reduced by use of the derived site-specific cleanup guidelines for natural uranium. Field screening of potential TRU proved to be effective. Removal of the glove boxes and fume hoods using minimal disassembly and having them super compacted proved to be an efficient approach, minimizing fieldwork involving radioactivity and reducing the waste volume.

The amount of mixed waste generated was larger than expected. Characterization for hazardous materials had been intentionally limited, and the HEPA filters were identified as mixed waste late in the project. Use of radioactivity concentration guidelines for filter media minimized mixed waste by allowing some filters with very low levels of natural uranium radioactivity to remain installed.

The interface between the project team and the Savannah River Site functioned smoothly. The waste liaison from Westinghouse Savannah River Company made frequent visits to the plant and to the consolidation facility. The use of biweekly telephone conference calls helped to keep all key people informed on the status of decommissioning and waste activities.

Chapter 29

THE DOE HANFORD SITE

29.1 INTRODUCTION

In 1942, the U.S. Army Corps of Engineers chose Hanford as the location for reactor, chemical separation, and related facilities for producing and purifying plutonium. A small farming and ranching community along the Columbia river, covering the towns of Richland, Hanford, and White Bluffs, was designated as site W of the Manhattan Project, the code name for the government's development of atomic and the nuclear weapons.. The land known as the Hanford Engineering Works was located in a sparsely populated area close to the Bonneville and Grand Coulee dams.

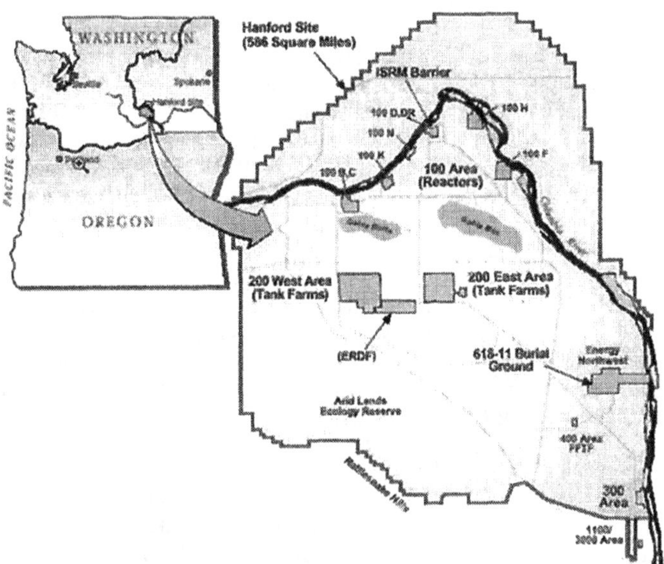

Figure 29.1 The Hanford Site

Hanford covers 586 square miles and is currently managed by DOE. The reservation was divided into areas according to their primary functions. The 100 area was for reactors and support structures; the 200 area comprised the separation facilities; the 300 area included testing and fuel fabrication facilities; and the 400 area contains the Fast Flux Test Facility (FFTF) and support structures.

The U Plant refined the plutonium produced at B and T Plants. These were shut down in 1958 and replaced with the C Plant, REDOX (Reduction-Oxidation), Plutonium Finishing Plant, and the

PUREX (Plutonium Uranium Extraction) for processing fuel. Eight of the original reactors were shut down by 1971, leaving only the N Reactor which produced both fuel grade plutonium and power until January 1987. The Plutonium Finishing Plant (2736Z plant) and the PUREX completed the final run of N Reactor fuel in December 1988, ending the Hanford Site production mission and entering a new era of environmental restoration. The B Reactor was cleaned up and is now a National Historic Site and museum.

Before the 1960s, scientists thought the process of separating isotopes and burning them up would be simple. They also thought the desert sand made an effective filter that would remove any contaminant from the process water, thereby rendering cleanup simple -- removal of the top layer of sand would remove all of the isotopes. This operational practice resulted in many "Past Practice" waste sites -- those where waste was dumped (intentionally or unintentionally) that was not subject to the regulations of a treatment, storage, and disposal (TSD) site.

In 1972, a decommissioning organization was first given the responsibility to remove unnecessary structures. By the late 1980s, the primary mission of a number of DOE sites, including Hanford, changed from weapons production to environmental cleanup. This clarified the need for changes in culture, management approach, and contracting strategies. New challenges were faced, including developing accurate baseline schedules and cost estimates, implementing effective project management, and initiating contract reforms to incorporate the best private industry practices.

In May 1989, the DOE, EPA, and the Washington State Department of Ecology signed the Hanford Federal Facility Agreement and Consent Order, also known as the Tri-Party Agreement, committing for site remediation within 30 yr. The agreement can be modified, depending on site needs and appropriated funding, by agreement of the involved parties. It is divided into a legal component and an action plan or schedule for cleanup. The site is divided into 78 operable units, each containing a number of past practices units for waste sites. Each unit is to be investigated and remediated separately, consistent with either CERCLA or RCRA.

In 1993, the DOE issued a Record of Decision with a preferred alternative to place the production reactors into safe storage, followed by a deferred one-piece removal of the reactor blocks and their transport to a specially-prepared burial facility within a plateau on the site. The C Reactor was selected as the first for action due to the advanced deterioration of building roof sections, which would have required extensive repair.

In 1994, as part of a contract reform effort, the DOE executed an Environmental Restoration Contract as one of two executed to bring best commercial practices into the complex. One of the original missions was to place the N Reactor's core in a secure safe enclosure and restore the Columbia River Corridor. The majority of this region has been returned to its natural state, with 110 acres already turned into a wildlife refuge. Given all the dams along the Columbia River, the Hanford Reach is the last free-flowing segment and provides spawning grounds for the autumn salmon runs and nesting areas for bald eagles. Elk, deer, and wild horses roam the site.

29.2 DECOMMISSIONING PLANNING

This decommissioning has three phases: initial preparation for facility closure, followed by surveillance/maintenance and transition, and final disposition. Preparation for facility closure includes all activities necessary for initial planning and decontamination. Transition includes documenting a project management plan and a facility end-point. Surveillance and maintenance will continue until final disposition is funded and executed.

The graphite core of the Hanford reactors is typically surrounded by a 3ft thick concrete wall. Beyond this wall are the control room, sample rooms, valve pit, accumulator room, control rods, ventilation rooms, fuel storage basin, office space and storage. An interim SAFSTOR enclosure can be achieved by removal of facilities beyond this wall, filling in the openings with concrete, and placing a 75 yr roof over the reactor. One set of doors will be left in place and welded shut, and radioactivity would decay to a level that the core can then be removed for final disposal.

Other priorities for closure under the Tri-Party Agreement include:

- Closure of the single-shell tanks and final disposal of all tank wastes
- Investigation and cleanup of all contamination at operable units
- Permitting and closure of TSD units
- Ceasing soil disposal of any contaminated liquids
- Operating the HLW vitrification plant.

More than 350 facilities are to be demolished and over 3,000 waste sites remediated. Some of the 55 TSD groups operate to facilitate cleanup. More than 1,000 Past Practice units have been categorized into 78 operable units. To date, over 150 units have been removed and the grounds remediated.

29.3 DECOMMISSIONING ACTIONS

29.3.1 Environmental Restoration Disposal Facility: serves as the primary repository for the bulk soil, demolition debris, and miscellaneous contaminated material from CERCLA remediation. This multi-celled burial trench is authorized to receive, stage, treat, and dispose of LLW, hazardous, and mixed waste. It will ultimately contain eight cells, with a total capacity of 6M yd^3 of Hanford waste.

29.3.2 The Reactor Facilities (100 Areas): Three production reactors (105B, 105D, and 105F) and two separation plants were built to produce plutonium for the first atomic bomb. B Reactor started operations as the world's first full-scale plutonium production in September of 1944, followed by the startup of the first plutonium processing plant (T Plant) in December. Starting in the late seventies, the DR-Reactor was used as a sodium-burn facility to test the effects of a fire on concrete. By 1994, the majority of ancillary facilities were removed from the 100 area, leaving the reactors and a few supporting structures. Six reactors were remediated, with ancillary facilities removed and the reactor structure reduced to a scale sufficient to contain the core. Of the remaining three reactors, one transitioned to disposition and two are in process.

29.3.3 Separation Facilities (200 Area): includes canyons, plutonium finishing, and tank farms.

Canyon Facilities: The original buildings were used for separation process until the late fifties and then reapplied for various new missions. The canyon facilities include the T, U, and B Plants, the REDOX, and the PUREX. The three original canyon facility plants are 600 ft long, 40 ft wide, and 60 ft tall. A 40 ft diameter by 60 ft high test vessel was added to the T-Plant for atmosphere testing and used by at least three contractors for different missions. The U and B Plants were also assigned different missions over the years. The REDOX and PUREX are longer and wider than the original three canyons. Cells in the bottom of the canyons contain the separation towers, tanks, and equipment necessary for separation of plutonium from the reactor fuel.

The canyon walls are approximately 8 ft thick. Removal of this much concrete is both expensive and time consuming. The alternatives being pursued to reduce the disposal cost are: 1) filling the canyon with mixed waste for long-term storage, and 2) burying the facility, perhaps collapsing it on itself before burial. In either case, the facility will remain disposed in place.

Plutonium Finishing: Plutonium slurry from the canyon facilities was processed into a finished product in plutonium finishing plants, PFP, Z, and 233S. They have the highest levels of loose contamination of any of the 350 site facilities.

Tank Farms: These were started in 1950 to collect liquid waste generated from the processing facilities. They include 177 USTs, with a total volume of 55M gal of liquid waste. The hazardous materials in the tanks can be divided into four categories. There are HLW tanks that produce a heat load requiring the routine replacement of water lost by evaporation to maintain acceptable limits. There are ferro-cyanide tanks that are explosive at elevated temperatures. Some of the tanks contain organics and are flammable, and some generate hydrogen. The Waste Vitrification Plant will separate the wastes into three separate streams, depending on the three latter categories.

29.3.4 Testing Facilities (300 Area): The process for every new facility was tested in this area, and at least three test reactors have been built and removed during the last 60 yr. Various facilities were used, cleaned up, rebuilt, and returned to other service. One old reactor building, used to test graphite blocks, was taken apart and reused as a garage, modified for isotope experiments, and again used as mockup for hot cell work. Another facility was reused as a fuel line for FFTF. Each building will therefore pose unique challenges. The lessons learned during facility removal will be essential in the characterization and planning required before removal and decontamination.

29.3.5 Fast Flux Test Facility (400 Area): The Fast Flux Test Facility (FFTF) is the newest of the reactors at the Hanford Site, placed into operation in 1979. As a test reactor, it was designed to test fuels and components for future fast flux reactors. Several missions have been proposed for the reactor, including generating power, burning weapons-grade plutonium, and manufacturing medical isotopes or tritium. Current plans are to remove the fuel from the facility, drain the sodium, shut down all nonessential systems, and eventual demolition.

29.4 REGULATORY AND STAKEHOLDER ISSUES

A major regulatory driver for the cleanup of the Hanford Site was the Tri-Party Agreement, whose cleanup milestones were used to develop the long-range plan integrating the technical scope, cost estimates, and detailed schedules with priority logic for identifying the funding levels necessary to complete milestones. By applying a project management approach to the long-range plan, planners reduced the baseline cost estimate by \$8 B (from \$20.4 to \$12.3 B) and accelerated the completion schedule by 12 yr (from 2047 to 2035).

Given the large number of facilities and waste sites to be removed, and the frequent changes to facilities during their operation, the critical regulatory compliance aspect is characterization before planning and starting decommissioning activities. It will be essential to determining the conditions of contamination before demolition, and to eventually verify completion of the cleanup effort.

The Yakima Indian Tribe settled the Hanford Site long before settlers reached the West Coast. They hunted on Gable Mountain and fished at White Bluffs. The 300 area and N Reactor are built on an ancient religious site. The tribe is involved as a key stakeholder in all aspects of the site. A major goal restoration is to return as much of the Hanford as possible to conditions existing before 1900.

29.5 CASE STUDY -- HANFORD C-REACTOR

Nine water-cooled, graphite-moderated production reactors were constructed by 1955. Construction of the C Reactor began in the 100 area in June 1952, with startup in late 1953. Reactor operations terminated and deactivation activities began in April 1969.

29.5.1 Decommissioning Planning: Planning for the C Reactor's interim safe storage project involved the signatories of the Tri-Party Agreement. Major objectives and milestones for the 24-month project were documented, and interim objectives and milestones then negotiated between the DOE and its decommissioning contractor.

Figure 29.2 Hanford Site C Reactor

A performance fee was based on achieving milestones. The project work plans and schedules were developed by representatives from all involved organizations, each of which then took ownership of their respective work elements.

The design effort incorporated long-term surveillance and maintenance needs into the plan for reactor storage. The protective structure that would remain had to be durable, while providing for safe access and practical follow-on maintenance operations. Some of the main features of the interim safe storage design were:

- Safe storage for up to 75 yr
- No credible releases of radionuclides to the environment under normal design conditions
- Required interim inspections on a five-year frequency basis
- Completion of a safe storage enclosure configuration that would not preclude or significantly increase the cost of any final decommissioning alternative

29.5.2 Decommissioning Actions: The interim safe storage project for the C Reactor included removing all portions of the facility outside of the reactor block shield walls. The areas removed included the fuel storage basin, the metal examination facility, outer rod room, control room, electrical room, switchgear room, lunch room, office space, fan supply and exhaust rooms, sample rooms, ready rooms, lift station, upper reactor framing and roofing, and other miscellaneous rooms and tunnels. The demolition reduced the original footprint of the facility by more than 80%, while the remaining portion (areas inside the concrete shield walls) became part of the enclosure.

After the upper reactor demolition was completed, new anchor bolts were grouted into the top of the concrete shield walls and new structural framing was installed. Galvalum-coated steel roofing and siding was then attached to the framework. Galvalum is a coating that contains 55% aluminum and 45% zinc. The resulting "cocoon" placed the reactor core in safe condition for at least 75 yr, while accommodating surveillance inspections scheduled for every five years.

Recycling of noncontaminated material and equipment averts large expenditures by reducing the cost of the waste disposal -- a major cost element in decommissioning. Therefore, recycling and waste minimization became key project objectives. Recycled materials are depicted in Table 29.1.

Table 29.1 Materials Recycled from the ISS of the C Reactor

Material	Quantity
Steel	362 metric tons (400 tons)
Copper and Aluminum	2,268 kg (2.5 tons)
Lead	36 metric tons (40 tons)
Mercury	3.8 L (1 gal)

The project was completed on schedule in September 1998, 24 months after decommissioning began, and with more than 260,000 hours of work without a lost-time injury or personnel contamination. The project successfully demonstrated 20 innovative technologies, which have since been deployed on other projects in both the U.S. and in the former Soviet Union. In recognition of these achievements, the project was selected as one of three international finalists in the Project Management Institute's 1998 "Project of the Year" competition.

29.5.3 Unique Issues: The C Reactor project was a testing ground to demonstrate new and innovative technologies with the potential benefit of lower life-cycle costs, accelerated schedules, and reduced worker exposure, among others. The participants were to identify and demonstrate new and innovative decontamination technologies that could benefit cost, schedule, and safety, and which could have potential applications for both DOE and private sector projects.

Innovative technologies were identified and evaluated in the areas of characterization, decontamination, dismantling, demolition, waste minimization and disposal, facility stabilization, and worker health and safety. The technologies were competitively selected using a "market search" approach, where the project presented problems to the industry, which responded with ideas for innovative technologies and/or new applications of existing technology. A team of international experts reviewed more than 200 identified technologies and selected 20 for demonstration at the C Reactor for comparison to existing baselines. Those chosen are shown in Table 29.2.

Table 29.2 Technologies Demonstrated During C Reactor Project
Laser-Assisted Ranging and Data (LARADS)
Concrete Shaving
Gamma-Ray Imaging
Concrete Diamond Grinder
Position-Sensitive Radiation Detector
Concrete Spaller
Self-Contained Pipe Cutting Shears
Reactor Stabilization
Heat Stress Monitoring System
Automatic Dust Suppression System
Mobil Integrated Temporary Utility System (MITUS)
System for Tracking Remediation, Exposure, Activities, and Materials (STREAM)
Seam-Seal Sack Suit
Residual Radioactivity Dose Model (RESRAD and RESRAD-BUILD)
Wireless Remote Monitoring
Nitrogen-Cooled Diamond Wire Cutting
2-D Linear Motion System
Compact Subsurface Discreet Sampler
High-Speed Clam Shell Pipe Cutter
Lead Decontamination (Chemical)

29.5.4 Outcome and Lessons Learned: In 1994, the DOE created its Contractor Purchasing Council to developing a cooperative government/industry relationship and to determine how the best commercial practices of private industry could be applied to the DOE's cleanup mission. The Council sponsored a number of studies, including a commercial procurement benchmarking study conducted in 1998. The primary purpose of the study was to learn from the experience and practices of others to address the pending procurement challenges throughout the DOE complex. Four nuclear power plants were used in the study: Oyster Creek, Trojan, Fort Saint Vrain, and Maine Yankee. A summary of the best practices identified and discussed is shown in Table 29.3.

Table 29.3 Lessons Learned and Best Commercial Practices Identified
Innovative contracting models solve old problems and save money/time
Performance-based incentives work
Competitive use of pre-qualified vendors
Unproven technologies are risky and time consuming - keep it simple
Finish line/closure culture
Property disposition
Planning/project controls
Safety, environmental compliance, quality expectations
Manage or eliminate risks
Project acceleration saves big money

Several results of the Council's work were applied at Hanford. To improve project communication and administration, the total C-Reactor project work force was located at the job site, including project management, engineering, planning and scheduling, waste management, radiological engineers, and safety personnel. Daily plan-of-the-day meetings were held with all craft, subcontractor, and non-manual work forces. Near-term schedule objectives and specific tasks for the day were discussed in detail. These daily meetings were an important factor in keeping project personnel aware of the risks in the work they were doing.

To address and work safely under conditions of radiological hazards, the project team applied the HSA approach, using historical data as a baseline and then updating work plans to reflect current sampling data. The baseline work plans were prepared with historical sampling records and knowledge of reactor operations that had occurred decades ago. Radiological sampling plans used the historical data as a starting point for obtaining current data profiles of conditions. The new data were then used to modify work packages and schedules to incorporate as-found conditions. Extreme precautions in personnel protection were used until actual plant conditions were verified.

The team reached the one million work-hour milestone without a lost work day three different times throughout the project, and the OSHA-recordable case rate was reduced from 6.46 cases per 200,000 hours worked in calendar year 1995 to 1.56 in 1999. The project was completed without any personnel skin contamination and with a total radiation exposure of 3.4 person-rem. The results of teamwork for success were also measured by reduced employee concerns and fewer grievances. From 1998 to 1999, the number of employee concerns diminished from 12 to 6, and grievances from 72 to 33.

REFERENCES

PART 1

AEA 1954 — Atomic Energy Act of 1954, as amended, Public Law 83-703.

BOCA 1996 — BOCA National Building Code, Building Officials and Code Administrators International, Albany, NY.

Bossart 1999 — Bossart, S.J. and Vagnetti, R., The DOE's Market for Deactivation and Decommissioning Services, *ANS Second Topical Meeting and Exhibition on DD&R*, Knoxville, TN.

CAA 1970 — Clean Air Act of 1970, as amended, 42 U.S.C. 7401.

CERCLA 1980 — Comprehensive Environmental Response, Compensation, and Liability Act, 42 U.S.C. 9601.

Compton 1956 — Compton, A., Atomic Quest: A Personal Narrative, Oxford University Press, Oxford, United Kingdom.

CWA 1972 — Clean Water Act of 1972, as amended, 33 U.S.C. 1251.

DBA 1964 — Davis-Bacon Act of 1964, as amended, 40 U.S.C. 276 a-b.

DOE 1995 a — Closing the Circle on the Splitting of the Atom, U.S. Department of Energy, Office of Environmental Management, Washington, DC.

DOE 1995 b — The Decommissioning Resource Manual, DOE/EM-0246, U.S. Department of Energy, Office of Environmental Management, Washington, DC.

DOE 1997 — Linking Legacies: Connecting the Cold War Nuclear Weapons Production Processes to their Environmental Consequences, U.S. Department of Energy, Washington, DC.

DOE 1998 a — Life Cycle Asset Management, DOE O 430.1A,(updated in Sept 2003, by DOE O 430.1B, Real Property Asset Management), U.S. Department of Energy, Washington, DC.

DOE 1998 b — Accelerating Cleanup: Paths to Closure, U.S. Department of Energy, Office of Environmental Management, Washington, DC.

DOE 2000 — Decommissioning Handbook: Procedures and Practices for Decommissioning, DOE/EM-0383, U.S. Department of Energy, Office of Environmental Management, Washington, DC.

DOE 2001 — Radioactive Waste Management, DOE Order 435.1, Change 1, U.S. Department of Energy, Washington, DC.

DOT 2003 a — General Information, Regulations, and Definitions, 49 CFR 171.

DOT 2003 b — Hazardous Materials Table, Special Provisions, Hazardous Materials Communications, Emergency Response Information, and Training Requirements, 49 CFR 172.

DOT 2003 c — Shippers - General Requirements for Shipment and Packaging, 49 CFR 173.

DOT 2003 d — Carriage by Rail, 49 CFR 174.

DOT 2003 e — Carriage by Aircraft, 49 CFR 175.

DOT 2003 f — Carriage by Vessel, 49 CFR 176.

DOT 2003 g — Carriage by Public Highway, 49 CFR 177.

DOT 2003 h — Specifications for Packaging, 49 CFR 178.

DOT 2003 i — Specifications for Tank Cars, 49 CFR 179.

DOT 2003 j — Continuing Qualification and Maintenance of Packaging, 49 CFR 180.

EPA 1976 — EPA Administered Permit Programs: The Hazardous Waste Permit Program, 40 CFR 270.

EPA 1999 — Land Disposal Restrictions for Hazardous Waste, EPA530-F-99-043, U.S. Environmental Protection Agency, Washington, DC.

EPRI 1999 — Comparison of Decommissioning Dose Modeling Codes for Nuclear Plant Use: RESRAD and D&D, *Technical Report 112874*, Electric Power Research Institute, Palo Alto, CA.

Groves 1962 — Groves, L., *Now It Can Be Told*, Harper and Row, New York, NY.

HMTA 1990 — Hazardous Material Transportation Act of 1975, as amended, U.S.C.1801-1819 and in 49 CFR 106.

Hewlett 1969 — Hewlett, R. G., and Duncan, F., *Atomic Shield: A History of the Atomic Energy Commission, Volume II 1947-1952*, 1990 Reprint, University of California Press, Santa Clara, CA.

IAEA 2000 — IAEA Safety Glossary, International Atomic Energy Agency, Vienna, Austria.

ICC 1997 — Uniform Building Code, International Code Council, Falls Church, VA.

NEPA 1982 — National Environmental Policy Act of 1969, as amended, 42 U.S.C. 4321.

NRC 1991 a — Disposal of High-Level Radioactive Waste in Geologic Repositories, 10 CFR 60.

NRC 1991 b	Requirements for Renewal of Operating Licenses for Nuclear Power Plants, 10 CFR 54.
NRC 1991 c	Duration of License Renewal, 10 CFR 50.51.
NRC 2000	Multi-Agency Radiation Survey and Site Investigation Manual, Rev. 1, NUREG-1575, Rev. 1 (also as EPA 402-R-97 and as DOE/EH-0624), U.S. Nuclear Regulatory Commission, Washington, DC.
NRC 2002 a	Definitions, 10 CFR 20.1003.
NRC 2002 b	Production, Storage, Disposal or Radioactive Waste, NUREG/BR-0216, Rev 2, U.S. Nuclear Regulatory Commission, Washington, DC.
NRC 2002 c	Information Digest, 2002 Ed., Vol. 14, NUREG-1350, U.S. Nuclear Regulatory Commission, Washington, DC.
NRC 2003 a	Standards for Protection against Radiation, 10 CFR 20.
NRC 2003 b	Rules of General Applicability to Domestic Licensing of Byproduct Material, 10CFR 30.
NRC 2003 c	Domestic Licensing of Production and Utilization Facilities, 10 CFR 50.
NRC 2003 d	Licensing Requirements for Land Disposal of Radioactive Waste, 10 CFR 61.
NRC 2003 e	Packaging and Transportation of Radioactive Material, 10 CFR 71.
NRC 2003 f	Licensing Requirements for the Independent Storage of Spent Nuclear Fuel, High-Level Radioactive Waste, and Reactor-Related GTCC Waste, 10 CFR 72.
NRC 2003 g	Domestic Licensing of Production and Utilization Facilities, Combining Licenses, 10 CFR 50.52.
OSHA 2003 a	Occupational Safety and Health Standards, 29 CFR 1910.
OSHA 2003 b	Safety and Heath Regulations for Construction, 29 CFR 1926.
PAA 1994	Price-Anderson Act of 1994, as amended, 42 U.S.C. 2210.
RCRA 1976	Resource Conservation and Recovery Act of 1976, as amended, 40 CFR 260-281 and 42 USC 6901.
Rhodes 1986	Rhodes, R., *The Making of the Atomic Bomb*, Simon & Schuster, Inc., New York, NY.
SARA 1986	Superfund Amendments and Reauthorization Act of 1986, as amended, 42 U.S.C. 9601.
SDWA 1974	Safe Drinking Water Act of 1974, as amended, 42 U.S.C. 300f.
TSCA 1976	Toxic Substances Control Act of 1976, as amended, 15 U.S.C. 2601.

PART 2

AAES 1998	Risk Analysis: The Process and its Application, American Association of Engineering Societies, Washington, DC.
AEA 1954	Atomic Energy Act of 1954, as amended, Public Law 83-703, Washington, DC.
AIChE 1992	Guidelines for Hazard Evaluation Procedures, Second Edition, American Institute of Chemical Engineers, New York, NY.
AID 1998	Engaging Customer Participation, U.S. AID's Organizational Change Experience, U.S. Agency for International Development, Washington, DC.
ANL 2001	Yu, C., et. al, User's Manual for RESRAD, Version 6, ANL/EAD-4 Argonne National Laboratory, Environmental Assessment Division, Argonne, IL.
ANSI 1992	Practices for Respiratory Protection, ANSI Z88.2-1992, American National Standards Institute, Washington, DC.
ANSI 2003	Minimum Requirements for Industrial Unit Type First Aid Kits, Z308.1-2003 American National Standards Institute, Washington, DC.
ASME/RSI 2001	Independent Peer Review of DOE Technologies Performed by the ASME, *Waste Management 2001 Conference*, American Society of Mechanical Engineers, Tucson, AZ.
ASME/RSI 2002	Report of the Technical Peer Review Panel: Strategy for Remediation of Groundwater Contamination at the Nevada Test Site, ASME/ Institute for Regulatory Science, New York, NY.
Cleland 1997	Cleland, D. I. (ed.), *Field Guide to Project Management*, John Wiley & Sons, New York, NY.
Coglianese 2004	Coglianese, Cary; Allen, Laurie; Does Consensus Make Common Sense? An Analysis of EPA's Common Sense Initiative, ENVIRONMENT, Jan,-Feb. 2004, Haldref Publications, Washington DC

Dixon 2003 — Dixon, E. and Peterson, K., Utilization of a Technical Peer Review to Support The Mission of the Nevada Test Site Community Advisory Board, *Waste Management '03 Conference*, Tucson, AZ.

DOE 1991 — Unreviewed Safety Questions, DOE O 5480.21, U.S. Department of Energy, Washington, DC. (Archived upon publication of Nuclear Safety Management, 10 CFR 830, on January 10 , 2001. Subpart Part B, Safety Basis Requirements, and in particular Section 10 CFR 830.203, Unreviewed Safety Question (USQ) Process, replaced the detailed requirements of DOE O 5480.21 regarding USQ.

DOE 1992 — Nuclear Safety Analysis Reports, DOE O 5480.23, U.S. Department of Energy, Washington, DC.

DOE 1993 — Radiation Protection of the Public and Environment, DOE Order 5400.5, Change 2, U.S. Department of Energy, Washington, DC.

DOE 1994 — Maintenance Management Programs, DOE O 4330.4B, U.S. Department of Energy, Washington, DC.

DOE 1995 a — Fire Safety Program, Implementation Guide for Use with DOE 420.1 and 440.1, DOE G 440.1-5, U.S. Department of Energy, Washington, DC.

DOE 1995 b — Planning and Conduct of Operational Readiness Reviews, DOE-STD-3006-95, U.S. Department of Energy, Washington, DC.

DOE 1996 a — Safety Management System Policy, DOE P 450.4, U.S. Department of Energy, Washington, DC.

DOE 1996 b — Firearms Safety Standard, DOE-STD-1091-96, U.S. Department of Energy, Washington, DC.

DOE 1996 c — Identifying, Implementing and Complying with Environment, Safety and Health Requirements, DOE P 450.2A, U.S. Department of Energy, Washington, DC.

DOE 1996 d — Technical Safety Requirements, DOE O 5480.22, U.S. Department of Energy, Washington, DC.

DOE 1997 a — Tailoring for Integrated Safety Management Applications, DOE G 450.3-3, U.S. Department of Energy, Office of Environment, Safety and Health, Washington, DC.

DOE 1997 b — Worker Protection Management for DOE Federal and Contractor Employees Guide, DOE G 440.1-1, U.S. Department of Energy, Washington, DC.

DOE 1997 c — Performance Indicators and Analysis of Operations Information, DOE O 210.1, U.S. Department of Energy, Washington, DC.

DOE 1997 d — Accident Investigations, DOE O 225.1, U.S. Department of Energy, Washington, DC.

DOE 1997 e — Contractor Occupational Medical Program Guide, DOE G 440.1-4, U.S. Department of Energy, Washington, DC.

DOE 1997 f — Construction Safety Management Guide for Use with DOE O 440.1, DOE G 440.1-2, U.S. Department of Energy, Washington, DC.

DOE 1997 g — Memorandum: Hazard Categorization for Environmental Management Activities Related to Stabilization, Deactivation, Decontamination and Decommissioning, and Environmental Restoration, U.S. Department of Energy, Office of Nuclear Safety Policy and Standards, Washington, DC.

DOE 1997 h — RCRA Corrective Measures Study: Area 6 Decontamination Pond Facility, Corrective Action Unit H 92, DOE/NV 493, U.S. Department of Energy, Nevada Operations Office, Mercury, NV.

DOE 1998 a — Life Cycle Asset Management, DOE O 430.1A,(updated in Sept 2003, by DOE O 430.1B, Real Property Asset Management), U.S. Department of Energy, Washington, DC.

DOE 1998 b — Integration of Environment, Safety and Health into Facility Disposition Activities, DOE-STD-1120-98, U.S. Department of Energy, Washington, DC.

DOE 1998 c — Statistical Evaluation of DOE D&D Occurrences, DOE/EH-578, U.S. Department of Energy, Washington, DC.

DOE 1998 d — DOE Explosives Safety Manual, DOE M 440.1-1, U.S. Department of Energy, Washington, DC.

DOE 1998 e — Occupational Exposure Assessment Implementation Guide, DOE G 440.1-3, U.S. Department of Energy, Washington, DC.

DOE 1998 f — Electrical Safety, DOE-HDBK-1092-97, U.S. Department of Energy, Washington, DC.

DOE 1999 a — Decommissioning Implementation Guide, DOE G 430.1-4, U.S. Department of Energy, Washington, DC.

DOE 1999 b — Deactivation Implementation Guide, DOE G 430.1-3, U.S. Department of Energy, Washington, DC.

DOE 1999 c Implementation Guide for Surveillance and Maintenance during Facility Transition and Disposition, DOE G 430.1-2, U.S. Department of Energy, Washington, DC.

DOE 2000 a Decommissioning Handbook: Procedures and Practices for Decommissioning, DOE/EM-0383, U.S. Department of Energy, Washington, DC.

DOE 2000 b Worker Protection Management for DOE Federal and Contractor Employees, DOE O 440.1A, U.S. Department of Energy, Washington, DC.

DOE 2001 a Transition Implementation Guide, DOE G 430.1-5, U.S. Department of Energy, Washington, DC.

DOE 2001 b Federal Employee Training, DOE Order 360.1B U.S. Department of Energy, Washington, DC.

DOE 2001c Personnel Selection, Qualification, and Training Requirements for Nuclear DOE Facilities, DOE O 5480.20A, U.S. Department of Energy, Washington, DC.

DOE 2001d Conduct of Operations Requirements for DOE Facilities, DOE O 5480.19, Change 2, U.S. Department of Energy, Washington, DC.

DOE 2001 e Employee Concerns Program, DOE O 442.1A, U.S. Department of Energy, Washington, DC.

DOE 2002 a Preparation Guide for U.S. Department of Energy Nonreactor Nuclear Facility Safety Analysis Reports, DOE-STD-3009-94, Change Notice No. 2, U.S. Department of Energy, Washington, DC.

DOE 2002 b Facility Safety, DOE O 420.1A, U.S. Department of Energy, Washington, DC.

DOE 2003 a Occupational Radiation Protection, 10 CFR 835, U.S. Department of Energy, Washington, DC

DOE 2003 b Startup and Restart of Nuclear Facilities, DOE Order 425.1C, U.S. Department of Energy, Washington, DC.

DOE 2003 c Environment, Safety, and Health Reporting, DOE O 231.1A, U.S. Department of Energy, Washington, DC.

DOE 2003 d The Excess Facility Transition to Deactivation & Decommissioning, Methods and Practices Handbook, Published by DOE/EM, U.S. Department of Energy, Washington, DC. (Replaced the Facility Deactivation Methods and Practices Handbook, DOE/EM-0318. Available at http://web.em.doe.gov.)

DOT 2003 a Hazardous Materials Table, Special Provisions, Hazardous Materials Communications, Emergency Response Information, and Training Requirements, 49 CFR 172.

DOT 2003 b General Information, Regulations, and Definitions, 49 CFR 171.

EPA 1976 EPA Administered Permit Programs: The Hazardous Waste Permit Program, 40 CFR 270.

EPA 1988 Limiting Values of Radionuclide Intake and Air Concentration and Dose Conversion Factors for Inhalation, Submersion, and Ingestion, *Federal Guidance Report No. 11*, EPA-520/1-88-020, U.S. Environmental Protection Agency, Washington, DC.

EPA 1989 Risk Assessment Guidance for Superfund Volume 1, Human Health Evaluation Manual, (Part A) Interim Final, EPA/540/1-89/002. Environmental Protection Agency, Washington DC

EPA 1990 National Oil and Hazardous Substances Pollution Contingency Plan, *Federal Register*, Vol. 55, p 8666-8865, (40 CFR 300), U.S. Environmental Protection Agency, Washington, DC.

EPA 1993 External Exposure to Radionuclides in Air, Water and Soil, *Federal Guidance Report No. 12*, EPA-402-R-93-081, U.S. Environmental Protection Agency, Washington, DC.

EPA 1994 a Estimating Radiogenic Cancer Risks, EPA 402-R-93-076, U.S. Environmental Protection Agency, Washington, DC.

EPA 1994 b Radiation Site Cleanup Regulations - Technical Support Document for the Development of Radiation Cleanup Levels for Soil, EPA 402-R-96-011A, U.S. Environmental Protection Agency, Washington, DC.

EPA 1999 a Cancer Risk Coefficients for Environmental Exposure to Radionuclides, *Federal Guidance Report No. 13*, EPA 402-R-99-001, U.S. Environmental Protection Agency, Washington, DC.

EPA 1999 b Project XL, Stakeholder Involvement: A Guide for Project Sponsors and Stakeholders, EPA-100-F-99-001, U.S. Environmental Protection Agency, Washington, DC.

EPA 2001 Health Effects Assessment Summary Tables, U.S. Environmental Protection Agency, Washington, DC.

EPA 2003 a Identification and Listing of Hazardous Waste, 40 CFR 261.

EPA 2003 b Chemical Accident Prevention Provisions, 40 CFR 68.

EPA 2003 c Public Involvement Policy, *Federal Register*, 68 FR 33946-33949, U.S. Environmental Protection Agency, Washington, DC.

EPRI 1996 Decommissioning Economics and Risk Advisor (DERAD), EPRI SW-106788, Electric Power Research Institute, Palo Alto, CA.

Fleming 1990 Fleming, J.R. and Knorr, B.R., History of the Clean Air Act, American Meteorological Society website at ttp://www.amet.soc.org/sloan/cleanair.cleanair_rightframe.html

HMTA 1990 Hazardous Material Transportation Act as amended by the Hazardous Materials Transportation Uniform Safety Act of 1990. PL 101-615

IAEA 1999 Decommissioning of Nuclear Power Plants and Research Reactors. IAEA Safety Guide, WS-G-2.1, International Atomic Energy Agency, Vienna, Austria.

Morris 2001 Morris, M., Life Cycle Analysis for Treatment and Disposal of PCB Waste at Ashtabula and Fernald, *Waste Management Symposium 2001 Conference*, University of Arizona, Tucson, AZ.

NAS 1990 Health Effects of Exposure to Low Levels of Ionizing Radiation: BEIR V, National Academy of Sciences, National Academy Press, Washington, DC.

NAS 2000 Natural Attenuation for Groundwater Remediation, National Academy of Sciences, National Academy Press, Washington, DC.

NAS 2001 A Risk-Management Strategy for PCB-Contaminated Sediments, National Academy of Sciences, National Academy Press, Washington, DC.

NEPA 1982 The National Environmental Policy Act of 1982, as amended, 42 U.S.C. 4321.

NFPA 2004 Standard for Electrical Safety Requirements for Employee Workplaces, 70E NFPA, National Fire Protection Association, Quincy, MA.

NIOSH 1975 List of Personal Hearing Protectors and Attenuation Data, Publication No. 76-120, U.S. Department of Health, Education, and Welfare, Morgantown, WV.

NRC 1978 Standard Format and Content of Safety Analysis Reports for Nuclear Power Plants, (LWR Edition), Regulatory Guide 1.70, U.S. Nuclear Regulatory Commission, Rockville, MD.

NRC 1988 10 CFR Parts 30, 40, 50, 51, 70, and 72, General Requirements for Decommissioning Nuclear Facilities, *Federal Register*, 53 FR 24018, U.S. Nuclear Regulatory Commission, Washington, DC.

NRC 1994 a Timeliness in Decommissioning of Materials Facilities, (10 CFR Parts 2, 30, 40, 70, and 72), *Federal Register*, 59 FR 36026, U.S. Nuclear Regulatory Commission, Washington, DC.

NRC 1994 b Site Decommissioning Management Plan, NUREG 1444, Supplement 1, U.S. Nuclear Regulatory Commission, Washington, DC.

NRC 1994 c Lessons Learned form the Three Mile Island-Unit 2 Advisory Panel, NUREG/CR -6252, August 1994, U.S. Nuclear Regulatory Commission, Washington, DC

NRC 1995 Clarification of Decommissioning Funding Requirements, (10 CFR Parts 30, 40, 50, 51, 70, and 72), *Federal Register*, 60 FR 38235, U.S. Nuclear Regulatory Commission, Washington, DC.

NRC 1996 a Decommissioning of Nuclear Power Reactors, (10 CFR Parts 2, 50, and 51), *Federal Register*, 61 FR 39728-39304, U.S. Nuclear Regulatory Commission, Washington, DC.

NRC 1996 b Instruction Concerning Risks from Occupational Radiation Exposure, Regulatory Guide 8.29, U.S. Nuclear Regulatory Commission, Washington, DC.

NRC 1997 a Radiological Criteria for License Termination, (10 CFR Parts 20, 30, 40, 50, 51, 70, and 72), *Federal Register*, 62 FR 39058-39092, U.S. Nuclear Regulatory Commission, Washington, DC.

NRC 1997 b NMSS Handbook for Decommissioning Fuel Cycle and Materials Licensees, NUREG/BR-0241, U.S. Nuclear Regulatory Commission, Washington, DC.

NRC 1998 Financial Assurance Requirements for Decommissioning Nuclear Power Reactors, (10 CFR Parts 30 and 50), *Federal Register*, 63 FR 50465, U.S. Nuclear Regulatory Commission, Washington, DC.

NRC 1999 Standard Format and Content of License Termination Plans for Nuclear Power Reactors, Regulatory Guide 1.179, U.S. Nuclear Regulatory Commission, Washington, DC.

NRC 2000 a Decommissioning of Nuclear Power Reactors, Regulatory Guide 1.184, U.S. Nuclear Regulatory Commission, Washington, DC.

NRC 2000 b Standard Format and Content Guide for Post-Shutdown Decommissioning Activities Report, Regulatory Guide 1.185, U.S. Nuclear Regulatory Commission, Washington, DC.

NRC 2000 c NMSS Decommissioning Standard Review Plan, NUREG 1727, U.S. Nuclear Regulatory Commission, Washington, DC.

NRC 2002 a Definitions, 10 CFR 20.1003.

NRC 2002 b Final Generic Environmental Impact Statement on Decommissioning of Nuclear Facilities, NUREG-0586, Supplement 1, U.S. Nuclear Regulatory Commission, Washington, DC.

NRC 2002 c Status of the Entombment Option for Nuclear Power Plants, SECY 02-191, U.S. Nuclear Regulatory Commission, Washington, DC.

NRC 2002 d 2002 Annual Update Status of Decommissioning Program, SECY 02-169, U.S. Nuclear Regulatory Commission, Washington, DC.

NRC 2003 a Citizens Guide to the NRC, NUREG/BR-0010, Rev. 4, U.S. Nuclear Regulatory Commission, Washington, DC.

NRC 2003 b Consolidated NMSS Decommissioning Guidance, NUREG 1757, Vol.1, Rev.1, Vol.2, and Vol.3, U.S. Nuclear Regulatory Commission, Washington, DC, 2003.

NRC 2003 c Standard Review Plan for Evaluating Power Reactor License Termination Plans, NUREG 1700, U.S. Nuclear Regulatory Commission, Washington, DC.

NRC 2003 d Domestic Licensing of Production and Utilization Facilities, 10 CFR 50.

NRC 2003 e Standards for Protection Against Radiation, 10 CFR 20.

OSHA 1989 Safety and Health Program Management Guidelines, 54 FR 3904-3916.

OSHA 1991 Occupational Exposure to Blood Borne Pathogens Standard: Safety and Health Program Management Guidelines, *Federal Register*, 56 FR 64175, Department of Labor, OSHA, Washington, DC.

OSHA 2002 Job Hazard Analysis: OSHA 3071, Department of Labor, OSHA, Washington, DC.

OSHA 2003 a Occupational Safety and Health Standards, 29 CFR 1910.

OSHA 2003 b Safety and Health Regulations for Construction, 29 CFR 1926.

OSHA 2003 c Recording and Reporting Occupational Injuries and Illnesses, 29 CFR 1904.

OSHA 2003 d Process Safety Management of Highly Hazardous Chemicals, 28 CFR 1910.119.

RSI 2003 Love, B., Straja, S., and Moghissi, A., Manual for Stakeholder Participation, RSI-03-04, Institute for Regulatory Science, Columbia, MD.

Schwartz 1996 Schwartz, N., Resource Book on Participation, Section IV, Participation and the Project Cycle, Inter-American Development Bank, Washington, DC.

SSS 1999 System Safety Analysis Handbook: A Source Book for Safety Practitioners, 2nd edition, System Safety Society, Unionville, VA.

Taboas 1997 Taboas, A.L., Restoration of the Site of the First Nuclear Reactors, *Environment International*, Vol. 23, Issue 4, p 565-585, Elsevier Science, Ltd.

WBank 2003 Source Book on Participation, The World Bank,www.worldbank.org/wbi/sourcebook/sbhome.htm

PART 3

AACE 1984 AACE International, Project and Cost Engineers' Handbook, Second Edition, Marcel Dekker, Inc., New York, NY.

AACE 2003 Association for the Advancement of Cost Engineering, Inc., *Certification Study Guide*, Second Edition, AACEI 1820-32, Association for the Advancement of Cost Engineering International, Morgantown, WV.

ANL 1993 a Yu, C. et al, Manual for Implementing Residual Radioactive Material Guidelines Using RESRAD: Version 5.0, ANL/EAD/LD-2, Argonne National Laboratory, Argonne, IL.

ANL 1993 b Yu, C. et al, Data Collection Handbook to Support Modeling the Impacts of Radioactive Material in Soil, ANL-EAIS-8, Argonne National Laboratory, Argonne IL.

ANL 1994 Yu, C. et al, RESRAD-BUILD: A Computer Model for Analyzing the Radiological Doses Resulting from the Remediation and Occupancy of Buildings Contaminated with Radioactive Material, ANL/EAD/LD-3, Argonne National Laboratory, Argonne, IL.

ANL 2001 — Yu, C. et al, July 2001, User's Manual for RESRAD Version 6, ANL/EAD-4, Argonne National Laboratory, Environmental Assessment Division, Argonne, IL.

ANSI 1999 — Surface and Volume Radioactivity Standards for Clearance, ANSI/HPS N13.12-1999, American National Standards Institute, Washington, DC.

ASTM 1996 — Rouhani, Srivastava, and Desbarats, Geostatistics for Environmental and Geotechnical Application, ASTM STP 1283, American Society for Testing and Materials, Philadelphia, PA.

ASTM 1997 a — Standard Guide for Preparing Characterization Plans for Decommissioning Nuclear Facilities, E 1892-97, American Society for Testing and Materials, West Conshohocken, PA.

ASTM 1997 b — Standard Guide for Selection and Use of Portable Radiological Survey Instruments for Performing In-Situ Radiological Assessments in Support of Decommissioning, E 1893-97, American Society for Testing and Materials, West Conshohocken, PA.

BOCA 1996 — BOCA National Building Code, Building Officials and Code Administrators International, Albany, NY.

Boing 2001 — Boing, L., Fueling Up for The Long Haul: Training for Decommissioning, *Radwaste Solutions Magazine*, Vol. 8, No. 1, American Nuclear Society, La Grange, IL.

Briggs 1997 — Briggs, M., Buck, S., and Smith, M., Decommissioning, Mothballing, and Revamping, Institute of Chemical Engineers, UK, Trowbridge, England.

DBA 1964 — Davis-Bacon Act of 1964, as amended, 40 U.S.C. 276a-b, et seq.

Demmer 2000 — Demmer, R., Decontamination Technologies, *IAEA/ANL Interregional Training Course on Decommissioning of Research Reactors and Other Small Nuclear Facilities*, Argonne, IL.

DOE 1980 — LaGuardia, T. and Manion, W., Decommissioning Handbook, DOE/EV/10128-1, U.S. Department of Energy, Washington, DC.

DOE 1993 — Radiation Protection of the Public and the Environment, DOE Order 5400.5, Change 2, U.S. Department of Energy, Washington, DC.

DOE 1994 — Decommissioning Handbook, DOE/EM-0142P, U.S. Department of Energy, Office of Environmental Restoration, Washington, DC.

DOE 1995 a — Policy on Decommissioning of Department of Energy Facilities under The CERCLA, U.S. Department of Energy, Washington, DC.

DOE 1995 b — Acceptance Priority Ranking and Capacity Report, DOE/RW-0457, U.S. Department of Energy, Office of Civilian Radioactive Waste Management, Washington, DC.

DOE 1998 a — Innovative Technology Summary Report: Expedited Site Characterization, DOE/EM-0420, Characterization, Monitoring, and Sensors Technology Crosscut Program and Subsurface Contaminants Focus Area, U.S. Department of Energy, Office of Environmental Management, Washington, DC.

DOE 1998 b — Accelerating Cleanup: Paths to Closure, DOE/EM-0362, U.S. Department of Energy, Office of Environmental Management, Washington, DC.

DOE 1999 a — Deactivation Implementation Guide, DOE O 430.1-3, U.S. Department of Energy, Washington, DC.

DOE 1999 b — Decommissioning Implementation Guide, DOE G 430.1-4, U.S. Department of Energy, Office of Field Integration, Washington, DC.

DOE 1999 c — Implementation Guide for Surveillance and Maintenance During Facility Transition and Disposition, DOE O 430.1-2, U.S. Department of Energy, Washington, DC.

DOE 1999 d — Quality Assurance Management System Guide for Use with 10 CFR 830.120 and DOE O 414.1, DOE G 414.1-2, U.S. Department of Energy, Washington, DC.

DOE 2000 a — Nuclear Safety Management, 10 CFR 830, Subpart A.

DOE 2000 b — Quality Assurance Requirements, 10 CFR 830.120-122.

DOE 2000 c — Decommissioning Handbook: Procedures and Practices for Decommissioning, DOE/EM-0383, U.S. Department of Energy, Office of Environmental Restoration, Washington, DC.

DOE 2001 a — Transition Implementation Guide, Draft DOE O 430.1-5, U.S. Department of Energy, Washington, DC.

DOE 2001 b — Quality Assurance, DOE Order 414.1A, Change 1, U.S. Department of Energy, Washington, DC.

DOE 2001 c — Adaptive Sampling and Analysis Programs, DOE/EM-0592, U.S. Department of Energy, Office of Environmental Management, Washington, DC.

DOE 2001 d — A Report to Congress on Long-Term Stewardship, DOE/EM-0563, U.S. Department of Energy, Office of Environmental Management, Washington, DC.

EPA 1979 — Moore, R. et al, AIRDOS-EPA: A Computerized Methodology for Estimating Environmental Concentrations and Dose to Man from Airborne Releases of Radionuclides, EPA 520/1-79-009 (Reprint of ORNL-5532), U.S. Environmental Protection Agency, Office of Radiation Programs, Washington, DC.

EPA 1987 — Rogers, V. and Hung, C., PATHRAE-EPA: A Low-Level Radioactive Waste Environmental Transport and Risk Assessment Code, Methodology and Users Manual, 520/1-87-028, U.S. Environmental Protection Agency, Office of Radiation Programs, Washington, DC.

EPA 1989 a — User's Guide for the COMPLY Code: Rev. 2, EPA 520/1-89-003, U.S. Environmental Protection Agency, Washington, DC.

EPA 1989 b — Guide for Determining Compliance with the Clean Air Act Standards for Radionuclide Emissions from NRC-Licensed and Non-DOE Federal Facilities, EPA 520/l-89-002, Rev. 2., U.S. Environmental Protection Agency, Washington, DC.

EPA 1992 — Parks, B., User's Guide for CAP88-PC, Version 1.0, EPA 402-B-92-001, U.S. Environmental Protection Agency, Office of Radiation Programs, Las Vegas, NV.

EPA 1994 — Radiation Site Cleanup Regulations - Technical Support Document for the Development of Radiation Cleanup Levels for Soil, EPA 402-R-96-011 A, U.S. Environmental Protection Agency, Washington, DC.

EPA 2000 a — Guidance for the Data Quality Objective Process, EPA QA/G-4HW, U.S. Environmental Protection Agency, Washington, DC.

EPA 2000 b — Test Methods for Evaluating Solid Waste, Physical/Chemical Methods, EPA SW-846, Third Edition, U.S. Environmental Protection Agency, Washington, DC.

EPA 2000 c — Hung, C-Y, Aug. 2000, User's Guide for the PRESTO-EPA-CPG/POP Operating System, Version 4.2, EPA 402-R-00-007, U.S. Environmental Protection Agency, Washington, DC.

EPA 2002 a — Guidance for QA Project Plans, EPA QA/G-5, EPA/240/R-02/009, Revised Dec. 2002, U.S. Environmental Protection Agency, Washington, DC.

EPA 2002 b — EPA Administered Permit Programs: The National Pollutant Discharge Elimination System, 40 CFR 122.

EPA 2003 a — National Primary Drinking Water Regulations, 40 CFR 141.16.

EPA 2003 b — National Primary Drinking Water Regulations, 40 CFR 141.15.

EPRI 1997 — Wood, C., Yankee Rowe Decommissioning Experience Record, Technical Report 107917-V1, Electric Power Research Institute, Palo Alto, CA.

EPRI 1998 — Fort St. Vrain Decommissioning: Public Relations and Human Resource Issues - Personnel Plans and Communications During Decommissioning of Nuclear Power Plants, Technical Report 109030, Electric Power Research Institute, Palo Alto, CA.

EPRI 2000 — Preparing for Decommissioning: The Oyster Creek Experience, Technical Report 1000093, Electric Power Research Institute, Palo Alto, CA.

EPRI 2001 — Decommissioning Pre-Planning Manual, Technical Report 1003025, Electric Power Research Institute, Palo Alto, CA.

Gilbert 1989 — Gilbert, T., et al, A Manual for Implementing Residual Radioactive Material Guidelines (the RESRAD Code), ANL/ES-160, DOE/CH/8901, Argonne National Laboratory, Argonne, IL.

IAEA 1992 — Monitoring Programmes for Unrestricted Release Related to Decommissioning of Nuclear Facilities, Technical Reports Series No. 334, STI/DOC/10/334, International Atomic Energy Agency, Vienna, Austria.

IAEA 1994 — Intervention Criteria in a Nuclear or Radiation Emergency, Safety Series No. 109, International Atomic Energy Agency, Vienna, Austria.

IAEA 1998 — Decommissioning of Nuclear Facilities Other than Reactors, Technical Report Series No. 386, STI/DOC/010/386, International Atomic Energy Agency, Vienna, Austria.

Lanais 1993 — Lanais, J., Baranowski, F., and Klevans, E., Sept. 1993, Final Report of Fast Flux Test Facility Independent Review Team, for U.S. Department of Energy, Richland, WA.

Kamboj 2000 — Kamboj, S. et al, Probabilistic Dose Analysis Using Parameter Distribution Developed for RESRAD and RESRAD-BUILD Codes, ANL/EAD/TM-89, Argonne National Laboratory, Argonne, IL.

LaGuardia 1986 — LaGuardia, T. et al., Guidelines for Producing Nuclear Power Plant Decommissioning Cost Estimates, Vol. 1 and 2, AIF/NESP-036, Atomic Industrial Forum / National Environmental Studies Project, Bethesda, MD.

Leigh 1993 — Leigh, C., Campbell, J., Thompson, B., Longsine, D., and Kennedy, R., User's Guide for GENII-S: A Code for Statistical and Deterministic Simulations of Radiation Doses to Humans from Radionuclides in the Environment, SAND91-0561A, Sandia National Laboratories, Albuquerque, NM.

MY 2000 — U.S. Court of Appeals for the Federal Circuit Decision, Docket No. 99-5138, -5139, -5140, Maine Yankee, Connecticut Yankee, and Yankee Atomic Power Companies, Washington, DC.

NCRP 1989 — Screening Techniques for Determining Compliance with Environmental Standards, NCRP Commentary No.3, NCRP Publications, Bethesda, MD.

NCRP 1991 — Calibration of Survey Instruments Used in Radiation Protection for the Assessment of Ionizing Radiation Fields and Radioactive Surface Contamination, Report No. 112, NCRP Publications, Bethesda, MD.

NCRP 1999 — Recommended Screening Limits for Contaminated Surface Soil and Review of Factors Relevant to Site-Specific Studies, Report No. 129, NCRP Publications, Bethesda, MD.

NRC 1971 — Quality Assurance Criteria for Nuclear Power Plants and Fuel Reprocessing Plants, 10 CFR 50, Appendix B, as amended at 36 FR 18301, Sept. 11, 1971; 40 FR 3210D, Jan. 20, 1975, U.S. Nuclear Regulatory Commission, Washington, DC.

NRC 1974 — Regulatory Guide 1.86, Termination of Operating Licenses for Nuclear Reactors, U.S. Nuclear Regulatory Commission, Washington, DC.

NRC 1983 — Licensing Requirements for Land Disposal of Radioactive Waste, 10 CFR 61.

NRC 1992 a — Manual for Conducting Radiological Surveys in Support of License Termination, NUREG/CR-5849, U.S. Nuclear Regulatory Commission, Washington, DC.

NRC 1992 b — Kennedy, W., Strenge, D., Meck, R., and Daily, C., Residual Radioactive Contamination from Decommissioning: Technical Basis for Translating Contamination Levels to Annual Total Effective Dose Equivalent, NUREG/CR-5512, Vol. 1., U.S. Nuclear Regulatory Commission, Washington, DC.

NRC 1995 — Konzek, G. et al, Revised Analysis of Decommissioning for the Reference Pressurized Water Reactor Power Station, NUREG/CR-5884, Battelle Pacific Northwest Laboratories, Richland, WA.

NRC 1996 — Smith, R. et al, Revised Analysis of Decommissioning for the Reference Boiling Water Reactor Power Station, NUREG/CR-6174, Battelle Pacific Northwest Laboratories, Richland, WA.

NRC 1997 a — Multi-Agency Radiation Survey and Site Investigation Manual (MARSSIM), G309, U.S. Nuclear Regulatory Commission, Washington, DC. Also published as EPA 403- R-97-016.

NRC 1997 b — Regulatory Analysis Technical Evaluation Handbook, NUREG/BR-0184, U.S. Nuclear Regulatory Commission, Washington, DC.

NRC 1998 a — Abelquist, E., Brown, W., Powers, G., and Huffert, A., Minimum Detectable Concentrations with Typical Radiation Survey Instruments for Various Contaminants and Field Conditions, NUREG-1507, U.S. Nuclear Regulatory Commission, Office of Nuclear Regulatory Research, Washington, DC.

NRC 1998 b — Demonstrating Compliance with the Radiological Criteria for License Termination, Draft Regulatory Guide DG-4006, U.S. Nuclear Regulatory Commission, Washington, DC.

NRC 1999 a — Haaker, R., Brown, T., and Updegraff, D., Comparison of the Models and Assumptions Used in the DandD 1.0, RESRAD 5.61, and RESRAD Build Computer Codes with Respect to the Residential Farmer and Industrial Occupant Scenarios, NUREG/CR-5512, Vol. 4, U.S. Nuclear Regulatory Commission, Washington, DC.

NRC 1999 b — Beyeler, W.E., et al., Residual Radioactive Contamination from Decommissioning, Parameter Analysis, NUREG/CR-5512, Vol. 3, Sandia National Laboratories, Washington, DC.

NRC 2000 a — Multi-Agency Radiation Survey and Site Investigation Manual, Rev. 1, NUREG-1575, Revision 1; U.S. Nuclear Regulatory Commission, Washington, DC. Also published as EPA 402-R-97 and as DOE/EH-0624

NRC 2000 b — NMSS Decommissioning Standard Review Plan, NUREG-1727, U.S. Nuclear Regulatory Commission, Washington, DC.

NRC 2001 a — McFadden, K. et al, Residual Radioactive Contamination from Decommissioning, User's Manual DandD, Version 2.1. NUREG/CR-5512, Vol. 2, U.S. Nuclear Regulatory Commission, Washington, DC.

NRC 2001 b — Domestic Licensing of Production and Utilization Facilities, Conditions of Licenses, 10 CFR 50, Subpart 50.54(bb).

NRC 2002 — Final Generic Environmental Impact Statement on Decommissioning of Nuclear Facilities, NUREG-0586, Supplement 1, U.S. Nuclear Regulatory Commission, Washington, DC.

NRC 2003 a — Radiological Criteria for Unrestricted Use, 10 CFR 20.1402.

NRC 2003 b — Licensing Requirements for Land Disposal of Radioactive Waste, 10 CFR 61.

NSP 1997 — Northern States Power Company, et al. vs. U.S. Department of Energy, U.S. Court of Appeals for the District of Columbia, Docket No. 97-1064, Northern States Power Company, Washington, DC.

ORISE 2002 — Compass Software Package Version 1.0, Oak Ridge Institute for Science and Education, Oak Ridge, TN.

ORNL 1981 — Begovich, C., Eckerman, K., Schlatter, E., Ohr, S., and Chester, R., DARTAB: A Program to Combine Airborne Radionuclide Environmental Exposure Data with Dosimetric and Health Effects Data to Generate Tabulations of Predicted Health Impacts, ORNL-5692/DE81030434, Oak Ridge National Laboratory, Oak Ridge, TN.

ORNL 1995 — ANISN, 1-D Neutron Transport and Gamma Transport in Slab, Cylindrical, Spherical Geometry with Anisotropic Scattering, CCC-0254, Oak Ridge National Laboratory, Oak Ridge, TN.

ORNL 1998 — TORT, 2-D 3-D (and DORT, 1-D 2-D) Discrete Ordinate Neutron and Photon Transport with Deep Penetration, CCC-0543, Oak Ridge National Laboratory, Oak Ridge, TN.

ORNL 2002 a — COMPASS: Implementation Software for Designing MARSSIM-Based Surveys, Oak Ridge Institute for Science and Education, Oak Ridge, TN.

ORNL 2002 b — Crofft, A., ORIGEN -2.2, Isotope Generation and Depletion Code Matrix Exponential Method, CCC-0371, Oak Ridge National Laboratory, Oak Ridge, TN.

PMI 2000 — A Guide to the Project Management Body of Knowledge, Project Management Institute, Newtown Square, PA.

Roctek 2000 — Rock Technologies, WinScale Software, Roctek Corporation, Chandler, AZ.

SAIC 1982 — *Proceedings of Alpha-Contaminated Waste Management Workshop,* Science Applications, Inc., Gaithersburg, MD.

Taboas 2000 — Taboas, A., et al., Decommissioning the World's Premier Facility for Radiological Research: The Janus Reactor, *Technology*, Vol. 7 p. 577-590, Cognizant Communication Corporation, Elmsford, NY.

Till 1988 — Till, J. and Moore, R., A Pathway Analysis Approach for Determining Acceptable Levels of Contamination of Radionuclides in Soil, *Health Physics Journal*, Vol. 55, 541-548, Charleston, SC.

TSCA 1976 — Toxic Substances Control Act of 1976, as amended, Washington, DC.

PART 4

Abrams 1966 — Abrams, C.S. and Salterelli, E.A., Decontamination of the Shippingport Atomic Power Station, USAEC Report WAPD-299, Bettis Atomic Power Laboratory.

AEA 1954 — Atomic Energy Act of 1954, as amended, Public Law 83-703.

Anspaugh 1973 — Anspaugh, L., et al, The Dose to Man Via Food-Chain Transfer Resulting from Exposure to Tritiated Water Vapor in Tritium, *Tritium*, Moghissi, A. and Cothern, C.R., ed., p 405-422, Messenger Graphics, Las Vegas, NV.

Ayres 1970a — Ayres, J.A., ed., *Decontamination of Nuclear Reactors and Equipment*, The Ronald Press Company, New York, NY.

Ayres 1970b — Ayres, J.A., Decontamination of Pressurized Water Reactors, in *Decontamination of Nuclear Reactors and Equipment*, The Ronald Press Company, New York.

Beaman 1982	Beaman, T.A. and Smee, J.L., Experience With Dilute Chemical Decontamination in Decontamination of Nuclear Facilities, September, 1982, ANS.
Berry 1979	Berry,W.E., Survey of Corrosion Product Generation, Transport, and Deposition in Light Water Reactors, NP-522, EPRI, 1979.
Bertholdt 1994	Bertholdt, H.O., Lejon, J., and Hermansson, X., A Full System Decontamination of the Oskarshamn 1 BWR, *International Conference Chemistry in WaterRreactors – Operating Experience and New Developments*, Nice, France.
Bertholdt 1998	Bertholdt, H.O., HP/CORD D UV – A New Decontamination Process for Decommissioning of NPPs, *1998 EPRI Chemical Decontamination Conference*, Greenville, SC, Report TR-110997.
Bertholdt 1998a	Bertholdt, H.O., A New Decontamination Process for Decommissioning of Nuclear Stations, *Nuclear Plant Journal*, Volume 12, No. 4.
Bradbury 1997	Bradbury, D. and Elder, G., EPRI DFD Decontamination for Decommissioning Process Evaluation, Electric Power Research Institute, TR-107707.
Bradbury 1998	Bradbury, D. and Elder, G., Review of Experience with the EPRI DFD Process: Decontamination for Decommissioning of Reactor Coolant Systems and Plant Components, EPRI, TR-109036.
Bradbury 1983	Bradbury, D. et al, Development of LOMI Chemical Decontamination Technology, EPRI Report NP-3177.
Cunnane 1978	Cunnane, J.C., Uhl, D.L., and Kennedy, A.J., Dynamic Corrosion Product Transport in Pressurized Water Reactors, Paper 155, *Corrosion '78*, NACE, Houston TX.
DOE/NV 1994	United States Nuclear Tests, July 1945 through September 1992, DOE/NV 209, U.S. Department of Energy, Nevada Operations Office, Mercury, NV.
DOT 2003a	Shippers - General Requirements for Shipment and Packaging, 49 CFR 173.
DOT 2003b	General Information, Regulations, and Definitions, 49 CFR 171.
EPA 2003	Identification and Listing of Hazardous Waste, 40 CFR 261.
EPRI 1991	*Full System Chemical Decontamination Workshop*, 91-WH-53, EPRI.
EPRI 1993	Proceedings: *Fifth Workshop on Chemical Decontamination*, NPD/M&C/93-1, EPRI.
EPRI 1995	*1995 Radiation Field Control and Decontamination Seminar*, TR-106009, EPRI.
EPRI 1998	Proceedings: *1998 EPRI Chemical Decontamination Conference*, TR-110997, EPRI.
Greenman 1986	Greenman, W., Nielsen, N., and Remark, J., LOMI Decontamination of Indian Point III, *AIChE Summer National Meeting*, Paper 63F, Boston, MA.
Irving 1985	Irving, B., Decontamination in Notes for ASME *Short Course on Radioactive Waste Management for Nuclear Power Plants and Other Facilities*, Alexandria, VA.
Irving 1986	Irving, B. and Allen, J.K., Mechanical Decontamination Techniques in Radioactive Waste Technology, The American Society of Mechanical Engineers, NY, NY.
Jeppeson 1999	Jeppeson, D., Evaluation of Tritium Removal Mitigation Technologies for Wastewater Treatment, DOE/RL-99-42 Rev. 0, U.S. Department of Energy, Washington, DC.
Kalyanam 1988	Kalyanam, K. and Sood, S., A Comparison of Process Characteristics for the Recovery of Tritium from Heavy Water and Light Water Systems, *Fusion Technology*, 14:524-528.
LeSurf 1982	LeSurf, J.E. and Weyman, G.D., Cost Effectiveness of Dilute Chemical Decontamination, *Nuclear and Chemical Waste Management*, Vol. 4.
LeSurf 1991	LeSurf, J.E., and Smee, J.L., Full-System Decontamination of a BWR Using the LOMI Process, Volume 4, Full-System Decontamination Experience and Cost-Benefit Analysis. EPRI Report TR-100049.
Lister 1979	Lister, D.H., The Accumulation of Radioactive Corrosion Products in Nuclear Steam Generators, *Corrosion*, Volume 35, No. 5, Page 219.
Lister 1989	Lister, D.H. and Davidson, R.D., Corrosion Product Release in Light Water Reactors, NP-6512, EPRI.
Lister 1996	Lister, D.H. et al, The Deposition of Cobalt on In-Core Surfaces During Boiling Heat Transfer, Paper 10, *VII International Conference on Water Chemistry of Nuclear Reactor System*, Bournemouth, UK.
Moghissi 1986	Moghissi, A. and Cothern, C. ed, Background Information for the Development of a Standard for Tritium in Drinking Water, *Radiation Protection Dosimetry* 16: 175-179, Oxford University Press.
NACE 2000	Laboratory Corrosion Testing of Metals, TM0169-2000, National Association of Corrosion Engineers, Houston, TX.

Nielson 1986 Nielsen, N., Indian Point Unit No. 3 Decontamination, *Seminar on PWR Water Chemistry and Radiation Field Control*, Paper 33, Berkeley, CA.

NRC 1991 Waste Form Technical Position, Rev. 1, January 19, 1991, Low Level Waste Management Branch, Division of Low Level Waste and Decommissioning, Nuclear Materials Safety and Safeguards NMSS, U.S. Nuclear Regulatory Commission, Washington, DC.

Osterhout 1980 Osterhout, M.M., ed, *Decontamination and Decommissioning of Nuclear Facilities*, Plenum Press, New York.

Parry 1998 Parry, J.O., Recontamination Rates at Indian Point 2 Two Years After a Full System Decontamination, *1998 EPRI Chemical Decontamination Conference*, EPRI, TR-110997.

Perrigo 1966 Perrigo, L.D., et al, Plutonium Recycle Test Reactor Decontamination Following a MgO-PuO2 Fuel Element Failure, USAEC Report BWNL-SA-636, Battelle-Northwest Laboratory.

RCRA 1976 Resource Conservation and Recovery Act of 1976, as amended, 40 CFR 260-281 and 42 USC 6901 et seq.

Remark 1981 Remark, J.F., Plant Decontamination Methods Review, NP-1168, EPRI.

Remark 1984 Remark, J.F., Quadrex HPS LOMI Decontamination of the Mendaciously Nuclear Power Station, *EPRI Seminar - BWR Corrosion, Chemistry, and Radiation Control*, Palo Alto, CA.

Remark 1986 Remark, J.F., Low Oxidation Metal Ion LOMI Chemical Decontamination of Indian Point III, *American Institute of Chemical Engineers 1986 Summer Meeting*, Paper No. 63F, Boston, MA.

Remark 1989 Remark, J.F., A Review of Plant Decontamination Methods-1988 Update, NP-6169, EPRI.

Remark 1993 Remark, J.F., and Bengel, T.G., US Patent 5,305,360, Process for Decontaminating a Nuclear Reactor Coolant System, Issued April 19, 1994.

SDWA 1974 Safe Drinking Water Act of 1974, as amended, 42 U.S.C. 300f et seq.

Segal 1983 Segal, M.G., and Swan, T., Water Chemistry-3 1983; 61-70, London England, British Nuclear Energy Society.

Sienkiewicz 1988 Sienkiewicz, C. and Lentz, J., Recovery of Tritium from Water, *Fusion Technology*, 14:444-449.

Smee 1999 Smee, J., Decontamination Handbook, TR-112352, EPRI, July 1999.

TSCA 1976 Toxic Substances Control Act of 1976, as amended, 15 U.S.C. 2601 et seq.

Van Hook 1968 Van Hook, W., Vapor Pressure of the Isotopic Waters and Ices, *The Journal of Physical Chemistry A*, Vol. 78, p 1234-1244, Washington, DC.

Weed 1968 Weed, R.D., Decontamination of the Plutonium Recycle Test Reactor PRTR Primary System, USAEC Report BNWL-711, Battelle Northwest Laboratory.

Wille 1986 Wille, H., et al, Development and Application of the OZOX/CORD Process, Paper 84, Fourth International Conference on Water Chemistry for Nuclear reactor Systems, British Nuclear Energy Society, London.

PART 5

AEA 1954 Atomic Energy Act of 1954, as amended, Public Law 83-703.

ANS 2003 DD&R Division Newsletter, American Nuclear Society, LaGrange Park, IL.

ANS 2004 DD&R Division Newsletter, American Nuclear Society, LaGrange Park, IL.

ASTM 1994 Standard Practice for Environmental Site Assessment: Phase I Environmental Site Assessment Process, E-1527-94, American Society for Testing Materials, West Conshohocken, PA.

ASTM 2000 Standard Practice for Environmental Site Assessment: Phase I Environmental Site Assessment Process, Rev. 1, E-1527-00, American Society for Testing Materials, West Conshohocken, PA.

MY 2000 U.S. Court of Appeals for the Federal Circuit Decision, Docket No. 99-5138, -5139, -5140, Maine Yankee, Connecticut Yankee, and Yankee Atomic Power Companies.

MY 2001 Maine Yankee License Termination Plan, Revision 1, Maine Yankee Atomic Power Company, Wiscasset, ME.

NHPA 1996 National Historic Preservation Act of 1966, as amended.

NRC 1974 Regulatory Guide 1.86, Termination of Operating Licenses for Nuclear Reactors, U.S. Nuclear Regulatory Commission, Washington, DC.

NRC 1992 Manual for Conducting Radiological Surveys in Support of License Termination, NUREG/CR-5849, U.S. Nuclear Regulatory Commission, Washington, DC.

NRC 1997 Multi-Agency Radiation Survey and Site Investigation Manual (MARSSIM), NUREG-1575, U.S. Nuclear Regulatory Commission, Washington, DC.

NRC 1999 Standard Format and Content of License Termination Plans for Nuclear Power Reactors, Regulatory Guide 1.179, U.S. Nuclear Regulatory Commission, Washington, DC.

NRC 2002 Final Generic Environmental Impact Statement on Decommissioning of Nuclear Facilities, NUREG-0586, Supplement 1, U.S. Nuclear Regulatory Commission, Washington, DC.

RS 2002a Plant Modifications - Defines and Controls Design Change Packages, RSAP-0303, Rev. 18, Rancho Seco Nuclear Plant, Sacramento, CA.

RS 2002b Removal of Tools and Equipment from Controlled Areas - Defines and Controls the Use of Inaccessible Surface Contamination Evaluation Forms, RP.305.09A, Rev. 12, Rancho Seco Nuclear Plant, Sacramento, CA.

RS 2003 Control of Decommissioning Projects - Defines and Controls Integrated Decommissioning Packages and Decommissioning Work Permits, RSAP-1900, Rev 8, Rancho Seco Nuclear Plant, Sacramento, CA.

ABOUT THE EDITORS

Anibal L. Taboas

Asst. Manager for Projects & Programs
US Department of Energy, Chicago Office
Argonne, IL 60439
(630) 252-2236
Anibal.Taboas@ch.doe.gov

Mr. Anibal L. Taboas conceived this handbook, and completed it as Lead Editor. His background includes nuclear facility operations, and executing major programs and projects that range from remediation of a Superfund site, to the production, disposition, and eventual civilian re-use of special materials from former nuclear weapons. Mr. Taboas serves on editorial boards of peer reviewed journals, and has published in topics from environmental policy, to cleanup standards post a radiological dispersal event. He maintains technical competence and independent thought, and frequently leads expert teams, such as to explore change in the regulatory scheme for nuclear and radiological safety, accident investigations, and international conferences. He has been recognized for significant accomplishments as senior executive in the U.S. Department of Energy. Mr. Taboas has had line responsibility for decommissioning projects, particularly nuclear research reactors, including Janus, the Brookhaven Graphite Research Reactor, and the Hammer Award winning Large Scale Technology Demonstration Project performed during decommissioning of the CP-5 Reactor. He has also led restoration of various sites, including Site A, where several of the nation's first nuclear reactors operated (a marker on the restoration of this site is depicted in the front cover). Prior activities include consulting on Monte Carlo calculations, and developing nuclear control systems at Argonne. Mr. Taboas earned degrees from the University of Dayton, Indiana State, and Northwestern University. He mentors minorities to develop superior technical qualifications, and was awarded a Doctorate degree, Honoris Causa, by the University of Puebla (UPAEP), Mexico.

A. Alan Moghissi
President, Institute for Regulatory Science
5457 Twin Knolls Road, Suite 312
Columbia, MD 21045
(301) 596-1700
RSI@nars.org

Dr. A. Alan Moghissi leads a non-profit organization dedicated to the principle that societal decisions must be based on best available scientific information. Previously he was Associate Vice President for Environmental Health at Temple University, and Assistant Vice President for Environmental Health and Safety, and professor in the School of Medicine of the University of

Maryland. He has held other academic affiliations and appointments, such as at the Georgia Institute of Technology, the University of Virginia, the University of Nevada, and the Catholic University of America. For several years, he was Principal Science Advisor for Radiation and Hazardous Materials at the U.S. Environmental Protection Agency. Dr. Moghissi worked closely with Mr. Taboas to foster sound risk management practices in federal rule-making, and in instituting a novel independent peer review process to determine the merit of further developing specific technologies for environmental remediation. Dr. Moghissi is Fellow of the American Society of Mechanical Engineers, and leads a significant related peer review effort. Dr. Moghissi has lectured on various topics at the Hahn Meitner Institute and at the Technical University of Munich (Germany), the Russian Academy of Sciences, the University of Puebla (UPAEP) Mexico, and the University of Kuwait. He has written numerous publications, including several books, and has served as Editor-in-Chief of peer reviewed technical journals. Dr. Moghissi studied at the University of Zurich and the Technical University of Karlsruhe, where he received a Doctorate degree in physical chemistry.

Thomas S. LaGuardia

President, TLG Services, Inc.
Vice President, Entergy Nuclear, Inc.
Bridgewater, CT 06752
(860) 355-2300
laguardia@tlgservices.com

Mr. Thomas S. LaGuardia founded TLG Services, Inc., in 1982 to specialize in decommissioning engineering and field services. In 2000, it then became a wholly-owned subsidiary of Entergy Nuclear, Inc., but maintains its proprietary relationship with all U.S. commercial utilities. Mr. LaGuardia's prior employment includes Nuclear Energy Services, Gulf Nuclear Fuels Corporation (formerly United Nuclear Corporation), and Combustion Engineering. Mr. LaGuardia earned mechanical engineering degrees from the Polytechnic Institute of Brooklyn and the University of Connecticut. A registered Professional Engineer in several states, he is also a Certified Cost Engineer by the Association for the Advancement of Cost Engineering International. Mr. LaGuardia has provided engineering, planning and management support for decontamination, decommissioning, and waste management for both commercial and government facilities. He is an industry-recognized expert in estimating decommissioning costs and schedules, and in facilitation by reducing time, cost, exposure, and/or radioactive waste. Mr. LaGuardia has overseen preparation of decommissioning feasibility and cost studies for over 128 nuclear power units and more than 200 fossil-fueled units, and has supported those estimates with expert testimony in over 125 rate hearings. He has organized expert teams to provide independent reviews of existing and proposed decommissioning projects at several government sites, and is heavily involved in establishing international guidelines for developing decommissioning cost estimates.

SYMBOLS AND UNITS RELEVANT TO DECOMMISSIONING

The International System of Units, known by its French abbreviation as SI, is an accepted and widely used system. Consisting of base and derived units, the system is coherent in that no multiplier is required for converting one to a second unit. Each unit, base or derived, is represented by a symbol (not an abbreviation or acronym). Both the SI and English systems make a distinction between a quantity and a unit. The quantity is an item such as length or electrical current. In contrast, a unit is how a quantity is measured, such as a meter or ampere. An important characteristic of SI is that each quantity has only one unit. For example, the unit for quantity of length is meter, with the symbol m. If one deals with small or large lengths, a prefix is used such as nm (for 10^{-9}m) or km (for 10^{3}m).

Decommissioning activities performed in the U.S. traditionally rely on the survey system of units, wherein a quantity may have several units. For example, this system uses inch, foot, yard, and mile for length. Although both systems contain extensive units for numerous quantities, the units used in decommissioning of nuclear facilities are limited. Since this handbook includes both English and SI units, a description of both and selected conversion factors are presented in the following table.

Prefixes: Both systems rely on prefixes to describe large or small units when writing exceedingly large or exceptionally small numbers (decimal places).

10^{24}	yotta	Y	10^{-1}	deci	d
10^{21}	zetta	Z	10^{-2}	centi	c
10^{18}	exa	E	10^{-3}	milli	m
10^{15}	peta	P	10^{-6}	micro	µ
10^{12}	tera	T	10^{-9}	nano	n
10^{9}	giga	G	10^{-12}	pico	p
10^{6}	mega	M	10^{-15}	femto	f
10^{3}	kilo	k	10^{-18}	atto	a
10^{2}	hecto	h	10^{-21}	zepto	z
10^{1}	deka	da	10^{-24}	yocto	y

Note that prefixes cannot be combined. For example, Mkm would not be accurate; Gm would be the proper notation.

Base Units: The following is a listing of the base and derived units of the SI system

Quantity	Unit	Symbol	Formula
Length	meter	m	base unit
Tine	second	s	base unit
Mass	kilogram	kg	base unit
Electric current	ampere	A	base unit
Temperature	kelvin	K	base unit
Frequency	herz	Hz	1/s
Force	newton	N	$kg\ m\ s^{-2}$
Pressure	pascal	Pa	$N\ m^{-2}$
Energy, work	joule	J	N m
Power	watt	W	$J\ s^{-1}$
Electric potential	volt	V	$W\ A^{-1}$
Activity (radionuclide)	becquerel	Bq	s^{-1}
Absorbed dose	gray	Gy	$J\ kg^{-1}$
Dose equivalent	sievert	Sv	$J\ kg^{-1}$

Other Units: Certain units are accepted by the SI system due to their practical usefulness. The following includes special names for certain SI units and those units used as a matter of tradition

Quantity	Unit	Symbol	Formula
Volume	liter	L or l	$1\ dm^3$
Surface	hectare	ha	$10^4\ m^2$
Mass	tone	t	10^3 kg
Activity (radionuclides)	curie	Ci	$3.7x10^{10}$ Bq
Absorbed dose (ionizing radiation)	rad	rad	0.01 Gy
Dose equivalent (ionizing radiation)	rem	rem	0.01 Sv
Temperature	celsius	^oC	$^oK + 273.15$

Conversion Factors: Quantities are presented in various units within this handbook. For the sake of readability, the equivalents are left to the reader where necessary. The following lists some of the most used equivalents for English and SI units.

Multiply	by	to Obtain
acres	0.4047	hectares (ha)
cubic feet (ft^3)	0.02832	cubic meters (m^3)
cubic yards (yd^3)	0.7646	cubic meters (m^3)
degrees Fahrenheit (°F) -32	0.5555	degrees Celsius (°C)
feet (ft)	0.3048	meters (m)
gallons (gal)	3.785	liters (L)
gallons (gal)	0.003785	cubic meters (m^3)
inches (in.)	2.540	centimeters (cm)
miles (mi)	1.609	kilometers (km)
pounds (lb)	0.4536	kilograms (kg)
short tons (tons)	907.2	kilograms (kg)
short tons (tons)	0.0972	metric tons (t)
square feet (ft^2)	0.09290	square meters (m^2)
square yards (yd^2)	0.8361	square meters (m^2)
square miles (mi^2)	2.590	square kilometers (km^2)
yards (yd)	0.9144	meters (m)
centimeters (cm)	0.3937	inches (in.)
cubic meters (m^3)	35.31	cubic feet (ft^3)
cubic meters (m^3)	1.308	cubic yards (yd^3)
cubic meters (m^3)	264.2	gallons (gal)
degrees Celsius (°C) + 17.78	1.8	degrees Fahrenheit (°F)
hectares (ha)	2.471	acres
kilograms (kg)	2.205	pounds (lb)
kilograms (kg)	0.001102	short tons (tons)
kilometers (km)	0.6214	miles (mi)
liters (L)	0.2642	gallons (gal)
meters (m)	3.281	feet (ft)
meters (m)	1.094	yards (yd)
metric tons (t)	1.102	short tons (tons)
square kilometers (km^2)	0.3861	square miles (mi^2)
square meters (m^2)	10.76	square feet (ft^2)
square meters (m^2)	1.196	square yards (yd^2)

SELECTED ACRONYMS AND NOTATIONS

The following lists the acronyms and abbreviations used most frequently in this handbook. Standard scientific notations aside, some have been created within the U.S. government's regulatory framework while others have developed over time at the country's nuclear reactor sites. In either case, certain terms have a universal recognition within the decommissioning industry and are therefore provided here to aid the reader in acclimating to the vocabulary, and to ease presentation in the text.

AACE	American Association of Cost Engineers
AEA	Atomic Energy Act
AGR	Advanced Gas Reactor
ALARA	As Low As Reasonably Achievable
ANS	American Nuclear Society
ANSI	American National Standards Institute
ASME	American Society of Mechanical Engineers
ASTM	American Society for Testing and Materials
BIO	Basis for Interim Operations
BWR	Boiling Water Reactor
CAA	Clean Air Act
CAD	Computer Aided Design
CEDE	Committed Effective Dose Equivalent
CERCLA	Comprehensive Environmental Response, Compensation, and Liability Act
CFR	Code of Federal Regulations (see http://www.gpoaccess.gov/cfr/index.html)
COE	US Army Corps of Engineers
cpm	counts per minute
CWA	Clean Water Act
D&D	Decontamination and Decommissioning
DCGL	Derived Concentration Guideline Level
DECON	NRC D&D alternative involving immediate dismantling
DF	Decontamination Factor
DOC	Decommissioning Operations Contractor
DOE	U.S. Department of Energy
DOE Order	A system of internal DOE directives
DOT	U.S. Department of Transportation
DP	Decommissioning Plan
dpm	disintegrations per minute
DQO	Data Quality Objectives
DTF	Decommissioning Trust Fund
EIS/EA	Environmental Impact Statement/Assessment
ENTOMB	NRC D&D alternative involving long-term encapsulation of a facility
EPA	U.S. Environmental Protection Agency
EPRI	Electric Power Research Institute
ES&H	Environment, Safety, and Health
FFCA	Federal Facility Compliance Act
FSAR	Final Safety Analysis Report

GCR	Gas Cooled Reactor
GIS	Geographical Information System
GTCC	Greater Than Class C (waste concentration)
HASP	Health and Safety Plan
HAZWOPER	Hazardous Waste Operations and Emergency Response
HEPA	High Efficiency Particulate Air (Filter)
HLW	High-Level (Radioactive) Waste
HMTA	Hazardous Materials Transportation Act
HP	Health Physics
HSA	Historical Site Assessment
HTGR	High Temperature Gas Cooled Reactor
HVAC	Heating, Ventilating, Air Conditioning
IAEA	International Atomic Energy Agency
INPO	Institute for Nuclear Power Operations
ISFSI	Independent Spent Fuel Storage Installation
ISMS	Integrated Safety Management System
JSA/JHA	Job Safety/Hazard Analysis
Kd	Distribution Coefficient
LLW	Low-Level (Radioactive) Waste
LLWPA	Low Level Waste Policy Act
LTP	License Termination Plan
MARSSIM	Multi-Agency Radioactive Site Survey and Site Investigation Manual
MCL	Maximum Contaminant Level
MCLG	Maximum Contaminant Level Goals
MPC	Multi-Purpose Canister
MSDS	Material Safety Data Sheet
MT	metric ton (t)
MWe	megawatt electric
MWt	megawatt thermal
NASE	National Association of Corrosion Engineers
NAAQS	National Ambient Air Quality Standards
NAS	National Academy of Science
NEI	Nuclear Energy Institute
NEPA	National Environmental Policy Act
NESHAP	National Emission Standards for Hazardous Air Pollutants
NFPA	National Fire Protection Association
NHPA	National Historic Preservation Act
NIOSH	National Institute for Occupational Safety and Health
NMSS	Office of Nuclear Materials Safety and Safeguards (NRC)
NPDES	National Pollutant Discharge Elimination System
NRC	U.S. Nuclear Regulatory Commission
NSSS	Nuclear Steam Supply System
NUREG	Nuclear Regulatory (NRC-sponsored document in use as a permanent reference)
NWPA	Nuclear Waste Policy Act
ODE	Occupational Dose Equivalent
OSHA	Occupational Safety and Health Administration
P&ID	Piping and Instrumentation Diagram

PCB	Polychlorinated Biphenyl
PHWR	Pressurized Heavy Water Reactor
POL	Possession-Only License
ppb	parts per billion
PPE	Personal Protective Equipment
ppm	parts per million
PSAR	Preliminary Safety Analysis Report
PSDAR	Post Shutdown Decommissioning Activities Report
PSv	Person-Sievert
PUC	Public Utility Commission
PWR	Pressurized Water Reactor
QA/QC	Quality Assurance/Control
RCA	Radiological Controlled Area
RCRA	Resource Conservation and Recovery Act
RESRAD	RESidual RADioactivity (a computer modeling code)
S&M	Surveillance and Maintenance
SAFSTOR	NRC D&D alternative involving placing a facility into safe and stable storage
SAR	Safety Analysis Report
SARA	Superfund Amendment and Reauthorization Act
SDWA	Safe Drinking Water Act
SNM	Special Nuclear Material
TEDE	Total Effective Dose Equivalent
TLD	Thermo Luminescent Dosimeter
TRU	Transuranic
TSCA	Toxic Substances Control Act
TSD	Treatment, Storage, and Disposal
TWA	Time Weighted Average
UCF	Unit Cost Factor
USC	United States Code
USQ	Unreviewed Safety Question
UST	Underground Storage Tank
VOC	Volatile Organic Compound
WAC	Waste Acceptance Criteria
WBS	Work Breakdown Structure
WIPP	Waste Isolation Pilot Plant

www.ingramcontent.com/pod-product-compliance
Lightning Source LLC
Chambersburg PA
CBHW050453190326
41458CB00005B/1259
9780791802243